In Vitro Aspects of Erythropoiesis

1. G. Rossi; 2. A. S. Gordon; 3. R. A. Rifkind; 4. C. Magli; 5. A. Urabe; 6. D. L. McLeod; 7. M. Ogawa; 8. M. J. Murphy, Jr.; 9. S. Sassa; 10. N. Testa; 11. A. P. Ball; 12. H. J. Seidel; 13. J. R. Zucali; 14. D. A. J. Ives; 15. J. W. Fisher; 16. S. Hasthorpe; 17. V. S. Gallicchio; 18. F. E. Trobaugh, Jr.; 19. Y. Moriyama; 20. A. J. Erslev; 21. P. Meyer; 22. C. Peschle; 23. G. Wagemaker; 24. E. D. Zanjani; 25. J. W. Adamson; 26. D. W. Golde.

In Vitro Aspects of Erythropoiesis

Edited by Martin J. Murphy, Jr.

Co-editors
Cesare Peschle
Albert S. Gordon
Edwin A. Mirand

With 192 illustrations

Springer-Verlag New York Heidelberg Berlin

Martin J. Murphy, Jr., Ph. D.
Director
Bob Hipple Laboratory for Cancer Research
Sloan-Kettering Institute for Cancer Research
1250 1st Avenue
New York, New York 10021

Cesare Peschle, M. D.
Istituto Patologia Medica
Nuovo Policlinico
Universita di Napoli
Via S. Pancini
80131 Napoli, Italy

Albert S. Gordon, Ph. D.
Research Professor
Dept. of Biology
New York University
Washington Square East
New York, New York 10003

Edwin A. Mirand, Ph. D., D. Sc.
Associate Institute Director and Head
Dept. of Biological Resources
Roswell Park Memorial Institute
Buffalo, New York

Library of Congress Cataloging in Publication Data
Main entry under title:
In vitro aspects of erythropoiesis.
   Bibliography: p.
   Includes index.
  1. Erythropoiesis. 2. Cell culture. I. Murphy,
Martin J. [DNLM: 1. Erythropoiesis. WH150 135]
QP92.V57   599'.01'13   78-16104

All rights reserved.

No part of this book may be translated or reproduced in any form without written permission from Springer-Verlag.

Copyright © 1978 by Springer-Verlag New York Inc.
Softcover reprint of the hardcover 1st edition 1978

9 8 7 6 5 4 3 2 1

ISBN-13: 978-1-4612-6303-6   e-ISBN-13: 978-1-4612-6301-2
DOI: 10.1007/978-1-4612-6301-2

Preface

The purpose of *In Vitro Aspects of Erythropoiesis* is to broaden our understanding of the regulation of normal and neoplastic hematopoietic cells by means of a synthesis of different techniques, divergent nomenclature as well as a closer standardization of reagents and methods which had heretofore produced vagaries of results among different laboratories. Hence the papers, discussions and appendices presented in this volume will be of value and interest to investigators and clinicians as well as to students and teachers of hematology.

In Vitro Aspects of Erythropoiesis is comprised of thirty papers delivered by twenty invited contributors at a three day conference which was held in Capri, Italy in October of 1977. The contributors were not only charged with a detailed accounting of materials and methods used in individual erythroid colony assay systems, but also with an itemization of those facets in common agreement and more importantly with an identification of those aspects where there was a divergence of technique between investigators.

The meeting was divided into six sessions and the text reflects this, the first section being devoted to *in vitro* methodology for the cultivation of erythroid stem cells. The second series of papers treats cellular and soluble factors which affect erythroid stem cell differentiation. Hormonal modulation of *in vitro* erythropoiesis is the theme of the third session followed by a series of papers in the fourth which focuses on erythropoiesis after transformation by either Friend or Rauscher viruses. The fifth session deals with serum inhibitors to erythropoiesis and the rôle of the macrophage in red blood cell production. Finally, the sixth session concentrates on the biogenesis of erythropoietin *in vivo* and *in vitro*.

Each of the six sections is followed by a discussion which was transcribed and initially edited at the meeting. In addition to the twenty-eight pages of these discussions which were further edited for relevance and clarity, a singularly important feature of this volume is the thirteen appendices which give the detailed materials and methodology employed by the authors in their research in the study of erythropoiesis *in vitro*.

New York, June 1978   Martin J. Murphy, Jr.

Acknowledgements

The thoughtful suggestions of Doctors A. S. Gordon and E. A. Mirand are appreciated and the participation in so many aspects of the meeting by Doctor Cesare Peschle is especially acknowledged, along with the hospitality of Doctor Sergio Gorini and his fine staff of the Fondazione Internazionale Menarini.

This workshop was also supported by contributions from Union International Contre le Cancer, Connaught Laboratories Limited, The Fogarty International Center, N.I.H., National Institute of Arthritis, Metabolism and Digestive Diseases, N.I.H., National Cancer Institute, N.I.H., National Heart, Lung, and Blood Institute, N.I.H., and The Bob Hipple Fund for Cancer Research.

We are indebted for the editorial assistance of Ann Murphy, Joanne Zisson and Sara E. Naughton. This volume could not have been so speedily published had we not had the cooperation of each of the contributors, for which we are sincerely thankful.

Contents

List of Contributors	xiii

1. Erythropoiesis *In Vitro:* A Brief Introduction — 1
   Martin J. Murphy, Jr.

2. Erythropoietin-Independence of Early Erythropoiesis and a Two-Regulator Model of Proliferative Control in the Hemopoietic System — 3
   N. N. Iscove and L. J. Guilbert

3. *In Vitro* Modulation of Human Erythropoiesis by Lymphocyte-Rich Fractions of Blood — 8
   Esmail D. Zanjani, Benet Nomdedeu, John Rinehart, Bobby J. Gormus, and Manuel E. Kaplan

4. Cryogenic Preservation of Human and Murine Erythropoietic and Granulopoietic Colony-Forming Cells — 21
   Hiroshi Hara, Makio Ogawa, and Harvey L. Bank

5. Miniaturization of Methylcellulose Erythroid Colony Assay — 28
   Akio Urabe and Martin J. Murphy, Jr.

6. Culture Systems *In Vitro* for the Assay of Erythrocytic and Megakaryocytic Progenitors — 31
   David L. McLeod, Mona M. Shreeve, and Arthur A. Axelrad

   Discussion (Chapters 1–6) — 37

7. Cellular and Soluble Factors Influencing the Differentiation of Primitive Erythroid Progenitor Cells (BFU-e) *In Vitro* — 44
   G. Wagemaker

8. The Effect of Sustained Hypertransfusion on Hematopoiesis — 58
   A. Erslev, R. Silver, J. Caro, S. Paist, and E. Cobbs

9. Inhibitors of Erythropoiesis: Effects of Red Cell Extracts on Heme Synthesis and Erythroid Colony Formation in Human Bone Marrow Cultures — 64
   Yoshiaki Moriyama

10. Production of Erythroid Precursor Cells in Long-Term Culture of Mouse Bone Marrow Cells — 68
    Martin J. Murphy, Jr., Akio Urabe, and Maureen E. Sullivan

11. Production of Erythroid Precursor Cells (BFU) *In Vitro* — 72
    N. G. Testa and T. M. Dexter

    Discussion (Chapters 7–11) — 75

12. Hormonal Modulation of Erythropoiesis *In Vitro*   81
    David W. Golde

13. Regulatory Mechanisms of Erythroid Stem Cell Kinetics   86
    C. Peschle, M. C. Magli, C. Cillo, F. Lettieri, F. Pizzella,
    G. Migliaccio, and G. F. Sasso

14. Hormonal Influences on Erythroid Colony Growth in Culture   95
    John W. Adamson, William J. Popovic, and James E. Brown

15. A Lack of Burst-Forming Unit (BFU-e) and Lymphocyte-Mediated Suppression of Erythropoiesis in Patients with Aplastic Anemia   100
    Yoshiaki Moriyama, Masatsugu Sato, and Yasutami Kinoshita

16. Pharmacological Agents and Erythroid Colony Formation: Effect of Beta-Adrenergic Agonists and Steroids   103
    James W. Fisher, Yasuhico Ohno, Bruno Modder,
    Franciszek Przala, Gregory D. Fink,
    and Dennis M. Gross

17. *In Vitro* Assessment of Similarities between Erythroid Precursors of Fetal Sheep and Patients with Polycythemia Vera   118
    Esmail D. Zanjani, Rona Singer Weinberg, Benet Nomdedeu,
    Manuel E. Kaplan, and Louis R. Wasserman

    Discussion (Chapters 12–17)   123

18. Murine Erythroleukemia: Cell Surface and Cell Cycle-Related Events of Induced Differentiation   129
    Richard A. Rifkind and Paul A. Marks

19. Regulation of Heme Biosynthesis in Mouse Friend Virus-Transformed Cells in Culture   135
    Shigeru Sassa, Joel L. Granick, Harvey Eisen, and
    Wolfram Ostertag

20. Comparative Studies on Hemopoietic Stem Cell Pools in Mice after Friend (FV-P) or Rauscher Virus Infection   143
    H. J. Seidel and Uta Opitz

21. Studies on Erythroid Differentiation in a Wild-Type and Dimethylsulfoxide-Resistant Clone of Friend Virus-Transformed Cells: Effects of Hemin, Hemoglobin, and Erythropoietin   149
    Akio Urabe, Martin J. Murphy, Jr., and Shigeru Sassa

22. Induction of Uroporphyrinogen-1 Synthase Activity in Mitogen-Stimulated Lymphocytes: Deficient Induction in Acute Intermittent Porphyria Cells   156
    Shigeru Sassa, Gregory L. Zalar, and Attallah Kappas

23. Potential for Differentiation, Virus Production, and
Tumorigenicity in Murine Erythroleukemic Cells Treated
with Interferon 159
*L. Cioé, A. Dolei, G. B. Rossi, F. Belardelli, E. Affabris,
R. Gambari, and A. Fantoni*

24. Kinetics of Erythroid and Myeloid CFU Following Rauscher
Virus Infection 172
Suzanne Hasthorpe

Discussion (Chapters 18–24) 177

25. The Role of Serum Inhibitors of Erythroid Colony-Forming
Cells in the Mechanism of the Anemia of Renal
Insufficiency 181
*James W. Fisher, Yasuhico Ohno, J. Barona, Maria Martinez,
and Arvind B. Rege*

26. Modulatory Effect of Macrophages on Erythropoiesis 189
Martin J. Murphy, Jr. and Akio Urabe

Discussion (Chapters 25–26) 192

27. Extrarenal Erythropoietin 194
*Brian A. Naughton, Albert S. Gordon, Sam J. Piliero, and
Philip Liu*

28. *In Vitro* Aspects of Erythropoietin Production 218
James R. Zucali and Edwin A. Mirand

29. Renal and Extrarenal Erythropoietin Production in Anemia 225
A. J. Erslev, J. Caro, and E. Kansu

30. Recent Advances in Erythropoietin Physiology 227
*C. Peschle, M. C. Magli, C. Cillo, F. Lettieri, F. Pizzella,
G. Migliaccio, A. Soricelli, G. Scala, G. Mastroberardino,
and G. F. Sasso*

Discussion (Chapters 27–30) 240

Appendices 1–13 243

Index 273

List of Contributors

John W. Adamson, M.D.
Associate Professor
Dept. of Medicine
Veterans Administration Hospital and
University of Washington
Seattle, Washington

S. S. Adler, M.D.
Assistant Professor of Medicine
Rush Medical College Chief
Special Morphology Laboratory
Dept. of Internal Medicine
Section of Hematology
Rush-Presbyterian-St. Luke's Medical Center
Chicago, Illinois

E. Affabris, Ph.D.
Postdoctoral Fellow
Section of Virology
Istituto Superiore de Sanità
Rome, Italy

Arthur A. Axelrad, M.D., Ph.D.
Professor of Anatomy (Histology)
Dept. of Anatomy
University of Toronto
Toronto, Ontario, Canada

Anne P. Ball, Ph.D.
Health Scientist Administrator
Blood Diseases Branch
Division of Blood Diseases and Resources
National Heart, Lung, and Blood Institute
National Institutes of Health
Bethesda, Maryland

Harvey L. Bank, Ph.D.
Assistant Professor
Dept. of Pathology
Medical University of South Carolina
Charleston, South Carolina

J. Barona, M.D.
Assistant Professor
Dept. of Pharmacology
Louisiana State University
New Orleans, Louisiana

F. Belardelli, Ph.D.
Postdoctoral Fellow
Section of Virology
Istituto Superiore di Sanità
Rome, Italy

Noelle Bersch, B.A.
Staff Research Associate
Dept. of Medicine
UCLA School of Medicine
Los Angeles, California

James E. Brown, M.D.
Instructor
Dept. of Medicine
Veterans Administration Hospital and
University of Washington
Seattle, Washington

J. Caro, M.D.
Clinical and Research Fellow in Medicine (Hematology)
Dept. of Medicine
Jefferson Medical College of
Thomas Jefferson University
Philadelphia, Pennsylvania

C. Cillo, Ph.D.
Associate Research Scientist of Internal Medicine
Istituto Patologia Medica
University of Naples–II Faculty
Naples, Italy

L. Cioé, Ph.D.
Instructor
Institute of Virology
University of Rome
Rome, Italy

E. Cobbs, B.A.
Research Technician
Dept. of Medicine
Jefferson Medical College of
Thomas Jefferson University
Philadelphia, Pennsylvania

T. M. Dexter, Ph.D.
Paterson Laboratories
Christie Hospital and Holt Radium Institute
Manchester, England

A. Dolei, Ph.D.
Instructor
Institute of Virology
University of Rome
Rome, Italy

Harvey Eisen, Ph.D.
Maitre de Recherche
Dept. of Molecular Biology
Institut Pasteur
Paris, France

xiv   List of Contributors

Allan J. Erslev, M.D.
Cardeza Research Professor of Medicine and
Director, Cardeza Foundation for Hematologic
  Research
Dept. of Medicine
Jefferson Medical College of
Thomas Jefferson University
Philadelphia, Pennsylvania

A. Fantoni, M.D.
Senior Investigator
Laboratorio di Dosimetria e Biofisica
CNEN–Casaccia
Rome, Italy

Eitan Fibach, Ph.D.
Research Associate
Cancer Center
Institute of Cancer Research
Columbia University
New York, New York

Gregory R. Fink, Ph.D.
Dept. of Pharmacology
Tulane University
New Orleans, Louisiana

James W. Fisher, Ph.D.
Professor and Chairman
Dept. of Pharmacology
Tulane University
New Orleans, Louisiana

Vincent S. Gallicchio, Ph.D.
Postdoctoral Fellow
The Bob Hipple Laboratory for Cancer Research
Sloan-Kettering Institute for Cancer Research
New York, New York

R. Gambari, Ph.D.
Research Associate
Istituto di Biologia Generale
University of Rome
Rome, Italy

David W. Golde, M.D.
Associate Professor of Medicine
Dept. of Medicine
UCLA School of Medicine
Los Angeles, California

Albert S. Gordon, Ph.D.
Research Professor
Dept. of Biology
New York University
New York, New York

Sergio Gorini, M.D.
President, Fondazione Internazionale Menarini
Milan, Italy

Bobby J. Gormus, Ph.D.
Assistant Professor
Dept. of Microbiology
University of Minnesota
School of Medicine
Minneapolis, Minnesota

Joel L. Granick, M.D.
Dept. of Medicine
Downstate Medical Center
State University of New York
New York, New York

Gisela Groothold, M.D.
Director for International Development
Menarini Laboratories
Florence, Italy

Dennis M. Gross, Ph.D.
Postdoctoral Fellow
Dept. of Pharmacology
Tulane University
New Orleans, Louisiana

L. J. Guilbert, Ph.D.
Friedrich Miescher Institute
Basel, Switzerland

Hiroshi Hara, M.D., Ph.D.
Visiting Professor
Dept. of Medicine
Medical University of South Carolina
Charleston, South Carolina

Suzanne Hasthorpe, Ph.D.
Visiting Investigator
Radiobiological Institute TNO
Rijswijk, Netherlands

N. N. Iscove, Ph.D.
Basel Institute for Immunology
Basel, Switzerland

David A. J. Ives, Ph.D., B.Sc., A.R.C.S., F.C.I.C.
Assistant Director, Research Administration
Connaught Laboratories Ltd.
Willowdale, Ontario, Canada

E. Kansu, M.D.
Cardeza Clinical and Research Fellow in Medicine
  (Hematology)
Dept. of Medicine
Jefferson Medical College of
Thomas Jefferson University
Philadelphia, Pennsylvania

Manuel E. Kaplan, M.D.
Professor of Medicine
Dept. of Medicine
University of Minnesota
School of Medicine
Minneapolis, Minnesota

Attallah Kappas, M.D.
Professor and Physician-in-Chief
Dept. of Metabolism and Pharmacology
The Rockefeller University
New York, New York

Yasutami Kinoshita, M.D., D.M.S.
Professor of Medicine
2nd Dept. of Medicine
Niigata University
School of Medicine
Niigata, Japan

F. Lettieri, M.D.
Associate Research Scientist of Internal Medicine
Istituto Patologia Medica
University of Naples–II Faculty
Naples, Italy

N. Lin, M.S.
Senior Research Technician
Hematology Research Laboratory
Veterans Administration Hospital
Seattle, Washington

Philip Liu, B.A.
Research Assistant
Dept. of Biology
New York University
New York, New York

David L. McLeod, M.D., F.R.C.P.
Professor of Anatomy (Histology)
Dept. of Anatomy
University of Toronto
Toronto, Ontario, Canada

M. C. Magli, Ph.D.
Associate Research Scientist of Internal Medicine
Istituto Patologia Medica
University of Naples–II Faculty
Naples, Italy

Paul A. Marks, M.D.
Frode Jensen Professor of Medicine
Director, Cancer Center
Dept. of Medicine
College of Physicians and Surgeons
Columbia University
New York, New York

Maria Martinez, M.D.
Assistant Professor
Dept. of Pharmacology
Louisiana State University
New Orleans, Louisiana

G. Mastroberardino, M.D.
Associate Professor of Internal Medicine
Istituto Semeiotica Medica III
University of Rome
Rome, Italy

G. Migliaccio, Ph.D.
Associate Research Scientist of Internal Medicine
Istituto Patologia Medica
University of Naples–II Faculty
Naples, Italy

Edwin A. Mirand, Ph.D., D.Sc.
Associate Institute Director and Head
Dept. of Biological Resources
Roswell Park Memorial Institute
Buffalo, New York

Bruno Modder, M.D.
Postdoctoral Fellow
Dept. of Pharmacology
Tulane University
New Orleans, Louisiana

Yoshiaki Moriyama, M.D., D.M.S.
Head of Hematology
2nd Dept. of Medicine
Niigata University
School of Medicine
Niigata, Japan

Martin J. Murphy, Jr., Ph.D.
Director of Hematology Training Program and
The Bob Hipple Laboratory for Cancer Research
Sloan-Kettering Institute for Cancer Research
New York, New York

Brian A. Naughton, B.S.
Assistant Research Scientist
Dept. of Biology
New York University
New York, New York

Benet Nomdedeu, M.D.
Visiting Scientist
Dept. of Medicine
University of Minnesota
School of Medicine
Minneapolis, Minnesota

Makio Ogawa, M.D., Ph.D.
Associate Professor
Dept. of Medicine
Medical University of South Carolina
Charleston, South Carolina

Yasuhico Ohno, M.D., D.M.S.
Postdoctoral Fellow
Dept. of Pharmacology
Tulane University
New Orleans, Louisiana

Uta Optiz, M.D.
Assistant Professor
Dept. of Clinical Physiology
University of Ulm
Ulm, West Germany

Wolfram Ostertag, Ph.D.
Professor
Dept. of Molecular Biology
Max Planck Institute for Experimental Medicine
Göttingen, West Germany

S. Paist, M.D.
Senior Resident
Dept. of Family Medicine
Highland Hospital
University of Rochester
School of Medicine
Rochester, New York

Cesare Peschle, M.D.
Associate Professor of Internal Medicine
Istituto Patologia Medica
University of Naples–II Faculty
Naples, Italy

Sam J. Piliero, Ph.D.
Professor and Chairman
Dept. of Histology
New York University
College of Dentistry
New York, New York

F. Pizzella, M.D.
Associate Research Scientist of Internal Medicine
Istituto Patologia Medica
University of Naples–II Faculty
Naples, Italy

William J. Popovic, M.D.
Research Fellow in Hematology
Dept. of Medicine
Veterans Administration Hospital and
University of Washington
Seattle, Washington

Franciszek Przala, Ph.D., D.V.M.
Postdoctoral Fellow
Dept. of Pharmacology
Tulane University
New Orleans, Louisiana

Arvind B. Rege, Ph.D.
Assistant Professor
Dept. of Pharmacology
Tulane University
New Orleans, Louisiana

Richard A. Rifkind, M.D.
Professor of Medicine and Human Genetics
Co-Director, Cancer Center
Dept. of Medicine and Human Genetics
College of Physicians and Surgeons
Columbia University
New York, New York

John Rinehart, M.D.
Assistant Professor
Dept. of Medicine
University of Minnesota
School of Medicine
Minneapolis, Minnesota

G. B. Rossi, M.D.
Associate Professor and Guest Investigator
Section of Virology
Istituto di Sanità
Rome, Italy;
Chairman, Dept. of Microbiology
University of Rome
Rome, Italy

Shigeru Sassa, M.D., D.M.S.
Associate Professor
Dept. of Metabolism and Pharmacology
The Rockefeller University
New York, New York

G. F. Sasso, M.D.
Associate Professor of Internal Medicine
Istituto Semeiotica Medica V
University of Rome
Rome, Italy

Masatsugu Sato, M.D.
2nd Dept. of Medicine
Niigata University
School of Medicine
Niigata, Japan

Umihiko Sawada, M.D.
Fellow in Hematology
Dept. of Internal Medicine
Section of Hematology
Rush-Presbyterian-St. Luke's Medical Center
Chicago, Illinois

G. Scala
Predoctoral Trainee of Internal Medicine
Istituto Patologia Medica
University of Naples–II Faculty
Naples, Italy

H. J. Seidel, M. D.
Professor
Dept. of Clinical Physiology
University of Ulm
Ulm, West Germany

Faith Shiota, B.S.
Research Technician
Hematology Research Laboratory
Veterans Administration Hospital
Seattle, Washington

Mona M. Shreeve
Laboratory Technician
Dept. of Anatomy
University of Toronto
Toronto, Ontario, Canada

R. Silver, D.Sc.
Research Associate in Medicine
Dept. of Medicine
Jefferson Medical College of
Thomas Jefferson University
Philadelphia, Pennsylvania

A. Soricelli
Predoctoral Trainee of Internal Medicine
Istituto Patologia Medica
University of Naples–II Faculty
Naples, Italy

Maureen E. Sullivan, B.S.
Research Assistant
The Bob Hipple Laboratory for Cancer Research
Sloan-Kettering Institute for Cancer Research
New York, New York

Nydia G. Testa, M.D., Ph.D.
Paterson Laboratories
Christie Hospital and Holt Radium Institute
Manchester, England

F. E. Trobaugh, Jr., M.D.
Professor of Medicine
Rush Medical College;
Director, Section of Laboratory Hematology
Dept. of Internal Medicine
Section of Hematology
Rush-Presbyterian-St. Luke's Medical Center
Chicago, Illinois

Akio Urabe, M.D.
Research Fellow
The Bob Hipple Laboratory for Cancer Research
Sloan-Kettering Institute for Cancer Research
New York, New York

Gerard Wagemaker, Ph.D.
Research Scientist
Radiobiological Institute TNO
Rijswijk, Netherlands

Lewis Wasserman, M.D.
Distinguished Service Professor
Dept. of Medicine
Mt. Sinai School of Medicine
City University of New York
New York, New York

Rona Singer Weinberg, Ph.D.
Research Associate
Dept. of Medicine
New York University
School of Medicine
New York, New York

Gregory L. Zalar, M.D.
Resident
Dept. of Dermatology
College of Physicians and Surgeons
Columbia University
New York, New York

Esmail D. Zanjani, Ph.D.
Professor and Director
Human Bone Marrow Diagnostic Research Unit
Dept. of Medicine and Physiology
University of Minnesota
School of Medicine
Minneapolis, Minnesota

James R. Zucali, Ph.D.
Senior Cancer Research Scientist
Dept. of Biological Resources
Roswell Park Memorial Institute
Buffalo, New York

In Vitro Aspects of Erythropoiesis

1 Erythropoiesis *In Vitro:* A Brief Introduction

Martin J. Murphy, Jr.

The *in vitro* culture of bone marrow cells began early in this century, with the work of Carrel and Burrows in 1910.[1] Most of the observations, related to bone marrow cells, were obtained employing short-term *in vitro* systems, with periods of observations limited to 6 to 24 hr. The classical methods of tissue culture such as coverslip preparations, hanging-drop cultures, coverslip chambers, and agar cultures were also applied using bone marrow cells. Among those methods, the cell suspension culture, introduced by Osgood and Muscovitz,[2] was advantageous because absolute cell counts and smear preparations could be performed.

Because erythroid cells make hemoglobin *in situ*, radioactive iron was an obvious choice as a label for the study of early erythroid differentiation at the cellular level. The recognition in serum and urine of anemic animals of a factor that specifically stimulates erythropoiesis, erythropoietin, led to the study of the effect of this glycoprotein hormone on erythroid cells in suspension culture. Krantz et al.[3] introduced a method using erythropoietin-stimulated heme-synthesis measured by radioactive iron incorporation into heme in suspension culture of marrow cells after 2 to 3 days of incubation. This method provided many results, especially in the study of responsiveness of marrow cells to erythropoietin in clinical cases.

As multipotential stem cells (i.e., CFU-s) had been detected by the method of spleen colony formation,[4] and granulocyte-macrophage progenitors had been assayed in semisolid agar,[5,6] the detection and assay of erythroid precursor cells became a high priority but elusive goal of many investigators. This impediment to the study of erythropoiesis was overcome by Stephenson et al.,[7] who introduced the plasma clot method for the formation of erythroid colonies *in vitro*.

Using plasma clot cultures, Matoth and Ben-Porath[8] reported that serum from anemic rabbits increased the mitotic index of normoblasts from rabbit bone marrow. In 1971, Stephenson et al.[7] reported the growth of erythroid cells in plasma clot culture using mouse fetal liver cells with exogenous addition of erythropoietin to the medium. Other early investigators reported the successful growth of erythroid cells using the plasma clot culture system, Gregory et al.[9] using mouse marrow and Tepperman et al.[10] using human marrow.

However, the plasma clot culture technique still possessed traits that hampered *in vitro* analysis of erythroid colony growth. These drawbacks were resolved with the development of the methylcellulose technique by Iscove et al.[11] Working primarily with agar, Ichikawa et al.[12] already used methylcellulose instead of agar in a brief series of experiments to examine the growth of granulocytic colonies. Worton et al.[13] found methylcellulose convenient for the evaluation of conditioned media in the granulocytic colony assay. Iscove et al.[11] introduced the methylcellulose method for the assay of erythroid colonies of mouse and human bone marrow. Methylcellulose is water soluble, which facilitates cytogenetic analysis of the colonies; furthermore, the colonies are scored directly in the culture dishes without need for further manipulation.

After the recognition of the rather late stage erythroid precursors (i.e., CFU-e) in plasma clot[9] and methylcellulose cultures,[11] Axelrad et al.[14] established the assay for the more primitive erythroid progenitors (i.e., BFU-e) using the plasma clot method. Later, the detection and quantitation of BFU-e was confirmed using the methylcellulose protocol.[15]

That was the "state of the art" in 1975. The advances made during the past two years will be considered in the following chapters. Although some artistry still lingers in the cultivation of erythrocytes, a great deal of hard science has made most of the technology reproducible and provided an ever increasing awareness of the complexity of erythropoiesis.

References

1. Carrel A, Burrows MT: Culture de moelle osseuse et de rate. *C R Soc Biol 69:*299, 1910.
2. Osgood EE, Muscovitz AN: Culture of human bone marrow. *JAMA 106:*1888, 1936.
3. Krantz SB, Gallien-Lartigue O, Goldwasser E: The effect of erythropoietin upon heme synthesis by marrow cells *in vitro. J Biol Chem 238:*4085, 1963.
4. Till JE, McCulloch EA: A direct measurement of the radiation sensitivity of normal mouse bone marrow cells. *Radiat Res 14:*213, 1961.
5. Pluznick DH, Sachs L: The cloning of normal

'mast' cells in tissue culture. *J Cell Comp Physiol 66:*319, 1965.
6. Bradley TR, Metcalf D: The growth of mouse bone marrow cells *in vitro. Aust J Exp Biol Med Sci 44:*287, 1966.
7. Stephenson JR, Axelrad AA, McLeod DL, Shreeve MM: Induction of colonies of hemoglobin-synthesizing cells by erythropoietin *in vitro. Proc Nat Acad Sci USA 68:*1542, 1971.
8. Matoth Y, Ben-Porath (Arkin) E: Effect of erythropoietin on the mitotic rate of erythroblasts in bone marrow cultures. *J Lab Clin Med 54:*722, 1959.
9. Gregory CJ, McCulloch EA, Till JE: Erythropoietic progenitors capable of colony formation in culture: State of differentiation. *J Cell Physiol 81:*411, 1973.
10. Tepperman AD, Curtis JE, McCulloch EA: Erythropoietic colonies in cultures of human marrow. *Blood 44:*659, 1974.
11. Iscove NN, Sieber F, Winterhalter KH: Erythroid colony formation in cultures of mouse and human bone marrow: Analysis of the requirement for erythropoietin by gel filtration and affinity chromatography on agarose-concanavalin A. *J Cell Physiol 83:*309, 1974.
12. Ichikawa Y, Pluznick DH, Sachs L: *In vitro* control of the development of macrophage and granulocyte colonies. *Proc Nat Acad Sci USA 56:*488, 1966.
13. Worton RG, McCulloch EA, Till JE: Physical separation of hemopoietic stem cells from cells forming colonies in culture. *J Cell Physiol 74:*171, 1969.
14. Axelrad AA, McLeod DL, Shreeve MM, Heath DS: Properties of cells that produce erythrocytic colonies *in vitro,* in Robinson WA: Hempoiesis in Culture. Washington, DC, US Government Printing Office, 1974, p 226.
15. Iscove NN, Sieber F: Erythroid progenitors in mouse bone marrow detected by macroscopic colony formation in culture. *Exp Hemat 3:*23, 1975.

2 Erythropoietin-Independence of Early Erythropoiesis and a Two-Regulator Model of Proliferative Control in the Hemopoietic System

N. N. Iscove and L. J. Guilbert

Introduction

A body of indirect evidence is accumulating that suggests that the earliest stages of erythropoiesis are governed by erythropoietin-independent mechanisms. In this chapter some of this evidence will be summarized and used as a basis on which to propose a two-stage model of humoral regulation in the hemopoietic system.

BFU-e and CFU-e represent cells at relatively early and relatively late points in erythropoiesis prior to hemoglobin synthesis. When mice are severely bled, the numbers of CFU-e in marrow increase threefold during the ensuing 4 days, whereas over the course of 8 days after hypertransfusion, they fall to one-third normal numbers (Figure 2-1). These changes are those expected for a cell population derived from erythropoietin-sensitive precursors. This type of response, however, is not shared by BFU-e. In the same time periods, marrow BFU-e decrease after bleeding and increase after hypertransfusion. The latter response pattern is also shown by the granulocyte/macrophage progenitors (CFU-c) as well as by the pluripotential stem cells (CFU-s), and appears to reflect anatomical relocations between bone marrow and spleen rather than significant net numerical change. This relative unresponsiveness of BFU-e suggests that physiological erythropoietin fluxes have little influence on the more primitive hemopoietic cells.[1]

This conclusion is supported by measurements of the sensitivity of BFU-e and CFU-e to the lethal effect of tritiated thymidine (^{3}HTdR) on S-phase cells (Figure 2-2). Seventy-five percent of CFU-e from control mice are killed by a 20-min pulse of ^{3}HTdR *in vitro*, indicating that all of these cells are in active cell cycle. The small but significant decrease in the sensitive proportion of CFU-e after hypertransfusion may suggest some dependence on erythropoietin for maintenance of the proliferative activity. Only 30% of BFU-e are ^{3}HTdR-sensitive in control, bled, and hypertransfused mice. However, this proportion increases to 63% during marrow regeneration in irradiated host mice, and the increase is not influenced by hypertransfusion of the host. The transition of BFU-e from a "low" to a "high" proliferative state during hemopoietic regeneration suggests that the cycling of BFU-e is subject to regulation that is erythropoietin-independent.[1]

The failure of BFU-e to sustain a similar high proliferative state in the normal mouse after bleeding indicates that these cells may not be able to respond to physiological erythropoietin concentrations. The fact that CFU-s and CFU-c also undergo similar transitions from low to high proliferative states during regeneration in irradiated hosts[2,3] suggests the possibility that the cycling of BFU-e, CFU-c, and CFU-s

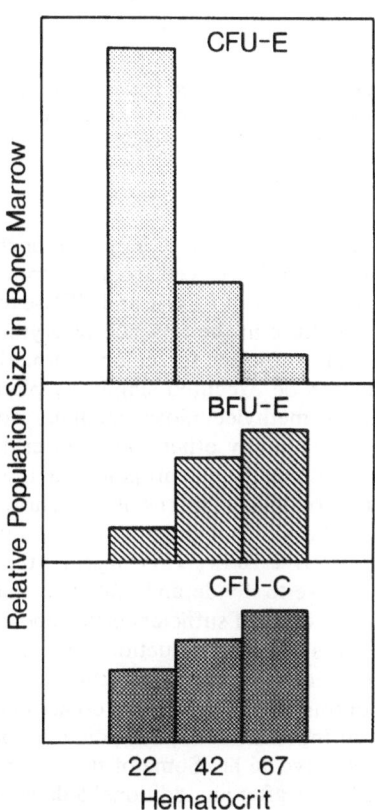

Figure 2-1 Relative changes in numbers of colony-forming cells in femoral marrow 4 days after bleeding (hematocrit, 22) or 8 days after hypertransfusion (hematocrit, 67). Normal values (hematocrit, 42) per two femurs are CFU-e, 84500, BFU-e (10 day), 3980, and CFU-e, 79300.

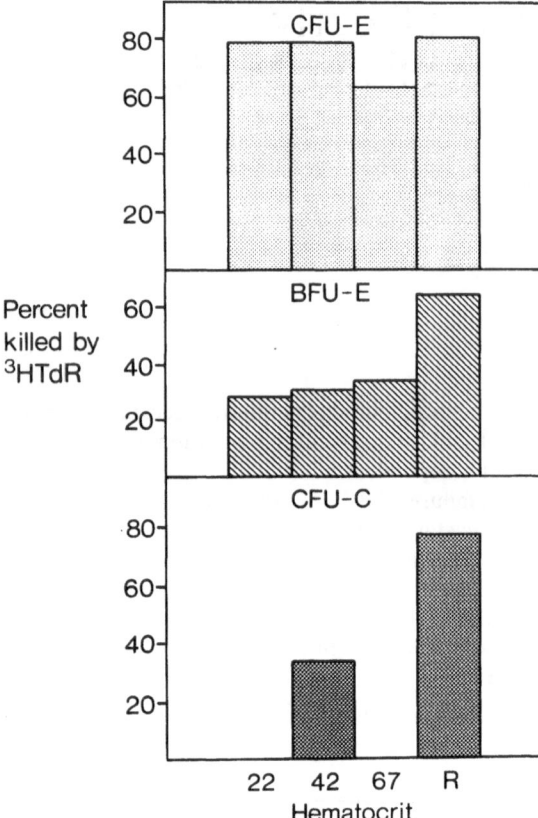

Figure 2-2 Sensitivity of colony-forming cells to a 20-min pulse of ^{3}HTdR *in vitro* after harvesting from marrows of normal, bled, and hypertransfused mice, and from marrow regenerating in irradiated hypertransfused (R) mice.

could be under the control of identical mechanisms.

Other evidence is derived from experiments that examine the capacity of BFU-e and CFU-e to survive a period of culture in the absence of erythropoietin concentrations sufficient for colony formation. The experiment can be executed simply by plating marrow cells in methylcellulose medium containing serum but without any other source of erythropoietin, and then adding the erythropoietin to the cultures after a period of delay. The results are shown in Figure 2-3. CFU-e survival drops from the outset and only a few survive 20 hr. This suggests that the capacity of CFU-e to divide and mature is maintained only in the presence of sufficient erythropoietin. Further, it appears that the production of CFU-e by their immediate precursors is also erythropoietin-dependent. In contrast, BFU-e numbers do not decline significantly during 48 hr of erythropoietin deprivation, and many survive 96 hr. Some of the latter go on to form large bursts after an additional 5 days of culture with erythropoietin. Since these bursts are as large as those in control cultures initiated with erythropoietin and incubated 9 days, it is possible that proliferation occurs during the initial 4 days of culture without erythropoietin. The results suggest that BFU-e do not need erythropoietin for survival, and may not even need it for proliferation.[4]

BFU-e and CFU-e require serum for colony formation in culture. It is now clear that the serum requirement of CFU-e can be explained, probably entirely, by their need for albumin, transferrin, selenium, and lipids (see ref. 5). Figure 2-4 indicates that when these substances are included in culture medium in addition to erythropoietin, large numbers of CFU-e form colonies in the absence of added serum. However, no bursts develop from BFU-e unless horse or fetal calf serum is added. These sera therefore appear to contain an activity essential for colony formation by BFU-e but not absolutely required by the later stage represented by CFU-e. The system provides a basis for an assay for burst-promoting activity (BPA) that can be expected to allow purification and characterization of the molecules responsible for the activity.

Data emerging from a number of other laboratories add support to the concept that there is an extra, nonerythropoietin requirement for BFU-e growth not shared by more mature erythroid precursors. Aye[6], Gregory and Eaves,[7] and Housman (personal communication) have recently reported that under certain conditions of culture, burst formation by human BFU-e is dependent on the addition of media conditioned by human embryo kidney cells or by human peripheral leukocytes. Wagemaker (Chapter 7) has confirmed and extended these findings in cultures of mouse bone marrow. Of particular interest are his observations that human leukocyte conditioned medium, in addition to exhibiting BPA in mouse marrow cultures, also stimulates DNA synthesis in mouse pluripotential stem cells (CFU-s) and promotes colony formation by a low-density subclass

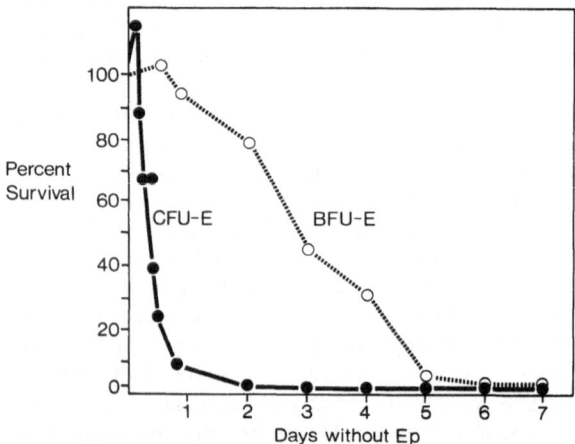

Figure 2-3 Survival of CFU-e and BFU-e in the absence of added erythropoietin.

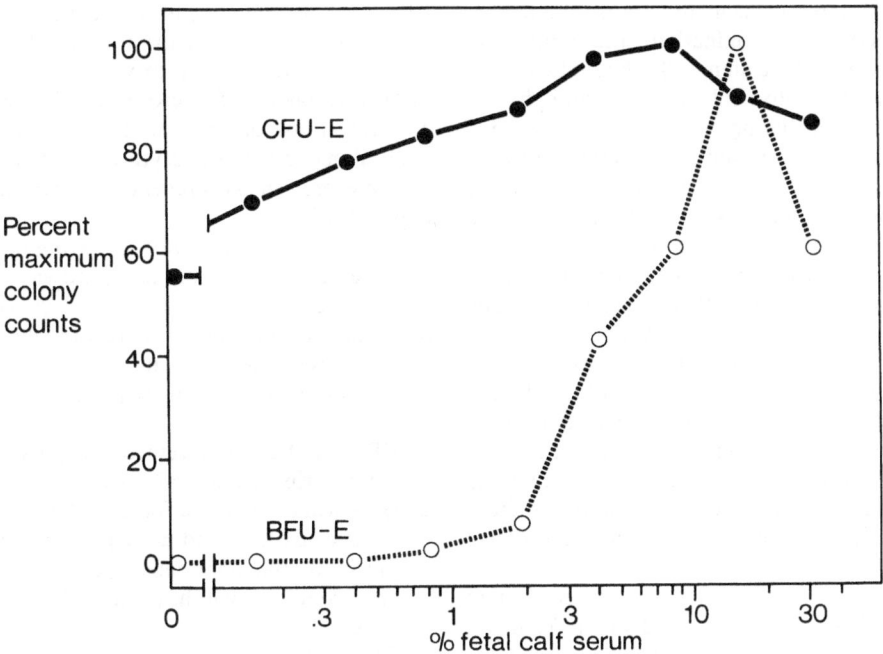

Figure 2-4 Dependence on serum for colony formation by CFU-e and BFU-e in cultures containing albumin, transferrin, selenite, soybean lipids, and cholesterol.

of CFU-c. Johnson and Metcalf have recently reported that large erythroid colonies develop in cultures of fetal liver or adult bone marrow (BM) when the cultures also contain Pokeweed Conditioned Medium (PWCM) conditioned by pokeweed mitogen-stimulated mouse spleen cells[8]. These colonies form in the absence of added erythropoietin. It appears likely that part of the action of both human leukocyte conditioned medium and mouse PWCM is to greatly decrease the amount of erythropoietin required for burst formation (Gregory and Axelrad, personal communications). In this way, the small amounts of erythropoietin in the serum component of the culture medium become sufficient to support burst development. PWCM also replaces serum in the BPA assay described above (Iscove, in progress).

The model illustrated in Figure 2-5 represents an attempt to integrate the foregoing data. It is proposed that pluripotential stem cells possess receptors for regulators that govern their proliferative activity.

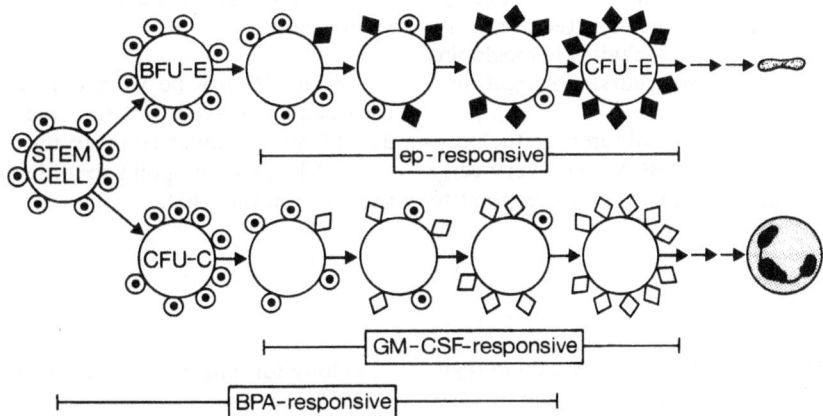

Figure 2-5 A two-stage model of humoral regulation of hemopoiesis. Early cells possess receptors (small dotted circles) to BPA, but not to pathway-specific factors. As the progeny of the earliest committed cells mature, they lose BPA receptors and acquire pathway-specific ones. Two examples, erythropoietin receptors (closed diamonds) and GM-CSF receptors (open diamonds), are illustrated.

Since time is required for cells to express the consequences of genetic determination, the immediate committed progeny of the stem cells should be identical to their parents in physical, chemical, and physiological properties, including size, density, metabolic and cell cycle activity, and surface structure. If this is so, then they should also retain responsiveness to the same stem cell regulators. As the earliest CFU-c and BFU-e mature, synthesis of receptors to stem cell regulators shuts down, and the receptors are diluted as the cells grow and divide. Sensitivity to stem cell regulators diminishes accordingly. At about the same time, synthesis of erythropoietin receptors is turned on in cells proceeding along the erythroid pathway. The first erythropoietin receptors appear and accumulate, conferring increasing sensitivity to that hormone. A parallel development of responsiveness to granulocyte/macrophage colony-stimulating factors (GM-CSF) occurs in the granulocyte/macrophage pathway.

It is proposed that the BPA assay detects a stem cell-type regulator or regulators, distinct from the pathway-specific factors erythropoietin and GM-CSF. The function of BPA, as well as of erythropoietin and GM-CSF, is to promote proliferation and survival of responsive cells. If BPA is present in high concentrations, then all BPA-responsive cells can survive and proliferate independently of erythropoietin, down to a point near but prior to CFU-e. Erythropoietin would then be required only for proliferation and survival of cells beyond the point where responsiveness to BPA is lost. Since cells at this point are very sensitive to erythropoietin, low concentrations of erythropoietin would be sufficient. Therefore, in the presence of high concentrations of BPA, proliferation and maturation of cells along the entire erythroid pathway is possible, with only a relatively low requirement for erythropoietin. BPA can provide most of the proliferation for burst development in culture, and erythropoietin is required only for the final maturation stages, including hemoglobinization, which render the burst recognizably erythroid.

The situation would be quite different in the presence of low concentrations of BPA. Only very early cells would have sufficient sensitivity to respond to BPA. Slightly later cells would be incapable of responding to low concentrations of BPA, but would now be able to respond to very high concentrations of erythropoietin. Therefore, in the presence of low concentrations of BPA, maturation and survival or cells along the entire erythroid pathway would be possible only in the presence of high erythropoietin concentrations.

The two-regulator model distinguishes between stem cell-type regulators and pathway-specific regulators. The existence of these two classes of regulator, and the persistence of responsiveness to stem cell-type regulators among early committed cells, would compel the following predictions:

1. BPA will be a requirement for growth of colonies from puripotential cells in culture.
2. BPA will be required for the efficient generation of CFU-c, BFU-e, and perhaps lymphoid cells from pluripotential stem cells in culture. GM-CSF and erythropoietin will have no influence on this process.
3. BPA will have identical effects on pluripotential stem cells and the earliest committed members of the erythroid, granulocyte/macrophage and perhaps lymphoid series.
4. The dose-response curve for erythropoietin dependence of burst formation will shift to the right or to the left depending on the concentration of BPA in the cultures.
5. BPA will reduce the requirement for GM-CSF by granulocyte/macrophage precursors by a mechanism similar to that for erythropoietin.
6. BPA will mimic GM-CSF in producing granulocyte/macrophage colonies. These will derive from a more primitive spectrum of cells than those produced by GM-CSF alone. The cells comprising BPA-stimulated colonies will cease growth before they have fully matured, unless GM-CSF is also present.

The model will be valid only to the extent that BPA can be assayed, isolated, and assigned a molecular identity distinct from erythropoietin and GM-CSF. All signs now point to an imminent and lively race toward this goal.

References

1. Iscove NN: The role of the erythropoietin in regulation of population size and cell cycling of early and late erythroid precursors in mouse bone marrow. *Cell Tis Kinet 10:*323–334, 1977.
2. Becker AJ, McCulloch EA, Siminovitch L, Till JE: The effect of differing demands for blood cell production on DNA synthesis by haemopoietic colony-forming cells of mice. *Blood 26:*296–308, 1965.
3. Iscove NN, Till JE, McCulloch EA: The proliferative states of mouse granulopoietic progenitor cells. *Proc. Soc. Exp Biol. Med 134:*33–36, 1970.
4. Iscove NN: Regulation of proliferation and mat-

uration at early and late stages of erythroid differentiation, in Saunders GF(ed): *Cell Differentiation and Neoplasia*. New York, Raven Press, 1978.
5. Guilbert LJ, Iscove NN: Partial replacement of serum by selenite, transferrin, albumin and lecithin in haemopoietic cell cultures. *Nature* 263:594–595, 1976.
6. Aye MT: Erythroid colony formation in cultures of human marrow: effect of leukocyte conditioned medium. *J Cell Physiol* 91:69–78, 1977.
7. Gregory CJ, Eaves AC: Human marrow cells capable of erythropoietic differentiation *in vitro*. Definition of three erythroid colony responses. *Blood* 49:855–864, 1977.
8. Johnson GR, Metcalf D: Erythropoietin independent erythroid colony formation *in vitro* by fetal mouse cells. *Exptl Hemat* 5(Supp. 2):75, 1977.

3 *In Vitro* Modulation of Human Erythropoiesis by Lymphocyte-Rich Fractions of Blood

Esmail D. Zanjani, Benet Nomdedeu, John Rinehart, Bobby J. Gormus, and Manuel E. Kaplan

Introduction

It is generally accepted that erythropoiesis in mammals is regulated primarily by the hormone erythropoietin (Ep).[1,2] Several lines of evidence suggest that cell–cell interactions also play an important role in the production of erythrocytes. Trentin,[3] for example, described an erythroid hematopoietic inductive microenvironment (HIM) that plays an important role in directing the differentiation of the multipotential stem cell or its progeny toward the erythroid line. A defect in the HIM is considered to be responsible for the genetically determined anemia seen in Steel mice.[4] Cells responsible for the HIM are thought to be a type of reticuloendothelial cell, since they exhibit resistance to ionizing irradiation.

The phenomenon of hybrid resistance or CFU-s repression described by Cudkowicz et al.[5] and McCulloch and Till[6] occurs in heavily irradiated animals, is not enhanced by immunization, and is unlikely to involve B-lymphocytes. Lotzova and Cudkowicz[7] have implicated the macrophages in hybrid resistance, since treatment of the host with silica, a macrophage toxin, abrogated hybrid resistance.[7] Additional evidence for cell–cell interactions in the regulation of erythropoiesis has been obtained from several types of *in vivo* and *in vitro* studies. Goodman and Shinpock[8] showed that proliferation of parent marrow in irradiated hybrid mice was enhanced by concomitant injection of parental thymus cells. The degree of this enhancement was related to the number of thymus cells injected. A recent study[9] demonstrates that the ability of the genetically normal (+/+) bone marrow or spleen cells to correct the macrocytic anemia of W/W^v mice may be T cell-dependent: treatment of these donor cells with complement and antisera to the T cell antigen Thy 1.2 abolishes their effect. Petrov et al.[10] have shown that the transfer of lymphocytes together with allogeneic bone marrow cells into lethally irradiated recipient mice resulted in the inactivation of spleen colony-forming units; this effect was not linked to the H-2 histocompatibility system.[11]

The availability of culture techniques for the study of proliferation and differentiation of erythropoietic precursors *in vitro*[12–16] has provided additional support for the existence of regulatory processes involving cell–cell interaction in erythropoiesis. In mice, Heath et al.[15] observed that co-cultivation of various fractions of marrow cells, previously separated by unit gravity sedimentation, resulted in a synergistic increase in the number of erythrocytic bursts (BFU-e) formed *in vitro*. It has also been shown that the plating efficiency of erythroid progenitor cells (CFU-e) in human marrow is influenced by an adherent cell population.[17] In a series of studies, we have examined the mechanisms underlying the anemia in patients with Diamond–Blackfan syndrome[18] and acquired aplastic anemia.[19] It was demonstrated in these states that human peripheral blood lymphocytes play a significant role in the proliferation and differentiation of human CFU-e *in vitro*. Additional studies revealed that in some cases this role can be attributed to T-lymphocytes, and that the effect is most probably mediated via direct cell–cell interaction. More recently, we have found that the formation of normal human peripheral blood BFU-e *in vitro* can be significantly affected by circulating monocytes. A description of these studies follows.

Materials and Methods

Informed consent was obtained from both patients and normal donors.

Bone Marrow Cultures (CFU-e Determinations)
Hematologically normal individuals served as sources of normal human bone marrow cells. All bone marrow aspirations from both patients and normal donors were performed at the posterior iliac crest. The bone marrow cells obtained were cultured in the plasma clot culture system by a modification of the procedure described by Tepperman et al.[14] In brief, 3 ml of bone marrow aspirate was placed in 5 ml of 2% FCS in MEM (see Appendix 1, p. 243) containing 10 units/ml heparin. The cells were spun at 1,000 rpm for 10 min at room temperature, resuspended in 5 ml of 2% FCS in MEM without heparin. After centrifugation (1,000 rpm, 10 min, room temperature), the supernatant was discarded and the buffy coat was placed in an appropriate volume of 2% FCS in MEM without heparin, and suspended by means of sterile Pasteur pipet. After determining the total numbers of nucleated cells present, desired concentrations of cells were prepared using 2% FCS in MEM without

heparin so that each cell concentration was in 0.1 ml volume of the medium. In all studies, the indicated concentration of Ep and bone marrow cells refers to the amounts present in 1.1 ml of culture medium. The culture medium consisted of beef embryo extract (0.1 ml), 10% bovine serum albumin (0.1 ml), NCTC-109 (0.1 ml), 0.02 mg L-asparagine in 0.1 ml NCTC-109, α-thioglycerol (0.1 ml; the final concentration in culture was $10^{-4}M$), heat-inactivated fetal calf serum (0.3 ml), Ep dissolved in NCTC-109 (0.1 ml), and citrated bovine plasma (0.1 ml). The total volume including the bone marrow cells was 1.1 ml. One-tenth ml aliquots of this mixture were then cultured in microtiter wells at 37°C in a humidified atmosphere of 5% CO_2 in air for 7 days. Studies designed to establish the Ep-dependent nature of erythroid colony formation involved the culture of dispersed bone marrow cells, obtained from marrow aspirates by centrifugation at 1,000 rpm for 10 min, at concentrations of 2, 4, 6, and 8 × 10^5 cells in the presence or absence of 2 IU Ep for 7 days. These studies, which were conducted in duplicates, involved bone marrow cells from five separate hematologically normal donors and included the following two additional culture conditions for each cell concentration from each donor: bone marrow cells were cultured in media containing, in addition to 2 IU Ep, 1 mg of either normal rabbit serum IgG or 1 mg of IgG obtained from pooled sera of rabbits that had been immunized with Ep. The immune IgG was prepared in rabbits according to the procedure outlined by Schooley et al.[20] The animals were initially immunized against a preparation of Ep (5 IU/mg protein) obtained from urine of a patient with aplastic anemia. Another preparation of human urinary Ep with a greater potency (630 IU/mg protein) was used to boost the antibody titers in these rabbits. One ml of the immune serum after absorption against normal human urinary proteins was found to neutralize approximately 90 IU of human urinary Ep. Absorption against human urinary protein was achieved by incubating 20 ml of the antiserum with an amount of urinary protein, separated by dialysis and ultrafiltration from 3 liters of normal human urine, for 1 hr at 37°C followed by an additional incubation for 5 hr at 4°C. The mixture was centrifuged at 20,000 g for 15 min and the supernatant was used in subsequent studies. In these studies, the IgG fraction of the immune serum, separated by DEAE-Sephadex (Pharmacia Fine Chemicals, Inc., Piscataway, NJ) chromatography,[21] was used. Similarly prepared gammaglobulin fraction from normal rabbit serum was used as control. One mg of the immune IgG was found to neutralize approximately 6 IU of human urinary Ep. The IgG was added to each cell concentration at the start of the culture. At 7 days of incubation, the clots were removed and transferred to glass slides, fixed in gluteraldehyde, and then stained with benzidine and hematoxylin. Under ×100 magnification, each clot was examined and only colonies consisting of 8 or more benzidine-positive cells were counted. No attempt was made to distinguish between smaller colonies (8 to 32 cells/colony) and large colonies (≥32 cells/colony) in the calculation of the results. A total of 16 separate clots were examined for each cell concentration and culture condition.

Peripheral Blood Mononuclear Cell Cultures
(BFU-e Determinations)

Mononuclear cells were obtained from peripheral blood of hematologically normal human donors by a combination of Ficoll-Hypaque density centrifugation, adherence to plastic plates, and use of carbonyl iron followed by Ficoll-Hypaque density centrifugation. Concentrations of 0.5, 1, 1.5, 2, 2.5, 3, 4, 5, and 6 × 10^6 cells of mononuclear cells were cultured in the presence of 4 IU Ep in the plasma clot as described by Clarke and Housman.[16] Except for the concentration of Ep used, the culture procedure for BFU-e differed from the CFU-e culture techniques in only two other aspects: semipurified thrombin (1 unit/1.1 ml culture) was used in place of beef embryo extract, and the cultures were incubated for 13 days instead of 7 days. Otherwise, the culture, as well as staining procedures, was similar to bone marrow cultures. These clots, however, were examined at ×10 magnification and areas of erythroid activity forming a distinct grouping of benzidine-positive subunits were considered as a single BFU-e colony. Although not all such subunits exhibited benzidine-positivity in every colony, generally 60 to 80% were stained with benzidine. In general, no difficulties in identifying such distinct areas of erythroid burst activity were experienced with concentrations of mononuclear cells up to 2 × 10^6 cells/1.1 ml of the culture. The determinations of BFU-e colonies, however, were more difficult with starting cell concentrations of 3 × 10^6 cells/1.1 ml culture medium and became almost impossible when 4 × 10^6 cells were used. In the latter case, the growth was confluent with no clear separation of areas of erythroid activity.

Separation of Peripheral Blood Mononuclear Cells

Mononuclear cells from peripheral blood were employed in two separate groups of studies. In the studies where these cells were co-cultured with human bone marrow cells, they were prepared as follows: 20 ml of heparinized blood was diluted 1:1 or 1:3 with Ca- and Mg-free Hank's balanced salt solution (HBSS). The diluted blood was layered on top of Ficoll-Hypaque in sterile glass test tubes and spun at 400 g for 30 min at 20°C. The interface cells were removed and transferred to sterile plastic tubes (Falcon Plastics) and washed three times with HBSS. The cells were resuspended at desired concentrations in MEM made 15% with fetal calf serum (FCS), and

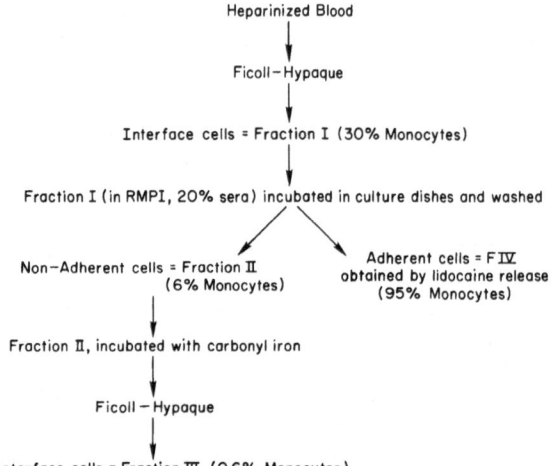

Figure 3-1 General scheme employed to prepare mononuclear cell fractions from human peripheral blood.

used in the co-culture studies with bone marrow cells. In experiments involving BFU-e determinations and the assessment of the role of monocytes in BFU-e proliferation, the procedure for obtaining mononuclear cells from blood is outlined in Figure 3-1 and detailed below. Heparinized blood, diluted 1:1 with cold Seligman's balanced salt solution (SBSS), was layered on Ficoll–Hypaque in 50 ml sterile conical polypropylene tubes and spun at 400 g for 25 min at 15°C. Cells at the interface, designated fraction I (F = I), were then removed, placed in 50 ml sterile polycarbonate tubes, and centrifuged at 400 g for 10 min at 4°C. The cells were then resuspended by vigorous agitation in ice cold SBSS and centrifuged at 200 g for 10 min at 4°C. This step was repeated and an aliquot of cells was resuspended at desired cell concentrations in SBSS containing 2% FCS. The percentage of monocytes was determined in the F = I cells on Giemsa stained, cytocentrifuge-prepared slides, and the remaining F = I cells were again centrifuged in ice cold SBSS at 200 g for 10 min and resuspended in RPMI-1640 containing 20% fresh autologous serum at a concentration of 1.5×10^6 monocytes/ml. Three-ml aliquots of this cell suspension were then placed in 60 mm plastic culture dishes and incubated for 1 hr at 37°C in a humidified atmosphere of 5% CO_2 in air. The nonadherent cells were then mobilized by swirling and decanted. The plates were then washed six times by placing 2 ml of HBSS onto each dish, swirling and decanting on each occasion. These washes and the original nonadherent cells were pooled and designated as fraction II (F-II). An aliquot of cells from this fraction was centrifuged at 200 g for 10 min and resuspended in HBSS in desired concentrations until used. The percentage of monocytes in F-II was determined as described. The remaining F-II cells were spun at 200 g for 10 min and resuspended to a final concentration of 10^6 cells/ml in autologous plasma in 50 ml polypropylene tubes. Carbonyl iron (25 mg/ml) was mixed with this cell suspension and the mixture incubated at 37°C with constant rotation (20 rpm) for 30 min. The mixture was then carefully layered on top of Ficoll–Hypaque and centrifuged at 400 g for 30 min at room temperature. Resultant interface cells (F-III cells) were then washed twice in RPMI-1640 by centrifuging at 200 g for 10 min at 4°C. Appropriate dilutions of these cells were achieved by using RPMI-1640.

The cells that had adhered to the plastic dishes when F-II cells were being prepared were immediately removed from washed culture dishes by the addition of 3 ml of SBSS containing 30 mM lidocaine, and 10% autologous serum and subsequent incubation at room temperature for 15 min. The cells (F-IV) thus removed were then pooled and placed in 35 ml polycarbonate tubes and immediately washed three times by centrifugation (200 g, 10 min) with cold SBSS. In general, F-IV cells were cultured within 10 min of availability. As with F-I and F-II cells, the percentage of monocytes was determined for F-III and F-IV cells.

T- and B-lymphocyte–enriched fractions from peripheral blood were prepared by incubating F-I cells with washed sheep red cells for 30 min at 37°C in the presence of heat-inactivated AB serum. Approximately 50 to 100 sheep red cells/lymphocytes were used. The mixture was incubated at 37°C for 5 min, spun at 500 rpm for 5 min at room temperature, incubated at 4°C for an additional 90 min, and resuspended by gentle shaking. The mixture was layered on top of Ficoll–Hypaque and centrifuged at 400 g for 30 min at 20°C. The interface cells were removed and washed twice with HBSS. These cells comprised the B-enriched fraction of lymphocytes. The "rosetted" cells were resuspended in a 1:5 solution of phosphate buffered saline and distilled water and allowed to stand at room temperature for 45 sec. The mixture was centrifuged at 4,000 rpm for 2 min, supernatant discarded, and the cells washed twice with HBSS. The cells were then resuspended in 15% FCS in MEM and designated as the T cell-rich lymphocyte fraction.

Patients

Two groups of patients with blood dyscrasias were studied.

Patients with Diamond–Blackfan syndrome (CHA)

Three groups of patients with confirmed diagnosis of Diamond–Blackfan syndrome [congenital-hypoplastic anemia (CHA)] were studied. These included: (a) Seven patients who failed to respond to steroid therapy at any time since first diagnosed and required frequent red blood cell transfusions. At the time of the study, all were receiving red cell transfusions and four were being given prednisone. No significant neutropenia and thrombocytopenia was present in these patients. (b) Five individuals who after diagnosis underwent complete remission in response to steroid therapy, were not receiving transfusions when studied, but were receiving low-dose prednisone; and (c) three patients who after a prolonged initial period of remission induced by prednisone were in relapse at the time of the study. Of these, all were receiving prednisone and red cell transfusions. Most of these patients were studied on more than one occasion and six were examined three times.

Patients with acquired aplastic anemia (AA)

A total of 17 patients with acquired severe aplastic anemia (AA) were studied. The underlying reasons were not determined in nine patients; three had a history of prior exposure to benzene-containing compounds, and the cause for the remaining five patients was suspected to be exposure to multiple antibiotics. Three of the patients exhibited moderately hypocellular bone marrow on biopsy, and in all other patients, severe bone marrow hypocellularity was present. The patients ranged in age from 7 to 52 years; seven were male, and all exhibited severe anemia, neutropenia, and thrombocytopenia. Serum Ep levels were elevated in all patients (1.5 to 2.8 IU/ml); the mean reticulocyte counts were 0.1 to 0.3% (11 patients) and 0.4 to 0.8% (six patients); mean platelet counts ranged from 2,000 to 21,000/mm^3, and the mean neutrophil counts were 22 to 321/mm^3. One patient was studied both before and after a successful bone marrow transplant. All but five of the patients were studied on at least two separate occasions; these five were studied only once.

Effect of Peripheral Blood Lymphocytes (PBL) on CFU-e Proliferation *in vitro*

Peripheral blood lymphocytes (PBL) were separated from freshly drawn heparinized venous blood (20 to 40 ml) by Ficoll–Hypaque density sedimentation as described above. Varying numbers of PBL (0.5 to 6 × 10^5 cells) from each patient were so-cultured in the plasma clot with 6 × 10^5 normal human bone marrow cells in the presence or absence of 2 IU Ep for 7 days. Each study involving patient PBL included two types of controls. Normal bone marrow cells (6 × 10^5 cells) were cultured with and without 2 IU Ep or 0.5 to 4 × 10^5 normal human PBL in the presence or absence of the hormone. At least two normal PBL donors were used in each study, and frequently more than one hematologically normal marrow was employed. In about 50% of all studies, at least one of the normal PBL donors was age and sex matched with the patient. In all other studies, the normal PBL donors were adult volunteers with no indications of hematological abnormalities or infections. Of the normal donors, 53% were male. In a significant number of studies, an additional control group was also included. Since a significant number of patients with CHA and AA were multiple transfused, PBL from three patients with hemolytic anemia, three adults with pure red cell aplasia, and four with refractory anemias who had received multiple red cell transfusions were studied. In all cases, each culture was prepared in duplicate.

Results

The potency of the anti-Ep preparation used in this study was determined in ex-hypoxic polycythemic mice[2] and is shown in Table 3-1. Formation of erythroid colonies by bone marrow cells from 18 hematologically normal individuals cultured for 7 days is demonstrated in Table 3-2. No colonies were formed in the absence of exogenous Ep. Addition of Ep to these cultures, however, resulted in the appearance of significant numbers of colonies at all cell concentrations examined. Addition of anti-Ep to cultures of bone marrow cells from five normal subjects resulted in the total inhibition of erythroid colony formation in response to Ep (Figure 3-2). Thus, whereas normal rabbit serum IgG did not exert a significant influence on colony formation, IgG from immune rabbit serum suppressed colony formation (Figure 3-2).

Table 3-3 demonstrates that bone marrow cells from patients with CHA respond to Ep by producing significant numbers of erythroid colonies. Similar concentrations of bone marrow cells from patients in

Table 3-1 Erythropoietin (Ep)-neutralizing activity of anti-Ep used in these studies.

Material Assayed[a]	% RBC-^{59}Fe Incorporation ± 1 SEM[b]
Saline	0.56 ± 0.15
0.5 IU Ep[c]	16.32 ± 1.73
+ 0.006 ml normal rabbit serum	15.87 ± 2.14
+0.006 ml anti-Ep serum	0.19 ± 0.06
+ 0.1 mg normal rabbit serum IgG	18.30 ± 1.47
+ 0.1 mg anti-Ep serum IgG	0.98 ± 0.26
+ (0.1 mg anti-Ep serum IgG + 0.2 mg GARGG[d] IgG)	17.01 ± 1.96

[a] Determined in the ex-hypoxic polycythemic mouse assay system

[b] Mean ± 1 SEM of four separate assays.

[c] Ep used in these studies was collected and concentrated by the Department of Physiology, University of Northeast, Corrients, Argentina, and further processed and assayed by the hematology Research Laboratories, Children's Hospital of Los Angeles, Los Angeles, California, under grant HE–10880 from the National Heart Lung and Blood Institute of the National Institutes of Health.

[d] Goat anti-rabbit gammaglobulin.

Table 3-2 Erythroid colony formation by normal human bone marrow cells *in vitro*.[a]

No. Bone Marrow Cells Cultured	Total No. BFU-e[b]	
	No Ep	+ Ep
2×10^5 cells	0	182 ± 29
4×10^5 cells	0	263 ± 36
6×10^5 cells	0	527 ± 52
8×10^5 cells	0	792 ± 63

[a] 2 IU Ep was used.
[b] Each value represents mean ± 1 SEM of 18 separate studies each involving a separate normal donor.

remission exhibited significantly greater numbers of CFU-e than those in relapse. However, in both instances the marrow content of CFU-e was substantially below normal (Table 3-3). Cultures of bone marrow cells from patients with CHA who had relapsed after an initial period of remission were not successful and therefore are not included in Table 3-3.

The results of erythroid colony formation by bone marrow of patients with AA are shown in Table 3-4. In general, it was difficult to obtain adequate numbers of bone marrow cells from AA patients, and except for three patients, we were unable to obtain sufficient marrow for study. It is apparent that the number of CFU-e's in AA bone marrow is below normal (Table 3-4).

The effect of different concentrations (0.025, 0.05, and 0.1 ml) of sera from eleven CHA and nine AA patients on erythroid colony formation by normal human bone marrow (6×10^5 cells) in the presence or absence of 2 IU Ep was examined. The serum was added at the start of the culture and the volume adjusted by the addition of appropriate amounts of NCTC-109. Controls consisted of (a) normal human serum and (b) serum from an adult patient with pure red cell aplasia (PRCA). The latter was previously shown to contain an inhibitor to erythropoiesis.[22] The results of these studies are shown in Table 3-5 and demonstrate that except for the PRCA serum no inhibitory activity was present in normal, CHA, and AA sera. It should be noted that these sera (except for normal serum) contained significant quantities of Ep, which may be responsible for the slight increase in CFU-e seen in these cultures.

The results of co-culture studies involving normal human PBL and bone marrow are shown in Table 3-6. The data from 115 normal PBL donors are included in this table. It can be seen that although the effect varied from donor to donor, PBL from a significant number of donors (49) enhanced the formation of erythroid colonies by normal bone marrow in the presence of Ep. No effect was seen when no exogenous Ep was present. Of the remaining donors, PBL from 25 failed to affect colony formation, 31 produced a moderate increase, and 10 caused a moderate decrease in erythroid colony formation (Table 3-6).

This enhancement of erythroid colony formation by normal PBL was not due to the presence of CFU-e in the PBL. The results in Table 3-7 demonstrate that when different numbers of normal PBL (0.5 to 12×10^5 cells) were cultured in the presence of absence of 2 IU Ep for 7 days, no erythroid colonies

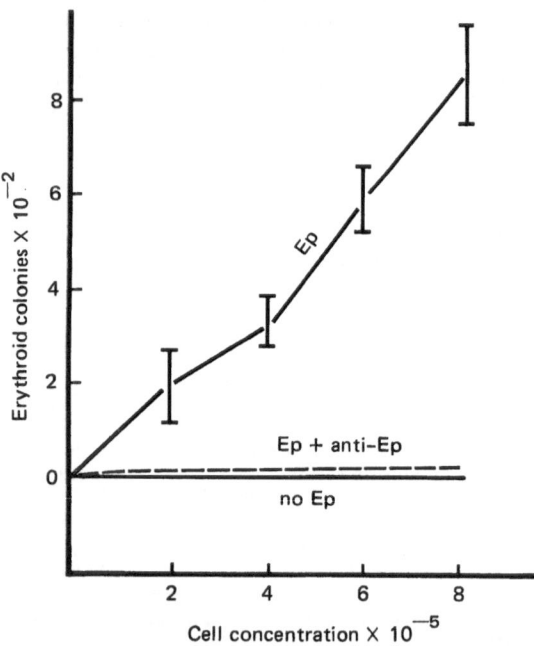

Figure 3-2 Effect of anti-erythropoietin (anti-Ep) on the formation of erythroid colonies by normal human bone marrow cells *in vitro*. Each point represents the mean ± 1 SEM of five separate studies each involving a different bone marrow donor. Two IU of erythropoietin was used for each cell concentration. Normal rabbit serum IgG was used as control.

Table 3-3 Erythroid colony formation by bone marrow cells from patients with Diamond–Blackfan syndrome (CHA) *in vitro*.[a]

Marrow Donors[b]	No. of Cells Cultured	Total No. CFU-e[c]
Normal (3)	4×10^5	311 ± 29
	6×10^5	482 ± 33
	8×10^5	702 ± 37
CHA in Relapse (4)	4×10^5	64 ± 8
	6×10^5	94 ± 26
	8×10^5	167 ± 19
CHA in Remission (3)	4×10^5	122 ± 17
	6×10^5	247 ± 43
	8×10^5	392 ± 31

[a] 2 IU Ep was used.
[b] Numbers in parentheses indicate the number of donors in each group.
[c] Each value represents mean ± 1 SEM of all studies.

Table 3-4 Erythroid colony formation by bone marrow cells from patients with acquired aplastic anemia (AA) in vitro.[a]

Bone Marrow Donors	Total No. CFU-e[b]		
	4×10^5 (Cell Conc.)	6×10^5 (Cell Conc.)	8×10^5 (Cell Conc.)
Normal	392 ± 44	585 ± 32	832 ± 53
Normal	287 ± 38	490 ± 25	614 ± 41
AA	50 ± 19	129 ± 14	157 ± 24
AA	51 ± 8	96 ± 7	120 ± 10
AA	63 ± 11	118 ± 12	148 ± 14

[a] 2 IU Ep was used.
[b] Each value represents mean ± 1 SEM of duplicate cultures.

were formed. By contrast, significant numbers of colonies were produced by normal marrow alone in the presence of Ep, and co-culture with PBL from some donors resulted in the enhancement of colony formation (Table 3-7).

The effect of PBL from patients with CHA on erythroid colony formation by normal bone marrow is shown in Table 3-8. Significant suppression of erythroid colony formation was observed when lymphocytes from six of the seven nonresponder patients were co-cultured with normal bone marrow in the presence of Ep. This inhibition was evident at all concentrations of PBL employed. However, when the ratio of lymphocytes to bone marrow cells approached one, a reversal of the inhibitory effect was seen (Table 3-8). Thus, whereas a 90 to 95% inhibition of colony formation occurred with concentrations of PBL ranging from 0.5×10^5 to 2×10^5 lymphocytes/6×10^5 bone marrow cells, the degree

Table 3-5 Effect of sera from patients with Diamond–Blackfan syndrome (CHA), acquired aplastic anemia (AA) or pure red cell aplasia (PRCA) on erythroid colony formation by normal human bone marrow cells in vitro.[a]

Serum Donors[b]	Serum Concentration (ml)	Total No. CFU-e/ 6×10^5 Cells[c]
Normal (8)	0.025	367 ± 29
	0.05	342 ± 32
	0.10	361 ± 21
CHA (7)	0.025	314 ± 39
	0.05	348 ± 36
	0.10	401 ± 10
AA (9)	0.025	397 ± 20
	0.05	382 ± 14
	0.10	412 ± 8
PRCA (1)[d]	0.05	46 ± 7

[a] 2 IU Ep was used.
[b] The numbers in parentheses indicate the number of donors in each group.
[c] Each value represents mean ± 1 SEM of all studies.
[d] Repeated with each study.

Table 3-6 Effect of peripheral blood lymphocytes (PBL) from normal humans on erythroid colony formation by normal human bone marrow cells in vitro.[a]

PBL Donors[b]	PBL Concentration	Total No. CFU-e/ 6×10^5 Cells[c]
Group 1 (49)	0	323 ± 10
	$0.5–2 \times 10^5$ cells	388 ± 21
	4×10^5 cells	482 ± 18
	6×10^5 cells	540 ± 12
Group 2 (25)	0	284 ± 7
	$0.5–2 \times 10^5$ cells	276 ± 13
	4×10^5 cells	302 ± 43
	6×10^5 cells	339 ± 34
Group 3 (31)	0	386 ± 17
	$0.5–2 \times 10^5$ cells	408 ± 8
	4×10^5 cells	437 ± 20
	6×10^5 cells	468 ± 11
Group 4 (10)	0	308 ± 14
	$0.5–2 \times 10^5$ cells	261 ± 15
	4×10^5 cells	283 ± 4
	6×10^5 cells	233 ± 13

[a] 2 IU Ep was used.
[b] The numbers in parentheses indicate the number of normal donors in each group
[c] Each value represents mean ± 1 SEM of all studies. Bone marrow cells from 55 separate donors were employed in these studies.

of suppression was approximately 80% for 4×10^5 lymphocytes and was decreased to 49% when 6×10^5 lymphocytes were used (Table 3-8). All concentrations of PBL from the remaining nonresponder CHA as well as from four of the patients who were in complete remission were without any significant effect on CFU-e proliferation (Table 3-8). PBLs from one patient in remission, and two who were in relapse following an initial prolonged period of remission caused inhibition of erythroid colony formation (Table 3-8). However, the degree of suppression approached statistical significance only when 4 to

Table 3-7 Erythroid colony formation by normal human peripheral blood lymphocytes (PBL) in the presence or absence of normal human bone marrow and/or erythropoietin in vitro.[a]

Cells Cultured	Total No. CFU-e[b]	
	No Ep	+ Ep
$0.5–12 \times 10^5$ PBL alone	0	0
6×10^5 bone marrow cells alone	0	418 ± 15
$0.5–2 \times 10^5$ PBL + 6×10^5 bone marrow cells	0	469 ± 13
$4–12 \times 10^5$ PBL + 6×10^5 bone marrow cells	0	583 ± 29

[a] PBL from 6 normal donors, and bone marrow cells from 9 donors were employed. The concentration of Ep used was 2 IU.
[b] Each value represents mean ± 1 SEM of all cultures.

Table 3-8 Effect of peripheral blood lymphocytes (PBL) from patients with Diamond–Blackfan syndrome (CHA) on erythroid colony formation by normal human bone marrow cells *in vitro*.[a]

PBL Donors[c]	Total No. CFU-e/6 × 10⁵ bone marrow cells[b]					
	0 (PBL Conc.)	0.5 × 10⁵ (PBL Conc.)	1 × 10⁵ (PBL Conc.)	2 × 10⁵ (PBL Conc.)	4 × 10⁵ (PBL Conc.)	6 × 10⁵ (PBL Conc.)
Normal (68)	395 ± 13	434 ± 9	484 ± 17	519 ± 8	573 ± 10	512 ± 14
CHA: nonresponder (6)		61 ± 3	47 ± 2	38 ± 4	110 ± 9	263 ± 5
CHA: nonresponder (1)		387 ± 63	413 ± 34	438 ± 53	400 ± 44	418 ± 60
CHA: in remission (4)		508 ± 26	439 ± 17	391 ± 11	487 ± 13	492 ± 48
CHA: in remission (1)		187 ± 36	206 ± 34	150 ± 28	310 ± 17	306 ± 43
CHA: in relapse (2)		211 ± 30	—	196 ± 23	201 ± 20	246 ± 14
CHA: in relapse (1)		343 ± 69	312 ± 16	444 ± 59	462 ± 32	441 ± 22

[a] 2 IU Ep was used.
[b] Each value represents mean ± 1 SEM of cultures.
[c] The numbers in parentheses indicate the number of donors in each group.

6×10^5 lymphocytes/6×10^5 bone marrow cells were employed. One responder patient who was in relapse when studied did not exhibit inhibitory activity (Table 3-8).

The results of the studies with AA patients are summarized in Table 3-9. Lymphocytes from 8 of the 17 patients with AA studied caused suppression of erythroid colony formation. The degree of suppression varied among these patients, with marked inhibition occurring with as few as 0.5×10^5 lymphocytes from some patients, whereas similar levels of inhibition were not observed in other patients until 2 to 6×10^5 lymphocytes were used (Table 3-9). PBLs from the remaining nine patients were without effect. The results presented in Table 3-10 demonstrate that (with one possible exception) PBL from patients with PRCA, hemolytic, and refractory anemias who had received multiple transfusions did not inhibit the formation of erythroid colonies by normal human bone marrow. A suggestion of suppressor lymphocyte activity was present in PBL from one patient with refractory anemia (Table 3-10).

The effect of T and B cell-enriched fractions of PBL from normal donors and patients with CHA on erythroid colony formation is shown in Table 3-11. No inhibitory effect was seen with normal human PBL, and T cell-enriched fractions from both normal and CHA patients. This was also true of T cell-enriched fractions of normal donors. On the other hand, both PBL and the B cell-enriched fraction of CHA patients caused a significant inhibition of colony formation (Table 3-11).

Table 3-11 also demonstrates that the inhibitory influence of PBL from CHA can be abolished by prior repeated freeze-thawing of these cells. Similar results have been obtained with heat-inactivated PBL. In addition, preliminary evidence suggests that treatment with antithymocyte globulin can abolish the suppressor effect of the PBL.

The cell separation procedure outlined in Figure

Table 3-9 Effect of peripheral blood lymphocytes (PBL) from patients with acquired aplastic anemia (AA) on erythroid colony formation by normal human bone marrow cells *in vitro*.[a]

PBL Donors[c]	Total No. CFU-e/6 = 10⁵ bone marrow cells[b]					
	0 (PBL Conc.)	0.5 × 10⁵ (PBL Conc.)	1 × 10⁵ (PBL Conc.)	2 × 10⁵ (PBL Conc.)	4 × 10⁵ (PBL Conc.)	6 × 10⁵ (PBL Conc.)
Normal (39)	268 ± 12	350 ± 11	342 ± 7	379 ± 14	403 ± 18	369 ± 13
AA (2)		44 ± 6	52 ± 3	—	38 ± 4	21 ± 2
AA (2)		169 ± 11	147 ± 8	39 ± 5	53 ± 8	—
AA (2)		311 ± 26	297 ± 38	202 ± 41	49 ± 6	16 ± 2
AA (1)		250 ± 26	275 ± 32	279 ± 18	80 ± 14	20 ± 4
AA (1)						
before transplant		110 ± 13	95 ± 21	27 ± 3	—	—
after transplant		297 ± 22	311 ± 19	315 ± 46	287 ± 14	346 ± 16
AA (9)		287 ± 10	300 ± 46	411 ± 52	396 ± 43	353 ± 36

[a] 2 IU Ep was used.
[b] Each value represents mean ± 1 SEM of all cultures.
[c] The numbers in parentheses indicate the number of donors in each group.

Table 3-10 Effect of peripheral blood lymphocytes (PBL) from multiply-transfused patients on erythroid colony formation by normal human bone marrow cells *in vitro*.[a]

PBL Donors[c]	Total No. CFU-e/6 × 10⁵ bone marrow cells[b]					
	0 (PBL Conc.)	0.5 × 10⁵ (PBL Conc.)	1 × 10⁵ (PBL Conc.)	2 × 10⁵ (PBL Conc.)	4 × 10⁵ (PBL Conc.)	6 × 10⁵ (PBL Conc.)
Normal (24)	422 ± 21	576 ± 18	612 ± 39	594 ± 17	572 ± 26	585 ± 31
Hemolytic anemia (3)		473 ± 32	—	512 ± 53	618 ± 49	—
Pure red cell aplasia (3)		486 ± 47	509 ± 55	617 ± 89	603 ± 77	591 ± 46
Refractory anemia (3)		572 ± 57	632 ± 81	594 ± 48	539 ± 63	528 ± 67
Refractory anemia (1)		390 ± 59	334 ± 51	301 ± 32	249 ± 19	206 ± 48

[a] 2 IU Ep was used.
[b] Each value represents mean ± 1 SEM of all cultures.
[c] The numbers in parentheses indicate the number of donors in each group.

3-1 resulted in progressively fewer monocytes in fractions I, II, and III. Fraction IV (adherent cells) consisted of 95% monocytes, which were shown to be functionally equivalent to the unseparated monocytes present in F-I. This separation scheme was undertaken primarily because of our inability to consistently grow BFU-e from F-I cells. Table 3-12 shows that the numbers of BFU-e formed in the presence of 4 IU Ep was greatest in F-III and smallest in F-I. When 2×10^6 cells in fractions I, II, III, and IV were cultured separately with 4 IU Ep, the number of BFU-e colonies formed at 13 days was inversely related to the number of monocytes present in each fraction (Table 3-12). The choice of cell and Ep concentrations (2×10^6 cells, 4 IU Ep) was dictated by dose-response studies employing F-III cells and was found to be optimal for these studies. To examine the possibility that monocytes were suppressing BFU-e proliferation, purified monocytes (F-IV cells) were admixed with 2×10^6 F-III cells in various concentrations (Table 3-13). Under these conditions, the number of colonies formed was inversely proportional to the numbers of monocytes added. In this regard, no colonies were formed when the number of monocytes present comprised 15% of the total cell population (Table 3-13).

Discussion

The results presented here support the view that an immunological basis may exist for reduced availability of circulating and bone marrow hematopoietic elements in some hematological disorders.[23–25] In addition, these findings provide strong evidence that cell–cell interactions have an important function in the overall proliferation and differentiation of erythrocytes.

An immunological basis for the anemia in some patients with PRCA was suggested by Krantz and his associates,[26,27] who demonstrated that a humoral antibody directed against the more differentiated members of erythroid precursors was present in these patients. However, in a majority of anemic individuals, a humoral inhibitor of erythropoiesis has not been demonstrated. In this regard, no inhibitory activity was present in sera of patients with CHA[18,28] or aplas-

Table 3-11 Effect of T and B cell-enriched fractions of peripheral blood lymphocytes (PBL) of normal humans and patients with Diamond–Blackfan syndrome (CHA) on erythroid colony formation by normal human bone marrow cells *in vitro*.[a]

PBL Donors[b]	Total No. CFU-e/6 × 10⁵ Bone Marrow Cells[c]
Normal (2):	
PBL (intact)	296 ± 32
PBL (freeze-thawed)	310 ± 42
T cell-rich	227 ± 31
B cell-rich	247 ± 29
CHA (2):	
PBL (intact)	53 ± 15
PBL (freeze-thawed)	328 ± 46
T cell-rich	243 ± 43
B cell-rich	32 ± 9

[a] 2 IU Ep was used.
[b] The numbers in parentheses indicate the number of donors in each group.
[c] Each value represents the mean ± 1 SEM of all cultures.

Table 3-12 Erythroid burst colony formation (BFU-e) by different fractions of peripheral blood mononuclear cells from normal humans *in vitro*.[a]

Fraction[b]	% Monocytes	BFU/10⁵ Cells[c]
I (9)	30 ± 3	1.1 ± 0.5
II (9)	6 ± 1	10.9 ± 2.5
III (9)	0.6 ± 0.2	20.7 ± 2.9
IV (5)	95	0

[a] 4 IU Ep was used. See text for definitions of fractions.
[b] The numbers in parentheses indicate the number of donors studied.
[c] Each value represents mean ± 1 SEM of all cultures.

Table 3-13 Effect of addition of purified monocytes (FIV cells) on the proliferation and/or differentiation of BFU-e by monocyte-free fraction (FIII cells) of normal human peripheral blood mononuclear cells *in vitro*.[a,b]

FIV Cells	BFU-e/10^5 Cells[c]
0	24.6 ± 2.1
40,000	18.3 ± 1.8
80,000	12.9 ± 2.3
120,000	9.7 ± 1.2
200,000	5.1 ± 0.8
300,000	2.6 ± 0.7
480,000	0

[a] 4 IU Ep was used. See text for description of fractions.
[b] Six normal donors were studied.
[c] Each value represents mean ± 1 SEM of all cultures.

tic anemia.[19] Similarly, in a patient with anemia associated with hypogammaglobulinemia we were unable to detect a humoral inhibitor of erythropoiesis.[29] Further investigation, however, revealed the presence in this patient of circulating lymphocytes capable of suppressing erythropoiesis *in vitro*.[29] This investigation was initiated because of recent developments in the understanding of mechanisms underlying the immunodeficiency states in man.

In general, deficient (at times absent) plasma cells have been demonstrated in patients with hypogammaglobulinemia.[30,31] A number of studies have demonstrated that gradient-isolated blood mononuclear cells of individuals with variable immunodeficiency with thymoma,[32-34] infantile X–linked immunodeficiency,[35] multiple myeloma,[36] and Hodgkin's disease[37] can block the differentiation of B-lymphocytes to immunoglobulin (Ig)-producing plasma cells *in vitro*. Waldmann et al.,[32] who first described the phenomenon using co-cultivation studies, suggested that suppressor T-lymphocytes in hypogammaglobulinemic subjects inhibited Ig synthesis in autologous and heterologous B-lymphocytes by interfering with its terminal differentiation.[32] He also proposed that in at least some patients, excess lymphocyte suppressor activity is a major pathogenic mechanism of the immunodeficiency.[32]

Cultured mononuclear cells from this patient produced much less Ig *in vitro* than cells of normals under similar conditions.[29] Co-cultivation studies involving the patient's PBL and normal cells demonstrated that Ig production by normal B-lymphocytes was significantly inhibited. Of interest was our finding that when lymphocytes from this patient were added to normal human bone marrow cells *in vitro*, a highly significant decrease in the number of erythroid colonies formed was seen.[29] Hydrocortisone did not affect the suppression phenomenon. Moreover, when adherent cells were removed from the patient's PBL, there was no loss of suppression in either the Ig or erythroid colony production. By contrast, isolated T cells did suppress cultivated cells in both systems.[29] Although these results raised the possibility that both phenomena, suppression of Ig synthesis and erythroid colony formation, were mediated by the same population of T-lymphocytes, additional studies have established the separate nature of the cells involved (ref. 29; unpublished observations). This is further supported by the observations that lymphocytes from patients with CHA can suppress erythroid colony formation. These patients, in general, do not exhibit immunodeficiency, and there is no evidence to suggest that cell-mediated suppression of Ig synthesis occurs in CHA.

Congenital hypoplastic anemia (CHA) is a normochromic, normocytic anemia beginning early in infancy and characterized by a selective absence of erythroid precursors in the bone marrow.[38] The production of Ep[28] and the numbers of erythroid progenitor cells[39] capable of proliferation and differentiation into red cells appear to be adequate. Since its initial description in 1938,[40] the etiology of the anemia in this syndrome has remained unknown. Several investigators have suggested that this disorder may have an immunological basis, since somewhat more than one-half of these patients respond to corticosteroid therapy. The results presented here lend support to such a possibility. The suppressive influence of lymphocytes from a majority of patients with CHA in relapse seen here suggests the involvement of immunocompetent cells in the production of red cells in this disorder. The fact that lymphocytes from patients in steroid-induced remission did not exhibit such a suppressor activity may be related to the overall steps operative in remission-induction in these patients. The possibility exists that, in some patients, treatment with immunoactive agents such as glucocorticosteroids dampens the suppressor activity of these lymphocytes. Alternatively, such cells may either be absent or not be present in sufficient numbers to exert any significant influence on erythropoiesis. In this regard, there is some evidence to indicate that the production of Ig in normal animals and man may be regulated by the relative concentrations of suppressor and helper lymphocytes.[41,42] Thus the ameliorative influence of the corticosteroid may be mediated through an effect on the existing populations of erythroid precursor elements. A number of studies have shown that steroid compounds can profoundly influence the proliferation and/or differentiation of hematopoietic precursors.[43-45] Another possibility is that the numbers and/or proliferative capacity of the erythroid stem cells in CHA may be inadequate. In support of this view, Nathan and his associates[46] have demonstrated that a significant deficiency in the numbers of circulating BFU-e exists even in patients in remission. Freedman and Saunders[47,48] were unable to detect the presence of suppressor lymphocytes in blood of patients with CHA both in relapse and during steroid induced remission. However, in the former, they showed the presence of a population of cells in patients' bone marrow that inhibited erythroid colony formation by autologous CFU-e.[48] It is not clear at present if the cells that suppress erythropoiesis reside in the bone marrow or arrive there as part of the recirculating lymphocyte pool that has free access into and out of the bone marrow parenchyma.[49]

The etiology of AA, like that of CHA, is unknown. Acquired aplastic anemia represents a group of disorders characterized by peripheral blood pancytopenia in association with varying degrees of diminished granulocytic, erythroid, and megakaryocytic elements in the bone marrow.[50] In this disease, there is a high degree of mortality as a result of hemorrhage and infection in the more severely involved patients.[51-53] Whether it is idiopathic,[50] drug-induced,[54,55] or associated with infection,[56,57] the pathogenesis of AA remains obscure.

In considering mechanisms of aplasia, one should recognize that bone marrow failure may result from a diversity of circumstances. Although known drugs and toxins have been implicated,[55] many of the patients seen have no such discernible association.[50] It is therefore appropriate that attention be called to the possibility that factors external to the pluripotential stem cell play an important role in the pathogenesis of the disease, at least in some of these patients. Animal models, of a genetic nature, exist that emphasize these differences. The W^v/W^v mouse has an inherently defective stem cell,[58] whereas the Steel, or Sl/Sl^d, strain of mice shows a defective microenvironment.[59] Thus, transplantation of the Sl/Sl^d or normal stem cells will engraft a W/W^v mouse, but neither normal nor W/W^v marrow will repopulate the Sl/Sl^d mouse. Successful marrow transplantation in patients with aplastic anemia appears to support the concept that environmental factors do not have an important role.[60] However, in many cases of transplantation with HLA-matched sibling marrow, failure of engraftment occurs.[61] In these cases, rejection has generally been attributed to undetected histocompatibility differences, but the possibility remains that an inhospitable environment could explain both the original marrow failure and that of the engrafted one.

In vitro bone marrow culture systems have come into increasing use as a means of assessing progenitor cells in aplastic anemia. When the patient's own marrow is studied, low numbers of both granulocytic and erythroid colonies have been found,[62] suggesting a reduction in number or an intrinsic defect in the committed cells of these lines. Recently, however, Ascensao et al.[24] reported that the addition of antilymphocyte globulin to the bone marrow culture of a patient with aplastic anemia increased the number of granulocytic colonies. In our investigations, we approached the question of environmental factors differently.

The role of serum and cellular inhibitors of erythropoiesis in patients with aplastic anemia was examined. The target cell against which these potential inhibitors were directed was the erythroid progenitor cell. Unfortunately, an easily performed assay for human pluripotent stem cells is not available. We therefore assume that factors affecting the erythroid-committed cell would do so by an effect on the pluripotent stem cell of the patient with aplastic anemia. In the patients whom we studied, we were unable to detect a serum borne inhibitor of erythropoiesis. However, the peripheral blood lymphocytes of a significant number of the patients studied were capable of inhibiting erythroid colony formation by normal human bone marrow. Since nearly all patients studied here had received at least one blood transfusion (patients with CHA had received multiple red cell transfusions), the possibility existed that the suppression of erythropoiesis by the patients' lymphocytes might be due to prior sensitization by foreign histocompatibility antigens. However, as reported here, when lymphocytes from several other patients who had been receiving prolonged transfusion of blood products as part of their therapy were cultured together with normal human bone marrow, no inhibition of colony formation was observed. Moreover, since not all patients with AA exhibited suppressor activity, it can be assumed that the observed phenomenon cannot be attributed to transfusion therapy. The persistence of lymphocytes capable of suppressing hematopoiesis in the bone marrow of patients with aplastic anemia could explain the difficulty seen with marrow engraftment of HLA-compatible cells.[61]

Recently, additional clinical evidence supporting this concept has been reported. Autologous bone marrow reconstitution has occurred in several patients with severe aplastic anemia after treatment with antilymphocyte globulin and HLA-semi-incompatible bone marrow transplants.[63] In addition, in one patient who had previously rejected an HLA-compatible transplant, administration of procarbazine and antihuman thymocyte globulin was followed by complete hematological recovery of the host marrow.[64] We can only assume that the immunosuppression that preceded the recovery of host marrow function in these cases may have disarmed a population of mononuclear cells capable of suppressing either the uncommitted stem cell or all three of the committed marrow progenitor cells. It is possible that the immunosuppressive therapy that precedes successful bone marrow transplants would also be directed against the suppressor cells and allow for a successful engraftment to occur. One of the patients studied here had received a successful bone marrow transplant and has been doing well more than 1 year later.

This patient's lymphocytes, which had exhibited strong inhibitory activity before, were no longer inhibitory after the transplant. Whether this loss of suppressor activity resulted from the immunosuppressive therapy or from replacement by the new cells remains to be determined. The presence or absence of suppressor lymphocytes was not related to age, sex, degree of aplasia, possible cause of the disease, or to the number of blood transfusions received by the patient.

An aspect of these studies that should be emphasized is that these findings provide strong evidence for the existence and relative importance of cell–cell interaction in erythropoiesis. Our findings that the proliferation and/or differentiation of normal human peripheral blood BFU-e can be influenced by relative concentration of accompanying monocytes indicate that cell-mediated regulation of erythropoiesis is exercised at the level of the most primitive functionally identifiable erythroid precursors. The mechanism underlying cell-mediated regulation of erythropoiesis is not known. Preliminary studies have shown that PBL-conditioned media (before and after tenfold concentration) do not affect red cell production in vitro. This was also true when these cells were cultured in the presence of mitogens. The results indicate that cellular integrity is a necessary component of the PBL effect; cells subjected to (a) repeated freeze-thawing, (b) heating at 56°C for 30 min, (c) ionizing irradiation, or (d) sonication failed to affect erythropoiesis. It is likely that the effect is exerted via short-range messages which may involve cell–cell contact. It should be noted, however, that although cell–cell interaction is an important short-range factor, further proliferation and differentiation of these precursor elements into mature erythrocytes requires the presence of the primary regulator of erythropoiesis, erythropoietin.

Acknowledgment

The studies reported here were supported by grants CA–18755 and CA–23021 from the National Cancer Institute, National Institutes of Health, and Veterans Administration Medical Research Fund.

References

1. Gordon AS: Erythropoietin. *Vit Horm 31*:105, 1973.
2. Zanjani ED, Lutton JD, Hoffman R, et al: Erythroid colony formation by polycythemia vera bone marrow in vitro dependence on erythropoietin. *J Clin Invest 59*:841, 1977.
3. Trentin JJ: Influence of hematopoietic organ stroma (hematopoietic inductive microenvironment) on stem cell differentiation, regulation of hematopoiesis, in Gordon AS (ed): Regulation of Hematopoiesis, vol. 1. New York, Appleton-Century Crofts, 1970, p 161.
4. Metcalf D, Moore MAS: Haemopoietic Cells. New York, American Elsevier, 1971.
5. Cudkowicz G, Stimpfling JH: Deficient growth of C57BL marrow cells transplated in F1 hybrid mice. Association with the histocompatibility-2 locus. *Immunology 7*:291, 1964.
6. McCulloch EA, Till JE: Repression of colony-forming ability of C57BL hematopoietic cells transplanted into non-isologous hosts. *J Cell Comp Physiol 61*:301, 1963.
7. Lotzova E, Cudkowicz G: Abrogation of resistance to bone marrow grafts by silica particles. Prevention of the silica effect by the macrophage stabilizer poly-2-vinyl pyridine N-oxide. *J Immunol 113*:798, 1974.
8. Goodman JW, Shinpock SG: Influence of thymus cells on erythropoiesis of parental marrow in irradiated hybrid mice. *Proc. Soc Exp Biol Med 129:* 417, 1968.
9. Wiktor-Jedrzejczak W, Sharkis S, Ahmed A, et al: Theta-sensitive cell and erythropoiesis: Identification of a defect in W/W^v anemic mice. *Science 196*:313, 1977.
10. Petrov RV, Kaitov RM, Aleinikova NV, et al: Factors controlling stem cell recirculation. III. Effect of the thymus on the migration and differentiation of hemopoietic stem cells. *Blood 49*:865, 1977.
11. Petrov RV, Seslavina LS, Pantelejev EI, et al: Inactivation of stem cells by lymphocytes non-linked to the H-2 histocompatibility system. *Transpl Proc 9*:555, 1977.
12. Stephenson JR, Axelrad AA, McLeod DL: Induction of colonies of hemoglobin-synthesizing cells by erythropoietin in vitro. *Proc Natl Acad Sci 68*:1542, 1971.
13. McLeod DL, Shreeve MM, Axelrad AA: Improved plasma clot culture system for production of erythrocytic colonies in vitro: quantitative assay method for CFU-E. *Blood 44*:517, 1974.
14. Tepperman AD, Curtis, JE, McCulloch EA: Erythropoietic colonies in cultures of human marrow. *Blood 44*:659, 1974.
15. Heath DS, Axelrad AA, McLeod DL, et al: Separation of the erythropoietin-responsive progenitors BFU-E and CFU-E in mouse bone marrow by unit gravity sedimentation. *Blood 47*:777, 1976.
16. Clarke BJ, Housman D: Characterization of an erythroid precursor cell of high proliferative ca-

pacity in normal peripheral blood. *Proc Natl Acad Sci 74:*1105, 1977.
17. Aye MT: Erythroid colony formation in cultures of human marrow: effect of leukocyte conditioned medium. *J Cell Physiol 91:*69, 1977.
18. Hoffman R, Zanjani ED, Vila J, et al: Diamond-Blackfan syndrome: lymphocyte-mediated suppression of erythropoiesis. *Science 193:*899, 1976.
19. Hoffman R, Zanjani ED, Lutton J, et al: Suppression of erythroid-colony formation by lymphocytes from patients with aplastic anemia. *N Engl J Med 296:*10, 1977.
20. Schooley JC, Garcia JF: Some properties of serum obtained from rabbits immunized with human urinary erythropoietin. *Blood 25:*204, 1965.
21. Sober HA, Gutter FJ, Wyckoff MM, et al: Chromatography of proteins. II. Fractionation of serum protein on anion-exchange cellulose. *J Am Chem Soc 78:*756, 1956.
22. Zalusky R, Zanjani ED, Gidari AS, et al: Site of action of a serum inhibitor of erythropoiesis. *J Lab Clin Med 81:*867, 1973.
23. Krantz SB: Pure red cell aplasia. *N Engl J Med 291:*345, 1974.
24. Ascensao JA, Pahwa R, Kagan W, et al: Aplastic anemia: evidence for an immunological mechanism. *Lancet 1:*669, 1976.
25. Kagan W, Ascensao JA, Pahwa RN, et al: Aplastic anemia: presence in human bone marrow of cells that suppress myelopoiesis. *Proc Natl Acad Sci 73:*2890, 1976.
26. Krantz SB, Kao V: Studies on red cell aplasia. I. Demonstration of a plasma inhibitor to heme synthesis and an antibody to erythroblast nuclei. *Proc Natl Acad Sci 58:*493, 1967.
27. Krantz SB, Morse WH, Zaentz SD: Studies on pure red cell aplasia. V. Presence of erythroblast cytotoxicity in γg globulin fraction of plasma. *J Clin Invest 53:*324, 1973.
28. Geller G, Krivit W, Zalusky R, et al: Lack of erythropoietic inhibitory effect of serum from patients with congenital pure red cell aplasia. *J Pediatr 86:*198, 1975.
29. Litwin SD, Zanjani ED: Lymphocytes suppressing both immunoglobulin production and erythroid differentiation in hypogammaglobulinemia. *Nature 266:*57, 1977.
30. Good RA: Studies on agammaglobulinemia. II. Failure of plasma cell formation in the bone marrow and lymph nodes of patients with agammaglobulinemia. *J Lab Clin Med 46:*167, 1955.
31. Creste PA, Heremans JF: The distribution of immunoglobulin containing cells along the human gastrointestinal tract. *Gastroentero 51:*305, 1966.
32. Waldmann TA, Broder S, Blaese RM, et al: Role of suppressor T-cells in pathogenesis of common variable hypogammaglobulinemia. *Lancet 2:*609, 1974.
33. Broom BC, DeLaconcha IG, Webster ADB: Intracellular immunoglobulin production in vitro by lymphocytes from patients with hypogammaglobulinemia and their effect on normal lymphocytes. *Clin Exp Immunol 23:*73, 1976.
34. Waldmann TA, Brode S, Durm M: T-cells in immunodeficiency, immune depression and cancer, in Siskind GW, Christian CL, Litwin SD (eds): *Proc 2nd Strassberger Symposium.* New York, Grune & Stratton, 1975.
35. Siegal FD, Siegal M, Good RA: Suppression of B-cell differentiation by leukocytes from hypogammaglobulinemic subjects. *J Clin Invest 58:*109, 1976.
36. Broder S, Humphrey R, Durm M, et al: Impaired synthesis of polyclonal (non-paraprotein) immunoglobulin by circulating lymphocytes from patients with multiple myeloma. *N Engl J Med 293:*887, 1975.
37. Twomey JJ, Laughter AH, Farrows, et al: Hodgkins disease: an immunodepleting and immunosuppressive disorder. *J Clin Invest 56:*467, 1975.
38. Diamond LK, Allen DM, Magil FB: Congenital (erythroid) hypoplastic anemia. *Am J Dis Child 102:*149, 1961.
39. Freedman MH, Amato D, Saunders EF: Erythroid colony growth in congenital hypoplastic anemia. *J Clin Invest 57:*673, 1976.
40. Diamond LK, Blackfan KO: Hypoplastic anemia. *Am J Dis Child 56:*464, 1938.
41. Gershon RK: T-cell control of antibody production. *Contemp Topics Immunobiol 3:*1, 1974.
42. Gershon RK: A disquisition on suppressor T-cells. *Transpl Rev 26:*170, 1975.
43. Mirand EA, Gordon AS, Wenig J: Mechanism of testosterone action in erythropoiesis. *Nature 206:*270, 1965.
44. Shahidi NT: Androgens and erythropoiesis. *N Engl J Med 289:*72, 1973.
45. Singer JW, Adamson JW: Steroids and hematopoiesis II. The effect of steroids on in vitro erythroid colony growth: evidence for different target cells for different classes of steroids. *J Cell Physiol 88:*135, 1976.
46. Clarke BJ, Nathan D, Hillman D, et al: Characterization of the burst forming erythroid stem cell (BFU-E) in peripheral blood mononuclear cells (PB-MNC) of normal individuals and Diamond-Blackfan syndrome (BD). *Blood 48:*65, 1976.
47. Freedman MH, Saunders EF: Diamond-Blackfan syndrome: in vitro analysis of the erythropoietic defect. *Pediatr Res 11:*471, 1977.
48. Freedman MH, Saunders EF: Diamond-Blackfan syndrome: types 1 and 2. *Clin Res 25:*339A, 1977.
49. Fauci AS, Dale DC: The effect of in vivo hydro-

cortisone on subpopulations of human lymphocytes. *J Clin Invest* 53:240, 1974.
50. Erslev AJ: Aplastic anemia, in Williams WJ, Beutler E, Erslev AJ (eds): *Hematology*. New York, McGraw-Hill, 1972, p. 207.
51. Davis S, Rubin A: Treatment and prognosis in aplastic anemia. *Lancet* 1:871, 1972.
52. Li FP, Alter BP, Nathan DG: The mortality of acquired aplastic anemia in children. *Blood* 40:153, 1972.
53. Lynch RE, Williams DM, Reading JC, et al: The prognosis in aplastic anemia. *Blood* 45:517, 1975.
54. Bithel TC, Wintrobe MM: Drug-induced aplastic anemia. *Sem Hematol* 4:194, 1967.
55. Williams DM, Lynch RE, Cartwright GE: Drug-induced aplastic anemia. *Sem Hematol* 10:195, 1973.
56. Camitta BM. Nathan DG, Forman EN, et al: Posthepatic severe aplastic anemia: an indication for early bone marrow transplantation. *Blood* 43:473, 1974.
57. Hagler L, Pastore RA, Bergin JJ: Aplastic anemia following viral hepatitis: report of two fatal cases and literature review. *Medicine* 54:139, 1975.
58. McCulloch EA, Siminovitch L, Till JE, et al: Spleen colony formation in anemic mice of genotype WW^v. *Science* 144:844, 1964.
59. McCulloch EA, Siminovitch L, Till JE, et al: The cellular basis of the genetically determined hemopoietic defect in anemic mice of genotype sl/sl^d. *Blood* 26:399, 1965.
60. Storb R, Thomas ED, Buckner CD, et al: Allogeneic marrow grafting for treatment of aplastic anemia. *Blood* 43:157, 1974.
61. Storb R, Buckner CD, Fefer A, et al: Marrow transplantation in aplastic anemia. *Transpl Proc* 6:335, 1974.
62. Kurnick JE, Robinson WA, Dickey CA: In vitro granulocytic colony-forming potential of bone marrow from patients with granulocytopenia and aplastic anemia. *Proc Soc Exp Biol Med* 137:917, 1971.
63. Jennet M, Speck M, Rubenstein B, et al: Autologous marrow reconstitutions in severe aplastic anaemia after ALG pretreatment and HLA semiincompatible bone marrow cell transfusion. *Acta Haematol* 55:129, 1976.
64. Thomas ED, Storb ER, Gilbert B: Recovery from aplastic anemia following attempted marrow transplantation. *Exp Hematol* 4:97, 1976.
65. Huang S-W, MacLaren NK: Insulin-dependent diabetes: a disorder of autoaggression. *Science* 192:64, 1976.

4 Cryogenic Preservation of Human and Murine Erythropoietic and Granulopoietic Colony-Forming Cells

Hiroshi Hara, Makio Ogawa, and Harvey L. Bank

Introduction

Recently, cryogenic preservation of mammalian cells has emerged as a useful technique in biomedical research as well as in clinical medicine. Development of a consistent method for cryopreservation of human marrow function would be of value in: (a) providing a constant source of viable human marrow cells for research in the pathophysiology of human hemopoiesis, and (b) facilitating transplantation of allogeneic marrow. Bone marrow transplantation is recognized as an effective treatment modality in such diseases as aplastic anemia and potentially in acute leukemia.

Over the past decade, the availability of semisolid clonal cell culture assays rendered it possible to quantitatively evaluate committed hemopoietic precursors in man and in mouse. The following is a model of hemopoiesis constructed on the basis of these functional assays. Committed granulopoietic precursors produce colonies when cultured in the presence of appropriate conditioned media,[1,2] and are termed colony-forming units in culture (CFU-c). Committed early erythropoietic precursors produce macroscopic colonies in culture and are termed erythropoietic burst-forming units (BFU-e).[3] The late erythroid precursors termed erythroid colony-forming units (CFU-e) produce small hemoglobinized colonies after short incubation time and are considered to be progenies of BFU-e.[4] Both CFU-c and BFU-e are direct descendants of hemopoietic stem cells that are measurable only *in vivo* in mice using the spleen colony techique developed by Till and McCulloch.[5]

Using CFU-c and the spleen colony techniques, a number of investigators reported successful cryopreservation of human and murine bone marrow functions.[6-14] Most investigators used variants of the freezing procedures originally described by Polge et al.[15] In the basic procedure, isolated bone marrow cells are suspended in a balanced salt solution with 10 to 15% (v/v) dimethylsulfoxide (DMSO) or glycerol as a cryoprotective agent. The cells are then sealed in ampules and cooled at 1 to 3°C/min to −79°C after which they are transferred directly into liquid nitrogen or a −79°C freezer. Cells are recovered by immersing the ampules in a 37°C water bath for 1 to 2 min.

Several investigators have explored the modifications of this basic procedure on murine bone marrow cells. Abrahams et al.[7] examined varying concentrations of DMSO and reported that 10 to 15% (v/v) are optimal for cryopreservation of murine marrow functions. Dobry and Livora[13] examined the effect of polyvinylpyrrolidone (PVP) and dextran. They found that the survival of cells protected with PVP was relatively independent of cooling rate, whereas those cells protected with dextran had a survival optima at 10°C/min. Lewis et al.[6] examined the effect of cooling rates between 0.8 and 20°C/min. They found that cell viability decreased as the cooling rate increased above 1°C/min. A comprehensive study was done by Leibo et al.[9] in which the interactions of cooling and warming rates on the survival of murine stem cells were examined in the presence of three cryoprotective agents. They found that the optimal rate of cooling varied with both the type and the concentration of cryoprotective agents used.

To our knowledge, there has been no previous report of the cryoprotection of marrow erythropoietic colony-forming cells. Here we report a simple and consistent method for cryopreservation of erythropoietic and granulopoietic precursors in human and murine marrows. We examined the effects of DMSO concentrations and cooling rates and found the highest survival for both granulopoietic and erythropoietic functions of marrow after cooling at 1°C/min in the presence of 10% DMSO.

Materials and Methods
Marrow Cells

Human marrow cells were obtained from a healthy volunteer and from eight patients during the course of their hematological assessment (Table 4-1). Marrow samples were aspirated from the posterior iliac crests and were transferred into a 6 ml Falcon plastic tube containing 200 units of heparin without preservatives (Chromalloy Pharmaceutical, Inc., Oak Park, Michigan). After centrifugation, the marrow buffy coat cells were removed by a Pasteur pipet and a single cell suspension was prepared by repeated pipetting.

Male BDF$_1$ mice, weighing 25 to 30 gm, were obtained from Simonsen's Laboratory, Gilroy, California. Bone marrow cells were flushed from femurs with a #23 gauge needle into α-medium (Flow La-

Table 4-1 Hematological findings in donors.

Diagnosis	RBC ($\times 10^6/mm^3$)	WBC ($\times 10^3/mm^3$)	Platelet ($\times 10^3/mm^3$)	Bone Marrow
Rectal carcinoma	4.27	5.7	205	Normal
Chronic alcoholism	2.90	6.0	150	Erythroid hyperplasia and sideroblastic changes
Chronic lymphocytic leukemia	2.64	60.0	84	Lymphocytosis
Microcytic anemia	4.11	8.1	300	Erythroid hyperplasia
Histiocytic lymphoma	3.76	6.6	132	Erythroid hyperplasia containing nests of tumor cells
Hodgkin's Disease	4.16	5.6	186	Normal
Chronic alcoholism	2.77	17.4	—	Normal
Splenomegaly	4.92	2.8	87	Erythroid hyperplasia

boratories, Inc., Rockville, Maryland). The cells were then dispersed by repeated pipetting until single cell suspensions were obtained.

Assay for Human and Murine CFU-c

Human marrow cells were cultured in methylcellulose (Fisher Scientific Co., Norcross, Georgia) by a modification of the technique first described by Iscove et al.[1] Single cells were suspended in 1 ml of α-medium containing 0.8% methylcellulose, 20% fetal calf serum (Flow Laboratories, Inc.) and 20% conditioned medium and plated into 35 mm Lux standard nontissue culture dishes (#5221R) (Flow Laboratories, Inc.). The plates were incubated at 37°C for 14 days in a humidified atmosphere containing 5% CO_2 in air. Murine CFU-c were also assayed in a methylcellulose system, but in this assay, the medium was supplemented with L cell-conditioned medium and the cells were cultured for only 7 days.[2]

Assay for Human and Murine CFU-e and BFU-e

Human erythropoietic precursors were assayed in methylcellulose using the technique originally described by Iscove et al.[16] and detailed elsewhere.[17] Nucleated cells were plated in a mixture containing α-medium, 0.8% methylcellulose, 30% fetal calf serum, 1% bovine serum albumin (Calbiochem, San Diego, California), and 1.0 unit/ml of the step III preparation of sheep plasma erythropoietin (EPO) (Connaught Labs, Ltd., Willowdale, Ontario, Canada). The plates were incubated at 37°C in a humidified atmosphere containing 5% CO_2 in air for 4 days for CFU-e assay[18] and for 14 days for BFU-e.[18,19] For the murine erythropoietic precursor assay, mercaptoethanol (Pierce Chemical Co., Rockford, Illinois) in a final concentration of $1 \times 10^{-4}M$ and dialyzed EPO (Connaught Labs, Ltd.) in a final concentration of 2.0 units/ml were used. The colonies and bursts were scored on an inverted microscope after 2 days[4] and 9 days[20] of incubation, respectively.

Cell Freezing

Marrow cells were frozen using a procedure similar to that described by Bank and Maurer.[21] Sterile 12 × 75 mm Falcon plastic tubes (#2003) (Becton Dickinson & Co., Oxnard, California) containing 1×10^7 murine marrow cells or 5 to 10×10^6 human marrow cells in 0.3 ml of α-medium were transferred to a 4°C water bath and the samples equilibrated for 5 min. Then 0.3 ml of double-strength DMSO in α-medium was added to each tube and the tubes were agitated to ensure adequate mixing. The tubes were attached to a thin rod with rubber bands to provide easy handling. A #30 gauge copper-constant thermocouple was placed in a tube containing 0.6 ml of the freezing medium. This tube was subsequently handled exactly like those tubes containing the bone marrow cells. After 20 min, the tubes were transferred to a −5°C bath and equilibrated for 5 min. Samples to be supercooled were seeded by touching the solution with isothermal ice crystals frozen in the first few millimeters of the tip of a Pasteur pipet. The tubes were held for an additional 5 min at −5°C to ensure that crystallization had reached equilibrium.

For cooling at 1°C/min, the tubes were transferred from the −5°C bath to a nonsilvered, nonevacuated 95 × 280 mm Dewar flask containing 80 ml of ethanol that had been precooled in liquid nitrogen to −5°C. The initial rate of cooling between −5 and −10°C was approximately 0.5°C/min. Subsequently, the rate of cooling was 1.0 ± 0.1°C/min between −10 and −60°C. A cooling rate of 0.5°C/min was obtained by using 200 ml ethanol in the Dewar flask. A cooling rate of 12 ± 3°C/min was obtained by substituting the Dewar flask with a plastic cup containing 100 ml of ethanol. After cooling to −60°C, the samples were transferred directly into liquid nitrogen and stored for up to 48 hr.

Liquid nitrogen was decanted from the samples to be thawed and the freezing tubes transferred to a 37°C water bath and rapidly agitated. Immediately after the last ice crystals in samples melted, 0.3 ml of α-medium equilibrated at 0°C was added. Subsequently at five minute intervals, 0.3 ml, 0.3 ml, 0.6 ml, and 0.6 ml of α-medium were added serially to further dilute DMSO. The tubes were centrifuged, washed twice with α-medium, and resuspended in α-medium.

Effect of DMSO at 37°C

Marrow cells were suspended in 1.0 ml of α-medium containing appropriate concentrations of DMSO and were incubated at 37°C in 17 × 100 mm Falcon tissue culture tubes for 30 min. Then the tubes were transferred to a 4°C bath and at 5 min interval, 0.5 ml, 0.5 ml, 1.0 ml, and 1.0 ml of α-medium equilibrated at 0°C was added to dilute DMSO in the medium. The tubes were then centrifuged for 10 min at 4°C, washed with cold α-medium twice, and resuspended in α-medium, and assayed for surviving fractions of colony-forming units.

Presentation of Data

In all experiments, the survival of precursors after freezing is expressed as percentage of assayable precursors in the original samples. Therefore, factors such as cell loss during centrifugation and washing, and direct deleterious effects of the cryoprotective agent are included in the final data. Data shown in Figures 4-1 to 4-4 indicate the mean and SEM of percent colony formation of three separate experiments, each consisting of four plates per group.

Results

In order to determine the optimal concentration of DMSO for both murine and human hemopoietic precursor cells, we assayed cell survival after freezing in different concentrations of DMSO. Results are presented in Figure 4-1. The survival of cells cooled at 1°C/min indicates that under the conditions tested, 10% DMSO is equal to or superior to other concentrations tested for cryopreservation of all hemopoietic precursors. In general, DMSO conferred somewhat less protective effect to BFU-e than CFU-c for both murine and human bone marrows. It also

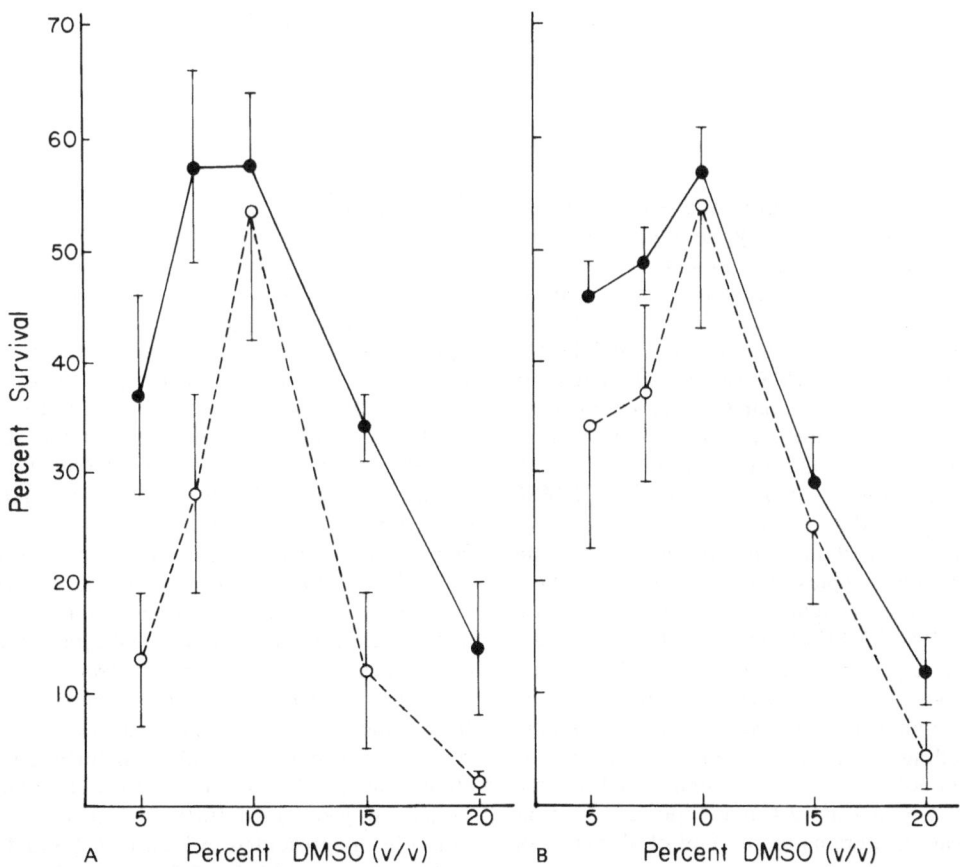

Figure 4-1 Cryoprotective effects of varying concentrations of DMSO on murine and human hemopoietic precursors. The cooling rate was 1°C per min. **A.** Murine Precursors; **B.** Human Precursors. Closed circles, CFU-c; open circles, BFU-e.

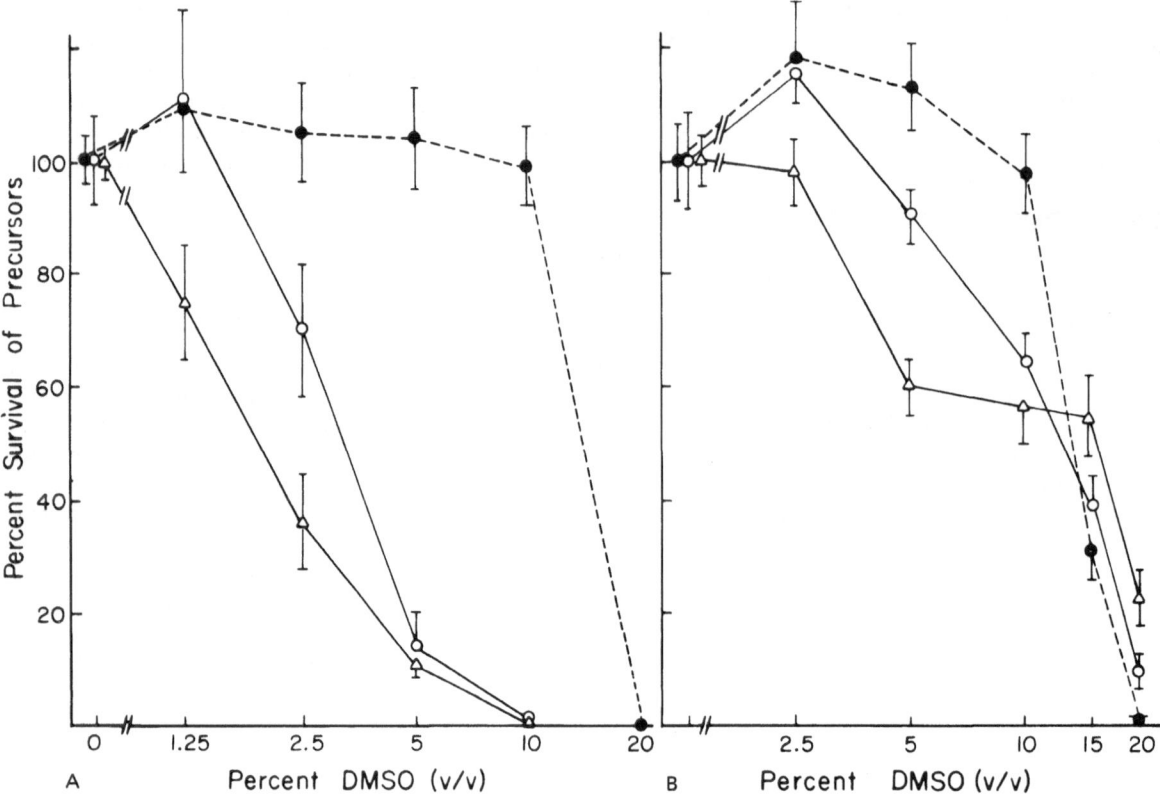

Figure 4-2 Effects of DMSO at 37°C on the erythropoietic and granulopoietic precursors. **A.** Murine Precursors; **B.** Human Precursors. Closed circles, CFU-c; triangles, CFU-e, open circles, BFU-e.

seemed possible that the differential survival of the erythropoietic precursors may be related to their stages of erythropoietic development. We exposed murine and human marrow cells to varying concentrations of DMSO for 30 min at 37°C and assayed the surviving fractions of CFU-c, CFU-e, and BFU-e (Figure 4-2). While 10% DMSO exhibited no deleterious effect to murine and human CFU-c, it completely inhibited murine erythropoietic colony formation. Furthermore, the results demonstrated that late erythroid precursors are more vulnerable to DMSO than early erythroid precursors.

We then evaluated different cooling rates and the effect of seeding samples with ice crystals. Survival of human and murine hemopoietic cells frozen in the presence of 10% DMSO are shown in Figures 4-3 and 4-4, respectively. Approximately 50% of CFU-c and BFU-e activities were preserved when the cooling rate of 1°C/min was employed. Except for human CFU-c, seeding with ice crystals did not improve survival when cooling rates of 0.5°C/min and 1.0°C/min were used. Cooling at 12°C/min followed by rapid warming was deleterious to survival of all hemopoietic precursors.

Discussion

We have determined the percentage of survival of three distinct types of human and murine bone marrow precursor cells by measuring the functional integrity of the cells. *In vivo* studies such as survival of an irradiated individual, subsequently transplanted with marrow cells or a spleen colony assay are not practical for quantitative studies of the survival of human marrow cells. Assay techniques that depend on the intactness of cell morphology, such as dye-exclusion or isotopic uptake do not necessarily reflect the functional integrity of frozen-thawed precursor cells, nor do they differentiate the survival of specific hematopoietic cell classes. Clonal cell culture techniques offer quantitative tests for the ability of precursors to proliferate and differentiate in culture.

Most of the human marrow samples were taken from patients during hematological assessment, and revealed morphological abnormalities on Wright's stain (Table 4-1). Despite these abnormalities, we consistently obtained 50 to 60% BFU-e and CFU-c activities as presented in Figures 4-1 and 4-3.

Cooling rates of 0.5 to 1.0°C/min followed by rapid

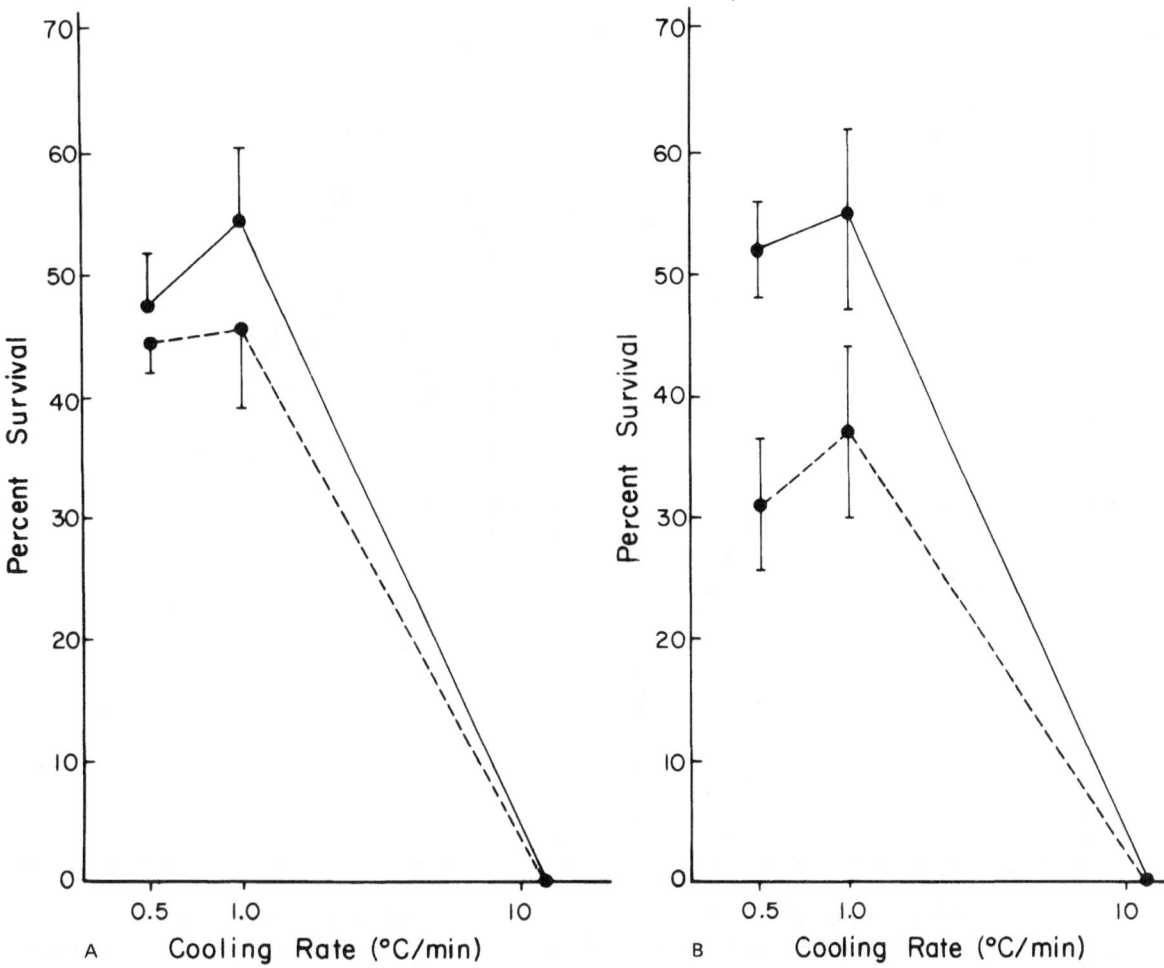

Figure 4-3 Effects of cooling rates and seeding samples with ice crystals on human hemopoietic precursors. Concentration of DMSO was 10% (v/v). **A.** Human BFU-e; **B.** Human CFU-c. Solid line, seeding; broken line, no seeding.

warming proved optimal for cryopreservation of both erythropoietic and granulopoietic precursors. Seeding with ice crystals did not confer additional protection. When the effect of different concentrations of DMSO were compared on human and murine precursor cells, 10% (v/v) proved to be the optimal for both erythropoietic and granulopoietic precursors (Figure 4-1). Our reported survival values for the cryopreservation of human CFU-c are somewhat lower than some of the previous reported data on human CFU-c.[10] The major reason for this discrepancy is due to differences in data calculations. Our data represent the actual percent survival of precursors compared to the number present in the original marrow samples. Our survival data thus include the injurious effects of the sample collection and cell washing procedures as well as the entire freezing sequence.

Exposure of the cells to DMSO at 37°C was deleterious to hemopoietic precursors (Figure 4-2). CFU-e were more sensitive to DMSO than were BFU-e. The deleterious effect of DMSO increased with increasing maturation of the erythropoietic precursors. This loss in viability may be due to a cytotoxic effect of the DMSO, an inability of the cells to automatically equilibrate during dilution of the DMSO, or to a change in the plating efficiency of the precursors. In addition, treatment of murine bone marrow cells with 10% DMSO for 30 min at 37°C caused a selective loss of erythropoietic activity. While we do not understand the underlying mechanisms, this differential sensitivity may be exploited further in the study of murine hemopoiesis.

Human bone marrow transplantation is widely recognized as having potential therapeutic value for a variety of pathological conditions. Technically, cryogenic storage of human bone marrow has been feasible for over 15 years and has been looked on as a means of circumventing difficulties encountered due

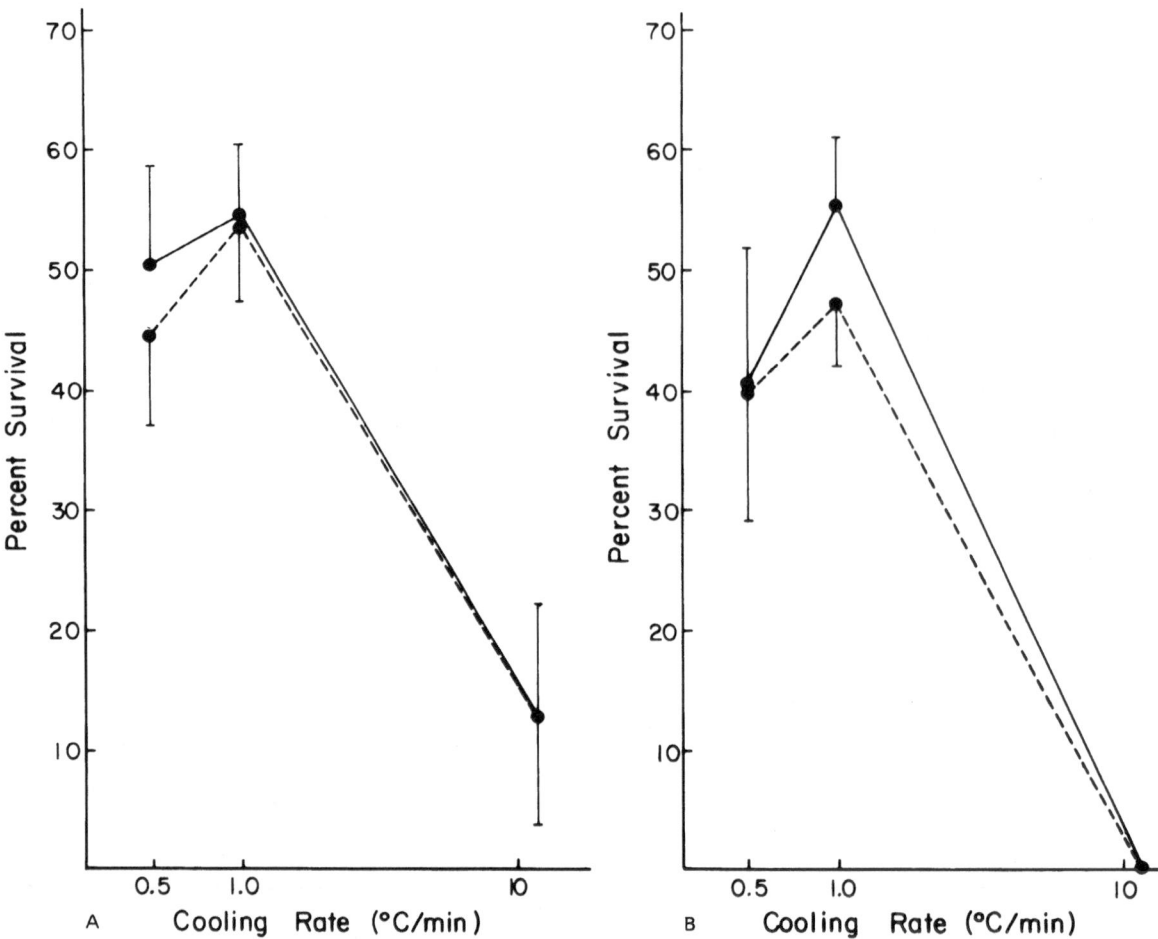

Figure 4-4 Effects of cooling rates and seeding samples with ice crystals on murine hemopoietic precursors. Concentration of DMSO was 10% (v/v). **A.** Murine BFU-e; **B.** Murine CFU-c. Solid line, seeding; broken line, no seeding.

to HLA incompatibilities. However, despite its great potential, this technique has had little clinical impact due to poor reproducibility of data obtained from different laboratories. We believe this is due, in part, to the lack of a direct reproducible survival assay for the precursor cells. With the advent of sophisticated cloning procedures for both granulocytic and erythrocytic precursor cells, it is now possible to systematically explore survival as a function of cryobiological variables.

Summary

We examined the effects of dimethylsulfoxide (DMSO) concentrations and cooling rates for cryogenic preservation of human and murine erythropoietic and granulopoietic colony-forming units using methylcellulose clonal cell culture techniques. Freezing in 10% (v/v) DMSO at a cooling rate of 1°C/min proved optimal and preserved approximately 50% of colony forming capacity of the original bone marrow samples. Prevention of supercooling by seeding samples with ice crystals did not improve survival except for human granulopoietic precursors. Exposure of the cells to DMSO at 37°C showed a preferential cytotoxicity to erythroid precursors. Further, the deleterious effect of DMSO increased with increasing maturation of the erythropoietic precursors.

Acknowledgments

We thank Mrs. Martha D. MacEachern, Ms. Alyce G. Martin, Mrs. Lobelia Avila, and Mr. Earl L. Alston for their excellent technical assistance, and Ms. P. Linda Skipper for her help in the preparation of this manuscript.

This investigation was supported by grants CA-17002 and HI-20913 from the National Institutes of Health, DHEW, and VA Basic Institutional Fund. Dr. Ogawa is a Leukemia Society of America Scholar. Dr. Hara is currently at the Second Department of Internal Medicine, Hyogo College of Medicine, Hyogo 663, Japan.

References

1. Iscove NN, Senn JS, Till JE, et al: Colony formation by normal and leukemic human marrow cells in culture: Effects of conditioned medium from human leukocytes. *Blood 37*:1–5, 1971.
2. Worton RG, McCulloch EA, Till JE: Physical separation of hemopoietic stem cells from cells forming colonies in culture. *J Cell Physiol 74*: 171–182, 1969.
3. Axelrad AA, McLeod DL, Shreeve MM, et al: Properties of cells that produce erythrocytic colonies in vitro, in Robinson WA (ed): *Hemopoiesis in Culture*. Washington DC, US Government Printing Office, 1974, pp 226–234.
4. Stephenson JR, Axelrad AA, McLeod DL, et al: Induction of colonies of hemoglobin-synthesizing cells by erythropoietin in vitro. *Proc Natl Acad Sci 68*:1542–1546, 1971.
5. Till JE, McCulloch EA: A direct measurement of the radiation sensitivity of normal mouse bone marrow cells. *Radiat Res 14*:213–222, 1961.
6. Lewis JP, Passovoy M, Trobaugh FE Jr: The transplantation efficiency of marrow cooled to −100°C at 2°C per minute. *Cryobiology 3*:47–52, 1966.
7. Abrahams S, Till JE, McCulloch EA, et al: Assessment of viability of frozen bone marrow cells using a cell culture method. *Cell Tissue Kinet 1*:255–261, 1968.
8. Carsten AL, Bond VP: Viability of stored bone marrow colony forming units. *Nature 219*:1082, 1968.
9. Leibo SP, Farrant J, Mazur P, et al: Effects of freezing on marrow stem cell suspensions: Interactions of cooling and warming rates in the presence of PVP, sucrose or glycerol. *Cryobiology 6*:315–332, 1970.
10. Gray JL, Robinson WA: In vitro colony formation by human bone marrow cells after freezing. *J Lab Clin Med 81*:317–332, 1973.
11. Goldman JM, Th'ng KH: The functional capacity of frozen mouse and human bone marrow cells in cryopreservation of normal and neoplastic cells, in Weiner RS, Oldham RK, Schwarzenberg L (eds): *Cryopreservation of Normal and Neoplastic Cells*. Paris, Inserm, 1973, pp 71–79.
12. Schaefer VW, Dicke KA: Preservation of haemopoietic stem cells. Transplantation potential and CFU-c activity of frozen marrow tested in mice, monkeys and man, in Weiner RS, Oldham RK, Schwarzenberg L (eds): *Cryopreservation of Normal and Neoplastic Cells*. Paris, Inserm, 1973, pp 63–68.
13. Dobry E, Livora J: Cryoprotective effect of polyvinylpyrrolidone and dextran in the preservation of murine marrow cells. *Blut 28*:282–287, 1974.
14. Ragab AH, Gilkerson E, Myers M: Factors in the cryopreservation of bone marrow cells from children with acute lymphocytic leukemia. *Cryobiology 14*:125–134, 1977.
15. Polge C, Smith AU, Parkes AS: Revival of spermatozoa after vitrification and dehydration at low temperatures. *Nature 164*:666, 1949.
16. Iscove NN, Sieber F, Winterhalter KH: Erythroid colony formation in cultures of mouse and human bone marrow: Analysis of the requirement for erythropoietin by gel filtration and affinity chromatography on agarose-concanavalin A. *J Cell Physiol 83*:309–320, 1974.
17. Ogawa M, Parmley RT, Bank HL, et al: Human marrow erythropoiesis in culture. I. Characterization of methylcellulose colony assay. *Blood 48*:407–417, 1976.
18. Ogawa M, MacEachern MD, Avila L: Human marrow erythropoiesis in culture. II. Heterogeneity in the morphology, time course of colony formation and sedimentation velocities of the colony forming cells. *Am J Hemat 3*:29–36, 1977.
19. Aye MT: Erythroid colony formation in cultures of human marrow: Effect of leukocyte conditioned medium. *J Cell Physiol 91*:69–78, 1977.
20. Iscove NN, Sieber F: Erythroid progenitors in mouse bone marrow detected by macroscopic colony formation in culture. *Exp Hemat 3*:32–43, 1975.
21. Bank HL, Maurer RR: Survival of frozen rabbit embryos. *Exp Cell Res 89*:188–196, 1974.

5 Miniaturization of Methylcellulose Erythroid Colony Assay

Akio Urabe and Martin J. Murphy, Jr.

Introduction

Semisolid culture methods for quantification of erythroid stem cells have recently been developed; a plasma clot method was originally described by Stephenson et al.[1], and later Iscove et al. introduced methylcellulose as a semisolid support medium.[2] Late stage erythroid precursors, colony-forming units-erythroid (CFU-e), form erythropoietin-dependent colonies in semisolid culture after 48 hr of incubation,[3] whereas more primitive erythroid stem cells, burst forming units-erythroid (BFU-e), give rise to macroscopic colonies after 9 to 14 days of incubation.[4]

The methylcellulose method discussed by Iscove et al.[2] has advantages in that the number of reagents required is less than in the plasma clot method and that cells can be easily taken for morphological analysis from the culture medium containing methylcellulose. The regular methylcellulose method, however, usually requires at least 10^5 cells for bone marrow culture and a total volume of 1 ml in a 35 × 10 mm petri dish.

In the plasma clot method, McLeod et al.[5] have already miniaturized their method using microtiter plates, reducing cell numbers and volumes of culture medium. We have modified and miniaturized the original methylcellulose method for quantification of erythroid colonies,[2] thus yielding distinct advantages in clinical and basic experimental studies of *in vitro* erythropoiesis.

Materials and Methods

Normal human bone marrow and normal human peripheral blood were obtained from volunteers. Marrow buffy coat and peripheral blood buffy coat were applied as a single cell suspension to a modification of the Ficoll–Isopaque technique for nucleated cell separation.[6,7] Namely, 108 ml of 9% (w/v) Ficoll–400 (Pharmacia Fine Chemicals, Piscataway, New Jersey) was added to 45 ml of 33.3% (w/v) of sodium metrizoate (Nyegaard & Co., Norway). Approximately 3 ml of buffy coat cell suspension in α medium (Flow Laboratories) was gently layered over 3 ml of this Ficoll–sodium metrizoate solution and was then spun at 2000 rpm for 30 min at room temperature. The nucleated cells located at the interface between Ficoll and medium were collected, and washed twice in α medium (2000 rpm, 5 min each wash). One to 2×10^4 nucleated bone marrow cells, or 0.5 to 2×10^5 nucleated peripheral blood cells were plated into small wells of flexible polyvinyl chloride flat-bottomed microtiter plates (Cooke Laboratory Products) in 0.1 ml of a mixture containing α medium, 0.8% methylcellulose (The Dow Chemical Co.), 1% bovine serum albumin (Calbiochem), 30% fetal bovine serum (Flow Labs), and 0.5 to 2 units of human urinary erythropoietin (38 IRP units/mg) per ml of medium. Microtiter plates were sterilized overnight by ultraviolet irradiation. Four to 6 microtiter wells were used for each CFU-e and for each BFU-e determination. The cell suspension within the microtiter wells was then incubated at 37°C in 5% CO_2 and 95% air with saturated humidity. Colonies were scored directly with an inverted microscope after staining with benzidine. The cells containing hemoglobin stained dark blue. The staining procedure was as follows: 0.02 ml of 3% hydrogen peroxide was added to 5 ml of 0.2% (w/v) benzidine dihydrochloride in 0.5 M acetic acid. A drop of benzidine-H_2O_2 solution was then added to each well. The benzidine-H_2O_2 solution was freshly prepared before each staining.

Bone marrow cells were also obtained from femora of 11 to 12 week old female BDF_1/Cum mice (Cumberland View Farms). Cell suspension was cultured as described above, except that 1×10^{-4} M β-mercaptoethanol (Sigma Chemical Co.) was added to the culture medium.

The numbers of CFU-e were counted after 2 days of incubation of mouse marrow and at 7 days of incubation for human marrow; the numbers of BFU-e were counted on day 9 for mouse marrow, and on day 14 for human marrow and peripheral blood. Only benzidine-positive colonies were enumerated. Morphology of cells within colonies was confirmed before benzidine-staining by cytocentrifugation (Shandon) and staining with Wright-Giemsa solution. Benzidine-stained colonies of eight or more cells were identified as CFU-e, and benzidine-stained bursts of macroscopic dimensions were identified as BFU-e.

Results and Discussion

The numbers of colonies, for human marrow CFU-e and BFU-e, and for human peripheral blood BFU-e,

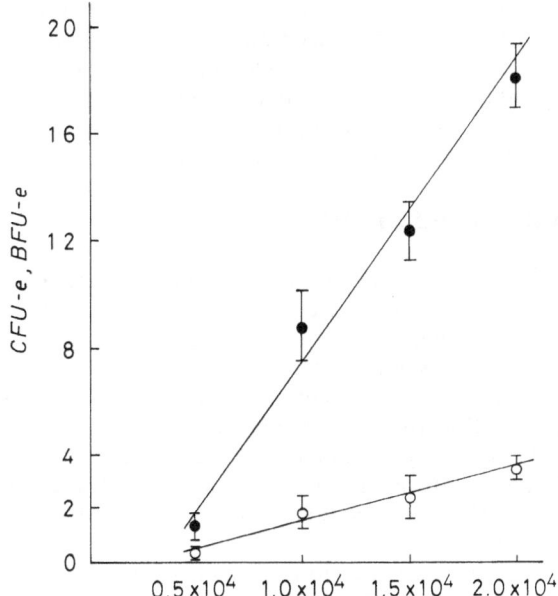

Figure 5-1 The numbers (mean ± standard error) of CFU-e and BFU-e corresponding to different concentrations of nucleated human bone marrow cells in the presence of erythropoietin (1 IRP unit/ml). Closed circles, CFU-e; open circles, BFU-e.

showed clear dose-response curves as a function of both the concentration of erythropoietin as well as the cell count (Figures 5-1 to 5-4). Each CFU-e revealed a colony consisting of 20 to 50 cells. BFU-e showed the characteristic configuration which was composed of 5 to 30 colonies, and each colony consisted of 50 to 1000 cells.

The numbers of CFU-e and BFU-e from mouse marrow also revealed erythropoietin dependence. The mean colony counts per 2×10^4 mouse bone marrow nucleated cells are given in Table 5-1.

These results demonstrate that this miniaturized methylcellulose culture system for CFU-e and BFU-e offers the same reliability as the method originally described by Iscove et al.[2] This simple technique has distinct clinical advantages, where, for example, only very few nucleated hematopoietic cells are sometimes available. Additionally, more cultures "per point" may easily be plated increasing the validity of

Table 5-1 Number of erythroid colonies from normal mouse bone marrow (2×10^4 Nucleated Cells/0.1 ml).

	CFU-e	BFU-e
Erythropoietin 2 units/ml	9.1 ± 0.5[a]	3.0 ± 0.5[a]
Without Erythropoietin	0.1 ± 0.1[a]	0

[a] Mean ± standard error.

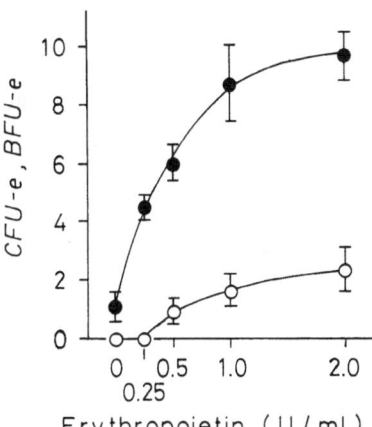

Figure 5-2 The numbers (mean ± standard error) of CFU-e and BFU-e in human marrow as a function of different concentrations of erythropoietin. The constant number of human bone marrow nucleated cells was 1×10^4 in a final volume of 0.1 ml. Closed circles, CFU-e; open circles, BFU-e.

statistical analysis, and an overall reduction of reagents, especially erythropoietin, makes for a more economical assay.

Summary

A methylcellulose culture technique for the enumeration of erythroid progenitor cells was miniaturized and standardized for both human and mouse bone marrow as well as for human peripheral blood. This

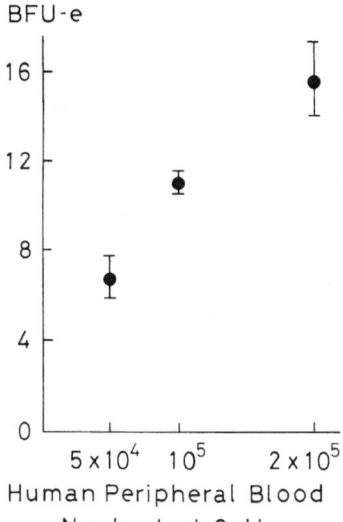

Figure 5-3 The number (mean ± standard error) of BFU-e in human peripheral blood as a function of different concentrations of human peripheral blood nucleated cells in the presence of erythropoietin (1 IRP unit/ml).

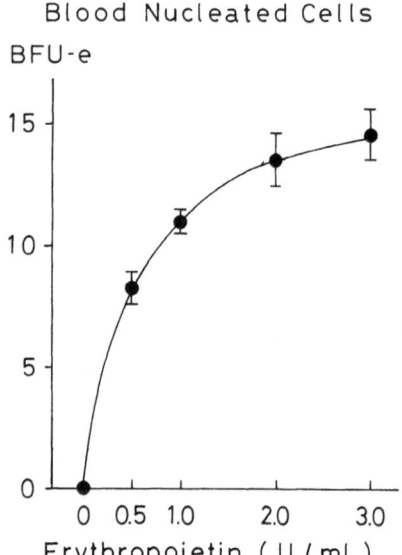

Figure 5-4 The number (mean ± standard error) of BFU-e in human peripheral blood as a function of different concentrations of erythropoietin. The constant number of human peripheral blood nucleated cells was 1×10^5 in a final volume of 0.1 ml.

miniaturized method has satisfactory reliability with respect to different concentrations of nucleated cells as well as erythropoietin. This method has distinct advantages in clinical and experimental studies of erythropoiesis, especially in clinical cases in which few marrow and blood cells are available.

Acknowledgments

This work was supported by a USPHS grant AM-19741 and by a grant from the Hipple Foundation.

The human urinary erythropoietin for the study described was collected and concentrated by Centro de Estudios Farmacologicos y de Principios Naturales, Buenos Aires, Argentina, further processed and assayed by Hematology Research Laboratories, Children's Hospital of Los Angeles, under Research Grant HL-10880 (National Heart, Lung and Blood Institute).

We thank Dr. Makio Ogawa of the Medical University of South Carolina for valuable technical advice. We also gratefully acknowledge Ms. M. Sullivan for her excellent technical help, and Ms. J. Zisson for her fine secretarial assistance.

References

1. Stephenson JR, Axelrad AA, McLeod DL, Shreeve MM: Induction of colonies of hemoglobin-synthesizing cells by erythropoietin in vitro. *Proc Natl Acad Sci USA.* 68:1542, 1971
2. Iscove NN, Sieber F, Winterhalter KH: Erythroid colony formation in cultures of mouse and human bone marrow: analysis of the requirement for erythropoietin by gel filtration and affinity chromatography on agarose-concanavalin A. *J Cell Physiol 83:*309, 1974.
3. Gregory CJ, McCulloch EA, Till JE: Erythropoietic progenitors capable of colony formation in culture: state of differentiation. *J Cell Physiol 81:*411, 1973.
4. Axelrad AA, McLeod DL, Shreeve MM, Heath DS: Properties of cells that produce erythrocytic colonies *in vitro,* in Robinson WA (ed): *Hemopoiesis in Culture.* Washington, DC, US Government Printing Office, 1974, p. 226.
5. McLeod DL, Shreeve MM, Axelrad AA: Improved plasma culture system for production of erythrocytic colonies *in vitro:* quantitative assay method for CFU-E. *Blood 44:*517, 1974.
6. Böyum A: Separation of leukocytes from blood and bone marrow. *Scand J Clin Lab Invest 21:*Suppl 97, 1968.
7. Hara H, Ogawa M: Erythropoietic precursors in mice under erythropoietic stimulation and suppression. *Exp Hemat 5:*141, 1977.

6 Culture Systems *In Vitro* for the Assay of Erythrocytic and Megakaryocytic Progenitors

David L. McLeod, Mona M. Shreeve, and Arthur A. Axelrad

Introduction

The production of erythrocytic colonies from murine fetal liver cells seeded in semisolid plasma culture medium was first accomplished in 1971.[1] The plasma culture system was then improved[2] and is now widely used along with the methylcellulose system[3] for the assay of the erythropoietic progenitors, CFU-e and BFU-e, of adult as well as fetal hemopoietic tissues of several species including man.

Recently our own work and that of other investigators has shown that megakaryocyte colonies can be produced in plasma[4] and in agar culture.[5] The growth of megakaryocyte colonies can be stimulated in these cultures by the addition of either erythropoietin or pokeweed-stimulated spleen cell conditioned medium (PWCM).[6] These colonies are usually found in association with either erythropoietic bursts or granulocytic colonies depending on the effector substance that is added to the cultures.

In pursuing these findings we have developed a modified agar culture system that supports the growth of megakaryocyte colonies when PWCM is added to the cultures and used routinely in our laboratory for the assay of megakaryocytic progenitors (CFU-M) in murine hemopoietic tissues.

This method, along with the plasma culture methods for the assay of CFU-e and BFU-e, is described in detail below. When we compared these culture systems we found that plasma cultures were more efficient for the production of murine erythrocytic colonies and erythropoietic bursts but less efficient than the agar system for the production of megakaryocyte colonies (Table 6-1).

Although the plasma culture system has the disadvantage that individual colonies cannot readily be picked from the medium for replating, it is the only system that permits colonies to be permanently fixed and stained *in situ* so that slides can be scored or reviewed at any time. It also provides a more reliable means of identifying cells and their products by light and electron microscopy[4] as well as by cytochemical[6,7] and immunochemical[8] methods. We have now successfully applied the methods used in the plasma culture system for the fixation and staining of agar cultures and this method, which we feel will be of advantage to those working with agar cultures, is also described.

Materials and Methods

The Plasma Culture System: Assay Method for Murine CFU-e

Preparation of Cell Suspensions

To prepare a suspension of mouse bone marrow cells, quickly remove femurs with the ends intact from at least three mice and wash in collection medium (see below) in a sterile petri dish. Clean each femur separately and cut off its ends. Flush out the contents of the marrow cavity into an empty plastic tube by passing 1 ml of collection medium through each cavity. Pool the marrow cells from all the femurs and pipet the suspension (avoiding frothing) to break up cell clumps. Measure the total volume and remove an aliquot for nucleated cell count. Centrifuge the cell suspension for 10 min at 160 g and resuspend the cells to the appropriate concentration in fresh collection medium.

To prepare spleen cell suspensions, remove spleen from at least three animals and wash in collection medium. Place on a coarse mesh (110/inch) stainless steel screen (Greening Wire Co., Hamilton, Ontario, Canada) and chop with scissors. Press the mince through the screen with a spatula and flush with 7 ml of collection medium into a petri dish. Scrape the material on the underside of the screen into the same medium and pipet several times and transfer to a tube. Wash the screen and petri with 3 ml of fresh medium and add these washes to the same tube. Pipet the contents of the tube once more, measure the total volume and take an aliquot for cell count. Centrifuge the cell suspension for 10 min at 160 g and resuspend the cells to the appropriate concentration in fresh collection medium.

Media

1. Supplemented Eagle's minimum essential medium with Hanks' balanced salt solution (called supplemented HMEM) is prepared as follows: Add to a flask 10 ml of MEM (Minimum Essential Medium) (Eagle) with Hanks' balanced salt solution (10× concentrated), 1 ml of MEM nonessential amino acid solution (100×), 1 ml of MEM sodium pyruvate solution (100 mM = 100×), 1 ml of L-glutamine (200 mM = 100×), 1.25 ml of 5% NaHCO$_3$ solution (autoclaved at 15 lb for 10 min) and make up to a final volume of 100 ml with

Table 6-1.

Source of cells	No. of cells seeded	Colonies or Bursts	No. of colonies or bursts (M ± S.E.)	
			Plasma	Agar
Bone marrow	5×10^4	Erythrocytic colonies	560.0 ± 42.0	51.6 ± 6.7
Bone marrow	1×10^5	Erythropoietic bursts	45.4 ± 2.4	21.3 ± 1.4
Bone marrow	1×10^5	Megakaryocyte colonies	20.3 ± 2.9	48.0 ± 4.6
Spleen	5×10^5	Erythropoietic bursts	20.5 ± 2.5	3.5 ± 1.2

Typical data obtained for number of erythropoietic bursts and erythrocytic colonies produced by C57BL/6Ut bone marrow or spleen cells in two culture systems.

water (deionized, distilled, sterilized by autoclave). Add penicillin and streptomycin to a concentration of 100 units/ml and 50 µg/ml, respectively. Store this supplemented medium after it has been made up, at 4°C and keep no longer than 3 weeks.

2. Bovine plasma (citrated): distribute in 10 ml amounts and store at −20°C.
3. Beef embryo extract (lyophilized, 50% by volume when reconstituted: reconstitute to the original volume with water, distribute in 1 ml amounts and store at −20°C. For use in the culture system add 5 ml of medium NCTC–109 to 1 ml of the reconstituted extract.
4. Fetal calf serum (FCS): heat inactivate at 60°C for 30 min, distribute in 10 ml amounts and store at −20°C.

   All of the above ingredients with the exception of NaHCO$_3$ and antibiotics are obtained from Grand Island Biological Co., Grand Island, New York, and used without further sterilization.

5. Erythropoietin Step III from anemic sheep plasma, 5 to 7 units/mg protein (Connaught Medical Research Laboratories, Willowdale, Ontario, Canada): Dilute in supplemented medium to a concentration of 10 units/ml, distribute in 1 ml amounts and store at −20°C. For use in the culture system dilute 1 ml of erythropoietin solution with NCTC–109 to give a final concentration of 2.5 units of erythropoietin per ml.
6. L-asparagine (Nutritional Biochemicals Corp., Cleveland, Ohio). Make up in supplemented medium to a concentration of 2 mg/ml, filter through a 0.45 µ Millipore filter (Millipore Ltd., Montreal, P.Q., Canada), distribute in 0.5 ml amounts, and store at −20°C. For use in the culture system, add 4.5 ml of medium NCTC–109 to the 0.5 ml so that 0.1 ml contain 0.02 mg of L-asparagine.
7. NCTC–109 (Microbiological Associates, Bethesda, Maryland.): Add penicillin and streptomycin to a concentration of 100 units/ml and 50 µg/ml, respectively and store at 4°C.
8. Bovine serum albumin fraction V (BSA) (Sigma Chemical Co., St. Louis, Missouri): Prepare a 10% solution of BSA by a modification of the technique of Worton et al.[9] Add 50 gm of BSA powder to 91 ml of sterile distilled water and dissolve at 4°C (stir with a glass rod). BSA will dissolve in approximately 2½ hr. To detoxify the BSA solution, stir 5 gm of AG–501–X 8(D) resin into the solution and leave at 4°C for 2 hr. Stir the mixture every 15 min for the first hour. At the end of the 2 hr decant the clear fluid at the top into a flask to which 5 gm of fresh resin has been added and repeat the previous procedure. During the final hour the solution is left at room temperature (this last step may have to be repeated several times if the final solution of BSA is toxic for cultures or if colony-forming efficiency for CFU-e is low). Decant the clear upper portion and measure. To make the solution isotonic, add 1.1 ml of 10× concentrated MEM (Eagle) with Hanks' salt solution to each 15 ml of BSA solution. This results in a 37% w/v BSA solution and to make the final 10% w/v stock preparation of BSA dilute this solution with 1× supplemented HMEM. Filter the 10% solution through a Whatman No. 1 filter paper and sterilize by filtering through a 0.45 µ Millipore filter. Store at −20°C overnight, thaw the next day, distribute in 20 ml amounts and store at −20°C. Before using in the culture system, thaw and add 0.5 ml 7% NaHCO$_3$ to each 20 ml of 10% BSA.
9. Penicillin (Connaught Medical Research Laboratories, Willowdale, Ontario, Canada).
10. Streptomycin sulphate (Allen and Hanbury's, Toronto, Canada).
11. Resin: AG–501–X8(D) 20 to 50 mesh (Bio-Rad Laboratories, Richmond, California).

Each new batch of bovine plasma, beef embryo extract, fetal calf serum, and bovine serum albumin is pretested for its ability to support the production of erythrocytic colonies and, as with preparations of L-asparagine and erythropoietin, can be reused if returned to the freezer.

Collection and Dilution Medium To prepare, add heat inactivated fetal calf serum to supplemented Eagle's minimum essential medium with Hanks' balanced salt solution to a final concentration of 2% (pH 7.2 to 7.3).

Culture Method

In our laboratory disposable microtiter plates, each well containing 0.1 ml of culture medium, are used routinely for erythrocytic colony culture. Plastic petri dishes 35 × 10 mm containing 1 ml of culture medium or titration trays, each well containing 0.5 ml of medium may also be used. The concentrations of cells and ingredients in the culture medium are kept the same. The use of microtiter plates reduces cell numbers, volumes of culture medium, including erythropoietin, and numbers of microscopic slides. It also facilitates removal of plasma clots for fixation and staining and increases the precision and ease of scoring colonies.

Disposable microtiter plates (MRC-96-clear) obtained from Linbro Scientific Inc., Hamden, Connecticut are rectangular plastic plates with 96 wells, each 6 mm in diameter. Cut into equal-sized sections, each containing six wells, and place the sections in a tissue culture hood and sterilize by UV irradiation for 1 hr (GE germicidal lamp Model G30T8 30 W. at a distance of 60 cm).

Prepare the culture medium in a plastic tube in the following quantities or some multiple of these quantities and keep in an ice bath: (1) 0.1 ml beef embryo extract; (2) 0.2 ml fetal calf serum; (3) 0.1 ml bovine serum albumin (10% solution); (4) 0.1 ml L-asparagine (0.02 mg); (5) 0.1 ml erythropoietin (0.25 U); (6) 0.1 ml cells in supplemented HMEM; and (7) 0.2 ml NCTC-109.

Check the mixture to ensure that it contains no cell clumps. Add 0.1 ml of bovine citrated plasma at room temperature to 0.9 ml of the mixture and mix the two by pipetting. Add 0.1 ml of this final mixture to each well and check to see that the cells are distributed singly in the clots (beef embryo extract furnishes a sufficient concentration of Ca^{2+} ions to overcome the effect of the citrate in the plasma). Place an uncovered 35 × 10-mm petri dish containing distilled water alongside the microtiter plates in each of the large petri dishes in order to maintain adequate humidity. Cover the large petri dish and incubate in a CO_2 water-jacketed incubator (National Appliance Company, Portland, Oregon) at 37°C in an atmosphere of high humidity (incubator window should be covered with moisture) with 5% CO_2 in air for 48 hr.

Fixation and Staining of Colonies

1. 3,3'-Dimethoxybenzidine (Practical) (Eastman Kodak Co., Rochester, New York), 1% solution in absolute methyl alcohol. The stain can be used for 1 month before a precipitate forms. Danger: highly carcinogenic. Use and store in a hood and wear gloves. Both vapor and direct contact are toxic.

2. Hydrogen peroxide, 30% (Fisher Scientific Company, Chemical Manufacturing Division, Fairlawn, New York): To make a 2.5% solution, add 5 ml of 30% hydrogen peroxide to 55 ml of 70% ethyl alcohol. This can be used for 1 month at room temperature.

3. Hematoxylin stain solution (Harris-Lillie) (Fisher Scientific). A precipitate forms with time on the surface of the solution. This can be swept away with the edge of a microscope slide.

Remove the plasma clot from each well of a microtiter plate by rimming the clot with the small end of a microstainless steel spatula (Fisher Scientific). Scoop the clot from the well, and drop it onto a 75 × 25-mm glass slide. Three clots are placed on one slide. Make sure that clots are left long enough to adhere to the slide before dehydration to prevent lifting of the clots (several minutes).

Partially dehydrate the plasma clot by placing a rectangular piece of Whatman No. 1 filter paper (slightly smaller than the slide) on its surface. Place a second piece of filter paper on the first and allow to remain just long enough for the paper to become moist. With the aid of a dissecting needle, remove the top piece of filter paper and drop 1 to 2 drops of 5% glutaraldehyde in 0.01 M phosphate buffer (pH 7.0 to 7.2) on the remaining filter paper over the clots with a Pasteur pipet. Leave the filter paper, moistened with the glutaraldehyde for 6 min. Absorb off the excess glutaraldehyde with filter paper and remove the remaining filter paper. Place the slides with the fixed plasma clots in distilled water for 8 min and dry completely with a gentle stream of forced air. No further fixation is necessary. The dehydration process is necessary to remove amorphous material which, if left, interferes with the fixation and staining. All steps are carried out at room temperature; fixation at 4°C is unnecessary.

We have found that fixation of the cultures in glutaraldehyde substantially reduces the degree of lysis of cells that occurs with methanol fixation.

Place the dried slides in staining racks in 1% benzidine solution for 2 min, drain and place in 2.5% hydrogen peroxide solution for 1 min, drain and place in distilled water for 1 min. Counterstain in Harris' hematoxylin for 1 to 2 min and finally "blue" in running tap water. Allow the stained preparations to air dry and cover with a cover slip using Permount (Fisher Scientific) as mounting medium.

Scoring of Colonies

A grid made from a piece of thin cellulose acetate can be ruled out in 1 mm squares. Place on the surface of the slide, and hold in place by pieces of tape. Examine the whole preparation at 100× magnification. Score colonies of eight or more cells showing a positive benzidine reaction (brown to orange).

For each experimental point in the titrations, four

to six plates are used, and the results are expressed as mean number of erythrocytic colonies per plate ± 1 SEM.

The Plasma Culture System: Assay Method for Murine BFU-e
Preparation of Cell Suspensions
Prepare suspensions of bone marrow cells and spleen cells as described under assay method for murine CFU-e.

Media The media used in this culture system are as described under assay method for murine CFU-e but with the following modifications.

1. Beef embryo extract: Omit from the culture.
2. L-asparagine with added $CaCl_2$. Make up L-asparagine in supplemented HMEM to a concentration of 2 mg/ml and filter through a 0.45-micron Millipore filter, distribute in 0.5 ml amounts and store as stock solution at −20°C. To NCTC–109 add $CaCl_2$ to a concentration of 280 mg/liter and sterilize by filtering through 0.45-micron Millipore filter and store at 4°C (no longer than 1 month). For use in the culture system, add 4.5 ml of NCTC–109 with added $CaCl_2$ to 0.5 ml of the stock solution of L-asparagine (the extra $CaCl_2$ promotes clotting of the plasma and increases the efficiency of the system for erythropoietic burst and megakaryocyte colony formation). This solution of L-asparagine with added $CaCl_2$ cannot be frozen because of precipitation and must be made up fresh each time.
3. Erythropoietin (Ep) Step III. Dilute in supplemented HMEM to a concentration of 60 units/ml, distribute in 1 ml amounts, and store at −20°C. For use in the culture system dilute 1 ml of the Epo solution with NCTC–109 to give a final concentration of 30 units per ml.
4. Bovine serum albumin (BSA): Prepare and use as described under the assay method for CFU-e. However, more than two treatments with Resin–AG–501–X8 (D) are rarely required for culture of erythropoietic bursts. Treatment with resin decreases the pH of the BSA solution and thus of the final culture medium. A lower pH favors growth of 48-hr erythrocytic colonies but decreases the efficiency of the culture medium for erythropoietic bursts. Addition of $NaHCO_3$ to adjust pH of the medium (up to 7.6) is helpful with certain batches of BSA.

Culture Method
In our laboratory titration trays containing 0.5 ml of medium are used for culture of erythropoietic bursts. Plastic petri dishes 35 × 10 mm containing 1 ml of culture medium may be used but again the cost in media (including erythropoietin), slides, and cells is doubled. In our hands microtiter plates with 0.1 ml of culture medium are unsatisfactory for BFU-e assay above 10 bursts/well.

Plastic titration trays (Disposo-Trays, Model 96V-TC, Linbro Scientific Inc., Hamden, CT) are rectangular plastic trays with 96 wells. Cut the trays into sections, each containing 6 wells. Place the wells in a tissue culture hood and sterilize by UV irradiation for 1 hr.

Prepare the culture medium in a plastic tube in the following quantities or some multiple of these quantities: (1) 0.7 ml of FCS; (2) 0.35 ml of BSA (10% solution); (3) 0.35 ml of L-asparagine (0.07 mg) containing added $CaCl_2$ (0/09 mg); (4) 0.35 ml of Epo (10.5 units); (5) 0.35 ml of cells in collection medium; and (6) 1.05 ml of NCTC–109. Mix the contents of the tube by gentle pipetting. Add 0.05 ml of plasma to each of six wells and pipet 0.45 ml of the mixture from the tube onto the plasma in each of the wells. Mix the contents of the wells by gently swirling. (The addition of plasma to the wells before adding the cells and remainder of the culture medium facilitates removal of the clots from the wells.) Place the six wells in a large petri dish with an extra tissue culture well containing sterile distilled water alongside. Cover the petri dish and incubate in a CO_2 water-jacketed incubator at 37°C in an atmosphere of high humidity with 5% CO_2 in air for 7 days.

Fixation, Staining, and Scoring of Erythropoietic Bursts
These procedures are the same as described for the assay method for murine CFU-e. Only two clots are mounted on each slide.

Agar Culture System: Assay Method for Murine Megakaryocytic Progenitors (CFU-M)
Preparation of Cell Suspensions
Prepare suspensions of bone marrow and spleen cells as described under assay method for CFU-e.

Media

1. FCS, BSA 10% solution, L-asparagine with added $CaCl_2$ and collection and dilution medium are prepared as described under plasma culture assay method for BFU-e.
2. Difco Bacto Agar (Difco Laboratories, Detroit, Michigan) 3% w/v solution. Make up fresh each time with deionized distilled water and sterilize by boiling in a glass beaker for 10 min.
3. Alpha modification of Eagle's medium with Earle's salts, with L-glutamine and without sodium bicarbonate 2× concentrated (Flow Laboratories Inc., Rockville, Maryland).
4. Pokeweed mitogen-stimulated spleen cell-conditioned medium (PWCM) prepared as follows: Add 6×10^7 murine nucleated spleen cells (C57

BL/6Ut spleen cells are used in our laboratory) to 18 ml of NCTC-109 plus 2 ml of FCS (inactivated at 60°C for 30 min) in a Falcon tissue culture Flask #3024. Add PWM to give a final concentration of 15 μg per ml of final culture medium. Incubate the mixture (with the top of the flask loose) in a water jacketed incubator for 4 days at 37°C in 5% CO_2 in air. Pour off the medium and cells and centrifuge to remove cells. Filter the medium through a 0.45-micron Millipore filter and store at −20°C (PWCM will keep at −20°C for 6 months with no loss of activity). Test each new batch of PWCM for megakaryocyte colony-stimulating activity against PWCM of known activity. The PWCM may be used at 5 or 10% concentration in the final agar culture medium.

Culture Method

Titration trays are used as described under plasma culture assay method for BFU-e. Prepare culture medium in a plastic tube in the following quantities to give a final volume of 3.5 ml. (1) 0.7 ml of FCS; (2) 0.35 ml BSA (10% solution); (3) 0.35 ml of L-asparagine with added $CaCl_2$; (4) 0.35 ml of alpha medium 2× concentrated; (5) 0.35 ml of PWCM; (6) 0.7 ml of NCTC-109 (Alpha medium 1 X concentrated may be substituted); and (7) 0.35 ml of 3% Bacto agar (that has been brought to 40 to 41°C in water bath). Mix the above ingredients by swirling the tube. Add 0.35 ml of cells in dilution medium and quickly mix the contents of the tube by gentle pipetting and transfer 0.5 ml aliquots of the mixture to each of the six wells. Place the wells in a large petri dish and add an extra well filled with sterile distilled water. Cover the cultures and leave out for 10 min at room temperature. Transfer to a water-jacketed incubator in an atmosphere of high humidity with 5% CO_2 in air for 7 days.

Fixation and Staining of Colonies

The method of harvesting the agar clots and fixing and staining the colonies is essentially the same as that used in the plasma culture method for CFU-e and BFU-e assay but some modifications are necessary.

Remove the cultures from the incubator and allow them to sit at room temperature for 10 min (this firms up the clots and makes them easier to handle). Cut the sections of trays with 6 wells into smaller sections of 2 wells each with scissors. Hold one of these smaller sections over a 75 × 25 mm glass slide and lightly tap the edge of the plastic section on the bench surface. Tilt the wells and let the agar clots slide gently onto the glass slide. Occasionally the clot has to be freed with a spatula. The two clots fit one glass slide. Place several rectangular pieces of Whatman No. 1 filter paper (slightly smaller than the slide) on the surface of the clots. Allow these to remain long enough to absorb most of the moisture but do not allow the filter paper in contact with the agar clot to dry. With forceps or a dissecting needle remove all but one piece of filter paper. Drop 1 to 2 drops of 5% glutaraldehyde in 0.01 M phosphate buffer (pH 7.0 to 7.2) on filter paper over each of the clots with a Pasteur pipette. Leave for 6 min and absorb off the excess of glutaraldehyde with filter paper leaving the original piece of filter paper on the clot. Drop 2 drops of water onto the filter paper over each of the clots. Leave for 3 min and absorb off the water. Place the slides in a staining rack and dry in an air stream long enough to dry the edge of the filter paper but not the portion of the paper in contact with the clot. Remove the remaining filter paper and dry completely in the air stream.

Stain and cover the preparations as described in the assay method for CFU-e.

Acknowledgement

This work was supported by the National Cancer Institute of Canada and the Medical Research Council of Canada.

References

1. Stephenson JR, Axelrad AA, McLeod DL, et al: Induction of colonies of hemoglobin-synthesizing cells by erythropoietin *in vitro*. *Proc Natl Acad Sci USA* 68:1542-1546, 1971.
2. McLeod DL, Shreeve MM, Axelrad AA: Improved plasma culture system for production of erythrocytic colonies *in vitro*: Quantitative assay method for CFU-e. *Blood* 44:517-534, 1974.
3. Iscove NN, Sieber F, Winterhalter KH: Erythroid colony formation in cultures of mouse and human bone marrow: Analysis of the requirement for erythropoietin by gel filtration and affinity chromatography on agarose-concanavalin A. *J Cell Physiol* 83:309-320, 1974.
4. McLeod DL, Shreeve MM, and Axelrad AA: Induction of megakaryocyte colonies with platelet formation in vitro. *Nature* 261:492-494, 1976.
5. Metcalf D, MacDonald HR, Odartchenko N, et al: Growth of mouse megakaryocyte colonies in vitro. *Proc Natl Acad Sci USA* 72:1744-1748, 1975.
6. Nakeff A, Daniels-McQueen S: *In vitro* colony assay for a new class of megakaryocyte precur-

sor: Colony-forming unit megakaryocyte (CFU-M). *Proc Soc Exp Biol Med 151:*587–590, 1976.
7. Strome JE, McLeod DL, Shreeve MM: Evidence for the clonal nature of erythropoietic bursts: Application of an *in situ* method for demonstrating heterochromatin in plasma cultures. *Exp Hematol:* 1977. (In print)
8. Papayannopoulou TH, Brice M, Stamatoyannopoulos G: Stimulation of fetal hemoglobin synthesis in bone marrow cultures from adult individuals. *Proc Natl Acad Sci USA 73:*2033–2037, 1976.
9. Worton RG, McCulloch EA, Till, JE: Physical separation of hemopoietic stem cells from cells forming colonies in culture. *J Cell Physiol 74:*171–182, 1969.

Discussion (Chapters 1–6)

Dr. Seidel: With respect to your remark on the increasing number of receptors, I would like to ask whether, after 4 days without Ep *in vitro*, you require lower amounts of Ep for BFU-e growth. If this is so, then it would support your last hypothesis.

Dr. Iscove: I have not tested that. My guess is that the surviving BFU-e mature to the point where they become able for the first time to respond to very high erythropoietin doses.

Dr. Zanjani: Do you see evidence of growth or proliferation during the first 5 days in culture without Ep? Is the beginning of colonies evident before the addition of erythropoietin?

Dr. Iscove: Yes. The problem is identifying them with certainty before they acquire erythroid markers.

Dr. Zanjani: Are they of respectable size? Do they look like CFU-e, for example?

Dr. Iscove: Yes, there are dispersed colonies, but they are composed of primitive cells that cannot be identified as being erythroid. However, it is important to be able to recognize these early colonies, because they provide a potential source of early cells at stages before CFU-e.

Dr. Zanjani: Just using the inverted microscope, do you know whether they are the same colonies that become erythroid following the addition of erythropoietin?

Dr. Iscove: Yes, one can mark the dish, for example, with a little scratch on the underside.

Dr. Pescle: Did you by any chance look at BFU-e kinetics early after hematopoietic manipulation such as bleeding or transfusion?

Dr. Iscove: No.

Dr. Gordon: Dr. Goldwasser has been able to remove receptors temporarily by treating marrow cells with proteolytic enzymes. Since Dr. Wagemaker has been able to separate CFU-c, CFU-e, and BFU-e, would it be feasible to subject pure populations of these cells to enzyme treatment and then look for changes in their capacity to respond to the various factors that you have studied *in vitro*?

Dr. Iscove: I think it would be interesting, but it has not been done yet with the colony forming cells, as far as I know.

Dr. Erslev: According to your hypothesis, the development of the receptors must be erythropoietin determined, since in the absence of erythropoietin, there is a marked decrease in the responsiveness of erythropoietin. This would give erythropoietin a new function, that of making receptors on the surface of BFU-e.

Dr. Iscove: It is a problem of the chicken or the egg, in a sense. I think that the delayed erythropoietin experiment suggests that erythropoietin sensitivity is developing in its absence.

Dr. Erslev: But you have shown that hypertransfusion, presumably causing a disappearance of erythropoietin, will lead to a decrease in the number of stem cells responsive to erythropoietin, or in other words, in CFU-e.

Dr. Iscove: That is right.

Dr. Erslev: Which means that the receptors or the cells did not develop in the absence of erythropoietin.

Dr. Iscove: My interpretation of the finding is not that the receptors fail to develop, but rather that the CFU-e themselves did not develop in the absence of erythropoietin. Erythropoietin is necessary for survival of CFU-e, and is also required for formation of CFU-e from their immediate precursors.

Dr. Ogawa: Have you examined serum from mice with regenerating marrow cells to see if there is a type 2 regulator present?

Dr. Iscove: No, not yet.

Dr. Testa: What was the degree of purity of the erythropoietin preparation that you used, and how can one distinguish between effects caused by nonerythropoietin impurities?

Dr. Iscove: That is always a problem with unpurified preparations. Mine is purified to the extent that I cannot measure any other activity in culture that I am aware of. With respect to the receptors for the early type of nonerythropoietin humoral regulation, I suggest that those receptors are already present in the pluripotent cells, and rather than being induced, they are, in fact, gradually lost. My guess would be that sensitivity to erythropoietin or to colony-stimulating factor is not induced, but is rather an inexorable consequence of the development program operating within the cell itself.

Dr. Sassa: The identification of the early CFU-e before the development of hemoglobin is obviously important. It is now possible, by means of the work of Harvey Eisen at the Pasteur Institute, to study some early markers of erythroid differentiation, such as spectrin or a chromatin protein IP25 in Friend virus transformed cells undergoing erythroid differentiation after treatment with DMSO. I have also demonstrated that early enzymes of heme biosynthetic pathway are induced long before the appearance of hemoglobin in these cells. Although this is not directly comparable to your study, we have also found evidence that goes along with your finding that erythropoietin acts on the late stages of erythroid differentiation. As Dr. Urabe will describe, he has identified a DMSO-resistant clone of Friend virus transformed cells that do, however, respond to erythropoietin to form hemoglobinized cells. Early enzymes of heme-synthetic pathway in these DMSO-resistant cells are induced normally as in the wild-type cells, yet they do not make hemoglobin. However, they can make hemoglobin when they are treated with erythropoietin or hemin.

Dr. Iscove: I think one of the key differences between transformed, "immortal" cells, and normal cells is that in the absence of erythropoietin, the normal cells undergo a programmed death in the unhemoglobinized state, whereas Friend cells do not. Some cell surface markers on colony-forming cells will be accessible by techniques such as cell sorting, for example, glycophorin, which can be detected without killing the cell. Spectrin and nuclear proteins will not be accessible to that kind of approach since their detection involves killing the cell.

Dr. McLeod: We have done similar experiments in plasma clot culture to those that Dr. Iscove described by adding pokeweed mitogen-stimulated spleen cell-conditioned medium (PWCM) to the cultures on day 1 and no erythropoietin until day 4, and obtained bursts on day 7. The bursts obtained were as large or larger than normally obtained by adding erythropoietin alone on day 1. It appears that PWCM causes proliferation of precursor cells and that erythropoietin hastens differentiation of these cells. One question I would like to ask Dr. Iscove is how he fits megakaryocyte colonies into his scheme. We get megakaryocyte colonies with our erythropoietic bursts and Dr. Metcalf also gets megakaryocyte colonies with granulocytic colonies. Looking at our cultures, I do not think this is just a matter of coincidence; I think there is a real relationship between them. We suspect that they may have a common progenitor.

Dr. Iscove: In my hands, addition of pokeweed conditioned medium at time zero increases the size of the bursts.

Dr. McLeod: If you add Ep on day 1 along with PWCM, does that influence the size of the bursts?

Dr. Iscove: Yes, the bursts are larger in size. The presence of megakaryocytes in both bursts and granulocytic colonies suggests a "mechanism" of determination. Decision by a pluripotent cell to differentiate along the erythroid pathway may involve simply the loss of granulocyte-macrophage-monocyte potential while retaining erythroid and megakaryocytic potential. Determination along the granulocytic pathway, conversely, may involve simply the loss of erythroid potential with retention of the other potentials. Only with further maturation would the capacity for megakaryocyte formation be lost.

Dr. Ogawa: How long do you incubate cultures for CFU-e assay?

Dr. Zanjani: Seven days.

Dr. Ogawa: The plating efficiency of BFU-e from peripheral blood seems to be about one-tenth of what we observe.

Dr. Zanjani: Well, actually, in fraction III cells we observe as many as 20 BFU-e's per 10^5 cells.

Dr. Ogawa: You stated that the unfractionated blood mononuclear cells yielded one burst per 10^5 cells.

Dr. Zanjani: Yes, however, often times we do not obtain BFU-e growth in fraction I cells.

Dr. Ogawa: Have you used Isopaque or sodium metrizoate, rather than Hypaque, which may be toxic?

Dr. Zanjani: No. Do you preincubate your cells?

Dr. Ogawa: No. We use the methylcellulose system and we simply plate 3×10^5 mononuclear cells per plate without any incubation.

Dr. Zanjani: It seems that when we remove the monocytes from the system or adherent cell populations, anyway, we get increasingly better results, and hence, we keep using the same fraction of cells all the time.

Dr. Ogawa: The reason why I asked about incubation time is because approximately 15% of human peripheral blood erythropoietic precursors produce large colonies, and you can begin to see the clear hemoglobinization on day 9 of incubation. I wonder if you are counting some colonies derived from your lymphocyte fractions in blood.

Dr. Zanjani: No erythroid colonies were found if we took peripheral blood lymphocytes along and incubated them with erythropoietin by day 7. Whether or not there is an interaction going on that brings about an earlier maturation remains to be determined. These studies were terminated on day 7, not day 9.

Dr. Ogawa: Since your plating efficiency of blood BFU-e is so low, I almost think you need additional independent evidence that the colonies in the co-culture experiments indeed come from the bone marrow.

Dr. Zanjani: One way I can try to answer this is that the number of colonies that are formed in our co-cultures are far greater than the number of BFU-e's one would get from a similar number of lymphocytes. When we add normal peripheral blood lymphocytes, for example, to 6×10^5 bone marrow cells, we get 600 or 700 colonies. If these colonies were coming from peripheral blood lymphocytes, how many would you anticipate you would get at that time? Obviously not that many. I believe that would eliminate the possibility of colonies coming from peripheral blood lymphocytes. More likely, the effect can be ascribed to some form of cell–cell interaction.

Dr. Adamson: The studies that you have published in aplastic anemia have generated a good deal of interest in Seattle, particularly at the Hutchinson Center. Have you had the opportunity to study any patients with aplastic anemia who have not been transfused prior to your working with their lymphocytes in culture, or patients who are new patients with congenital hypoplastic anemia, who have neither received treatment nor relapsed, or who are being maintained on treatment, or being transfused?

Dr. Zanjani: No. Most of our patients had received a blood transfusion before we studied them, and most of them received many transfusions.

Dr. Adamson: In the canine system, which has been looked at by Drs. Beverly and Rainer Storb over the last year, deliberate and selective cross-transfusion in dog histocompatibility locus-matched animals indicates that such exposure to cells has a profound influence on the survival of a marrow graft. This is also reflected in culture as either nonstimulation or some degree of suppression of erythroid colony formation by donor lymphocytes when co-cultured with recipient marrow cells. There has been a major effort in Seattle to look at both transfused and untransfused patients with aplastic anemia, and to look at the effects of co-cultured aplastic anemia lymphocytes on marrow cell growth from both normal individuals and from HLA-matched donors. Some suppression is observed when one crosses both histocompatibility and transfusion exposure barriers, so that both of these may in fact have some influence on the results of co-culturing experiments. The experience that Drs. Jack Singer and Jim Brown have accumulated in the last year shows that in five untransfused individuals with aplastic anemia, patients' lymphocytes will only enhance the growth of CFU-e type colonies from the marrow of HLA-matched siblings who are to be used as donors.

Dr. Zanjani: The possibility exists that some of our observations are caused by transfusion requirements of the patient. The problem there is that this is not true of all individuals we have looked at. It may be that some patients react to transfusion differently than others; but, whatever the reason, a point is reached where the circulating cells in these individuals can bring about suppression of red cell production, at least in this *in vitro* environment. I think there is abundant evidence that under certain circumstances, treatment with immunosuppressive therapy may be sufficient to bring about a remission in some patients; but then one also has to consider that some of these patients undergo spontaneous remission. I think perhaps it is not so important that there is an immunological basis for certain of these diseases, but that the cell–cell interaction is occurring at some level that regulates production of red blood cells.

Dr. Moriyama: I have a similar result in the aplastic anemia. I wonder if the same inhibition happens if you use the HLA-matched or histocompatibility lymphocytes in co-culture systems, so that there is no damage between lymphocytes?

Dr. Zanjani: I really cannot tell you that. In some studies we have taken the patient's bone marrow and co-cultured it with the patient's own peripheral lymphocytes, and observed suppression, but we have always had trouble getting a sufficient amount of marrow to do the kind of studies we want to do. I think if histoincompatibility was responsible, then we should have picked up significant inhibition with the multitude of normals we have looked at. I think it is more specific than that.

Dr. Sassa: Would you comment on the effect of rates of increase in temperature during cell thawing?

Dr. Ogawa: We have not studied that. We use a standard rapid warming.

Dr. Sassa: Viability of cells changes as a function of both the rate of cooling and the rate of increase of temperature. Both effects should be studied.

Dr. Zanjani: Were similar numbers of viable cells cultured before and after the freezing process?

Dr. Ogawa: We do not even count the cells after thawing. We cultured a fraction of the original sample and froze the remainder.

Dr. Zanjani: Therefore, one cannot be certain that during processing a significant portion of cells may not have simply died. How long were the cells kept frozen before they were thawed and cultured?

Dr. Ogawa: As long as 48 hrs.

Dr. Zanjani: Rather than loss of response to erythropoietin or CSF, could cell survival to freeze-thawing lose its ability to respond to the usual physiological regulators?

Dr. Ogawa: I agree that this is a possibility.

Dr. Trobaugh: Concerning your cooling curves, when you say 1° or 2°C per min, is that the average temperature change from ambient to −60°C, or does it represent the rate of change from ambient to freezing and then from freezing to −60°C?

Dr. Ogawa: We keep the marrow cells at 4°C before freezing.

Dr. Trobaugh: Several years ago we showed that for cryopreservation of CFU-s, the rate of cooling from ambient to freezing and the temperature at which freezing was induced was unimportant in influencing viability of the cells, but that it was important to get the heat of fusion back to the freezing temperature as quickly as possible. We also found that the longer the cells remained at freezing temperature, the greater their loss of viability. Could you give us details of the temperature changes?

Dr. Ogawa: The cells were kept at 4°C before freezing. They were then exposed to DMSO at 4°C and transferred to a −5°C water bath.

Dr. Gallicchio: Dr. Ogawa, I would like to ask you if you performed the same series of experiments without ice crystallization?

Dr. Ogawa: Yes, and in this case, the temperature probably went down to about −15°C and rose briefly on crystallization.

Dr. Wagemaker: Did you measure temperature decline directly inside the ampule?

Dr. Ogawa: Yes, the thermocouple was placed in a separate tube containing 0.6 cc. of alpha medium and handled exactly the same as the experimental tubes.

Dr. Wagemaker: Did you try other cryoprotective agents like hydroxyethyl starch, which is a bit less toxic than DMSO?

Dr. Ogawa: We studied glycerol only in concentrations less than 20% which proved to be a poor cryoprotective agent. Apparently, high concentrations of glycerol used in fast cooling are effective in cryopreservation of mature red cells; however, we did not examine the latter conditions.

Dr. Adamson: I may have missed this point in the technical details of your presentation, but what was the serum concentration?

Dr. Ogawa: We used no serum.

Dr. McLeod: Dr. Ogawa do you count BFU-e on day 14?

Dr. Ogawa: That is right.

Dr. McLeod: One more question. You showed us pictures of what you are calling bursts, that is a large multiple colony. It it true that not all of your bursts will appear as one large mass of cells?

Dr. Ogawa: That is right. I think the distinction between a burst and a colony in the mouse is getting fuzzier and fuzzier. I believe that probably the most important factor is the incubation time, and when you try to segregate certain classes of colonies like, "this is a colony" so I count it separately, and "this is a big colony" so I classify it a burst, then I think we are probably adding more error. What I have been doing, whether it is acceptable or not, is counting everything that is hemoglobinized red on a particular date, and CFU-e on day 4 is the latest erythroid that I can identify in methylcellulose, which I think is very similar to a murine CFU-e. Whether it is better to count day 17 or 21 or 14, for the earlier BFU-e I do not know. The only reason I am using day 14 is that in peripheral blood, they are all saturated with hemoglobin at that time.

Dr. McLeod: Talking of the mouse model, of course, the CFU-e (i.e., the erythrocytic colony) lyses very early. At least in the plasma clot assay it lyses very early, so there is no overlap between erythrocytic colonies and erythropoietic bursts, and no problem distinguishing one from the other. But I have always had this problem in trying to distinguish CFU-e and BFU-e in cultures of human hematopoietic cells.

Dr. Rifkind: Have you, or has anyone, by serial photographic means, actually established the precise natural history of the sequential colonies, by photographic mapping techniques? I am not so much concerned about the clonal nature of the colonies; I accept that, but I would like to know a little bit more, for example, about the cellular origins of a colony which goes on to have as its components many subcolonies. Can one map the cell(s) that go to form the burst? What was the relationship of the precursor cell to precedent colonies that may have existed in the same area?

Dr. Testa: The speed with which CFU-e colonies disappear is surprising when one considers that red cells have a life span of 40 days in mice and 100 days in humans. Do the red cells produced *in vitro* have a shorter life span?

Dr. Ogawa: I do not know. One of my colleagues, a long time ago, examined the colonies and the whole methylcellulose culture fluid, and we found many matured red cells without nuclei but with a lot of Heinz bodies. I don't know when the colonies lyse and, I do not know whether the cells are dead or not. In fact I think many of them just simply float in the media, and do not know how long they survive in methylcellulose.

Dr. Golde: I was wondering how you deal with the statistical problem of colony counting when you are using such a small dish?

Dr. Urabe: In each experiment, 4 to 6 small wells are used for each CFU-e determination, and also 4 to 6 wells are used for each BFU-e determination.

Dr. Hasthorpe: If I am not mistaken, it appears that the CFU-e level is rather low. In this system you have about one hundred colonies per 10^5, whereas the BFU-e are in quite reasonable numbers, for mice. Is there a problem in detection when you're counting CFU-c in the microdishes?

Dr. Urabe: We stain with benzidine-H_2O_2 solution, then add one drop of the solution to each well, and count directly using an inverted microscope. There is no problem counting the colonies since one can see all the colonies in each well clearly.

Dr. Adamson: Does the cloning efficiency just alluded to have about 100 CFU-e per 10^5 cells per plate, for human or murine cells?

Dr. Urabe: For human marrow, there are approximately 100 CFU-e per 10^5 cells.

Dr. Adamson: In the table which you presented, you showed the growth of nine colonies per 2×10^4 cells from a murine source. This is a number of CFU-e on the order of 45 colonies per 10^5 cells. That is approximately one-eighth of what we would ordinarily expect cloning efficiency for murine CFU-e to be.

Dr. Urabe: The table was the number of colonies in mouse bone marrow culture that in that series of experiments yielded 9 CFU-e from 2×10^4 mouse marrow cells.

Dr. Iscove: I noticed a wine bottle up near the top of your slide showing the microwells. I just wondered whether that was your idea or whether it was Dr. Murphy's?

Dr. Urabe: It is not an integral reagent in our culture medium.

Dr. Rifkind: There is another problem that the small assay may introduce; there is a greater amount of side wall surface in these little culture wells than in bigger wells. The edge has always given us problems, even with Friend cells, in that colonies may accumulate there and are difficult to score. By decreasing culture volume, you have increased that indeterminate zone where defraction and phase changes make viewing very difficult. I wonder whether you have experienced this problem?

Dr. Urabe: The number of colonies in this miniaturized method is not always exactly one-tenth of the usual method. But using this miniaturized method, one gets clear dose response curves corresponding to different concentration of Ep and cell count.

Dr. Murphy: Just a comment to Dr. Rifkind's question. We have overcome this optical problem by using modulation interference optics, which are a type of Nomarski interference optical system rendering a planar, very flat field which I think you can appreciate, especially from our color photographs of BFU-e. It makes it very easy to score colonies at the interface of the medium and well wall.

Dr. Trobaugh: I was interested in Dr. Rifkind's comments about the colonies being at the edge of the culture dishes. Our CFU-e colonies are more numerous at the edge of the culture dish. They seem to prefer the edge and it is difficult to visualize them. Do others have this same problem or observation?

Dr. Peschle: Yes, we do.

Dr. Fisher: Can I ask both Drs. Urabe and Iscove a question because it deals with the methylcellulose system? I think Dr. Urabe is in a good position to use in his system the very highly purified erythropoietin because only small amounts would be required. Have you used the very low specific activities erythropoietin, e.g., one unit per mg and go on up to the very high specific activity, e.g., the 10,000 units per mg, material to determine whether the very high specific activity Ep supports growth of the CFU-e and the BFU-e. Most of us have used the very impure erythropoietin, but you would be in a good position to do this with your microsystem, because it would not require such a high amount of pure erythropoietin. Have you done any work of this type with increasing specific activity and determined whether it supports CFU-e and BFU-e growths in the same way?

Dr. Urabe: I used only one human urinary erythropoietin preparation that was 38 IRP units/mg of protein.

Dr. Iscove: Fritz Sieber did an experiment in our lab that bears on this point. He subjected human urinary erythropoietin to a sequential purification by charge, size, and adherence to calcium phosphate, concanavalin A-sepharose and PHA-sepharose. The relative positions of the dose response curves for CFU-e and BFU-e or semilog plots did not change at any stage of the purification. The specific activity at the final stage was approximately 4500 units/mg. I think the results show that erythropoietin itself is likely to be a requirement for the formation of mature hemglobinized bursts by BFU-e in culture. Other factors may also be required, but are probably supplied by the serum or other cells. The Connaught sheep erythropoietin gives bursts which are larger than those obtained with human urinary material. This could be due to the presence of an additional factor in the sheep preparation. It would be instructive to compare the activity of very highly purified sheep erythropoietin with that of cruder preparations.

Dr. Iscove: Can you form clots with purified fibrinogen and thrombin?

Dr. McLeod: No.

Dr. Iscove: Which system is easier to handle, plasma clot or agar, for producing stained, glass-mounted preparations?

Dr. McLeod: I think that the plasma clot is a little easier to fix and mount on glass slides. The agar clot is a little softer and a little more difficult to remove from the dish, but once it is mounted on glass it is just as easy to handle and stain.

Dr. Iscove: I think that the methods for preparing and specifically staining entire cultures with the relationships intact are going to become extremely important. They will probably prove essential to developing the emerging clonal assays for pluripotential cells in culture.

Dr. McLeod: I think that it may not be possible to do labeled antibody studies in agar and that the plasma system may be the method of choice for these sorts of studies.

Dr. Peschle: You mentioned a difference in the pH values for maximum growth of BFU-e and CFU-e. Could you give us some exact figures? Dr. S. Suzuki in our laboratory has been adding sodium bicarbonate to his cultures to bring the pH to 7.4 to 7.6, and finds that the efficiency for burst formation is increased. In our laboratory the method of preparing the BSA solution raises the pH and gives similar results.

Dr. Gordon: Dr. McLeod, when we were first beginning studies with the plasma clot system, we examined the relative effects of the human urinary preparation and the sheep material. We worked with preparations that had approximately the same unitage per mg but both were relatively crude, namely about 5 units per mg. We found rather marked differences with them in the plasma clot system. We felt that these different effects were perhaps attributable to differences in the type and amount of impurities in the two preparations. Another explanation may have related to differences in cell response to Ep of human as compared to sheep origin. Can you enlighten me further on this difference, if you have encountered it?

Dr. McLeod: In our experience, the human urinary erythropoietin gives the same results as sheep erythropoietin. I am not sure that is true for human erythroid colonies. But at least for murine CFU-e and BFU-e, we have not found a difference between the human and the sheep erythropoietin preparations.

Dr. Gordon: Is that even with relatively crude human and sheep material?

Dr. McLeod: With some of the early preparations we encountered some toxicity when we used more than 3 units per ml for the BFU-e assay and had a drop in the number of bursts. I think this was strictly a toxic effect of the material. The erythropoietin preparations we now use run 5 to 7.5 units per mg and we see no difference between that and the human urinary erythropoietin preparations which are about 74 units per milligram.

Dr. Gordon: Have you had some experience with this problem, Dr. Iscove?

Dr. Iscove: Sheep erythropoietin produces larger bursts than human erythropoietin under some conditions of culture. I want to make a comment about pH. Larry Guilbert has carefully measured the pH in culture dishes within the incubator. He found the optima for both CFU-e and BFU-e to be 7.4. A word of caution about adding things to cultures: keep osmolarity in mind. Osmolarity titration curves have broad plateaus but with very steep shoulders. If the osmolarity of the culture is near a shoulder, the addition of anything which isn't iso-osmotic may have drastic effects on plating efficiency. This would have to be kept in mind when, for example, the pH of the medium is changed by addition of sodium bicarbonate. The same precaution applies to the use of hepes buffer which is not itself toxic.

Dr. Peschle: Using the same system as Norman, we have observed that the optimum pH for both BFU-e and CFU-e growth is between 7.3 and 7.4.

Dr. Ogawa: The comment on the species of mice, I think, is very important. We found that the plating efficiency of BALB/c BFU-e is about one-half that of BDF_1 mice. In addition, Dr. Hara noted that the age of the mice is important to consider. The plating efficiency in young mice is considerably higher than older mice.

Dr. Fisher: I would like to ask Dr. McLeod a technical question about the quality control of the media used in our systems. For example, the bovine citrated plasma has been a problem in our laboratory in which some batches will not support colony growth. In fact, several batches from GIBCO have failed to support colony growth. When we do get a good batch, we use it for a prolonged period of time and compare any new batches of media with this standard.

The second question is how you sterilize your erythropoietin. How much loss of erythropoietin do you get by passage through a Millipore filter?

Dr. McLeod: Well, in answer to the first comment, I agree that one can have bad batches of plasma. And I stressed that we pretest every batch and I think this is essential. With any culture system one must pretest all the ingredients. As for sterilization of erythropoietin, I think the loss with Millipore filtering is a myth. We have only had to filter erythropoietin two or three times, but didn't have any loss of the hormone with this procedure.

Dr. Golde: First, I would like to respond to the point made about comparing sheep plasma erythropoietin with human urinary in human cultures. We find them quite different. In human cultures, the step III preparation gives us a peak colony number at a half unit per ml. With the human urinary erythropoietin, we have not found a peak, going up to 10 units.

Dr. Peschle: But how much does the CFU-e number increase when you add large amounts of Ep, say 5 or 10 units per plate?

Dr. Golde: We reach no plateau with human urinary erythropoietin and human bone marrow target cells?

Dr. Peschle: But I meant on a percentage basis. Assuming that with 3 units per plate you get 100% of colony numbers, how much can you increase it over 100% by plating 10, 15 or 20 units?

Dr. Golde: It keeps going up but the slope may change at very high erythropoietin concentrations. The other thing with the loss of erythropoietin by sterilization with Millipore, I can just quote what Dr. Peter Dukes tells us. He says that he loses 20 to 30% of the activity using Millipore filters, and I hope he is right, because we are now using silver filters. We do not lose any Ep activity with these silver filters.

Dr. Iscove: Millipore filters bind protein, but have a finite binding capacity. Therefore the percentage loss depends on the total amount of protein going through a given size of filter. Effectively this means that if you want to filter a preparation of very high specific activity, you will have to protect the erythropoietin from binding by including a competing substance, such as albumin or polyethylene glycol (6,000 daltons, 0.1% final concentration).

Dr. Trobaugh: How much loss is there with a Nucleopore filter?

Dr. Wagemaker: They do not cause any loss of Ep activity in our laboratory.

7 Cellular and Soluble Factors Influencing the Differentiation of Primitive Erythroid Progenitor Cells (BFU-e) *In Vitro*

G. Wagemaker

Introduction

The mammalian regulation of red blood cell production appears to be mediated mainly by the hormone erythropoietin (Ep), produced by involvement of the kidney in response to changing oxygen demand. Ep has been characterized as a sialic acid containing glycoprotein, with a relative molecular mass of 46,000 daltons as estimated from highly purified sheep plasma preparations.[1] Its biological activity has been standardized in units established by an international reference preparation (IRP).[2-4] On the assumption that preparations of 70,000 units/mg^{-1} are pure circulating levels of Ep, which maintain steady state erythropoiesis, can be calculated to amount to a few pmol/liter^{-1}, concentrations of approximately 2 pmol/liter^{-1} can elicit a significant response of erythroid progenitor cells *in vitro* and maximal responses are obtained at concentrations lower than 1 nmol/liter^{-1}.

The mechanism of action of this very potent molecule has not been clarified. Early observations[5] characterized its target cells, later termed erythropoietin-responsive cells[6] (ERC), as a morphologically unidentified cell population that responds to Ep by giving rise in 24 hr to a wave of proerythroblasts. ERC appear to differ from pluripotent stem cells in kinetics of recovery following perturbation of hemopoiesis[7-12] and in sensitivity to cell cycle-specific cytotoxic agents.[10,13-15] In contrast to pluripotent hemopoietic stem cells (CFU-s) ERC seem to be in rapid cell cycle, even in mice made polycythemic to suppress erythropoiesis to a level of virtually nill.[13,16,17] The mechanisms regulating the size of the ERC population are difficult to investigate, since a change in *in vivo* Ep responsiveness does not distinguish between a change in ERC number and a change in sensitivity of ERC toward Ep. Some inferences can be made from the appearance of ERC in spleen colonies under conditions of suppressed Ep production by induced polycythemia, indicating that Ep is not involved in the differentiation of CFU-s into ERC.[18-21] It has, on the other hand, been demonstrated that repeated administration of Ep can restore Ep responsiveness in busulphan-treated mice without restoring CFU-s numbers, suggesting that the population size of ERC can be controlled by Ep in the absence of an influx derived from differentiating CFU-s.[12] This apparent dual regulation of ERC population size might be considered as a characteristic of stages of erythroid differentiation intermediate between CFU-s and the morphologically recognizable erythroid precursor cells. Lack of a clonal assay for direct enumeration of ERC, however, has been a serious drawback and has effectively prevented detailed studies.

Thus, much potential progress was recently made when, in addition to a clonal assay for granulocyte/macrophage progenitor cells (CFU-c), clonal assays for cells in the erythroid series became available by the observation that some primitive hemopoietic cells differentiate into clusters and colonies consisting of erythroid cells in semisolid cultures containing Ep. Stephenson et al.[22] demonstrated the formation of clusters of hemoglobin-synthesizing cells in Ep-stimulated cultures and termed the cells from which these clusters originated erythroid colony-forming units (CFU-e). A cell of much larger erythroid colony-forming ability was detected by Axelrad et al.[23] and termed erythroid burst-forming unit (BFU-e) because of the disperse appearance of the colonies containing abundant numbers of erythroid cells. BFU-e can form colonies consisting of up to 10^4 cells.[24] Eventually these cells will differentiate into reticulocytes, giving the bursts a bright red appearance due to their hemoglobin content.

We have previously demonstrated that erythropoietic stress by intermittent isobaric hypoxia greatly increases the number of CFU-e in the bone marrow, but does not significantly affect the number of BFU-e.[25] The time course followed by BFU-e during exhypoxic polycythemia was characterized by a small but significant transient rise. A similar pattern was shown not only for ERC,[26] but also for CFU-s and CFU-c, indicating that the mechanism by which it occurred was not specific for erythropoiesis. CFU-e, on the other hand, rapidly declined during exhypoxic polycythemia, in a manner paralleling and preceding peripheral blood ^{59}Fe uptake. It can be concluded from these data that BFU-e population size in the bone marrow is not regulated by mechanisms specific for erythropoiesis, but rather by as yet poorly defined mechanisms that also influence the compartment size of CFU-s and CFU-c. Recently, Iscove[27] supplemented these data by demonstrating that physiological stimulation and suppression of erythropoiesis does not affect the BFU-e cycling state. The *in vivo* unresponsiveness to Ep of BFU-e has led to the suggestion that the early stage of erythroid burst formation *in vitro* is also independent of Ep.[27,28]

The CFU-e population is, in contrast to BFU-e and, by definition, to ERC, suppressable[25,29] by induced polycythemia, indicating that its population size in the bone marrow is largely dependent on circulating Ep. Several authors, however, demonstrated the existence of a small subpopulation of CFU-e resisting suppression by polycythemia.[23,27,30] Recently, Reissmann et al.[30] demonstrated an increased number of splenic CFU-e following administration of endotoxin, which might indicate that the population size of CFU-e is not exclusively controlled by Ep. These data, similar to observations on ERC, may indicate a dual control of the population size of CFU-e.

It is apparent from these considerations that Ep dependency gradually increases during erythroid differentiation, and appears to be absolute for the morphologically recognizable erythroid precursor cells in the mouse bone marrow. BFU-e appears to be the most primitive cell population of the erythroid series and is, as far as current evidence shows, neither *in vivo*, nor *in vitro* (this chapter) directly responsive to Ep. Its population size seems to be controlled by an as yet poorly defined, perhaps a specific mechanism. ERC and CFU-e are best characterized as intermediate, probably overlapping, cell populations that are not exclusively being maintained by Ep.

In contrast to any other clonable hemopoietic progenitor cell the sedimentation rate[31] and density[32] distributions of BFU-e completely coincide with those of CFU-s.[33] By virtue of its physical properties, proliferative potential, and cycling state, BFU-e appears to be the erythroid progenitor cell population closest to CFU-s, which is consistent with its unresponsiveness to Ep *in vivo*. It can therefore be presumed that factors modulating the Ep responsiveness of BFU-e *in vitro* may be relevant to the regulation of the earliest differentiation of erythroid progenitor cells. The present study is directed at cellular and soluble factors that *in vitro* profoundly influence the number of BFU-e detected and their sensitivity to Ep.

Materials and Methods

For all experiments inbred ND2 female mice 8 to 12 weeks of age were used.

Preparation of cell suspensions, assay for erythroid burst forming units and erythropoietin (Ep) preparations were as previously described[25,34] and as extensively described in Appendix 4.

Individual erythroid bursts were harvested from the cultures using a capillary pipette and transferred to glass microscope slides for appropriate staining.

Colonies harvested from cultures were stained with orcein in 70% acetic acid and covered by a glass coverslip. Reticulocytes were stained by 3.3% brilliant cresyl blue in 100% ethanol. This solution was dried on glass slides; harvested colonies were transferred to such a glass slide and squeezed by a glass coverslip.

Mouse bone marrow cells were separated on the basis of differences in their buoyant density using bovine serum albumin (BSA) gradient equilibrium centrifugation as described by Shortman.[32,35] BSA (Sigma Chemical Co., St. Louis, Missouri, Cohn fraction V, cat. nr. A4503) was prepared as reported by Shortman et al.[36] The final osmolarity of the gradient was iso-osmotic with mouse serum,[37] and the pH was maintained at 5.1. The bone marrow cells were suspended into the dense medium, and a continuous linear gradient was prepared at 4°C with a single chamber and a peristaltic pump driven generator. The stirring rate of the paddle mixer was 75 rpm^{-1}, the gradient being generated over 20 min. Gradients of 15 ml volume containing 1 to 2×10^8 cells were centrifuged at 4000 g for 30 min at 4°C in a JS–13 swinging bucket rotor of a Beckmann Model J–21B centrifuge to bring the cells in the position of their buoyant density. Fractions of fixed volume were collected by upward displacement with bromobenzene. Over a density range of 1.060 to 1.090 gm/cm^{-3} approximately 20 fractions were collected. Density measurements were performed on each fraction by use of a linear bromobenzene/petroleum ether gradient, calibrated by standard BSA solutions, in turn calibrated by a PAAR DMA 40 digital density meter. The cells in each fraction were recovered by dilution and centrifuged at 4°C for 10 min at 500 g. The cell pellets were resuspended in Hank's Balanced Salt Solution (HBSS) and kept on melting ice until assayed. Nucleated cell counts were made by hemocytometer after dilution by Türk's solution. The density distributions for the various cell types were prepared by relating cells per fraction per density increment to density, with peak values adjusted to 100%.

Cells were separated on the basis of essentially size differences by sedimentation at unit gravity as described by Miller and Phillips.[31] Bone marrow cells (0.5 to 1×10^8) in 0.3% BSA in HBSS, covered by a layer of 50 ml HBSS, were loaded on top of a shallow (1 to 2%) BSA gradient in a conically based, cylindrical, siliconized, sterile glass chamber ($\alpha = 11.4$; Glasapparatebau Weil, Germany), and allowed to sediment for 4 hr at 4°C. Fractions were collected in a volume of 15 ml, centrifuged at 4°C for 10 min at 500 g, resuspended in a fixed volume of α-medium, and kept on melting ice until assayed. Cell counts were made by hemocytometer after dilution with Türk's solution. Sedimentation rates per fraction were calculated as outlined by Miller.[38] The sedimentation rate distributions for the various cell types were prepared by relating cells per fraction to sedimentation rate, with peak values adjusted to 100%. Approximate cell diameters were calculated by rewriting Stokes' law for the sedimentation rate of spheres in a viscous medium as[38]:

$$r = \left\{\frac{s}{k(\rho - \rho_0)}\right\}^{\frac{1}{2}}$$

where r represents the radius of the sedimenting sphere, k a constant (5.0×10^9 mm²/gm^{-1}/hr^{-1}), ρ the density of the sphere, and ρ_0 of the surrounding medium.

Cells were irradiated in a well-aerated chamber at a rate of 1.15 Gy/min^{-1} using a ^{137}Cs source.

Serum from polycythemic mice was prepared as follows: Polycythemia was induced by intermittent isobaric hypoxia essentially as previously described.[39] Briefly, 100 mice were kept in a closed cage ($0.70 \times 0.50 \times 0.30$ m³) flushed with a gas mixture of nitrogen and air at a rate of 3 liters/min^{-1}, the oxygen inflow being adjusted to 8%. The mice were exposed to this mixture 8 hr/day^{-1} for 9 days. Food and drinking water were, as in all experiments, provided *ad libitum*. The mice were bled by orbital puncture under ether anesthesia on the 9th day of hypoxia and at 2 days after termination of hypoxic conditions.

Serum from anemic mice was prepared as follows: Anemia was induced by intraperitoneal injection of neutralized phenylhydrazine (80 mg/kg^{-1}) on 2 consecutive days, and bled by orbital puncture under ether anesthesia 3 days after the first injection. Bone marrow aplasia was induced by lethal irradiation (9.5 Gy); the mice were bled on day 7 following irradiation. A separate group of mice received in addition phenylhydrazine on days 4 and 5 following irradiation. The collected sera were sterilized by filtration through Millipore membranes with a pore size of 0.22 µm.

Colony-stimulating factor (CSF) from pregnant mouse uteri was prepared according to the methods of Bradley et al.[40] and Stanley et al.[41] The biological activity of this CSF has been investigated in detail by Van den Engh[42] and is identical to CSF from human urine. It was further purified by gel filtration on Sephadex G150, its distribution coefficient[43] K_{av} being 0.22, and subsequently by Con A-Sepharose chromatography. It is designated as CSF-pmue. In all experiments described in this report it is added to the cultures at such a concentration that no increase in colony number occurs on further increasing the CSF-pmue concentration. This is referred to as stimulation by a "plateau" concentration.

Human leukocyte conditioned medium (HLCM) was prepared as previously described.[34]

Results
Characteristics of erythroid bursts

The appearance of erythroid bursts in methylcellulose cultures of murine bone marrow cells is dependent on the concentration of Ep, on the concentration of nucleated cells, and on multiple serum components. Optimal stimulation of erythroid burst formation was found to occur at Ep concentrations of 1 to 2 units/ml.$^{-1}$ Commercially available Ep preparations from sheep serum tend to inhibit at higher concentrations, whereas addition of higher concentrations of human urinary Ep to the cultures did not induce significant changes in the number of bursts obtained, up to a concentration of 10 units/ml^{-1}.

Erythroid bursts are recognizable in unstained cultures starting from day 6 of incubation. They can be easily detected by abundant numbers of small cells, characteristically growing in many separate clusters. Under optimal conditions individual bursts obtain a red appearance due to hemoglobin synthesis. Bursts tend to lyse between days 14 and 20 of incubation; the predominant cell types present at the beginning of lysis are reticulocytes, and to a lesser extent mature erythrocytes. Figure 7-1 shows reticulocytes from a burst formed at an Ep concentration of 1 unit/ml^{-1} at day 15 of incubation. Various stages of maturation can be distinguished, and this apparent lack of synchronization of the final stage of burst formation can also be demonstrated in earlier stages. Figure 7-2 shows orcein-stained cells from a day 9 burst. Multiple stages of erythroid differentiation are detectable, which seem to be grouped in separate clusters. Except for small erythroid cells, bursts generally contain a number of large cells with multiple nuclei. These cells resemble megakaryocytes, in accordance with data from McLeod et al.[44] on platelet-producing megakaryocytes in Ep-stimulated cultures containing erythroid bursts. The clonal nature of colonies containing both erythroid cells and megakaryo-

Figure 7-1 Reticulocytes from an erythroid burst at day 15 of incubation.

tration suggests that an active principle associated with bone marrow cells is required for BFU-e to be able to respond to Ep. This burst feeder activity (BFA) appeared to be an absolute requirement; no erythroid bursts were detected at cell concentrations lower than 2.5 to $5 \times 10^4/\text{ml}^{-1}$. Not only the incidence of BFU-e but also the size of the bursts was dependent on the cell concentration of the cultures (unpublished observation). Under identical culture conditions the incidences of CFU-e and CFU-c were found to be independent of the concentration of nucleated cells.

BFA appeared not to be abolished by irradiation up to 20 Gy. Figure 7-3 shows irradiation dose-effect curves for BFU-e and for the enhancing effect of irradiated bone marrow cells on unirradiated BFU-e. In agreement with the nonlinear relationship of burst formation and nucleated cell concentration the addition of $5 \times 10^4/\text{ml}^{-1}$ bone marrow cells to $2 \times 10^5/\text{ml}^{-1}$ bone marrow cells resulted in an increase in the number of bursts disproportional to the number obtained with $5 \times 10^4/\text{ml}^{-1}$ bone marrow cells alone. Increasing doses of irradiation resulted in a rapid disappearance of erythroid bursts (the D_0

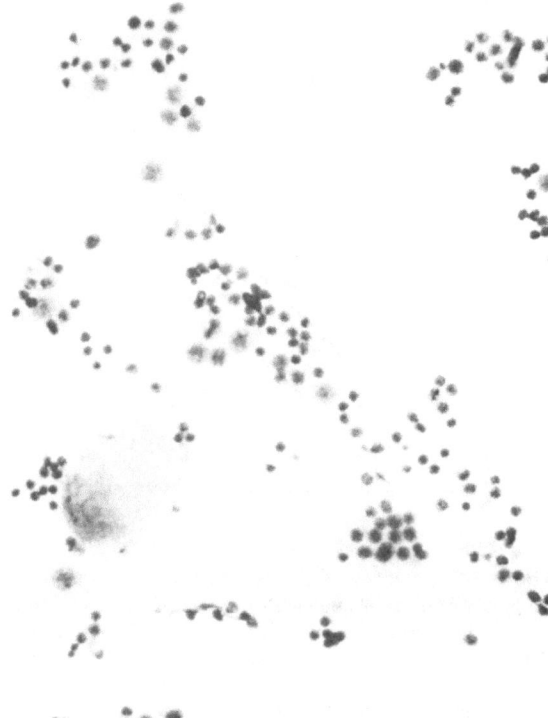

Figure 7-2 Orcein stained cells from a burst at day 9 of incubation. The culture was stimulated by 1 unit/ml^{-1} Ep.

cytes has as yet not been proven. In this series of experiments, 31 of 40 bursts investigated at day 10 of incubation contained the type of large cells shown in Figure 7-2, in 12 of them in multiple clusters. The number of these large cells ranged from 2 to 29, with a mean of 11 per erythroid burst. Since the occurrence of these cells was investigated in bursts transferred to glass slides, the values represent minimal numbers.

Induction of Ep-responsiveness in BFU-e by distinct population of bone marrow cells

Detection and biological effects of bone marrow-associated burst feeder activity (BFA)

Under our culture conditions the number of BFU-e detected was, irrespective of the EP concentration, consistently dependent on the concentration of nucleated cells in the cultures, in agreement with data from Heath et al.[45] but in contrast to data from Iscove and Sieber[24] and Gregory.[16] The degree of dependency, however, varied with different serum batches used to supplement the culture medium, in general agreement with the notion that a specific serum component is necessary for burst formation.[28] The dependence of the differentiation of BFU-e on cell concen-

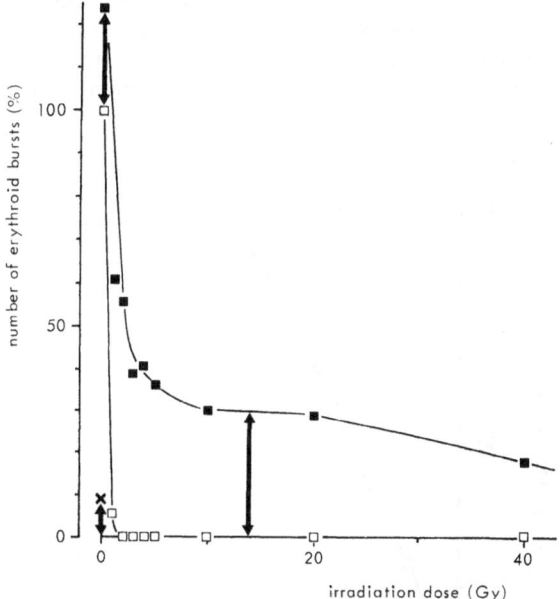

Figure 7-3 Effect of adding irradiated bone marrow cells to Ep-stimulated cultures on the number of bursts detected, as a function of irradiation dose. x: $5 \times 10^4/\text{ml}^{-1}$ unirradiated bone marrow cells; open squares: $2 \times 10^5/\text{ml}^{-1}$ irradiated bone marrow cells; closed squares: $2 \times 10^5/\text{ml}^{-1}$ irradiated bone marrow cells $+ 5 \times 10^4/\text{ml}^{-1}$ unirradiated bone marrow cells. The vertical arrows emphasize the increased number of bursts induced in $5 \times 10^4/\text{ml}^{-1}$ bone marrow cells by addition of $2 \times 10^5/\text{ml}^{-1}$ bone marrow cells, irrespective of irradiation. 100% indicates 66 bursts. See text for details.

value of BFU-e was found to be 0.73 ± 0.08 Gy; unpublished data). In contrast, the enhancing effect of $2 \times 10^5/\text{ml}^{-1}$ bone marrow cells on the number of BFU-e detected in $5 \times 10^4/\text{ml}^{-1}$ unirradiated bone marrow cells appeared to resist irradiation up to a dose of 20 Gy.

Subsequently, the number of detected BFU-e was determined as a function of the number of irradiated (15 Gy) bone marrow cells added. Figure 7-4 shows that graded doses of irradiated bone marrow cells added to Ep-stimulated cultures containing $5 \times 10^4/\text{ml}^{-1}$ bone marrow cells results in increasing numbers of bursts induced, approaching plateau levels at approximately 100 BFU-e per 10^5 unirradiated cells. This incidence is 3 to 8 times more than previously reported,[25,27,46] which implies that characteristics of BFU-e reported previously might apply to a subpopulation rather than to the full population of BFU-e. No bursts were obtained in the absence of unirradiated cells. Not only the number of bursts detected but also their size increased by addition of irradiated bone marrow cells. The ascending part of the curve of Figure 7-4 shows a linear relationship between the number of bursts induced and the concentration of irradiated cells. This particular feature indicates that the number of bursts induced can be employed as a direct measure of the concentration of BFA in the cultures, which is relevant for studies directed at the physical characteristics of the bone marrow cells associated with BFA (this chapter), and

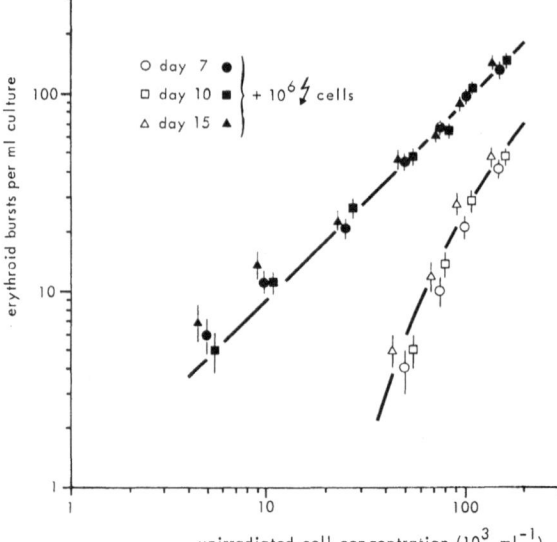

Figure 7-5 Number of erythroid bursts as a function of the concentration of nucleated bone marrow cells, both in the absence (open symbols) and presence (closed symbols) of $10^6/\text{ml}^{-1}$ irradiated (15 Gy) bone marrow cells. Open and closed circles: scored at day 7 of incubation; open and closed triangles: at day 10; open and closed squares: at day 15. The cultures were stimulated with 1 unit/ml^{-1} Ep. Vertical bars indicate standard errors.

for studies measuring changes in BFA in response to *in vivo* perturbation of hemopoiesis.[47]

The results of the previous experiments predicted that addition of irradiated cells to a plateau concentration would result in an independence of the number of bursts formed on the cell concentration. Figure 7-5 shows a double logarithmic plot of burst number as a function of unirradiated cell concentration, showing, in the absence of irradiated cells, curvilinearity characteristic for cellular interaction phenomena, and, in the presence of irradiated cells, burst incidence on unirradiated cell concentration. In fact, addition of irradiated cells resulted in a slope of 0.97 ± 0.07 for the relationship of burst number and unirradiated cell concentration, which is not significantly different from the predicted value of 1.00. The incidence of BFU-e was again found to be approximately 100 per 10^5 nucleated cells. The experiment was scored at three different time intervals to ascertain that the higher number of bursts induced by the irradiated cells was not due to a higher appearance rate of the bursts. No significant changes, however, were observed as a function of incubation time. A noteworthy result of addition of irradiated cells is the fact that not only BFU-e incidence but also burst size is independent of cell concentration.

The absolute dependency of Ep responsiveness of BFU-e on BFA was expected to influence the dose-response curve relating Ep concentration and num-

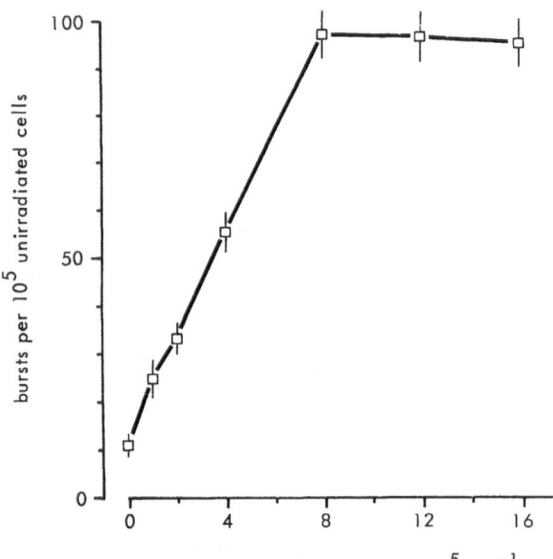

Figure 7-4 Number of erythroid bursts as a function of the concentration of irradiated (15 Gy) bone marrow cells, added to cultures containing $5 \times 10^4/\text{ml}^{-1}$ unirradiated cells and a plateau concentration of Ep. Mean of three independent experiments. Vertical bars indicate standard errors.

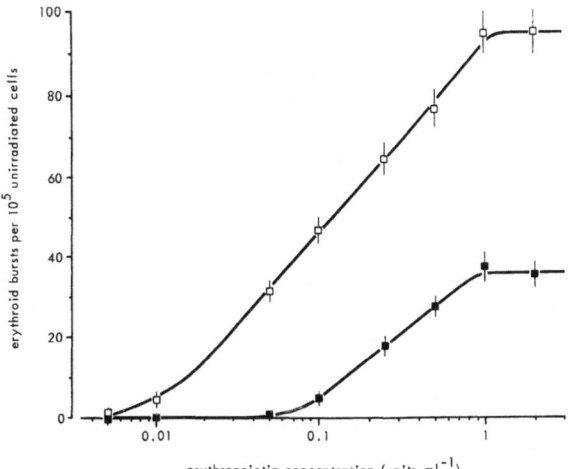

Figure 7-6 Dose-response relationship relating burst number to the Ep concentration, in the absence (closed squares) and presence (open squares) of $10^6/ml^{-1}$ irradiated (15 Gy) bone marrow cells. The concentration of unirradiated cells was $10^5/ml^{-1}$. Bursts were scored at day 10. Vertical bars indicate standard errors.

ber of BFU-e. Figure 7-6 shows that addition of a plateau concentration of irradiated cells to the cultures results in an increased range of sensitivity to Ep, and a greatly decreased threshold concentration of Ep for burst formation. The data thus imply that the number of BFU-e detected and their sensitivity to Ep is directly related to the BFA concentration of the cultures. It might be emphasized that the threshold concentration of Ep for BFU-e approximately equals the threshold concentration previously reported for CFU-e.[25] The fact that the Ep sensitivity of BFU-e is directly related to the BFA concentration also shows the hazards of considering Ep sensitivity measured *in vitro* as a specific attribute of the differentiation stage of the target cells.[46] Rather, the lack of capacity of BFU-e to respond to Ep in the absence of BFA characterizes BFU-e as a cell population in which the early differentiation step leading to Ep responsiveness has not yet taken place. It remains to be determined if the large range in sensitivity to EP can be explained by a heterogeneity of BFU-e and reflects subpopulations with different thresholds for stimulation. The homogeneity, however, of the dose-response relationship, and the dependency of Ep responsiveness of BFU-e on BFA suggest that the wide range in sensitivity is not due to subpopulations with different thresholds. Rather, this wide range in sensitivity might reflect an intrinsic property of Ep responsiveness. It is of interest that the dose-response relationship for Ep of BFA-induced BFU-e resembles closely the dose-response curve of polycythemic mice of the same strain (unpublished observation).

In summary, these data demonstrate that the Ep responsiveness of BFU-e is induced by a bone marrow-associated activity, which has been termed BFA. The data imply that *in vitro* the number of BFU-e detected and their sensitivity to EP is a direct function of the concentration of BFA.

Physical characterization of the bone marrow cells associated with BFA

The cellular nature of BFA was demonstrated by cell separation employing continuous density gradient centrifugation and velocity sedimentation at unit gravity. The experiments were based on the linearity of the ascending part of the dose-response curve Figure 7-4 relating induction of burst formation and concentration of irradiated bone marrow cells, from which it was concluded that induction of burst formation may be employed as a direct measure of the BFA concentration of the cultures, and thus as a relative measure of the concentration of cells associated with BFA. The cell separation experiments were designed in such a way that the peak values of the density or sedimentation rate distributions of the cells associated with BFA represented a number of bursts induced close to but lower than the number induced by a plateau concentration of irradiated cells. Irradiated (15 Gy) cells of a fixed proportion of the fractions were cultured with unseparated bone marrow cells cultured at a concentration that by itself resulted in an absence or in a very low number of bursts. In this way it was ascertained that the number of bursts induced was linearly related to the BFA concentration of the fractions.

It was previously demonstrated that cells associated with BFA can be almost completely separated from BFU-e by density centrifugation. BFU-e appeared to belong to low density cells, whereas cells associated with BFA were restricted to density regions higher than the modal density of all nucleated cells.[48] The buoyant density distribution of BFA associated cells appeared to be characterized by a modal density of 1.083 gm/cm^{-3}, with a band width at 50% of the peak value of 0.005 gm/cm^{-3}. The latter feature suggests that BFA is associated with a population of cells that is homogeneous with respect to density.

The separation of BFU-e and BFA appeared to have important implications for the density distribution of BFU-e. In initial experiments directed at comparison of the density distributions of BFU-e and other clonable hemopoietic progenitor cells[33] a very low recovery of BFU-e was noted. The recovery of BFU-e from the gradient varied between 9 and 20% (four experiments), in contrast to a recovery ranging from 70 to 90% for all nucleated cells and for simultaneously assayed hemopoietic progenitor cells such as CFU-s, CFU-c, and CFU-e. This low recovery might indicate either a high loss of BFU-e, or, in view of the previous data, might be due to separation of BFU-e

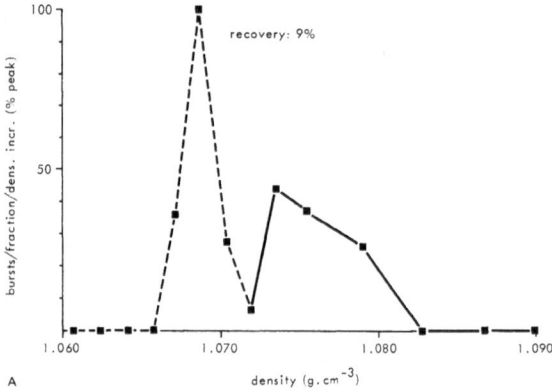

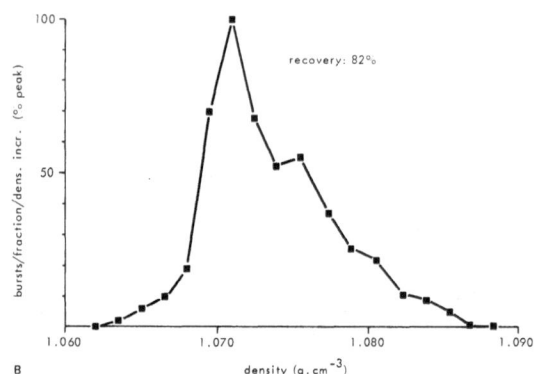

Figure 7-7 Density distributions of bone marrow BFU-e cultured in the absence of irradiated cells (**A**) and in the presence of $10^6/ml^{-1}$ unseparated irradiated (15 Gy) bone marrow cells (**B**). Solid line, normal-sized bursts; broken line, small bursts. (**B** shows the mean of 2 independent experiments). The cultures were stimulated by 2 units/ml^{-1} Ep. The interrupted line indicates bursts of extremely small size. Points connected by drawn lines represent bursts with a size distribution not significantly different from unseparated bone marrow.

from cells associated with BFA. The upper panel of Figure 7-7 shows the result of a representative density centrifugation experiment directed at the distribution of BFU-e. The distribution obtained, representing only 9% of the number of BFU-e loaded on the gradient, characteristically consisted of a low density peak of extremely small bursts, and a second peak of bursts with a normal size distribution. This density distribution was repeatable over three other experiments. The size of the bursts composing the low density peak agreed with burst sizes obtained at low concentrations of unseparated bone marrow cells, i.e., at low concentrations of BFA. The low recovery of BFU-e and the heterogeneous distribution of burst size over the gradient appeared indeed to be due to a relative deficiency of BFA in density separated fractions compared to unseparated bone marrow. It was mentioned before that BFU-e and BFA are almost completely separated by density centrifugation, the distribution of BFU-e being characterized by a modal density of 1.070 gm/cm^{-3}, and the distribution of BFA by a modal density of 1.083 gm/cm^{-3}. In agreement with this observation the lower panel of Figure 7-7 shows that culturing BFU-e from density separated fractions in the presence of a plateau concentration of BFA results in a recovery of 82%. The density distribution of BFU-e cultured in the presence of a plateau concentration of BFA appeared to coincide with the distribution of CFU-s.[33] The modal density of BFU-e was 1.070 gm/cm^{-3} (mean of four experiments), whereas the distribution was found to be heterogeneous with a shoulder at 1.075 gm/cm^{-3}. The significance of this apparent heterogeneity of BFU-e will be discussed in detail in a separate publication.

A similar dependency of burst formation on cell concentration was noted by Heath et al.,[45] who, however, demonstrated that the sedimentation rate distribution of BFU-e was not influenced by cell concentration. We have previously reported on the sedimentation rate distribution of BFU-e[25] and have noted that the recovery of BFU-e did not deviate from recoveries obtained for other hemopoietic progenitor cells. It was consequently predicted that separation of BFU-e and cells associated with BFA does not occur to the extent noted for density centrifugation. The upper panel of Figure 7-8 shows the sedimentation rate distribution of BFA associated cells compared to the distribution of all nucleated cells. The cells associated with BFA appeared to sediment among the more rapidly sedimenting cells. The sedimentation rate distribution of BFA-associated cells appeared, in contrast to the density distribution, to be heterogeneous, being characterized by a modal sedimentation rate of 4.7 mm/hr^{-1}, and a second peak of activity at 6.1 mm/hr^{-1}. This heterogeneity was reproducible. On the basis of a buoyant density of 1.083 gm/cm^{-3} the diameter of BFA-associated cells was calculated to range from 6.7 to 8.4 μm. A possible inaccuracy in the buoyant density would maximally introduce an error of ±5%.

As was predicted the distribution of the cells associated with BFA and the distribution of BFU-e appeared to considerably overlap, which is shown in the lower panel of Figure 7-8. BFU-e appeared to sediment at a modal rate of 4.0 mm/hr^{-1}, which, combined with a buoyant density of 1.070 gm/cm^{-3}, would result in a modal diameter of 7.0 μm. The difference in sedimentation rate between BFU-e and the slowly sedimenting band of cells associated with BFA can therefore be considered due to the large difference in density rather than to a difference in cell size. The small difference in sedimentation rate is compatible with the independence of cell concentration of the sedimentation rate distribution of BFU-e.

It is thus concluded that expression of the potential of BFU-e to form erythroid progeny is depend-

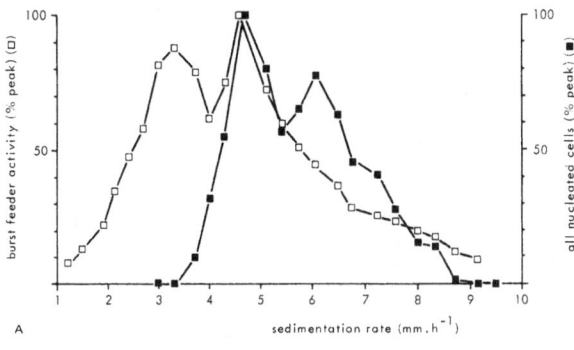

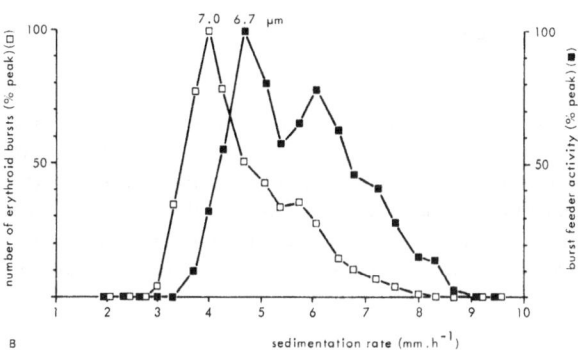

Figure 7-8 **A:** sedimentation rate distribution of BFA associated cells (closed squares) in comparison to the distribution of all nucleated cells (open squares). **B:** sedimentation rate distribution of BFA associated cells (closed squares) in comparison to the distribution of BFU-e (open squares). Mean values of two independent experiments. BFU-e was cultured in the presence of $10^6/ml^{-1}$ irradiated (15 Gy) cells. All cultures were stimulated with 2 units/ml^{-1} Ep. Recoveries were 91% for all nucleated cells, 83% for BFA and 89% for BFU-e.

ent, in addition to Ep, on a distinct population of bone marrow cells. This population is characterized by a buoyant density of 1.083 gm/cm^{-3} and a modal sedimentation rate of 4.7 mm/hr^{-1}. Since the incidence of CFU-e did not appear to be dependent on the cell concentration of the cultures, the culture conditions employed can be considered as sufficient for support of Ep-induced erythroid differentiation. It is thus concluded that cells associated with BFA probably play a role in the early differentiation of BFU-e, leading to responsiveness to the specific regulator Ep. Following the detection of BFA-associated cells we initiated a search for soluble factors with a biological action similar or identical to BFA.

Soluble Factors Enhancing Erythroid Burst Formation

In addition to the search for a molecular form of BFA, other studies have indicated that serum probably contains an activity that is important for *in vitro* differentiation of BFU-e. It is possible, by adding albumin, transferrin, phospholipid, and selenite to the cultures to restrict the serum requirements of CFU-c and CFU-e to 1% of fetal calf serum,[49] but this does not apply to BFU-e.[28] It has therefore been suggested that serum contains an activity specific for BFU-e[28] involved in the early phases of its differentiation. In view of these studies we have analyzed the serum requirements of BFU-e in terms of the presence of an activity similar to the biological activity of BFA. The data further warranted a pilot study aimed at the detection of burst-enhancing activities in serum of mice following perturbation of hemopoiesis, or, more specifically, erythropoiesis. Since the physical characteristics of BFU-e are very similar to those of CFU-s, the possible BFU-e enhancing activity of a substance known to initiate DNA synthesis in CFU-s was also investigated. Such an activity is expressed by medium conditioned by human peripheral blood leukocytes (HLCM), which also induces a distinct subpopulation of early murine CFU-c to become responsive to CSF.[50,51]

Dependence of Erythroid Burst Formation on Fetal Calf Serum and Horse Serum

It is well known that individual batches of serum may vary widely in their capacity to supplement culture media supporting hemopoietic differentiation *in vitro*. It was noted that this variation is much larger for erythroid bursts than for granulocyte/macrophage colonies. It was further observed that horse serum (HS) is consistently more conducive to burst formation than fetal calf serum (FCS), but that mixing the two types of serum in equal volumes gives results superior to those obtained with double the concentration of individual sera. Table 7-1 summarizes results with five randomly chosen serum batches, in comparison with "standard" sera, especially selected for their capacity to support burst formation in combination with HS/FCS. Except for the already mentioned additive effect of mixing HS and FCS, indicating that these two serum sources contain different activities important for burst formation, the data demonstrate that the effectiveness of HS compared to FCS is partially due to a BFA-type activity. This can be inferred from comparison of the ratios of burst numbers induced by a plateau concentration of BFA over those obtained with Ep alone. The smallest ratios were obtained in those cultures supplemented with "standard" HS, whereas supplementing the medium with only HS results in a much smaller ratio than that obtained with only FCS. It is further demonstrated that the quality of "standard" FCS to support growth of erythroid bursts in combination with HS is not due to an activity that by itself supports burst formation, not to another activity present in HS, and not to an activity that is similar to BFA. The data thus allow us to attribute the serum requirements of bursts to multiple, partly interdependent activities, of which one ap-

Table 7-1 Dependence of erythroid burst formation on fetal calf serum (FCS) and horse serum (HS).

Culture medium[a] supplemented with	Erythroid bursts/10^5 cells ± s.d.		Ratio[c]
	Ep	Ep + BFA[b]	
20% FCS	0.6 ± 0.5	5.2 ± 2.9	8.7 (5)
20% HS	4.1 ± 3.2	14.4 ± 6.7	3.5 (5)
10% HS + 10% FCS	7.9 ± 5.0	25.4 ± 3.1	3.2 (25 combinations)
10% HS + 10% "standard" FCS	11.6 ± 4.7	46.8 ± 11.5	4.0 (5)
10% FCS + 10% "standard" HS	21.2 ± 3.3	44.8 ± 10.6	2.1 (5)
10% "standard" FCS + 10% "standard" HS	30 ± 6	78 ± 9	2.6
20% "standard" HS	9 ± 3	23 ± 5	2.6
20% "standard" FCS	0	5 ± 2	>10

[a] Already supplemented with albumin, transferrin, lecithin and selenite.
[b] 2 units/ml^{-1} human urinary Ep + 10^6/ml^{-1} irradiated (10 Gy) bone marrow cells.
[c] Numbers in parentheses represent number of serum batches.

pears to be nonadditive to the action of FBA. This pattern may predict that biochemical isolation of the activities will not be easily accomplished.

Detection of a Burst-enhancing Activity in Mouse Serum Following Perturbation of Hemopoiesis

One of the attributes of a humoral substance regulating a hemopoietic differentiation event might be a change in its peripheral blood levels in response to a perturbation of hemopoiesis. We have been able to detect a burst-enhancing activity in serum of mice subjected to various treatments resulting in profound changes in hemopoiesis. These treatments included lethal irradiation either combined with or without hemolytic anemia induced by administration of phenylhydrazine, induced hemolytic anemia alone, intermittent hypoxia, and exhypoxic polycythemia. Details and time schedules for collecting serum were described under Materials and Methods. Graded doses of the sera were added to Ep-stimulated cultures containing 2 × 10^5/ml^{-1} bone marrow cells. Results are given in Figure 7-9 for bursts scored on day 7 of incubation. All sera appeared to inhibit burst formation at relatively high concentrations. At lower concentrations a significant ($p < 0.005$) burst-enhancing activity became apparent in the sera collected from mice exposed to phenylhydrazine treatment, either combined with or without irradiation, and in the serum collected from exhypoxic polycythemic mice. No enhancement was observed in sera from normal, irradiated, and hypoxic mice. This pattern was not changed by scoring bursts on days 10 and 15. The nonenhancing sera did not seem to contain a higher inhibiting activity than the enhancing sera. None of the sera exerted a significant enhancing influence on the number of CFU-c scored, as is shown for serum derived from phenylhydrazine treated mice in Figure 7-10. None of the sera stimulated burst formation in the absence of added Ep, as is also shown in Figure 7-10.

It can be concluded from these data that some forms of perturbation of hemopoiesis in mice result in the appearance of a burst-enhancing activity in the peripheral blood, which cannot be detected in serum from untreated control animals. This activity differs from Ep in not stimulating but rather enhancing

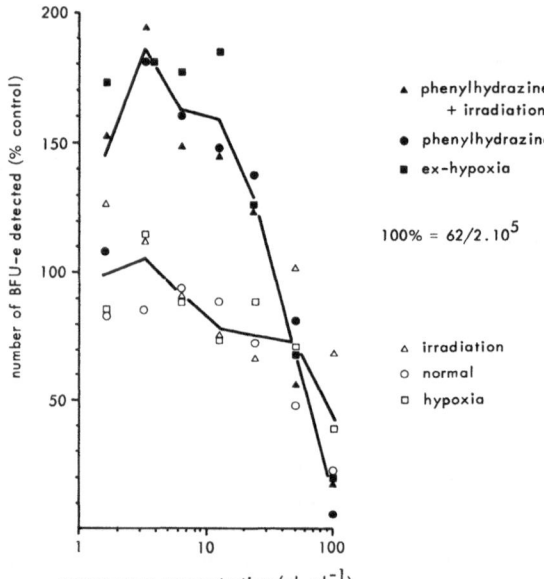

Figure 7-9 Effect of adding graded volumes of mouse serum to Ep-stimulated cultures on the number of BFU-e detected. The sera were collected from mice exposed to various treatments: hypoxia, lethal irradiation, administration of phenylhydrazine, and phenylhydrazine treatment combined with irradiation. Also included were serum from exhypoxic polycythemic mice (closed squares) and serum from untreated mice (open circles). To avoid crowding of graphs, lines connecting mean values are given for sera significantly ($p < 0.005$) enhancing burst formation (closed symbols) and nonenhancing sera (open symbols). All cultures contained 2 × 10^5/ml^{-1} nucleated bone marrow cells and were stimulated with 1 unit/ml^{-1} Ep. 100% indicates 62 bursts obtained in control cultures to which no mouse serum was added.

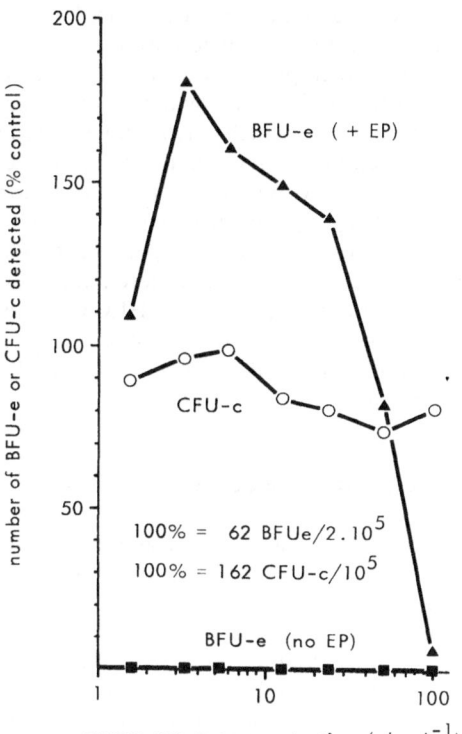

Figure 7-10 Effect of serum from mice with hemolytic anemia induced by phenylhydrazine on the numbers of BFU-e and CFU-c detected. Closed symbols represent BFU-e (closed triangles: 1 unit/ml^{-1} Ep; closed squares: no Ep); o: CFU-c. 100% indicates 66 bursts per 2×10^5/ml^{-1} nucleated cells or 81 granulocyte/macrophage colonies per 5×10^4/ml^{-1} nucleated cells, in control cultures to which no mouse serum was added.

erythroid differentiation *in vitro*. This difference is confirmed by its detection in the serum from exhypoxic polycythemic mice, which does not contain detectable concentrations of Ep as measured by an *in vivo* assay[39] using peripheral blood ^{59}Fe-incorporation in polycythemic mice (unpublished observation). Its detection in serum from exhypoxic rather than from hypoxic mice is in principle in agreement with our previous observations, showing that no significant change in BFU-e numbers was induced by exposure to intermittent hypoxia, whereas during exhypoxic polycythemia BFU-e numbers transiently rose.[25] Obviously, however, a more rigorous experimentation is required to demonstrate a relationship between BFU-e population size *in vivo* and circulating levels of burst-enhancing activity. From the experiments no conclusions can be drawn as to which particular signal induces the presence of detectable levels of burst-enhancing activity in the peripheral blood.

The data thus demonstrate the detection of a novel type of humoral activity formed in response to changes in hemopoiesis, possibly only in erythropoiesis. Since the activity can be detected in mouse serum concentrations of a few μl/ml^{-1} and was not detected in control mouse serum, its action is probably not due to a general improvement of the nutrient qualities of the culture medium. Studies on the nature of this activity and its possible physiological role are in progress.

Detection of a Burst-enhancing Activity in Human Leukocyte-conditioned Medium (HLCM)

HLCM has been observed to stimulate human CFU-c and to influence profoundly hemopoietic progenitor cells of the mouse, including initiation of DNA synthesis in CFU-s *in vitro*. A major action of HLCM is induction of CSF responsiveness in a distinct subpopulation of murine CFU-c, which appeared to require at least two divisions.[34,50,51] In addition, it was demonstrated that HLCM contains an activity-enhancing murine BFU-e in semisolid cultures stimulated by Ep.

Table 7-2 summarizes data showing independence of the actions of HLCM on BFU-e and CFU-c. The incidences of CFU-c and BFU-e were scored in independent cultures and in cultures containing both Ep and CSF, and compared with a similar set of cultures in addition stimulated by HLCM. From the results shown in Table 7-2 no evidence can be obtained that HLCM influences the direction of differentiation. The additional erythroid burst formation induced by HLCM did not occur at the expense of the additional number of granulocyte/macrophage colonies. The occurrence of mixed erythroid/macrophage colonies was not more frequent than would be predicted by chance spatial coincidence of CFU-c and BFU-e.

Figure 7-11 shows that the enhancing effect of HLCM on the number of bursts is dependent on the concentration of bone marrow cells of the cultures, a decreasing ratio of bursts induced by Ep + HLCM over bursts obtained by Ep alone becoming apparent with increasing cell concentrations. This was due to

Table 7-2 Effect of HLCM on erythroid burst formation.

	Detected BFU-e^a/10^5		Detected CFU-c/10^5	
Stimuli	−HLCM	+HLCM	−HLCM	+HLCM
Ep	23 ± 2^b	35 ± 3	n.d.^{c}	n.d.
Ep + CSF-pmue	22 ± 2	33 ± 3	110 ± 5	146 ± 6
CSF-pmue	0	0	109 ± 5	145 ± 6
—	n.d.	0	n.d.	n.d.

^{a} Stimulated by 1–2 units Ep, and scored at day 8 of incubation.
^{b} Standard error, computed on the assumption that colony counts are Poisson distributed.
^{c} Not done or not scored.

54 Factors Influencing the Differentiation of Primitive Erythroid Progenitor Cells (BFU-e) In Vitro

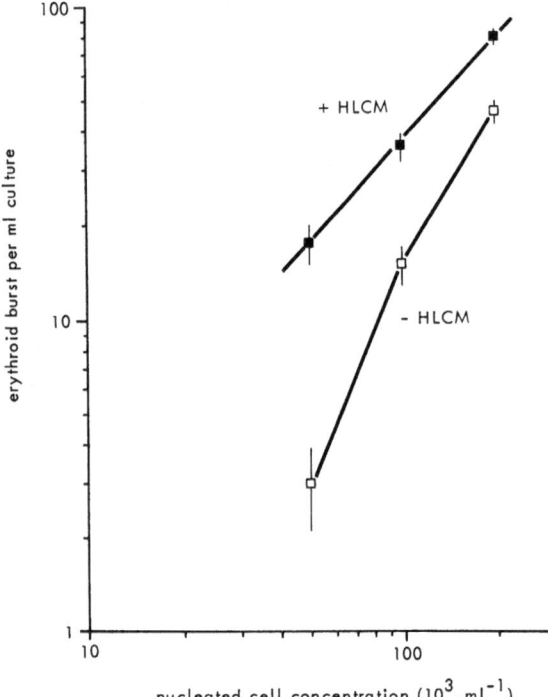

Figure 7-11 Number of erythroid bursts as a function of nucleated cell concentration in the absence (open squares) or presence (closed squares) of 10% HLCM. All cultures were stimulated by 1 unit/ml^{-1} Ep. Vertical bars indicate standard errors.

murine CFU-c resisted separation from human CSA by gel filtration, ion-exchange chromatography, and affinity-chromatography.[50] The available evidence thus points to a new factor or family of factors, related to human CSA, which *in vitro* have profound effects on the earliest hemopoietic progenitor cells of the mouse. It may be noted that the actions of HLCM, inducing Ep responsiveness in BFU-e and CSF responsiveness in a subpopulation of CFU-c, can be interpreted as parallel differentiation events. Combined with the capacity of HLCM to initiate DNA-synthesis in CFU-s, this opens interesting prospects for more detailed understanding of the earliest differentiation of hemopoietic cells.

Summary and Discussion

Physical characteristics and proliferative potential characterize BFU-e as the most primitive erythroid progenitor cells as yet detectable *in vitro*. The experiments described in this chapter confirm, in agreement with *in vivo* observations, that BFU-e lacks the capacity to respond directly to Ep *in vitro*. Both soluble and cellular activities have been detected that apparently control the *in vitro* responsiveness of BFU-e to EP. The concentration of a distinct bone marrow cell population appeared to determine the number of bursts detected and their sensitivity to Ep. The activ-

burst numbers in cultures of Ep + HLCM being directly proportional to the cell concentration of the cultures in contrast to cultures lacking HLCM. This feature of the action of HLCM on BFU-e demonstrates that a soluble substance can induce independence of burst formation of the cell concentration of the cultures, without, however, inducing the high numbers obtained with a plateau concentration of BFA. This probably explains earlier mentioned discrepancies in results concerning the dependence of burst formation on cell concentration. HLCM, as a consequence, seems to contain an activity with an action on BFU-e similar to the action of BFA. This notion was confirmed by direct experiments, summarized in Figure 7-12, which demonstrated a decrease in the relative effect of HLCM when increasing concentrations of irradiated (15 Gy) cells were added to the cultures. It is therefore concluded that the biological activity of HLCM on BFU-e is nonadditive to the action of BFA.

The nature of this activity in HLCM has as yet not been investigated in detail. An enhancing activity on human BFU-e in preparations of human CSA has been reported by Aye.[52] We have as yet not investigated the occurrence of murine burst-enhancing activity in purified HLCM preparations, but have earlier noted that the activities of HLCM on CFU-s and

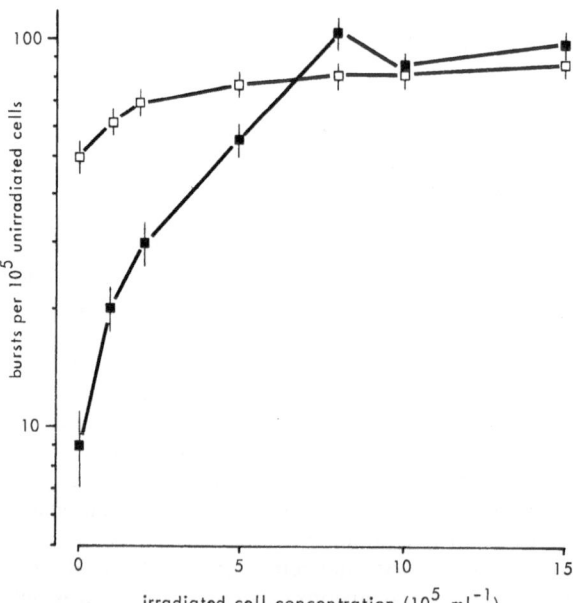

Figure 7-12 Number of erythroid bursts as a function of the concentration of irradiated (15 Gy) bone marrow cells added to 5 × 10^5/ml^{-1} unirradiated bone marrow cells in the absence (closed squares) or presence (open squares) of 10% HLCM. The cultures were stimulated by 2 units/ml^{-1} Ep. Vertical bars indicate standard errors.

ity exerted by this cell population has been termed BFA. BFA-like activities have been traced in serum, and in medium conditioned by human peripheral blood leukocytes. A burst-enhancing activity has also been detected in serum from mice following perturbation of hemopoiesis. It is presumed that these activities, *in vitro* controlling Ep-dependent differentiation of BFU-e, are relevant to the regulation of the earliest differentiation of erythroid progenitor cells.

The detection of a cell population in mouse bone marrow, on which the *in vitro* Ep responsiveness is dependent, might imply that the early differentiation of erythroid progenitor cells is controlled by a local, cellular mechanism. No specific data are as yet available about the nature, specificity, mechanism of action, incidence, and physiological role of the cells associated with BFA. From the cell separation experiments it can be concluded that this cell population is composed of relatively small cells of high buoyant density; the size distribution appeared to be heterogeneous. From the cell concentrations required for a plateau concentration of BFA it may be inferred that the cells associated with BFA are either not numerous or that their action is restricted to a short range. It might be interesting to investigate whether BFU-e and BFA-associated cells are distributed in the bone marrow according to a specific spatial pattern, as has been described for CFU-s and CFU-c.[53] Recent experimentation demonstrated that infusion of latex particles in mice induces Ep-independent changes in the numbers of erythroid progenitor cells. These changes are most pronounced in BFU-e and appeared to be gradually damped in more differentiated populations, the change in CFU-e numbers occupying an intermediate position between BFU-e and morphologically recognizable erythroid cells. Depending on the dose, the change may either be an increase or a decrease. In either case the change in erythroid progenitor cells was characteristically preceded by a corresponding change in BFA.[47] Essentially, these data provide the first bit of evidence indicating a physiological role of BFA-associated cells.

The detection of an activity similar to BFA in serum and in medium conditioned by peripheral blood cells presumably implies that neither BFA-associated cells nor their product is restricted to the bone marrow. Elevated levels of a burst-enhancing activity have been demonstrated in the serum of mice under conditions of perturbed hemopoiesis. Although the similarity of this enhancing activity and BFA remains to be investigated, these observations also point to a more systemic presence of activities modulating erythroid differentiation *in vitro*. Since elevated levels of burst enhancing activity were detected exclusively following profound changes in hemopoiesis, or, more specifically, erythropoiesis, this activity might well reflect a physiological mechanism involved in the early regulation of erythropoiesis.

As yet the physiological role of activities modifying Ep responsiveness of BFU-e can be discussed only in speculative terms. Erythropoiesis, however, can be subjected to extensive *in vivo* experimentation, indicating that decisive data on the physiological significance of these activities can probably be obtained on relatively short notice. Irrespective of this question, the detection of an activity which *in vitro* controls the sensitivity of BFU-e toward Ep implies that tools have become available to investigate the earliest differentiation of primitive erythroid cells. It might be indicative for the organization of hemopoiesis that this activity exerts a stimulatory rather than an inhibitory influence.

Acknowledgments

This investigation is part of a study on the regulation of hemopoiesis that is supported by a program grant of the Netherlands Foundation for Medical Research (FUNGO), which is subsidized by the Netherlands Organization for the Advancement of Pure Research (ZWO). Special thanks are due to Dr. S. J. L. Bol and Dr. D. Zipori for their scientific contributions to this work, and to Mr. A. Brouwer, Mrs. M. F. Peters, Mrs. V. E. Ober and Miss T. P. Visser for their expert technical assistance.

References

1. Goldwasser E, Kung CKH: Molecular weight of sheep plasma erythropoietin. *J Biol Chem* 247:5159, 1972.
2. Cotes PM, Bangham DR: The international reference preparation of erythropoietin. *Bull World Health Org* 35:751, 1966.
3. Cotes PM: Quantitative estimation of erythropoietin. *Ann NY Acad Sci* 149:12, 1968.
4. Annable L, Coates PM, Mussett MV: The second international reference preparation of erythropoietin, human urinary, for bio-assay. *Bull World Health Org* 47:99, 1972.
5. Jacobson LO, Goldwasser E, Plzak LF, Fried W: Studies on erythropoiesis. IV. Reticulocyte response of hypophysectomized and polycythemic rodents to erythropoietin. *Proc Soc Exp Biol Med* 94:243, 1957.
6. Gurney CW, Fried W: The regulation of numbers

of primitive hemopoietic cells. *Proc Natl Acad Sci 54:*1148, 1965.
7. Blacket NM, Roylance PJ, Adams K: Studies of the capacity of bone-marrow cells to restore erythropoiesis in heavily irradiated rats. *Br J Haematol 10:*453, 1964.
8. Schooley JC, Cantor LN, Havens VW: Relationship between growth of colony-forming cells in mice irradiated with 200 R of ^{60}Co gamma rays. *Exp Haematol 9:*55, 1966.
9. Porteous DD, Lajtha LG: On stem cell recovery after irradiation. *Br J Haematol 12:*177, 1966.
10. Porteous DD, Lajtha LG: Restoration of stem cell function after irradiation. *Ann NY Acad Sci 149:*151, 1968.
11. Reissmann KR, Samorapoomipichit S: Effect of erythropoietin on proliferation of erythroid stem cells in the absence of transplantable colony-forming units. *Blood 36:*287, 1970.
12. Reissmann KR, Udupa KB: Effect of erythropoietin on proliferation of erythropoietin-responsive cells. *Cell Tissue Kinet 5:*481, 1972.
13. Lajtha LG: Bone marrow stem cell kinetics. *Sem Hematol 4:*293, 1967.
14. Hodgson G: Effect of vinblastine and 4-amino-N^{10} methyl pteroylglutamic acid on the erythropoietin responsive cell. *Proc Soc Exp Biol Med 125:*1206, 1967.
15. Morse BS, Rencricca NJ, Stohlman F, Jr: Relationship of erythropoietin effectiveness to the generative cycle of erythroid progenitor cells. *Blood 35:*761, 1970.
16. Lajtha LG, Pozzi LV, Schofield R, Fox M: Kinetic properties of haemopoietic stem cells. *Cell Tissue Kinet 2:*39, 1969.
17. Lajtha LG, Gilbert CW, Guzman E: Kinetics of hemopoietic colony growth. *Br J Haematol 20:*343, 1971.
18. Feldman M, Bleibert I: Studies on the feedback regulation of haemopoiesis, in de Reuck AVS, Knight J (eds): *Ciba Foundation Symposium on Cell Differentiation.* p. 79, London, Churchill, 1967.
19. Bleiberg I, Liron M, Feldman M: Reversion by erythropoietin of the suppression of erythroid clones caused by transfusion induced polycythemia. *Transplantation 3:*706, 1965.
20. O'Grady LF, Lewis JP, Trobaugh FE, Jr: Effect of timed doses of erythropoietin on the development of hematopoietic tissue. *Exp Hematol 12:*62, 1967.
21. O'Grady LF, Lewis JP, Trobaugh FE, Jr: Further studies on the effect of plethora in the formation of hematopoietic colonies. *Exp Hematol 12:*70, 1967.
22. Stephenson JR, Axelrad AA, McLeod DL, Shreeve MM: Induction of colonies of hemoglobin synthesizing cells by erythropoietin *in vitro*. *Proc Natl Acad Sci 68:*1542, 1971.
23. Axelrad AA, McLeod DL, Shreeve MM, Heath DS: Properties of cells that produce erythrocytic colonies *in vitro*, in Robinson WA (ed): *Hemopoiesis in Culture.* p. 226, DHEW Publ No (NIH) 74:*205, 1974.
24. Iscove NN, Sieber F: Erythroid progenitors in mouse bone marrow detected by macroscopic colony formation in culture. *Exp Hematol 3:*32, 1975.
25. Wagemaker G, Ober-Kieftenburg VE, Brouwer A, Peters-Slough MF: Some characteristics of *in vitro* erythroid colony and burst-forming units, in Baum SJ, Ledney GD (eds): *Exp Hematology Today.* p. 103, New York, Heidelberg, Berlin, Springer-Verlag, 1977.
26. Wagemaker G: Erythropoietine, enkele aspecten van de humorale regulatie van de erythropoiese. PhD dissertation, Erasmus University, Rotterdam, 1976.
27. Iscove NN: The role of erythropoietin in regulation of population size and cell cycling of early and late erythroid precursors in mouse bone marrow. *Cell Tissue Kinet 10:*323, 1977.
28. Iscove NN: Erythropoietin independent regulation of early erythropoiesis *Exp Hematol 5* (Suppl 2) (abs):6, 1977.
29. Gregory CJ, Tepperman AD, McCulloch EA, Till JE: Erythropoietic progenitors capable of colony formation in culture: response of normal and genetically anemic W/W^4 mice to manipulation of the erythron. *J Cell Physiol 84:*1, 1974.
30. Reissmann KR, Udupa KB, Labedzki L: Induction of erythroid colony forming cells (CFU-e) in murine spleen by endotoxin. *Proc Soc Exp Biol Med 153:*98, 1976.
31. Miller RG, Phillips RA: Separation of cells by velocity sedimentation. *J Cell Physiol 73:*191, 1969.
32. Shortman K: The separation of different cell classes from lymphoid organs. II. The purification and analysis of lymphocyte populations by equilibrium density centrifugation. *Aust J Exp Biol Med Sci 46:*375, 1968.
33. Wagemaker G, Peters MF, Bol SJL: Comparison of the physical characteristics of CFU-s, BFU-e, CFU-c and CFU-e by density centrifugation and sedimentation at unit gravity. (submitted for publication, 1978).
34. Wagemaker G, Peters MF: Effects of human leukocyte conditioned medium on mouse hemopoietic progenitor cells. *Cell Tissue Kinet 11:*1, 1978.
35. Shortman K: Physical procedures for separation of animal cells. *Ann Rev Biophys Bioengineer 1:*93, 1972.
36. Shortman K, Williams N, Adams P: The separation of different cell classes from lymphoid

organs. V. Simple procedures for the removal of cell debris, damaged cells, and erythroid cells from lymphoid cells. *J Immunol Meth 1:*273, 1972.
37. Williams N, Kraft N, Shortman K: The separation of different cell classes from lymphoid organs. VI. The effect of osmolarity of gradient media on the density distribution of cells. *Immunology 22:*885, 1972.
38. Miller RG: Separation of cells by velocity sedimentation, in Pain RH, Smith BJ (eds): *New Techniques in Biophysics and Cell Biology.* p. 87, London, Wiley, 1973.
39. Wagemaker G, Van Eijk HG, Leijnse B: A sensitive bio-assay for the determination of erythropoietin. *Clin Chim Acta 36:*357, 1972.
40. Bradley TR, Stanley ER, Sumner MA: Factors from mouse tissues stimulating colony growth of bone marrow cells *in vitro. Aust J Exp Biol Med Sci 49:*595, 1971.
41. Stanley ER, Metcalf D, Maritz JS, Yeo GF: Standardized bio-assay for bone marrow colony stimulating factor in human urine: levels in normal man. *J Lab Clin Med 79:*657, 1972.
42. Van den Engh GJ: Quantitiative *in vitro* studies on stimulation of murine haemopoietic progenitor cells by colony stimulating factor. *Cell Tissue Kinet 7:*537, 1974.
43. Laurent FC, Killander J: Theory of gel filtration and its experimental verification. *J Chromatogr 14:*317, 1964.
44. McLeod DL, Shreeve MM, Axelrad AA: Induction of megakaryocyte colonies with platelet formation *in vitro. Nature 261:*492, 1976.
45. Heath DS, Axelrad AA, McLeod DL, Shreeve MM: Separation of erythropoietin-responsive progenitors BFU-e and CFU-e in mouse bone marrow by unit gravity sedimentation. *Blood 47:*777, 1976.
46. Gregory CJ: Erythropoietin sensitivity as a differentiation marker in the hemopoietic system: studies of three erythropoietic colony responses in culture. *J Cell Physiol 89:*289, 1976.
47. Ploemacher R: Work in progress.
48. Wagemaker G, Peters MF, Bol SJL: Induction of erythropoietin responsiveness *in vitro* by a distinct population of bone marrow cells. (submitted for publication)
49. Guilbert LJ, Iscove NN: Partial replacement of serum by selenite, transferrin, albumin and lecithin in haemopoietic cell cultures. *Nature 263:*594, 1976.
50. Wagemaker G, Peters MF, Brouwer A, Bol SJL, Ober VE: Isolation and biological characteristics of colony stimulating factors from human leukocyte conditioned medium *Exp Hematol* 5(Suppl 2) (abs): 16, 1977.
51. Wagemaker G, Peters MF, Bol SJL: Detection of a distinct subpopulation of mouse bone marrow CFU-c responding to medium conditioned by human peripheral blood leukocytes. (submitted for publication)
52. Aye MT: Erythroid colony formation in cultures of human marrow: effect of leukocyte conditioned medium. *J Cell Physiol 91:*69, 1977.
53. Lord BI, Testa NG, Hendry JH: The relative spatial distributions on CFU-s and CFU-c in the normal mouse femur. *Blood 46:*65, 1975.

8 The Effect of Sustained Hypertransfusion on Hematopoiesis

A. Erslev, R. Silver, J. Caro, S. Paist, and E. Cobbs

Introduction

Hypertransfusion polycythemia suppresses the production of erythropoietin, but in most mammals considerable erythropoietic activity remains despite elevated hematocrits. In rats, this residual activity can be reduced substantially by infusing anti-erythropoietin,[1] indicating that it is not autonomous but erythropoietin-dependent and that in these animals hypertransfusion alone is incapable of eliminating circulating erythropoietin. This may be due in part to the fact that polycythemic blood is viscous and may cause local or general hypoxia, offsetting the effect of an increase in oxygen carrying capacity.[2]

In mice, however, transfusion polycythemia results in an almost complete cessation of red cell production as assessed by reticulocyte counts, bone marrow examinations, and utilization of radioactive iron. Consequently, it seems likely that the increase in oxygen-carrying capacity has suppressed erythropoietin production maximally. This near absence of erythropoietin affords us an opportunity to examine the effect of a deletion of one cellular element in the bone marrow on the proliferation of other elements. The effect on certain cell types by short time suppression of red cell production has been reported,[3-7] but in the present study transfusion polycythemia was maintained for 6 weeks in order to evaluate both immediate and delayed effects on differentiated and undifferentiated hemic cells.

Materials and Methods

Random bred female Swiss-Webster mice were used throughout. They were transfused intraperitoneally with 0.9 ml of a suspension of washed packed isologous red cells (hematocrit: 85%) on days 1, 0, and then every 6 days for 6 weeks. Hematocrit, white blood cells count, granulocyte count, and platelet count were determined by routine laboratory techniques.

The utilization of radioiron was measured by injecting 0.5 μCi ^{59}Fe as ferrous citrate i.p. and determining radioactivity of 0.2 ml whole blood obtained 66 hr later.[8] In the calculation of total radioiron utilization the blood volume was taken as 7% of body weight. The *in vivo* response to human urinary erythropoietin was measured by injecting exactly 1 unit subcutaneously and measuring 66-hr radioiron utilization 48 hr later. The erythropoietin was obtained from the NIH Subcommittee on Erythropoietin and was derived from urine of patients with hookworm anemia. It was collected and concentrated by the Department of Physiology, University of the North-East Corrientes, Argentina, and further processed until it attained a specific activity of 60 units/mg of protein by the Haematology Research Laboratories, Children's Hospital, Los Angeles.

The number of CFU-e, CFU-c and CFU-M ion spleen and bone marrow were determined on semisolid clot cultures. The mice were killed by sudden neck luxation. Both femurs and spleen were removed. Marrow was flushed from the femurs with a 25-gauge needle using 5 ml MEM with 2% fetal calf serum as flushing and suspending solution. A spleen cell suspension was prepared with a hand tissue grinder in 10 ml of above solution. The total number of nucleated bone marrow and spleen cells was determined by a hemocytometer and cellular stock suspensions containing 1×10^7 bone marrow cells per ml and 2.5×10^7 spleen cells per ml were prepared.

CFU-e enumeration (modified from refs. 9 and 10)

0.9 ml cellular suspension for culture was prepared as follows:

- 0.2 ml Fetal calf serum
- 0.1 ml Thrombin in MEM (10 units/ml)
- 0.1 ml 2-mercaptoethanol in MEM (1 mM/ml)
- 0.1 ml Erythropoietin in MEM (10 μ/ml)
- 0.2 ml MEM
- 0.2 ml Bone marrow cell stock suspension (1×10^7/ml) or 0.2 ml Spleen cell stock suspension (2.5×10^7/ml)

To this was added 0.1 ml bovine citrated plasma with immediate pipetting of 0.1 ml of the final suspension, containing either 2×10^5 bone marrow cells or 5×10^5 spleen cells, into each of 8 microtiter wells. The trays were placed in a CO_2 incubator under 5% CO_2 and 21% O_2 at 37° for 3 days. The clots in the well were transferred to glass slides, fixed with 5% glutaraldehyde, stained with a 1% solution of 3'-3' dimethoxybenzidine in methanol for 5 min, a 1% solution of hydrogen peroxide in methanol for 1 min, and then counterstained with 1% Giemsa solution for 20 min. Colonies containing eight or more benzidine-positive cells were enumerated and the total number of colonies per femur or per spleen was calculated.

CFU-c enumeration (modified from refs. 10 and 11)

0.9 ml cellular suspension for culture was prepared as follows:

- 0.2 ml Fetal calf serum
- 0.1 ml thrombin in MEM (10 units/ml)
- 0.1 ml 2-mercaptoethanol in MEM (1 mM/ml)
- 0.1 ml Colony-stimulating factor (CSF)*
- 0.3 ml MEM
- 0.1 ml Bone marrow cell stock suspension (1×10^7/ml), or 0.1 ml Spleen cell stock suspension (2.5×10^7/ml)

To this was added 0.1 ml bovine citrated plasma with immediate pipetting of 0.1 ml of the final suspension, containing either 1×10^5 bone marrow cells or 2.5×10^5 spleen cells, into each of 8 microtiter wells.

The trays were placed in a CO_2 incubator under 5% CO_2 and 21% O_2 at 37° for 8 days. The clots in the wells were transferred to glass slides, fixed in 5% glutaraldehyde, and stained with a 1% Giemsa solution for 20 min. Colonies containing 50 or more cells were enumerated and the total number of colonies per femur or per spleen was calculated.

CFU-M enumeration (modified from ref. 12)

0.9 ml cellular suspension for culture was prepared as follows:

- 0.5 ml L-15
- 0.2 ml Horse serum
- 0.1 ml $CaCl_2$ in L-50 (13 mg%)
- 0.1 ml Bone marrow cell stock suspension (1×10^7/ml), or 0.1 ml Spleen cell stock suspension (2.5×10^7/ml)

To this was added 0.1 ml bovine citrated plasma with immediate pipetting of 0.1 ml of the final suspension, containing either 1×10^5 bone marrow cells or 2.5×10^5 spleen cells, into each of 8 microtiter wells.

The trays were placed in a CO_2 incubator under 5% CO_2 and 21% O_2 at 37° for 4 days. The clots were then transferred to glass slides, fixed in 5% glutaraldehyde, and stained for acetyl cholinesterase with acetylthiocholine iodate as described by Karnovsky and Roots.[14] Clusters of megakaryocytes containing four or more cells were enumerated in cultures of the spleen, cholinesterase staining was very intense and it was difficult to differentiate colonies from single cells.

CFU-s enumeration (modified from ref. 15)

Bone marrow and spleen cells were pooled from three mice and suspensions in MEM were prepared. Bone marrow cells in numbers varying from 1×10^4 to 3×10^4 nucleated cells or spleen cells varying from 1×10^4 to 3×10^5 nucleated cells were injected intravenously into ten mice, irradiated 3 hr before injection with 115 rads/min for 7 min. Eight days after the injection, the recipient mice were sacrificed, their spleens removed and placed in Bouin's solution. The splenic surface was examined under low magnification ($3\times$) and the number of surface colonies enumerated. The number was then recalculated to express the total number of colonies (CFU-s) per femur or per spleen, assuming that only 20% of the injected CFU-s would lodge in the spleen of the recipient mouse.

Results and Discussion

Erythrocytes

The hematocrit of the recipient mice increased to about 70% after two injections of isologous red cells and was maintained between 65 and 70% for 43 days (Figure 8-1). The 66 hr utilization of radioactive iron was reduced from 44% to 0.2% within 5 days and remained below 0.2% during the total period of hypertransfusion polycythemia. This would suggest that the production of red blood cells ceased almost completely.

Erythropoietin Responsive Stem Cells

The response to an injection of one unit of erythropoietin increased from about 6% (49 to 43%) before transfusion to 10% on day 5 (the day usually used for routine assays) and then leveled off at 6.5% from day 9 on (Figure 8-1). However, the 6% response in normal animals is a mean of data with tremendous variations and unacceptable standard deviations. The choice of the polycythemic mouse as a recipient in erythropoietin assays rather than the normal mouse is actually based on the observation that reproducible data cannot be expected unless the base line is low and steady.[16] It is not based on a greater sensitivity of the polycythemic animals to erythropoietin. On the contrary, there is a decreasing response after transfusion reaching its lowest level from day 9 on. This suggests a gradual decrease in the number of erythropoietin responsive stem cells until a certain base line has been reached. The gradual decrease may explain the considerable difference in erythropoietin response reported by investigators using different days for their injection of erythropoietic material.

---

* Colony-stimulating factor (CSF) was prepared from the serum of mice given 50 µg endotoxin 3 hr before bleeding.[13] The CSF consisted of 50% such serum + 50% MEM.

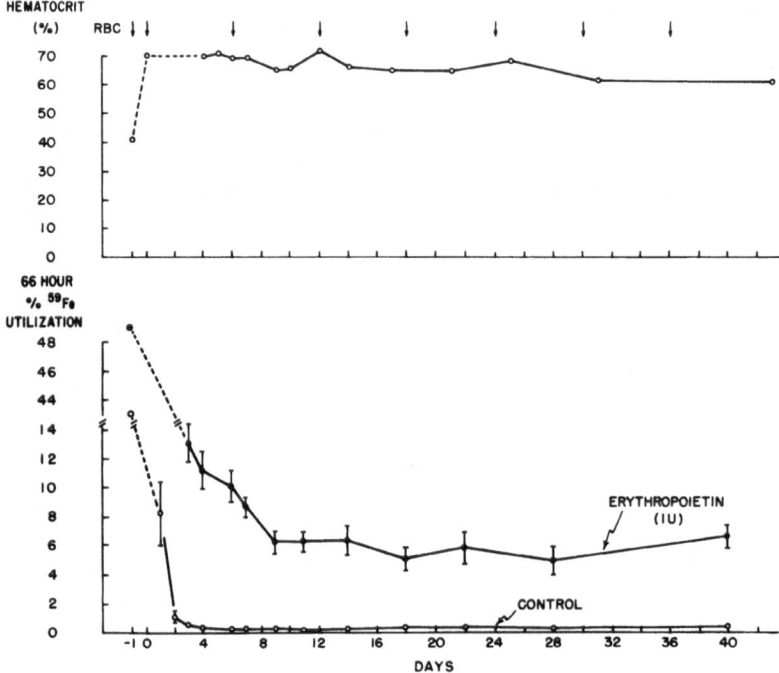

Figure 8-1 The upper panel shows the effect of intermittent transfusions on the hematocrit. The lower panel shows the effect on the 66 hr utilization of radioiron as well as on the response to a single injection of 1 unit of erythropoietin. The data are listed as mean ± 1 SEM.

CFU-e

The total number of CFU-e per femur decreased from 96,000 to approximately 1,000 and per spleen from 388,000 to approximately 50,000 (Figure 8-2). Similar reductions have been reported by Axelrad et al.,[5] Iscove,[6] and Hara and Ogawa[7] in short-term polycythemic mice. The more complete suppression of femoral CFU-e probably reflects the well-known erythropoietic preference for the spleen.

BFU-e

These large colonies that appear in response to high doses of erythropoietin have been reported in polycythemia to remain unchanged in number in the femur[5,7] but to decrease moderately in the spleen.[7] In the studies reported here, one unit of erythropoietin was found to be inadequate to induce reproducible numbers of BFU-e.

Granulocytes

The white cell count as well as the total granulocyte count increased moderately during the 6-week experimental period (Figure 8-3). A number of reports have suggested the existence of competition between the granulocytic and erythrocytic components[17-20] and the present observations did suggest that erythropoietic suppression resulted in a slight facilitation of the production of mature granulocytes.

CFU-c

After an initial decrease, the number of CFU-c in both femurs and spleen increased moderately (Figure 8-3). A similar increase in the femurs was observed in short-term transfusion polycythemia by Iscove,[6] but the increase was not enough to be considered suggestive of stem cell competition. In the present study the increase in spleen CFU-c are more suggestive but need to be substantiated by further CFU-c enumeration after more prolonged hypertransfusion polycythemia.

Thrombocytes

The platelet count decreased abruptly after transfusion from 1.5×10^6 to 1.1×10^6/cu mm of whole blood and remained at that level throughout the polycythemic period (Figure 8-4). This change in platelet count probably reflects the fact that platelets traditionally are expressed per cu mm of whole blood rather than per cu mm of plasma.[21] Functionally, platelets are plasma components interacting with coagulation factors that also depend on their activity on their concentration in plasma rather than on their concentration in whole blood. If platelets were expressed per cu mm of plasma or in terms of total number of circulating platelets, they were found to be unaffected by hypertransfusion.

erythroid stem cells was associated with an increase in megakaryocyte stem cells. However, this increase was not reflected by higher circulating platelet counts.

CFU-s

Exogenous colonization of the spleen of irradiated recipients of bone marrow or spleen showed some transfusion related changes similar to those reported by Guzman and Lajtha.[3] The number of CFU-s in the femur increased gradually while the splenic CFU-s tripled in number right after transfusion and remained at that level throughout (Figure 8-5).

Conclusion

The observations reported here would suggest that a hypertransfusion induced decrease in circulating erythropoietin causes a decrease in CFU-e and a reduced competition for stem cells resulting in an increase in the number of parent stem cells (CFU-s)

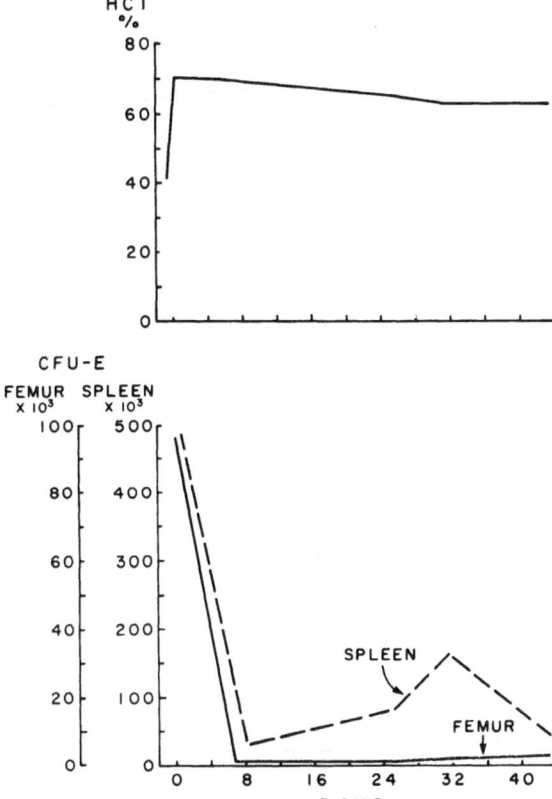

Figure 8-2 The upper panel shows the hematocrit response to red cell transfusions. The lower panel shows the number of CFU-e per femur and per spleen of the transfused animals.

CFU-M

Splenic and bone marrow cultures on agar or on fibrin clots have disclosed the appearance of acetylcholine positive colonies made up by giant cells, structurally similar to megakaryocytes.[13] Recently, mature platelets have been observed to originate from these cells in plasma clots and it appears that we are observing the cloning of specific megakaryocyte committed stem cells (CFU-M).[22] The addition of a conditioned medium consisting of either a lymphocytic derived culture medium[23] or of erythropoietin[22] has been reported to be required for maximal growth and development of CFU-M. However, in our hands, a fibrin clot without the addition of specific conditioned medium has supported the growth of megakaryocytic colonies equally well. In this study, the number of CFU-M in bone marrow had decreased moderately on the seventh day of transfusion polycythemia and then returned to a normal pretransfusion value (Figure 8-4). However, the CFU-M in the spleen increased gradually and became significantly elevated toward the end of the experimental period. This observation would suggest that a decrease in

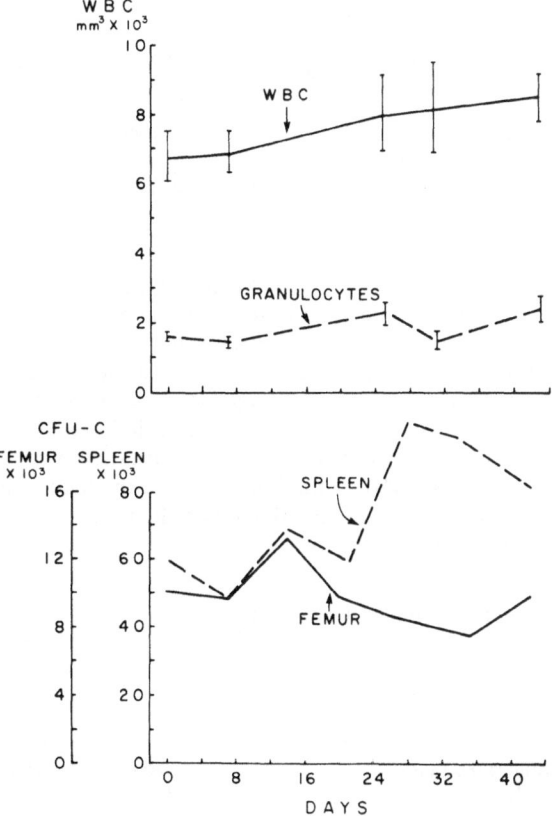

Figure 8-3 The upperpanel shows the white blood cell count and total granulocyte count of intermittently hypertransfused mice. The lower panel shows the number of CFU-c per femur and per spleen of the transfused animals.

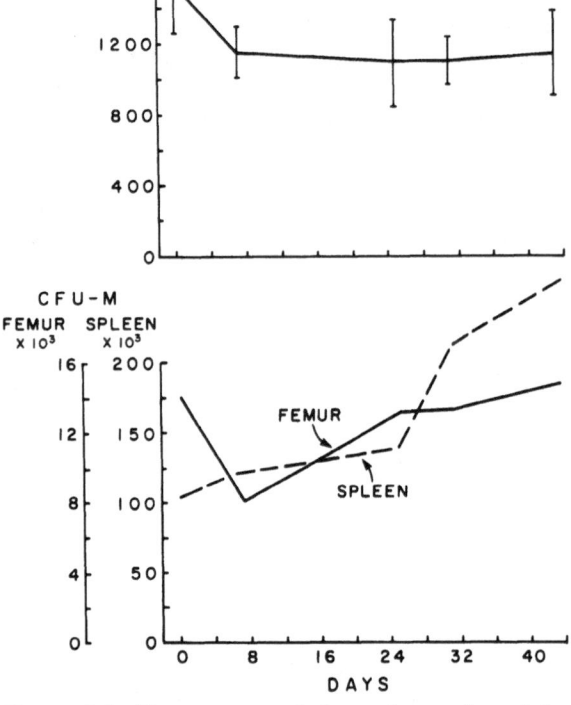

Figure 8-4 The upper panel shows the number of the platelet count in hypertransfused mice. The lower panel shows the number of CFU-M per femur and per spleen of the corresponding mice.

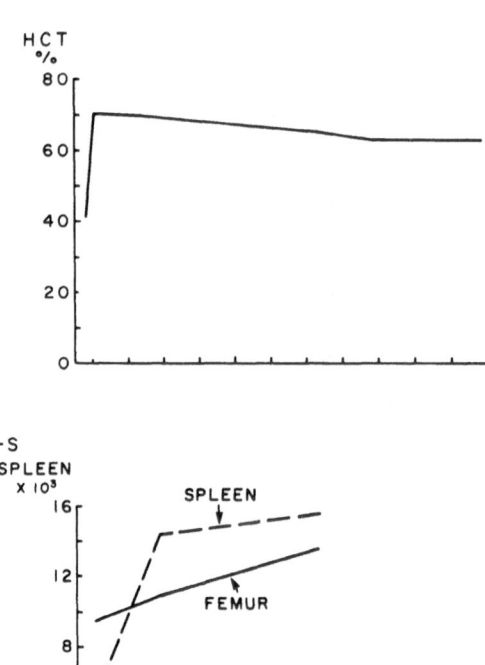

Figure 8-5 The upper panel shows the hematocrit response to red cell transfusions. The lower panel shows CFU-s per femur and per spleen of the corresponding mice.

and sister stem cells (CFU-c and CFU-M) (Table 8-1). However, this increase in the number of stem cells did not result in a significant increase in the differentiation toward myeloid or megakaryocytic cells. The animals, however, already had an adequate number of differentiated cells in their circulation and it is possible, as described by several investigators,[19,20] that increased differentiation from the expanded stem cell pools occurs when radiation or chemotherapy induced cytopenia exists.

Summary

In order to evaluate the effect of a sustained cessation of erythropoiesis on the kinetics of bone marrow cells, normal Swiss-Webster mice were kept polycythemic for 43 days by i.p. transfusion of packed red blood cells every 6 days. The 66 hr ^{59}Fe utilization was reduced from 44% to 0.2% within 5 days and remained at this level throughout. The response to an injection of 1 unit of erythropoietin decreased from 48% to 6.5% from day 9 on. CFU-e determined on fi-

Table 8-1 Stem cell compartments.

	Femur × 10³		Spleen × 10³	
	Controls	Day 25	Controls	Day 25
Total Nucleated Cells	27,000 ± 3,400	25,000 ± 2,200	277,000 ± 64,000	330,000 ± 24,000
CFU-s	40 ± 10	55 ± 18	102 ± 23	313 ± 34
BFU-e	—	—		—
CFU-e	97 ± 21	0.8 ± 0.4	388 ± 135	89 ± 33
CFU-c	10 ± 1.2	11 ± 0.4	48 ± 9	65 ± 18
CFU-M	14 ± 4	13 ± 1.5	104 ± 6	142 ± 22

Mean ± 1 SEM

brin clot decreased from 96,000 per femur to approximately 1,000 from day 7 on and from 500,000 per spleen to approximately 50,000 from day 7 on. The granulocyte count increased slightly as did the CFU-c in the femur. However, the CFU-c in the spleen appeared to increase more significantly. The platelet count decreased abruptly per cu mm of whole blood but when expressed in terms of total circulating number, the platelets were not changed by hypertransfusion. The CFU-M in femur determined on a fibrin clot remained fairly unchanged whereas the CFU-M in spleen increased throughout the experimental period. The CFU-s determined by exogenous assay per femur remained unchanged but in the spleen tripled in number. These observations show that exogenous suppression of erythropoiesis causes a pronounced reduction in the number of CFU-e in both bone marrow and spleen. The numbers of CFU-c and CFU-M in spleen increased as if released from competitive inhibition by the CFU-e. The CFU-s in spleen also increased, but despite this availability of parent stem cells and committed stem cells, the production of granulocytes and thrombocytes was not altered significantly.

References

1. Schooley JC, Garcia JF: Suppression of erythropoiesis in the plethoric rat by antierythropoietin. *Proc Soc Exp Biol Med 133:*953–954, 1970.
2. Thorling EB, Erslev AJ: The tissue tension of oxygen and its relation to hematocrit and erythropoiesis. *Blood 31:*332–343, 1968.
3. Guzman E, Lajtha LG: Some comparisons of the kinetic properties of femoral and splenic haemopoietic stem cells. *Cell Tissue Kinet 3:*91–98, 1970.
4. Bradley TR, Robinson W, Metcalf D: Colony production *in vitro* by normal, polycythaemic and anaemic bone marrow. *Nature 214:*511, 1967.
5. Axelrad AA, McLeod DL, Shreeve MM, Heath DA: Properties of cells that produce erythrocytic colonies *in vitro*, in Robinson W (ed): *Hemopoiesis in Culture*. Washington DC, US Government Printing Office, 1974.
6. Iscove NN: The role of erythropoietin in regulation of population size and cell cycling of early and late erythroid precursors in mouse bone marrow. *Cell Tissue Kinet 10:*323–334, 1977.
7. Hara H, Ogawa M: Erythropoietic precursors in mice under erythropoietic stimulation and suppression. *Exp Hemat 5:*141–148, 1977.
8. Erslev AJ: Erythropoietin assay, in Williams WJ, Beutler E, Erslev AJ, Rundles RW (eds): New York, McGraw-Hill, 1977.
9. Stephenson JR, Axelrad AA, McLeod DL, et al: Induction of colonies of hemoglobin-synthesizing cells by erythropoietin *in vitro*. *Proc Natl Acad Sci (USA) 68:*1542–1546, 1971.
10. Silver RK, Erslev AJ: The action of erythropoietin on erythroid cells *in vitro*. *Scand J Haemat 13:*338–351, 1974.
11. Steinberg HN, Handler ES, Handler EE: Assessment of erythrocytic and granulocytic colony formation in an *in vivo* plasma clot diffusion chamber culture system. *Blood 47:*1041–1051, 1976.
12. Nakeff A, Daniels-McQueen S: *In vitro* colony assay for a new class of megakaryocyte precursor: Colony-forming unit megakaryocyte (CFU-M). *Proc Soc Exp Biol Med 151:*587–590, 1976.
13. Quesenberry P, Morley A, Stohlman F Jr, et al: Effect of endotoxin on granulopoiesis and colony-stimulating factor. *New Engl J Med 286:*227–232, 1972.
14. Karnovsky MJ, Roots L: A "direct-coloring" thiocholine method for cholinesterases. *J Histochem Cytochem 12:*219–221, 1964.
15. Till JE, McCulloch EA: A direct measurement of the radiation sensitivity of normal mouse bone marrow cells. *Rad Res 14:*213–222, 1961.
16. Erslev AJ: The clinical usefulness of erythropoietin measurement, in Fisher JW (ed): *Kidney Hormones, Vol. 2 Erythropoietin*. London, Academic Press, 1977.
17. Hellman S, Grate HE: Haemopoietic stem cells: Evidence for competing proliferative demands. *Nature 216:*65–66, 1967.
18. Lawrence JS, Craddock CG Jr.: Stem cell competition: the response to antineutrophilic serum as affected by hemorrhage. *J Lab Clin Med 72:*731–738, 1968.
19. Morley A, Howard D, Bennett B, et al: Studies on the regulation of granulopoiesis. II. Relationship to other differentiation pathways. *Br J Haematol 19:*523–532, 1970.
20. Firkin FC, Hays EF, Cline MJ: Effect of hypertransfusion on granulopoiesis in bone marrow depression: Studies in the irradiated mouse. *Br J Haematol 35:*225–231, 1977.
21. Shaikh B, Erslev AJ: Dilution thrombocytopenia in hypertransfused mice. *Clin Res 22:*405a, 1974.
22. McLeod DL, Shreeve MM, Axelrad AA: Induction of megakaryocyte colonies with platelet formation *in vitro*. *Nature 261:*492–494, 1976.
23. Metcalf D, MacDonald HR, Odartchenko N, et al: Growth of mouse megakaryocyte colonies *in vitro*. *Proc Natl Acad Sci (USA) 72:*1744–1748, 1975.

9 Inhibitors of Erythropoiesis: Effects of Red Cell Extracts on Heme Synthesis and Erythroid Colony Formation in Human Bone Marrow Cultures

Yoshiaki Moriyama

Introduction

The hormone erythropoietin is one of the important factors that primarily regulates erythropoiesis in man. However, the control mechanisms of red blood cell (RBC) production involving negative feedback factors are not understood. Evidence that RBC mass has a regulating effect on erythropoiesis can be given by the use of methemoglobinized erythrocytes[1] and by the studies on nephrectomized humans, in which a transfusion of erythrocytes caused a depression of bone marrow erythroblast numbers.[2] In addition, Kivilaakso and Rytömaa[3] first reported, measuring tritiated thymidine incorporation in short-term bone marrow cell cultures, the existence of an inhibitor (erythrocyte chalone) of RBC precursor cell proliferation that was obtainable directly from mature erythrocytes. Using a similar *in vitro* technique, several investigators[4,5] confirmed their findings. We have also reported the presence of an erythropoietic inhibitory substance, which is of low-molecular weight, in sera of normal and hypertransfused-polycythemic rabbits.[6,7] These findings suggest the existence of RBC-specific inhibitors that may control the proliferation of erythroid precursor cells as negative feedback factors. However, little is known of the mode of action of such inhibitors and how RBC-specific inhibitors take part in erythropoiesis in patients with various hematological disorders.

The present studies were undertaken to investigate the ability of human red blood cell extracts (RCE) to modify the proliferation of erythroid precursor cells *in vitro* and to attempt to clarify the role of RCE in the mechanism of the anemia of various diseases.

Materials and Methods

The sources of RCE were normal human volunteers and patients with hematological disorders (iron deficiency anemia, hemolytic anemia, aplastic anemia, and polycythemia vera).

Heparinized blood from these patients was collected by venous puncture, and centrifuged at 1000 g for 10 min at 20°C, and the buffy coat was removed. The remaining erythrocytes were then washed twice in NCTC–109 solution and incubated in saline at 50% hematocrit for 2 hr at 37°C, then centrifuged 1000 g, 10 min, 5°C. The conditioned supernatant was used as the crude extract and stored at −20°C or partially purified by passing through a series of Amicon-filtration membranes. The fraction with a molecular weight range of 500 to 10,000 daltons was lyophilysed and stored at −20°C. Before it was used in the bone marrow cultures RCE was diluted again with NCTC–109 solution and was made equal to the initial volume of erythrocytes. The bone marrow culture technique was the method of Krantz et al.[8] for measuring heme synthesis *in vitro* and a modification of the method of McLeod et al.[9] for the growth of erythroid colonies. Bone marrow was obtained from normal human volunteers with their proper consent. Approximately 2 ml of bone marrow was aspirated and transferred immediately to a sterile plastic heparinized tube containing 5 ml supplemented Eagle's minimum essential medium (HMEM, Gibco, Grand Island, New York) and 2% fetal calf serum (Gibco). After centrifugation, the marrow buffy coat cells were washed once, and adjusted for plating.

For cultures where ^{59}Fe incorporation into heme was determined, 2×10^6 washed marrow cells were plated in 35×10 mm tissue culture dishes in 2 ml NCTC–109 solution containing 20% fetal calf serum and varying concentrations of RCE diluted in NCTC–109 solution. The cultures were incubated with and without the addition of human urinary erythropoietin with a specific activity of 5.29 units/mg protein at 37°C in a humidified atmosphere of 5% CO_2 in air. After 29 hr of incubation, ^{59}Fe bound to homologous transferrin was added (0.5 μCi/plate), and the incubation continued for an additional 16 hr. After the end of incubation the harvested cells were washed twice and heme was extracted with cyclohexanone and radioactivity determined in the cultured cells.

Cultures of erythroid colonies were carried out according to a modification of the method of McLeod et al.[9] as follows: 1 to 2×10^4 nucleated marrow cells were plated on wells of disposable microtiter plates in 0.1 ml culture media consisting of 10% beef embryo extract (Gibco), 20% fetal calf serum, 10% human AB serum, 10% bovine serum albumin (Sigma Chemical Co., St. Louis, Missouri), 10% supplemented HMEM, 10% bovine citrated plasma (Gibco), and 30% NCTC–109 solution containing L-asparagine 0.02 mg/ml and human urinary erythropoietin (0.5 to 1.0 units). Cultures were incubated with and without RCE diluted in NCTC–109 solution for 7 to 10 days at 37°C. After incubation the cultures were removed from the wells and placed onto glass microscope slides. The plasma clots were flattened with filter paper, fixed with 5% glutaraldehyde in 0.01 M phosphate buffer for 10 min, and washed in distilled water. After they were stained with benzidine, colonies containing eight or more erythroblasts were counted using 100× magnification.

Burst-forming units (BFU-e) were counted depending on the spatial orientation of erythroid colonies in which individual colonies were grouped into discrete aggregates consisting of three or more identifiable erythroid colonies.

Results

Figure 9-1 shows the effect of RCE from normal donors at various concentrations on heme synthesis in erythroid cells in normal human bone marrow cultures. The doses of RCE 0.2, 0.5, and 1.0 ml indicate those made from 0.2, 0.5, and 1.0 ml RBC, respectively. Total heme synthesis per plate decreased with increasing concentrations of RCE with and without stimulation with erythropoietin (0.2 units) in cultures. RCE at all the concentrations in cultures significantly ($p < 0.05$) inhibited heme synthesis in erythroid cells *in vitro* when compared with controls with no addition of RCE.

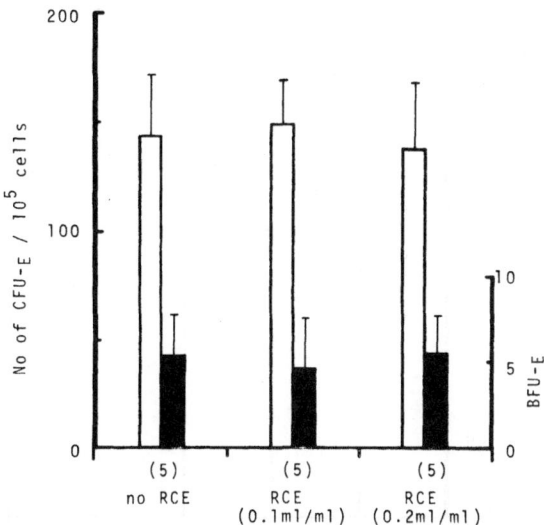

Figure 9-2 Effects of red cell extracts from normal donors on erythroid colony growth in normal human bone marrow cultures in the presence of erythropoietin (0.5 units/ml). White bars, CFU-e; black bars, BFU-e.

In order to determine whether RCE obtained from normal donors affects the *in vitro* growth of human marrow bursts (BFU-e) including more mature (CFU-e) erythroid precursor cells, cultures were performed with and without the addition of RCE. As illustrated in Figure 9-2, none of the concentrations of RCE modified the ability of erythroid committed stem cells to form erythroid colonies *in vitro* in response to erythropoietin (0.5 units/ml).

The effects of RCE obtained from normal donors on cell survival in cultures are shown in Table 9-1. After 45 hr of incubation, the numbers of viable cells in cultures were slightly reduced. However, there was no difference in the numbers of viable cells with and without the addition of RCE *in vitro*, indicating that RCE is nontoxic to the cells *in vitro*.

To investigate the role of RCE as a negative feedback factor in the control of erythropoiesis, RCE obtained from patients with various blood disorders was tested by measuring its inhibitory effect on heme syn-

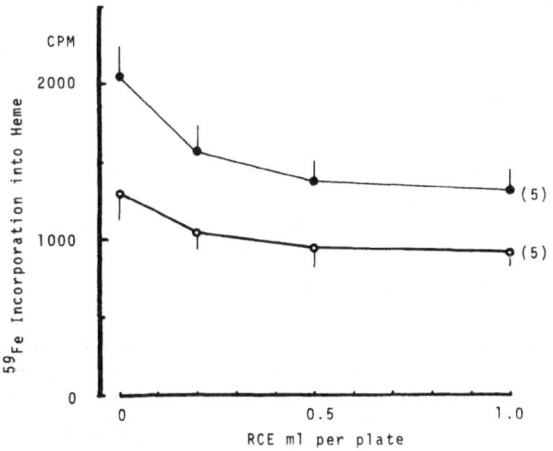

Figure 9-1 Inhibitory effects of red cell extracts from normal donors at various concentrations on heme synthesis in erythroblasts in normal human bone marrow cultures. Closed circles, with ESF; open circles, without ESF.

Table 9-1 Effects of red cell extract (RCE) on nucleated cells plated *in vitro* before and after incubation.

Treatment	No. of Experiments	No. of Cells[a] Before	After
no RCE (controls)	5	1.96 ± 0.11	1.68 ± 0.18
RCE 0.2 ml	5	1.99 ± 0.17	1.71 ± 0.24[b]
RCE 0.5 ml	5	1.96 ± 0.16	1.69 ± 0.22[b]

[a] × 10^6.
[b] Not significantly different from controls.

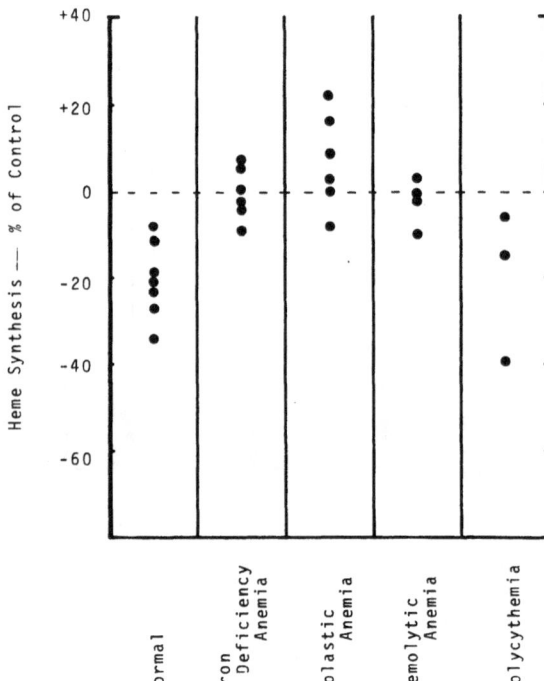

Figure 9-3 Inhibitory effects of red cell extracts obtained from patients with various blood disorders on heme synthesis in erythroblasts in normal human bone marrow cultures in the presence of erythropoietin (0.2 units/plate).

thesis in erythroblasts in normal human bone marrow cultures in the presence of erythropoietin (0.2 units/plate). The amount of all RCE tested here is 0.2 ml, which is equivalent to 0.2 ml RBC of each case. As shown in Figure 9-3, RCE obtained from normal donors and patients with polycythemia markedly inhibited heme synthesis in erythroblasts *in vitro* whereas no effects were seen in patients with iron deficiency anemia and hemolytic anemia. On the other hand, RCE made from patients with aplastic anemia slightly stimulated heme synthesis in human bone marrow cultures.

Discussion

The results of the present studies indicate that RCE obtained from normal donors produced a significant inhibition of ^{59}Fe incorporation into heme in erythroblasts in normal human bone marrow cultures with and without the addition of erythropoietin. In addition, this inhibitory activity was increased with increasing concentrations of RCE in the cultures. However, this inhibitor did not affect the numbers of marrow cells inoculated *in vitro* during incubation, indicating that this inhibitory effect on heme synthesis is unlikely to be due to cytotoxic factors.

In studies of erythroid colony formation with the use of normal human bone marrow in plasma clot culture system, RCE did not produce a change in the numbers of erythroid colonies formed *in vitro*. This suggests that the inhibitor made from mature erythrocytes does not affect the ability of the erythroid stem cells to form erythroid colonies in response to erythropoietin *in vitro* but probably acts more specifically on the differentiated erythroid cells (heme synthesizing cells) to alter their cell cycles.

The RCE observed in our system is a low molecular weight substance between 500 and 10,000 daltons. This suggests that it is not an antibody to erythropoietin and also that its molecular weight is similar to that of the inhibitor that we have reported in normal, uremic, and polycythemic rabbit sera.[6,7,10] It is also quite possible that this inhibitor may be related to the inhibitor (erythrocyte chalone) reported by several investigators.[3-5] However, whether RCE represents a natural physiological process in the control of erythropoiesis and also how the inhibitor may be related *in vivo* are unknown. In the present report, in order to clarify the role of RCE as a negative feedback factor in the control of erythropiesis in humans, the effect of RCE obtained from patients with various blood disorders on heme synthesis in erythroblasts was studied using normal human bone marrow culture system in the presence of erythropietin (Figure 9-3). The inhibitory activity of RCE made from mature RBC was found to be different in different conditions. In normal and polycythemic conditions the RCE showed some inhibitory effect on heme synthesis in erythroblasts *in vitro* whereas no inhibition was seen in anemic conditions. This suggests that the inhibitor that comes from mature RBC may have a role in the control of erythropoiesis as a negative feedback factor. However, most of extracts made from RBC of patients with aplastic anemia slightly stimulated ^{59}Fe incorporation into heme in normal human bone marrow cultures. The reason for this is now not clear.

Summary

Extracts (RCE) made from human mature erythrocytes were assayed *in vitro* in order to study their ability to modify the proliferation of maturing erythroid cells. The RCE obtained from normal donors was found to significantly inhibit ^{59}Fe incorporation into heme in erythroblasts in normal human bone marrow cultures with and without the addition of erythropoietin. This inhibitor is a low molecular weight substance and nontoxic to the cells *in vitro*. In addition, the inhibitory activity of RCE was different in normal, anemic, and polycythemic conditions. However, there was no effect on the committed erythroid stem cells. These data suggest that the inhibitor extracted from mature erythrocytes acts more specifically on the differentiated erythroid cell compartment (rather than on the stem cell compartment) as a negative feedback factor of red cell production.

Acknowledgment

Erythropoietin used here was kindly shared by Prof. James W. Fisher, Tulane University School of Medicine, New Orleans, Louisiana.

References

1. Kilbridge TM, Fried W, Heller P: The mechanism by which plethora suppresses erythropoiesis. *Blood 33:*104, 1969.
2. Naets JP, Wittek M, Toussaint C, Van Geertryyden, J: Erythropoiesis in renal insufficiency and anephric man. *Ann NY Acad Sci 149:*143, 1968.
3. Kivilaakso E, Rytömaa T: Erythrocytic chalone, a tissue-specific inhibitor of cell proliferation in the erythron. *Cell Tissue Kinet 4:*1, 1971.
4. Bateman AE: Cell specificity of chalone-type inhibitors of DNA synthesis release by blood leucocytes and erythrocytes. *Cell Tissue Kinet 7:*451, 1974.
5. Lord BI, Cercek L, Cercek B, Shah GP, Dexter TB, Lajtha LG: Inhibitors of hematopoietic cell proliferation: Specificity of action within the hematopoietic system. *Br J Cancer 29:*168, 1974.
6. Moriyama Y, Lertora JJL, Fisher JW: Studies on an inhibitor of erythropoiesis I. Effects of sera from normal and polycythemic rabbits on heme synthesis in rabbit bone marrow cultures. *Proc Soc Exp Biol Med 147:*740, 1974.
7. Moriyama Y, Rege A, Fisher JW: Studies on an inhibitor of erythropoiesis II. Inhibitory effects of serum from uremic rabbits on heme synthesis in rabbit bone marrow cultures. *Proc Soc Exp Biol Med 148:*94, 1975.
8. Krantz SB, Gallien-Lartique O, Goldwasser E: The effect of erythropoietin upon heme synthesis by marrow cells *in vitro*. *J Biol Chem 238:*4085, 1963.
9. McLeod DL, Shreeve MM, Axelrad AA: Improved plasma culture system for production of erythrocytic colonies *in vitro:* Quantitative assay method for CFU-e. *Blood 44:*517, 1974.
10. Moriyama Y, Rege A, Fisher JW: Inhibitory effects of uremic serum on heme synthesis and erythroid colony formation in bone marrow cultures, in Nakao K, Fisher JW, Takaku Y (eds): *Erythropoiesis*. Tokyo, University of Tokyo Press, p 15, 1975.

10 Production of Erythroid Precursor Cells in Long-Term Cultures of Mouse Bone Marrow Cells

Martin J. Murphy, Jr., Akio Urabe, and Maureen E. Sullivan

Introduction

The establishment of murine bone marrow cultures in which proliferation of pluripotent hematopoietic stem cells[1] and committed progenitor cells of granulocytes, macrophages,[1] and megakaryocytes[2] enables us to study the mechanisms governing the commitment of pluripotent stem cells into the various blood cell lineages *in vitro*. It is likely that this *in vitro* system is subject to many of the same inductive and regulatory mechanisms as are present *in vivo,* since cultured and freshly isolated granulocytes, macrophages, and their clonable precursor cells have been shown to have common biological and physical properties.[3] A significant deficiency of the system, however, has been the consistent inability to generate and maintain erythroid cells in the cultures. In this chapter, we report that although no differentiated erythroid cells persist in the long term cultures, early erythroid precursor cells (i.e., BFU-e) are actively produced.

Materials and Methods
Long-Term Marrow Culture
The original method of Dexter et al.[1] was modified as described elsewhere.[3] Femoral bone marrow cells from (DBA/2 × C57Bl/6) F_1 mice (Cumberland View Farms) were flushed directly into 25 ml sterile plastic flasks (Corning). No attempt was made to break up marrow aggregates. The cells were incubated at 33°C in 10 ml of Fischer's medium (Gibco Chagrin Falls, Ohio) containing 20% horse serum (Flow). After 3 or 4 days of incubation, half of the medium and nonadherent cells were removed and the cultures were fed with an equal volume of fresh medium. After 7 days, all of the nonadherent cells were removed, and 5×10^6 freshly isolated, syngenic bone marrow cells in fresh medium were added to the flask. One-half of the suspension cells in the medium were harvested each week from five to ten culture flasks, pooled, and assayed every week up to 6 weeks of incubation for the determination of pluripotent stem cells (CFU-s), and committed erythroid precursor cells (BFU-e and CFU-e), granulocyte-macrophage precursor cells (CFU-c), B-lymphocyte precursor cells (CFU-B), and megakaryocyte precursor cells (CFU-M). The differential morphology of the nonadherent cells was assessed using standard Wright-Giemsa staining and light microscopy. The protocol is sketched in Figure 10-1.

Pluripotent Stem Cell Assay
Spleen colony forming units (CFU-s) were quantitated using the technique originally described by Till and McCulloch.[4] Pooled, suspension culture cells or normal bone marrow cells were injected into groups of no less than 6 lethally irradiated (900 R) BDF_1 recipient mice. After 9 days, spleens were removed, fixed in Bouin's solution, and scored for macroscopic colonies.

Megakaryocyte Precursor Cell Assay
Megakaryocyte precursor cells (CFU-M) were cloned in semisolid agar using a modification of the technique of Metcalf et al.[5] as described elsewhere.[2] Cells from femoral bone marrow (7.5×10^4) and suspension cells (2.5×10^4) were cultured in 35 mm petri dishes (Falcon 3001), in 1 ml McCoy's 5A modified medium containing 15% fetal calf serum, 0.3% bactoagar, supplemented with essential and nonessential amino acids, 200 mM glutamine, asparagine, sodium pyruvate, and 10^{-4} M 2-mercaptoethanol.

Two activities have been shown to be required for quantitation of CFU-M in mouse marrow.[2] The two entities were obtained for these studies from conditioned media from cultures of the WEHI-3 murine myelomonocytic leukemic cell line[2,6] and from long-term cultures of bone marrow cells.[2] CFU-M were grown in cultures containing 50 μl of an eight-fold concentration of conditioned medium (approximately 700 μg protein/ml) from cultures of the WEHI-3 conditioned medium and 200 μl of bone marrow conditioned medium. Cultures were scored at 40× magnification for megakaryocyte colonies (>2 cells) after 7 days of incubation at 37°C in a humidified incubator in an atmosphere of 7% CO_2 in air.

Colonies of megakaryocytes were readily identifiable by the size of the majority of cells contained in them. WEHI-3 conditioned medium also stimulated the growth of macrophage-granulocyte colonies[7]; however, the cells comprising the megakaryocyte colonies were markedly larger than the cells that comprise either granulocyte or macrophage colonies.

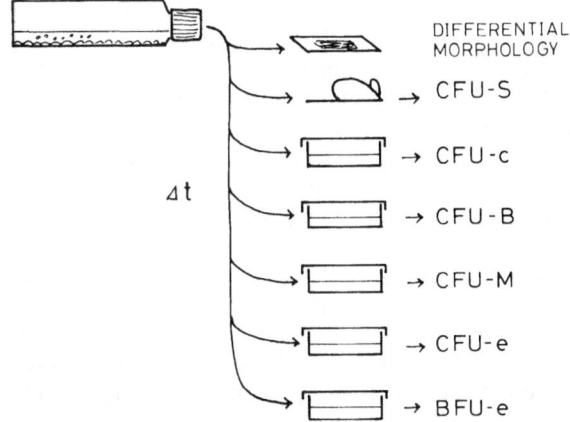

Figure 10-1 A schematic representation of the various assays performed weekly on nonadherent cells from long-term mouse bone marrow cell cultures.

Each megakaryocyte colony was generally made up of fewer cells (<50 cells) than the average, typical granulocyte or macrophage colony (>100 cells). Cytochemical identification of both large and small megakaryocytes was obtained with isolated colony cells by staining for the presence of acetylcholinesterase.[8,9]

Granulocyte-Macrophage Precursor Cell Assay

Progenitors that give rise to granulocytes or macrophages in semisolid agar were assayed using the same conditions as the CFU-M assay with the following exceptions: 1.0×10^4 suspension cells from long-term marrow cultures or 2.5×10^4 normal bone marrow cells were plated in each 1 ml of culture; granulocyte-macrophage colony-stimulating factor was derived from L cell-conditioned medium.[10,11] Supernatant medium was taken biweekly from L cells growing continuously in Dulbecco's modified medium containing 2% fetal calf serum. The conditioned medium was dialyzed extensively against distilled water, concentrated fivefold by Amicon filtration and then titrated against normal mouse bone marrow colony-stimulating activity for routine assay.

B-Lymphocyte Precursor Cell Assay

Assay for CFU-B was performed as described previously.[12,13] From 1 to 5×10^4 nonadherent cultured cells were plated in 1 ml cultures containing 5×10^{-5} M 2-mercaptoethanol and 10 μg/ml lipopolysaccharide (LPS) from *Salmonella typhosa* WOD01 (Difco Labs, Detroit, Michigan). Colonies of greater than 20 cells were scored after 6 days of incubation at 37°C with 7% CO_2 in air, in a humidified incubator.

Erythroid Precursor Cell Assay

The suspension cells from the marrow cultures were washed five times with alpha medium (Flow) and assayed for CFU-e and BFU-e using the methylcellulose culture method of Iscove and Sieber[14] as modified and described in Appendix 10.[15] Briefly, 2×10^5 nucleated cells were suspended in 1 ml of alpha medium containing 0.8% methylcellulose (Dow, Midland, Michigan), 1% bovine serum albumin (Calbiochem, San Diego, California), 30% fetal calf serum (Flow, Rockville, Maryland), 1×10^{-4} M 2-mercaptoethanol (Sigma Chemical Co., St. Louis, Missouri), and 2 units of human urinary erythropoietin in 35×10 mm petri dishes (Lux, Newbury Park, California). The cultures were incubated at 37°C in 5% CO_2 in air with a humidified incubator. The plates were examined for the presence of CFU-e at 2 days and BFU-e at 9 days using an inverted microscope ($\times 100$ magnification) and a benzidine stain.[16] Cytocentrifuge preparations were made of individual colonies plucked from the plates and stained with Wright-Giemsa.

Results

The suspension cells from the cultures never contained any identifiable erythroid precursors. Once established, the composition of the suspension cells remained rather stable, containing primarily unclassifiable blasts, myeloid cells, and macrophages.

Table 10-1 shows the numbers of BFU-e that were detected after 3 weeks of incubation in long-term marrow cultures. In the four individual experiments listed, CFU-e were never detected.

CFU-s and CFU-c were maintained throughout the course of the 6 weeks of observations of the long-term cultures; however, the numbers of CFU-M, BFU-e and CFU-B gradually diminished.

The committed erythroid progenitor cells detected were actively produced in the long-term marrow cultures, since they contained more BFU-e than would be expected if the BFU-e were simply the survivors of the initial population of precursor cells added when the cultures were established (Table 10-1). The maximum number of BFU-e after 3 weeks would be one-eighth of the input number because one-half the suspension cells were taken each week for assay.

Table 10-1 Sustained proliferation of erythroid precursors (BFU-e) after 3 weeks in long-term marrow culture.

Culture Number	Cells Harvested	Observed BFU-e
167	1.67×10^6	928
218	10.00×10^6	770
221	5.50×10^6	413
294	2.35×10^6	219

Table 10-2 Numbers of BFU-e in long-term mouse marrow cultures (BFU-e/10^5 Nucleated Cells).

Culture Number	Weeks in Culture					
	1	2	3	4	5	6
167	—	—	58	—	16.3	0
218	15	—	7.7	—	—	—
221	—	—	7.5	—	—	—
239/252	11.4	—	—	—	—	—
267	20.5	6	—	—	—	—
294	13	6.7	9.3	1.2	0	0
Mean (±S.E.)	15 (±2)	6.4 (±0.4)	20.6 (±12.5)	1.2 —	8.2 (±8.2)	

This is the maximum number possible without self-renewal or recruitment from more primitive cells, and assumes that no loss of BFU-e resulted from either cell death or by differentiation into more mature cells.

As shown in Table 10-2, BFU-e colonies were observed through the fifth week of culture, supporting the proposition that not only was there maintenance of BFU-e, but moreover, proliferation of BFU-e in long-term mouse marrow cultures.

Discussion

Precursor cells of all the hematopoietic cell lineages, with the exception of lymphocytes, appear to be actively sustained over many weeks in mouse long-term bone marrow cultures. It is noteworthy that the *in vitro* cloning assay for B-lymphocytes detects a proportion of B-lymphocytes with surface immunoglobulin, but not immunoglobulin negative, Ia positive pre-B-lymphocytes.[17] Unlike the cloning assays for the other hematopoietic cell lineages, the clonable B-lymphocytes are not immature cells. It remains possible, therefore, that in the long-term cultures, pre-B cells, like the BFU-e, are actively produced in the cultures, but mature cells are not formed.

The sustained proliferation of BFU-e in the absence of added erythropoietin is an intriguing phenomenon. It is impossible to exclude that the long-term cultures containing 20% horse serum may contain small but immeasurable amounts of erythropoietin. Nevertheless, these observations support the suggestion that recruitment into the BFU-e pool is relatively independent of detectable erythropoietin.[18] The absence of added erythropoietin in the long-term marrow cultures might explain the absence of CFU-e in the nonadherent cell population as well, since recruitment into the CFU-e pool seems to be erythropoietin-dependent.[18]

This culture system permits sustained proliferation of the BFU-e compartment, but no differentiation into morphologically recognizable hemoglobinizing cells is observed. Therefore these cultures represent an important approach for distinguishing the different mechanisms regulating the movement of stem cells into committed progenitor cells, and their further proliferation and differentiation to erythrocytes.

Summary

Pluripotent stem cells (CFU-s), granulocyte-macrophage precursor cells (CFU-c), early erythroid precursor cells (BFU-e), megakaryocyte precursor cells (CFU-M), and B-lymphocyte precursor cells (CFU-B) were quantitatively assayed from suspension cells obtained from long-term mouse bone marrow cultures.

CFU-s and CFU-c were observed throughout the observation period of 6 weeks. Distinct colonies of CFU-c were observed during 4 to 5 weeks of incubation; however, BFU-e and CFU-M diminished gradually during the course of the experiments. No CFU-e were observed. These results indicate that CFU-s, CFU-c, BFU-e, and CFU-M are not only maintained but also proliferate in long-term cultivation of mouse marrow.

Acknowledgments

The technical assistance of Heather Jackson is gratefully acknowledged. This work was supported by grants AM-19741 and CA-17353 from the USPHS, American Cancer Society, CH-3, The Hipple Foundation, and The Gar Reichman Foundation.

References

1. Dexter TM, Allen TD, Lajtha LG: Conditions controlling the proliferation of haemopoietic stem cells *in vitro*. *J Cell Physiol 91*:335, 1977.
2. Williams N, Jackson H, Sheridan APC, Murphy MJ Jr, Elste A, Moore MAS: Regulation of megakaryopoiesis in long term murine bone marrow cultures. *Blood 51*:245, 1978.
3. Williams N, Jackson H, Rabellino EM: Proliferation and differentiation of normal granulopoietic cells in continuous bone marrow cultures. *J Cell Physiol 93*:435, 1977.
4. Till JE, McCulloch EA: A direct measurement of the radiation sensitivity of normal mouse bone marrow cells. *Radiat Res 14*:213, 1961.

5. Metcalf D, McDonald HR, Odartchenko N, Sordat B: Growth of mouse megakaryocyte colonies in vitro. Proc Natl Acad Sci (USA) 72:1744, 1975.
6. Ralph P, Moore MAS, Nilsson K: Lysozyme synthesis by established human and murine histiocytic lymphoma cell lines. J Exp Med 143:1528, 1976.
7. Bradley TR, Metcalf D: The growth of mouse bone morrow cells in vitro. Aust J Exp Biol Med Sci 44:287, 1966.
8. Jackson CW: Cholinesterase as a possible marker of early cells of the megakaryocyte series. Blood 42:413, 1973.
9. Nakeff A, Daniels-McQueen S: In vitro colony assay for a new class of megakaryocyte precursor: Colony forming unit megakaryocyte (CFU-M). Proc Soc Exp Biol Med 151:587, 1976.
10. Austin PE, McCulloch EA, Till JE: Characterization of the factor in L cell conditioned medium capable of stimulating colony formation by mouse marrow cells in culture. J Cell Physiol 77:121, 1971.
11. Stanley ER, Cifone M, Heard PM, Defendi V: Factors regulating macrophage production and growth: Identity of colony stimulating factor and macrophage growth factor. J Exp Med 143:631, 1976.
12. Metcalf D, Nossal CJV, Warner NL, Miller JFAP, Mandel TE, Layton JE, Gutman GA: Growth of B-lymphocyte colonies in vitro. J Exp Med 142:1532, 1975.
13. Kincade PW, Ralph P, Moore MAS: Growth of B-lymphocyte clones in semi-solid culture is mitogen-dependent. J Exp Med 143:1265, 1976.
14. Iscove NN, Sieber F: Erythroid progenitors in mouse marrow detected by macroscopic colony formation in culture. Exp Hemat 3:32, 1975.
15. Murphy MJ Jr, Sullivan ME: Culture of erythroid stem cells from murine and human marrow and blood, in: Murphy MJ Jr (ed): In Vitro Aspects of Erythropoiesis, New York, Springer-Verlag, p. 262.
16. Gallicchio V: A modified benzidine method for the staining of bone marrow. Lab Med 6:15, 1975.
17. Kincade PW: personal communication, 1977.
18. Iscove NN: The role of erythropoietin in regulation of population size and cell cycling of early and late stage erythroid precursors in mouse bone marrow. Cell Tissue Kinet 10:323, 1977.

11 Production of Erythroid Precursor Cells (BFU) *In Vitro*

N. G. Testa and T. M. Dexter

Introduction

A long-term bone marrow culture that allows sustained proliferation of multipotential stem cells of mice (CFU-s)[1] has been useful for the study of some of the mechanisms controlling haemopoiesis.[2,3,4] In this system, production of granulocytic[2,3,5] and megakaryocyte[6] precursor cells occurs concomitantly with stem cell proliferation. These cells, in turn, undergo proliferation and maturation to fully mature granulocytes, macrophages, and megakaryocytes.[2,3,4,6] Morphologically recognizable erythroid cells are absent. However, since *in vitro* colony assay systems are available that permit the study of erythroid progenitors, it was of interest to determine whether such precursor cells are present in the cultures, and if so, at what stage erythropoiesis was halted.

The present study was carried out to determine if BFU (burst-forming unit),[7] which has extensive proliferation capacity[7,8,9] and is thought to be closely related to the stem cells,[8] and CFU-e, a more differentiated cell,[7,8] are present in these cultures.

Materials and Methods

Femora were removed from 8-week-old BDF_1 female mice, and the contents of a single femur were flushed into each of 10 culture bottles (United Glass, London, England) containing 10 ml of Fischer's medium (Gibco Chagrin Falls, Ohio) plus 25% horse serum (Flow Labs Rockville, Maryland) with 500 units/ml of benzyl penicillin and 50 μg/ml of streptomycin sulphate.

The cultures were gassed with 5% CO_2 in air and were kept at 33°C. They were fed at weekly intervals by removing one-half of the growth medium, and adding an equal volume of fresh growth medium. After 3 weeks, all the growth medium was removed and 10^7 fresh bone marrow cells were inoculated into each culture. This is considered day 0 of the culture. At weekly intervals thereafter, the cultures were fed as described above, and the pooled suspension cells removed were assayed for colony-forming cells.

Colony Assays

The CFU-s assay was performed as described previously.[3] The BFU assay was performed according to the technique described by Iscove and Sieber.[9] Before plating, the cells were washed twice in alpha medium (Gibco) with 2% fetal calf serum (Tissue Culture Associates) to remove traces of horse serum which may be inhibitory for BFU growth. Cells were cultured at 2×10^5 cells per plate in one ml of alpha medium containing 0.8% methylcellulose, and supplemented with 30% fetal calf serum, 10^{-4} M 2-mercaptoethanol, 1% deionized bovine serum albumin (Fraction V, Sigma Chemical Co., St. Louis, Missouri), 3.4×10^{-6} M human transferrin (Sigma Chemical Co.) half-saturated with iron, and 10^{-7} M sodium selenite.[10] Each culture contained 2U erythropoietin (Step III sheep plasma Epo, Connaught). The cultures were incubated for 9 days at 37°C in a sealed box in a humidifed atmosphere of a mixture of air plus 5% CO_2.

The same technique was used for the CFU-e assay, but only 0.5 units Epo per culture was used routinely and the cultures were scored after 2 days. Both types of colonies were recognized by their characteristic morphology[7,8,9] and were scored under 75× magnification. As a check, some were picked, placed on slides and stained with benzidine. They were found to contain hemoglobinized cells.

Results

Figure 11-1 shows the results of weekly determinations of BFU in a set of 10 replicate cultures. BFU were present until the cultures were terminated after 9 weeks. Their numbers, however, were consistently low after 6 weeks. However, a single experiment in which the numbers of BFU were determined 1 and 7 days after feeding (Table 11-1) indicates that immediately after feeding there is a marked increase in BFU, which is not detected if the assays are carried out 7 days after feeding (as was the case in the data shown in Figure 11-1).

Assays for CFU-e in this, and in four other sets of cultures investigated after 5 or more weeks in culture were consistently negative (Table 11-2). Also, repeated morphological examinations of the harvested cells failed to detect recognizable erythroid cells. As the long-term cultures are kept at 33°C (a temperature shown to be optimal for stem cell maintenance) and grandulopoiesis[11] control experiments were carried

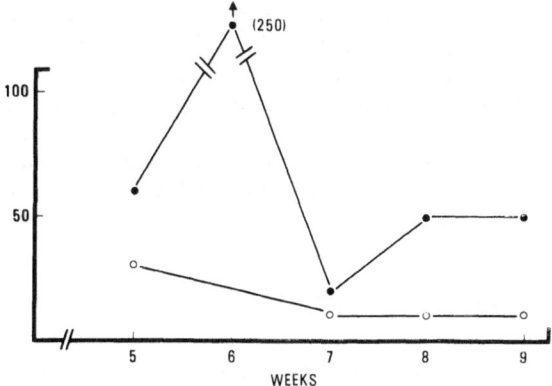

Figure 11-1 Numbers of BFU and CFU-s per culture determined in pooled cells from a set of ten replicate cultures. Closed circles, BFU; open circles, CFU-s.

out to determine if erythroid differentiation in methylcellulose can take place at that temperature. The results in Table 11-3 indicate that similar numbers of BFU and CFU-e were detected in fresh bone marrow cells cultured at 37°C or at 33°C. Furthermore, there was no obvious difference either in the size of the colonies or the degree of hemoglobinization.

Addition of Epo *directly* to the long-term cultures, however, did not result in the appearance of morphologically recognizable erythroid cells. In fact, the addition of 2 units of Epo per ml resulted in decreased granulopoiesis and increased numbers of mononuclear cells. This was so whether the Epo was added dissolved in Fischers' medium plus horse serum (the growth medium used for the long-term cultures) or in the growth medium used for the BFU assay (i.e., alpha medium plus fetal calf serum, bovine serum albumin, transferrin, sodium selenite, and 2-mercaptoethanol).

Discussion

The present results confirm and extend earlier work[12,13] that showed that BFU are present for several weeks in long-term bone marrow cultures. Although the numbers detected were not high, the plating efficiency in control assays using fresh normal bone marrow was only 5 to 10 per 10^5 cells. This, together with the fact that the dose of Epo used gives a

Table 11-1 Effect of feeding on BFU numbers.

Weeks in Culture	Cell Count × 10^{-6}	Total BFU
8[a]	1.14	11
8 + 1 day[b]	2.06	47
9[a]	2.06	10

[a] Cells assayed 7 days after feeding.
[b] Cells assayed 1 day after feeding.

Table 11-2 BFU and CFU-e in long-term bone marrow cultures.

Age of Cultures (weeks)	Number of Cultures Assayed[a]	BFU	CFU-e
5	2	Present	Absent
6	2	Present	Absent
5–9[b]	1	Present	Absent
11	1	Present	Absent

[a] Pooled cells from 4 to 10 bottles per point.
[b] Weekly determinations (See Figure 11-1.)

constant, but not necessarily maximum plating efficiency,[8] indicates that the real numbers of BFU in the cultures are likely to be higher.

The data in Figure 11-1 indicate that the BFU can be maintained in the cultures for as long as CFU-s,[11,12] although their eventual decline as the culture ages is more noticeable. However, all the assays were carried out 7 days after feeding, a time that is not optimal to detect the marked increase in numbers that the feeding induces. More frequent feeding and earlier sampling after feeding is likely to result in higher numbers of BFU being detected.

As each weekly feeding halves the numbers of suspension cells in the cultures, the maintenance of BFU numbers indicates that new BFU are produced. That this occurs over a period of several weeks suggest that they arise from differentiation from the CFU-s.

One striking observation concerns the size that individual bursts achieve: most of those derived from cultured cells are comparable to the biggest derived from fresh bone marrow, which suggest that the BFU population in long-term cultures are comparable not to the whole BFU compartment in the bone marrow but only to the most immature (i.e., with the highest proliferation capacity). The fact that they are detected in long-term cultures grown without exogenous Epo (and with Epo levels in the horse serum of the order of 0.01 to 0.05 units/ml) indicates that *in vitro*, as *in vivo*[13] differentiation from stem cells into early BFU is Epo independent. The absence of

Table 11-3 BFU and CFU-e in bone marrow cultures kept at 37°C or 33°C.

	No. 10^5 Cells	
	37°C	33°C
BFU	5	5
CFU-e	207	179
	161	184
	206	207

Each number represents the average count of two dishes. Results of four separate experiments are shown.

CFU-e agrees with the notion that they are relatively mature, Epo-dependent cells[8], and further indicates that the block in maturation, along the erythroid pathway, takes place at some stage between the BFU and CFU-e. The uniformly big size of the bursts suggests that the maturation block is probably at the level of the earliest (primitive) BFU stage.

This indicates that there may be a differentiation sequence (between the BFU stage and the beginning of Epo dependence) that does not take place under the present culture conditions. Further support for this view is given by the observation that replacement of the horse serum plus Fishers' medium for the growth medium used for the BFU assay plus Epo (2 units/ml) failed to induce recognizable erythroid differentiation when added directly to the long-term cultures. This suggests that the adherent cells may modify the response of the BFU to the Epo preparation. Whether it is by interaction with the Epo itself, or with the other molecules present in the Epo preparation (which has a specific activity of only 2 units/mg), remains to be elucidated. Using the culture system described, many of the questions raised are now amenable to study.

Acknowledgment

This work was supported by grants from the Cancer Research Campaign and the Medical Research Council.

References

1. Till JE, McCulloch EA: A direct measurement of the radiation sensitivity of normal mouse bone marrow cells. *Radiat Res 14*:213, 1961.
2. Dexter TM, Lajtha LG: Proliferation of haemopoietic stem cells *in vitro*. *Br J Haemat 28*:525, 1974.
3. Dexter TM, Testa NG: Differentiation and proliferation of haemopoietic cells in culture, in Prescott DM (ed): *Methods in Cell Biology Vol. XIV*. New York, Academic Press, p. 387, 1976.
4. Allen TD, Dexter TM: Cellular interrelationships during *in vitro* granulopoiesis. *Differentiation 6*:191, 1976.
5. Bradley, TR, Metcalf D: The growth of mouse bone marrow cells *in vitro*. *Aust J Exp Biol Med Sci 44*:287, 1966.
6. Williams N, Jackson H, Sheridan APC, Murphy MJ, Elste A, Moore MAS: Regulation of megakaryopoiesis in long-term bone marrow cultures. *Blood 51*:245, 1978.
7. Axelrad AA, McLeod DL, Shreeve MM, Heath DS: Properties of cells that produce erythrocytic colonies *in vitro*, in Robinson WA (ed): *Proceedings of the Second International Workshop on Haemopoiesis in Culture*. Washington DC, US Government Printing Office, p. 226, 1974.
8. Gregory CJ: Erythropoietin sensitivity as a differentiation marker in the haemopoietic system: studies of three erythropoietic colony responses in culture. *J Cell Physiol 89*:289, 1976.
9. Iscove NN, Sieber F: Erythroid progenitors in mouse bone marrow detected by macroscopic colony formation in culture. *Exp Hemat 3*:32, 1975.
10. Gilbert LJ, Iscove NN: Partial replacement of serum by selenite, transferrin, albumin and lecithin in haemopoietic cell cultures. *Nature 263*:594, 1976.
11. Dexter TM, Allen TD, Lajtha LG: Conditions controlling the proliferation of stem cells *in vitro*. *J Cell Physiol 91*:335, 1977.
12. Testa NG, Dexter TM: Long term production of erythroid precursor cells (BFU) in bone marrow cultures. *Differentiation* (in press).
13. Iscove NN: The role of erythropoietin in regulation of population size and cell cycling of early and late erythroid precursors in mouse bone marrow *Cell Tissue Kinet 10*:323, 1977.

Discussion (Chapters 7–11)

Dr. Gordon: I wondered whether any consideration, Dr. Moriyama, has been given to changes in the specific activity of the labeled iron that you use for measuring heme synthesis. Iron could conceivably be liberated from some of the red cell extracts you have used. This "cold" iron would dilute the radioactive iron you inject, thus yielding falsely low values for ^{59}Fe incorporation into heme. Has this been considered in the interpretation of your data?

Dr. Moriyama: I have measured the levels of cold iron in our red cell extracts and I have not found detectable cold iron levels in these extracts when measured by the method of Landers and Zak.

Dr. Sassa: I think that Dr. Gordon has made a good point. Since ^{59}Fe uptake has been done as a tracer study, its uptake is highly dependent on cold iron concentrations in the extract. It can be very variable, particularly when you use the extract from hemoglobin or red cells as in this study. Thus, it is necessary to establish how much iron is present in the extract.

Dr. Murphy: Dr. Moriyama, is your red cell extract that comes from human red cells also effective when incubated with mouse bone marrow?

Dr. Moriyama: No, I have not tested it with mouse marrow.

Dr. Murphy: It might be very interesting to try this red cell extract in such a system as Friend virus-transformed erythroleukemia, in which the enzymes of heme synthesis are well characterized in their sequence of activities. It would be very interesting to see at which point the red cell extract did have its inhibitory influence.

Dr. Sassa: I would like to comment about the effects of heme on ^{59}Fe uptake, reported by Ponka and Neuwirt in Prague, that small concentrations of heme inhibit the iron uptake by the reticulocytes or erythroid precursor cells. The effect of heme on the iron uptake is that heme inhibits the release of iron from the iron binding sites on the cell membrane. It is thus complicated to study ^{59}Fe uptake into heme in the presence of exogenously added hemin. One could follow in such a case the fractional uptake of ^{59}Fe in hemoglobin heme as percent of total ^{59}Fe bound to the cells.

Dr. Fisher: I might mention that when Dr. Moriyama reported this inhibitor in serum of transfused rabbits while he was working in our laboratory, he did in fact measure the serum iron levels in both transfused and normal rabbits and they were not significantly different.

Dr. Murphy: A singularly unique feature of Dr. Moriyama's presentation and certainly a refreshing one is the demonstration of a factor that actually has no effect on CFU-e and BFU-e!

Dr. Ogawa: Dr. Hutchinson in my laboratory carried out the following experiment: He injected endotoxin into mice and measured BFU-e and CFU-c kinetics. The changes of BFU-e and CFU-c were very similar. BFU-e and CFU-c declined in the marrow and increased into the spleen. It seems that no matter what hemopoietic stimulation is given to mice, BFU-e and CFU-c migrate from marrow to spleen. Data from many kinetic studies may simply reflect the fact that murine marrow is packed and that it has to release precursors to spleen to accomodate hemopoietic expansion. And yet, we don't quite know the relative importance of the spleen, marrow, and blood compartment in terms of their contribution to the total hemopoietic system. I don't know how to approach this problem, but I have been somewhat disappointed with the kinetic approach in the study of so-called "stem-cell competition."

Dr. Erslev: I cannot disagree with what you have said. However, we did find an increase in spleen CFU-s and CFU-M without a concomitant decrease in bone marrow CFU-s, suggesting a total increase in these cells, again suggesting the elimination of a competitive effect of CFU-e.

Dr. Peschle: I would like to go back to the interesting comment by Dr. Ogawa in regard to the migration of BFU-e from marrow to spleen. We have observed that a polycythemic mouse injected with either exogenous Ep or testosterone shows an early wave of amplification of the BFU-e pool. This is followed by a decrease down to lower than normal values: the phenomenon is at least partially due to BFU-e migration from marrow to spleen. This seems to be a general mechanism: when the BFU-e population in marrow is expanded beyond a certain critical point, BFU-e migration from marrow to spleen initiates, thus allowing erythropoiesis to expand further in the spleen, which acts as a large reservoir of erythroid microenvironment for stress erythropoiesis.

Dr. Fisher: I wonder if you have any idea about the mechanism for your inhibition of CFU-e and radioiron incorporation in red cells? Do you think that it is in the red cell itself? Have you ever taken serum or plasma, for example, in the transfused animal and reintroduced it into a recipient animal and measured the CFU-e to determine whether the inhibition is due to the serum or the red cells?

Dr. Erslev: I have spent too much time trying to look

for this inhibitor in polycythemic plasma to want to go through it once more. I have never observed an inhibitory effect of polycythemic plasma on peripheral blood counts, and I am somewhat reluctant to try to look for an effect on stem cells. However, it ought to be done by someone else. I still have great trouble fitting our present concept as expressed by Dr. Iscove, to the reduction in CFU-e's found regularly after transfusion polycythemia. If CFU-e's are in a continuous turnover as suggested by a constant suicide rate regardless of the erythropoietin concentration in circulating blood, there should be a high death rate of CFU-e's in hypertransfused animals but still a normal number of responsive cells available. However, that is not so since there is a very striking reduction in the total number of CFU-e. I think there's something wrong with the hypothesis suggesting a continuous, autonomic, non-erythropoietin dependent turnover of CFU-e. I believe that the rate of the turnover must be dependent on the presence of erythropoietin.

Dr. Ogawa: I agree with Dr. Erslev's comment completely. Dr. Iscove, have you studied suicide rates of 3 or 4 day BFU-e?

Dr. Iscove: I haven't but Connie Gregory has. In her abstract for the Basel meeting, the day 3 BFU-e percentage killed with ^{3}HTdR was intermediate between the eight day BFU-e and CFU-e. It was about 50%.

Dr. Ogawa: Have you examined, or did Dr. Gregory examine if the suicide rate is changed in anemic or polycythemic models?

Dr. Iscove: I do not recall.

Dr. Adamson: Those of you who attended the meeting in San Francisco have seen some work that was presented on the influence of anemia on the class of intermediate burst-forming cells that grow from mouse marrow on days 3 to 5. In our laboratory, the suicide rate of that particular colony-forming unit in the normal animal, which is an LAF-1 female, is 50% and rises in response to a single bleed, to over 80% by day 3.

Dr. Murphy: May we end on a nonkinetic question, Allan, with regard to the data on CFU-M. We monitor not only for megakaryocytic colonies, but also examine whether these colonies are indeed producing and secondly shedding platelets.

Dr. Erslev: We did not look. Dr. McLeod had just published a paper in which he saw platelet production in these colonies, and certainly morphologically they look like perfectly normal megakaryocytes. By the way, we have not needed to add any stimulating factors, erythropoietin or some factors related to lymphocytes; in our hands simple semisolid culture was enough to support the growth of CFU-M.

Dr. Iscove: Do you have kinetics for the appearance of the burst-promoting activity, after phenylhydrazine for example? And do you have any molecular characterization?

Dr. Wagemaker: Not yet. We only demonstrated its existence in some of these mice sera and we thought that worth communicating because it apparently is quite difficult to find such substances.

Dr. Iscove: Can you rule out that you are looking at erythropoietin?

Dr. Wagemaker: The evidence that we are not looking at erythropoietin is (1) that a few μl, of serum in the cultures will give optimal enhancement of bursts, (2) that the mouse serum does not stimulate bursts in the absence of added Ep, and (3) that Ep is already present in the cultures in saturating amounts.

Dr. Golde: Do you know which cell is making the feeder activity. It is adherent or phagocytic?

Dr. Wagemaker: I don't know. The burst-enhancing activity in MLCM seems to be closely related to human CSA.

Dr. Golde: If that's the case, then it may implicate the monocyte.

Dr. Wagemaker: Exactly, yes.

Dr. Zanjani: We were talking about human peripheral blood BFU-e; these studies are done in bone marrow.

Dr. Testa: Can you separate the burst-feeding activity from the colony-stimulating activity for granulocytic colonies in mouse lung conditioned medium?

Dr. Wagemaker: As far as the enhancement produced by mouse lung conditioned medium over CSF-PMUE type stimulators is concerned, MLCM and lung conditioned medium factor are nonadditive but we didn't try it for BFU-e, so I can't give you specific data about that.

Dr. Iscove: David Houseman presented some relevant data at Cold Spring Harbor a few weeks ago. When he depleted human peripheral leukocytes of E-rosette-forming cells, presumably T cells, then bursts did not form in erythropoietin-containing cultures of the residual cells. The supernatant of tetanus toxoid-stimulated peripheral blood cells (added) to the T cell-deficient cultures restored burst formation. That evidence suggests that the activity is produced by T cells.

Dr. Wagemaker: The physical data I presented do not exclude that at all. Actually the size distribution of burst feeder cells is quite compatible with lymphocytes. But we don't have information about the morphological appearance of burst feeder cells.

Dr. Zanjani: Perhaps this will complicate matters a bit more. It turns out, at least in our hand, that if you deplete the peripheral blood mononuclear cells of its adherent cell population, following both adherence in a plastic dish and after ingestion of carbonyl iron, and then subject the remaining cells to SRBC-rosetting so that you remove most of the T cells, then the remaining population can still form erythroid colonies (BFU-e). So that there may be an interaction between the T cell population and the adherent cells to bring about the effect that was mentioned by Dr. Iscove.

Dr. Murphy: Have you tried freeze-thawing your bone marrow, or any other kind of cell disruption, and if so, do you also get BFA?

Dr. Wagemaker: No, we did not try freeze-thawing on the bone marrow. We irradiate the bone marrow just to be able to study BFA independent of BFU-e, or, more directly, to get rid of BFU-e.

Dr. Murphy: A philosophical question. In your first figure did lines indicate an action or a putative action of erythropoietin on BFU-e, and then between BFU-e and CFU-e you place the ERC. Are you implicating, then, that there is a cell which is responsive to Ep but otherwise indistinguishable between a BFU-e and a CFU-e?

Dr. Wagemaker: Well, as far as current evidence goes, BFU-e do not seem to be responsive directly to erythropoietin, so from that point of view I think we should exclude BFU-e from the *in vivo* ERC population. Referring now to CFU-e, that population is to a large extent, but not completely, suppressed by experimental polycythemia and, by definition that does not apply to ERC. The model I showed is for descriptive purposes and I cannot exclude considerable overlap between these cell populations.

Dr. Murphy: It's interesting that you bring this up, because if it really is an erythropoietin responsive cell, it is to be called an ERC, by definition. And if the CFU-e are the only cells responsive to erythropoietin, then they should not be equatable to the BFU-e.

Dr. Peschle: I noticed that you have a plateau for BFU-e growth starting from one unit of erythropoietin per plate. This is of course distinctly less than that reported by most investigators. What do you think is the cause of this difference?

Dr. Wagemaker: We have always found optimal levels of erythropoietin to be approximately 1 to 2 units and when using most of the Connaught preparations, we see a decline in burst number with higher concentrations. With human urinary erythropoietin, we found the same value of 1 to 2 units, but a stable plateau up to 10 units. So I think these discrepancies depend on different culture conditions used by different investigators, rather than a difference in Ep-responsiveness by itself. I don't know which particular conditions are important.

Dr. Iscove: This is a comment. CFU-e and 10 day BFU-e are not the only cells in the system. There appears to be a continuum of cells, at least in terms of the observed colony formation. There are colonies which stop proliferating at 48 hours, others at 3 days, some at 4 days, and so on, right out to the largest ones that are just beginning to hemoglobinize on the tenth day. So I think that there's a maturational continuum of colony-forming cells. The traditional "ERC" are likely to be a spectrum of cells located in that continuum somewhere between CFU-e and the most primitive BFU-e.

Dr. Gordon: Is there not a remote possibility that the factor you are dealing with is a product of hemoglobin destruction.

Dr. Wagemaker: Yes.

Dr. Gordon: Because even in the exhypoxic mouse, which obviously has an excess of red cells, these cells are undergoing rapid destruction, and some of the products are entering the circulation. Have you tested products of hemoglobin destruction?

Dr. Wagemaker: I have not tried that. The only thing we know is that the activity of burst feeder cells is changed by infusion of latex particles, which might indicate, but this is only speculation, that it's a phagocytic cell. I can't exclude that burst enhancing activities are a product of hemolysis. In our laboratory hemolysate is used to stimulate a specific subpopulation of BFU-e, but it did not appear to be active on all bursts.

Dr. Murphy: Is the burst feeder cell a medium to large cell?

Dr. Wagemaker: The size ranged from 6.7 to 2.6 mm.

Dr. Murphy: It is therefore a small cell.

Dr. Wagemaker: Yes.

Dr. Murphy: But it seems to be a phagocytic cell, is that right?

Dr. Wagemaker: I know for sure that it is a rather small cell. I don't know whether it is a phagocytic cell, but there is a suggestion that it is a phagocytic cell. But I agree that the size does not agree with a phagocytic cell.

Dr. Zanjani: The question is essentially directed to the group as a whole, and I guess I can first direct it to Dr. Murphy. His finding that what looked like CFU-e in methylcellulose turned out to be something different may indicate a real difference between the plasma clot culture system and the methylcellulose assay, and this may explain some of the differences we are beginning to observe. For example, there have been some reports that normal human bone marrow cells form colonies in the absence of exogenous erythropoietin in methylcellulose and I don't think that's the case with the plasma clot culture system. And then there is our own inability to grow what we call fraction I cells in the plasma clot to form a consistently sufficient number of BFU-e's, whereas of course Dr. Ogawa and others do get, with the routinely separated peripheral blood cells, formation of BFU-e. I wonder if we shouldn't begin to define the two systems that are involved, that care should be taken to describe the procedures and then the culture system, and begin to, under certain circumstances, define different population of cells.

Dr. Murphy: That's the reason why we're all here. The introduction to each of the grant proposals I wrote for this meeting stated precisely that. With regard to methylcellulose and the alleged appearance in the absence of exogenously added erythropoietin of CFU-e in methylcellulose we do not find it. These colonies that actually arose from the continuous marrow culture are definitely different from any other colony we have ever seen. This is very important and one of the reasons why we've asked all of you to be as fastidious as possible in writing the materials and methodology section which will appear as an appendix to this volume.

Dr. Zanjani: Perhaps I can direct this question to Dr. Golde. Unless I'm mistaken, you did publish that in methylcellulose, normal human bone marrow cells, and mouse too, form erythroid colonies in the absence of erythropoietin?

Dr. Golde: No, I didn't say that. I said in the absence of added erythropoietin. Fetal calf serum which as you know contains erythropoietin might be the reason.

Dr. Zanjani: Obviously that is not the same amount of erythropoietin we usually add to the culture.

Dr. Golde: That's clear, but there's erythropoietin present in the dish. Even if we neutralize it we still get a few colonies.

Dr. Zanjani: That was the point. Is the methylcellulose system therefore more sensitive than the plasma clot culture system?

Dr. Golde: First of all, with regard to the staining, there must be other kinds of cells because you can't even demonstrate peroxidase in the macrophage cytochemically with the peroxidase staining, so this is another problem. The usual cytochemical difficulty with benzidine is slight staining of the eosinophil, the brown peroxidase, and I guess if you had a lot of esosinophil colonies and you weren't aware of the problem, you could be confused. In the methylcellulose system, as you know, you can pick the colonies out; you can even take the whole dish and spin it down with a cytocentrifuge and there's absolutely no doubt that those colonies are hemoglobinized colonies. About the concept of growing without erythropoietin, first of all we take the cells from an individual human or mouse that has circulating erythropoietin, and if you like the receptive theory, and I do, you can postulate that erythropoietin is bound to a certain number of these receptors at the time you remove the cell from the host. That is why we cannot eliminate this background cloning.

Dr. Zanjani: But the fact remains that in plasma clot you don't get those colonies appearing as frequently. We don't get any endogenous colonies with normal marrow, so there appears to be a difference between the two culture systems.

Dr. Iscove: I doubt that there's any essential difference in this point. I think whether or not you see "spontaneous" colonies is a function of the particular batch of serum that you use. In my hands, erythroid colonies in methylcellulose without erythropoietin are extremely rare, and certainly do not occur in a serum-free system. The other point that I'd like to make concerns the acidic benzidine reaction with unfixed cells *in situ*. This reaction is not really specific enough to be useful. Although the blue color develops rapidly with hemoglobin, it also forms with myeloperoxidase. To use it is to ask for trouble. In the case of human marrow in methylcellulose, there's much less of a problem than with mouse cells. Human erythroid colonies hemoglobinise very well in culture. Colonies from CFU-e can be scored by counting obviously orange or red colonies at 6 or 7 days. Colonies from primitive BFU-e can be scored between 14 and 21 days by placing the dish on a white background and counting the red spots with the unaided eye.

Dr. Testa: The number of background colonies depends on the batch of serum one uses. If you change a batch of sera, you may have a different background.

Dr. Peschle: I would like to extend this concept on the presence of Ep in serum. Ep is also present in the serum which you are using in your liquid phase culture system. Thus, BFU-e proliferation there may not be totally Ep-independent.

Dr. Testa: That's what I said. There is a small amount of erythropoietin in the culture, but there is not an amount of erythropoietin which will allow proliferation of the BFU-e in the standard assay.

Dr. Iscove: Since horses are bled repeatedly for serum, it is quite possible that some are anemic and that the serum contains elevated levels of erythropoietin.

Dr. Tests: We've actually measured the erythropoietin levels in this batch of horse serum and it's of the order of 0.05 to 0.10 of a unit per ml. Now if it's used at 25%, that will put the level of erythropoietin something around 0.02 units per ml.

Dr. Fisher: I would like to ask Dr. McLeod and Dr. Iscove about the appearance of colonies in the absence of erythropoietin. David, did I understand you correctly that even when you add anti-erythropoietin to your culture that you still get a few colonies? I just wonder if there are cells that have been triggered into cycle *in vivo* and when you carry them over into the culture they go on to proliferate and form colonies. Dr. Iscove, have you ever abolished the low basal level that you see in the methylcellulose system with anti-erythropoietin.

Dr. Iscove: I'd like to stress again that the background in methylcellulose is usually zero in my hands.

Dr. Golde: I'd like to make a comment about the use of antibody. It's very easy to get the appropriate results with the antibody because many of them are cytotoxic. If you want to get rid of the colonies, many antibodies will work. Yours, Jim, is not cytotoxic, but Esmail has another technique of doing it that I think gets around the problem of cytotoxic antibodies.

Dr. Zanjani: We use IgG from a relatively potent antibody preparation, and then remove the excess antibody with goat anti-rabbit gamma globulin, and use the supernatant. Like David said, the data on the polycythemic patients we published show that there are some colonies still around after anti-Ep treatment, which really indicates that we are not destroying cell *per se* by cytotoxicity, but actually preventing formation of cells.

Dr. Ogawa: Dr. Mike Dexter was very certain that even the best fetal calf serum which supports large burst formation in clonal culture, is not adequate for suspension culture. Did you find the same?

Dr. Testa: We have a very unfortunate experience with fetal calf serum. When practically everybody was growing CFU-c in fetal calf serum, we couldn't grow them. We had to use horse serum, and with horse serum we got, as everybody else, similar results with regard to plating efficiency. Of course I don't think Dr. Dexter has tested as many batches of fetal calf serum as of horse serum, but he certainly tested several batches, and none of the batches tested supported growth. I cannot tell you if one batch in a hundred will support CFU-s proliferation but so far every batch we've tested, or rather that Dr. Dexter tested, has been negative. I don't know why, but we can never find a batch of fetal calf serum that will support CFU-c growth. It's very difficult to find out how the serum samples are collected and stored. It may depend entirely on the manufacturer or the supplier of sera. Suppose that in England, for example, the fetal calf serum comes all the way from Australia and you don't know what's happened during transport.

Dr. Ives: I think I should stress at Connaught we've made fetal calf serum and bovine serum and have tested other manufacturers', because of our cell culture work for vaccines. There are very many differences between manufacturers, for instance the way of collection and the testing of the serum. Do you test for adventitious agents such as viruses, do you get protocols from your suppliers as to what's there, and what isn't there? Are you sure of the age of your animals that the serum is taken from, for instance: fetal calf serum can come from a truly fetal calf, or it can come from a calf that's some months old? I think that you're trying to standardize your methods here, and I think that one of the problems that you're having is possibly in your materials, even the Ep can't be standardized completely. I'd like to ask Dr. Murphy a question, and that is that he has continuous cultures of bone marrow, and he is harvesting the cells in suspension. I had always been told by my cell culture friends that a suspension culture is a transformed culture and it's no longer a true cell. Can you check that?

Dr. Murphy: When we say a "cell suspension," actually there is a very intimate association between the already adherent layer of cells and the cells that are not necessarily on top. They are often interdigitated within this adherent layer, and it requires agitation to really bring off sufficient cells for assay. But if you're asking the question whether they are morphologically differentiated, I believe I showed you one of my slides that throughout the period of our cultures there are 30 to 40% unidentifiable cells which we classified as blasts. Maybe Dr. Testa would like to comment as well?

Dr. Testa: I agree with what Dr. Murphy said. Usually cell lines which grow in suspension are thought to be transformed. Now, the cells in the long term bone marrow culture are better described as loosely attached, if you like, but they depend for proliferation and differentiation on the adherent layer. Without the adherent layer, they won't proliferate and differentiate, so it's not comparable to suspension cultures of cell lines.

Dr. McLeod: I was going to comment on the plasma culture system. Certainly with murine bone marrow or spleen cells, you get some background of erythroid colonies without adding erythropoietin. We rarely see it with the human bone marrow cells, but there can be quite a variation in background with murine cells. And the number of colonies depends at least in part on the serum used in the cultures. Dr. Ives was talking about fetal calf and I might add that I believe fetal calf serum is collected from a number of fetuses and kept for some time before being processed. Some of the serum electrolyte concentrations can be 3 to 4 times normal levels.

Dr. Trobaugh: Concerning identification of colonies, I would like to ask if any of you have used a 415 nm filter to identify hemoglobin containing cells in the colonies? We have examined a number of cultures using the Soret band and it is fascinating to see how nice and black the red cell colonies appear while the leukocyte colonies are clear. One of our biomedical engineers, Dr. Bacus, uses the Soret band to study red cells and has a very complicated set up including a large computer and a television viewing screen. However, one can view the cultures directly through the microscope. The field is very dim, dark blue, but the eyes adapt quickly and the red cell colonies appear as a solid black against this blue field. There's just no question as to whether a particular cell or colony contains hemoglobin since the hemoglobin absorbs all the light.

Dr. Testa: We did try that, and certainly you see black cells. The problem was that total light absorption was so marked that the field was too dark. I would like to know what percentage of incident light was absorbed by the filter you used.

Dr. Trobaugh: I'm sorry, I can't answer that.

Dr. Murphy: I believe I can perhaps give some indication of this. This should appropriately be a question directed to Marcel Bessis because he is one of the few hematologists who has long remembered Dr. Soret who initially described this wavelength in 1876. The filter is a $70 item, but the important thing probably for our studies is to use a xenon lamp which has a very high-intensity beam, so that your noise-to-signal ratio is very low. In other words, you have a bright background, and when you see a colony, it's not dim. So you just need a stronger source of illumination with a light transmission of 40%.

Dr. Ogawa: In the Experimental Society Meeting in Basel, Switzerland, Dr. Dexter said that he's now removing the entire suspensions, and still continues to see growth of CFU-c and BFU-e. That means these cells are at least loosely attached to the bottom. I wonder if there are any confirmation that CFU-c and BFU-e in your culture are coming from the second inoculum or are partly derived from the first marrow inoculum?

Dr. Testa: This is Dr. Dexter's work, really, but I will try to answer. That depends on the batch of serum, if the batch is good, there will be CFU-s in the adherent layer at the end of a 3 weeks period.

Instead of removing half of the growth medium, all the growth medium is removed, the adherent layers rinsed and after that, the CFU-s content of the adherent layer is measured and the finding is that CFU-s are present among the attaching cell population. It makes a lot of sense to find CFU-s in the adherent layer, because if one postulates cell-cell interactions, then one expects to find the stem cells in close contact with the adherent cells. The CFU-c are also there. Now after feeding CFU-s and CFU-c are leased into the liquid phase. If again all the growth medium is removed and replaced they are found again in the liquid phase. It means that cells within the adherent layer are capable of regenerating the culture. I expected to find BFU-e also in the adherent layer, and we did a couple of experiments in cultures in which we found BFU-e in suspension, and in fact we didn't find any in the adherent layer. Now, I don't think they're not there, I think that we weren't able to detect them. But there was a very marked proliferation of cells in the culture, even if we plated very low numbers of cells. There were lots of fibroblastoid cells, and lots of macrophage-mononuclear cells. We have heard today about macrophages stimulating and macrophages inhibiting BFU-e growth. What we saw in those experiments was a lot of macrophage proliferation, and no BFU-e growth. It may be that for some reason that marked proliferation was hindering the development of BFU-e.

Dr. Golde: On that point about the fetal calf serum, I think that those comments are very cogent. These batches have been tested by other investigators. We've tested for certain hormone content. You mention that the potassium level goes anywhere from 2 milliequivalents up to about 8. That's not the worst of it. The calcium levels are all over the place; many have endotoxin, and the hormone levels of everything from gonadoptropins to growth hormones, to insulin and what-not, are wildly disparate from batch to batch.

Dr. Peschle: We all hope that soon we shall be working with serum-free media.

12 Hormonal Modulation of Erythropoiesis In Vitro

David W. Golde

Introduction

Hormonal modulation of cell growth and differentiation is a fundamental mechanism of intercellular communication and an invariable accompaniment of multicellular organization in both plants and animals. Because of the importance of hormones in mammalian physiology, the study of their interaction at the cellular level constitutes a substantial proportion of investigations in the field of cell biology. The role of humoral factors in the regulation of blood cell production is widely appreciated. Since the discovery of erythropoietin as the major regulator of erythropoiesis, a wide body of knowledge has developed concerning the modulation of red cell production by this hormone under physiological and abnormal circumstances.[1] Erythropoietin is the primary and specific regulator of erythropoiesis; however, other hormones have important effects on red cell production.[2]

The effect of endocrine hormones on erythropoiesis has been investigated in animals and man for many years. Physiological studies on the erythropoietic effects of endocrine hormones in animals provided the knowledge on which much of our understanding of hormonal modulation of erythropoiesis was based. Although considerable information was acquired from *in vivo* experiments, recently developed clonal culture methods have permitted a clearer definition of the direct effects of various hormones on erythroid cell proliferation. In this chapter we present our recent findings relative to the hormonal modulation of normal and neoplastic erythropoiesis *in vitro* and summarize some of the pertinent work of other investigators in this area.

Materials and Methods

Bone marrow was obtained from young male white Swiss-Webster mice and from appropriately informed healthy adult human volunteers.[3,4] The bone marrow cells were isolated by centrifugation in Wintrobe tubes and cultures established for erythroid colonies using the methylcellulose technique originally described by Iscove.[5] Our detailed methodology is presented in the appendix. Basically, marrow cells were cultured in 0.8% methylcellulose with α medium, 30% fetal calf serum, and 10^{-4} M α-thioglycerol. For murine experiments, anemic sheep plasma erythropoietin (Connaught Willowdale, Ontario, Canada) was used. In human cultures, we used human urinary erythropoietin (approximately 73 units/ml per mg protein). Although various concentrations of erythropoietin were employed, the routine culture system contained 0.5 units/ml sheep erythropoietin and 1 unit/ml human urinary erythropoietin. Mouse erythroid colonies were counted with an inverted microscope at 48 hr, and human colonies were counted after 8 days. Each colony contained a minimum of eight hemoglobinized cells and the erythroid nature of these colonies was confirmed by removal of random colonies and staining with benzidine reagent. Hormones were dissolved in alcoholic, acidic, or alkaline solutions as dictated by their solubility properties, and further diluted with phosphate-buffered saline. Control cultures were included in each experiment using concentrations of diluent material equivalent to those present over the range of concentrations of hormone employed.

Friend erythroleukemia cells (clone 745) were maintained in continuous suspension culture with α medium and 20% fetal calf serum. The erythroleukemia cells were grown in T-25 flasks and divided and fed with fresh medium 1 to 3 days prior to cloning in methylcellulose. The methylcellulose culture system was the one previously described; however, 0.5% bovine serum albumin was substituted for the fetal calf serum in the culture dishes. The cloning system was serum free and no erythropoietin or other stimuli were used. The plates contained 10^4 cells and were incubated at 37°C in a humidified atmosphere of 7.5% CO_2 and air. Clusters of eight or more cells were enumerated at 72 hr using an inverted microscope.

Results and Discussion

Table 12-1, derived from our data and those of other investigators, summarizes the identified effects of nonhematopoietic hormones on erythroid colony formation *in vitro*. Androgenic and related steroids stimulate erythropoiesis *in vitro*[6,7] and the structure–activity relationships of these hormones have been clearly delineated *in vitro*.[7]

The effect of glucocorticosteroids on erythropoiesis has been studied *in vivo* and *in vitro* and the

Table 12-1 Effect of hormones on erythroid colony formation *in vitro*.

Adrenergic agonists	↑
Androgens	↑
cAMP	↑
Dexamethasone	↑
Estrogen	−
Growth hormone	↑
Human chorionic gonadotropin	−
Human chorionic somatomammotropin	±
Progesterone	−
Prolactin	−
Prostaglandin E	↑
Prostaglandin F$_\alpha$	−
Thyroid	↑

results are controversial[3,7,8] Using dexamethasone as the prototype glucocorticosteroid for *in vitro* studies because of its potency and lack of serum protein binding, we found potentiation of erythroid colony formation.[3] Dexamethasone caused a consistent increase in cloning of murine and human erythroid progenitors *in vitro* and this potentiation was blocked by appropriate concentrations of progesterone, suggesting that dexamethasone was operating through the usual glucocorticosteroid receptor mechanism. The potentiation of erythroid colony formation by dexamethasone was most prominent at suboptimal concentrations of erythropoietin. It is possible, therefore, that glucocorticosteroids may modulate the sensitivity of the response to erythropoietin at the precursor cell level. A similar permissive role for glucocorticoids has been described in other systems.[9,10] We also investigated the effect of dexamethasone on neoplastic erythroid cell proliferation. Dexamethasone potentiated human erythroid colony formation *in vitro* in patients with polycythemia vera, preleukemia, and erythroleukemia.[11,12] On the other hand, dexamethasone was a potent inhibitor of cloning of murine erythroleukemia cells and the inhibition was blocked by progesterone and deoxycortisol. This latter finding may relate to the glucocorticosteroid induction of virus in these cells, a process mediated via the usual steroid receptor mechanism.

We confirmed that adrenergic agonists potentiate erythroid colony formation *in vitro* and extensive and elegant experiments by others have shown β2 specificity for this receptor mechanism.[13,14] Since adrenergic receptor activity is linked to the cyclic adenosine monophosphate system, we looked at the effect of other modulators of intracellular cAMP concentration. In confirmation of other reports,[15,16] we also found potentiation of erythroid colony formation by dibutyryl cAMP and phosphodiesterase inhibitors. The effects of various prostaglandins are also associated with cyclic nucleotide activity. We found that prostaglandins of the E series (E$_1$ and E$_2$) potentiated murine erythroid colony formation, whereas those of the F$_\alpha$ series did not (Table 12-2). Similar results have previously been reported by Dukes.[16] Thus, it seems that hormones and pharmacological agents that cause a rise in intracellular cAMP potentiate erythroid colony formation.

We performed studies examining the effect of thyroid hormones on hematopoiesis *in vitro*. A series of thyroid analogs were tested and potentiation of erythroid colony formation was regularly noted.[17] Surprisingly, we could not find a clear correlation between calorigenicity or metamorphosing potential and stimulation of erythropoiesis *in vitro*. Thus, T$_4$ was slightly more potent than T$_3$ in the assay system, and D-T$_4$ also showed potentiation. We also found a stimulating effect of "reverse" T$_3$, which presently has no known biological activity. The observed stimulation of erythropoiesis by thyroid hormones is consistent with other *in vitro* and *in vivo* studies.[18,19] Popovic and co-workers have recently reported that thyroid hormone potentiation of erythroid colony formation *in vitro* is blocked by β-adrenergic antagonists, suggesting that the β receptor is necessary for expression of this activity of thyroid hormone.[20]

A strong relationship between pituitary function and erythropoiesis in mammals has been known for many years and there is considerable evidence suggesting that growth hormone is necessary for normal mammalian erythropoiesis *in vivo*. We studied the effect of growth hormone *in vitro* using purified hormones. Highly purified growth hormone was found to stimulate erythroid colony formation *in vitro*.[4] The growth hormone effect was demonstrable in nanogram concentrations and was shown to be species specific. Thus, both human and bovine growth hormone stimulated murine erythroid colonies, but only the human growth hormone potentiated human erythropoiesis *in vitro*. Figure 12-1 shows the effect of bovine growth hormone on murine erythroid colony formation. Clear enhancement was detected at a concentration of added hormone of 5 ng/ml and peak activity occurred between 50 and 100 ng/ml. Thereafter there was a decrease in stimulation of colony formation. Figure 12-2 shows the effect of human growth hormone on human erythroid colony formation *in vitro*. The highly purified growth hormone is compared to a growth hormone preparation obtained from the National Institutes of Health. Clear potentiation was detectable at a concentration of 50 ng/ml and peak activity was seen at approximately 100 ng/ml. Prolactin had no effect on erythroid colony

Table 12-2 Effect of prostaglandins on mouse CFU-e.

PGE$_1$	123 ± 8% of control at 5 × 10^{-7} M
PGE$_2$	129 ± 11% of control at 5 × 10^{-7} M
PGF$_{2\alpha}$	No effect
PGF$_{2\alpha}$	No effect

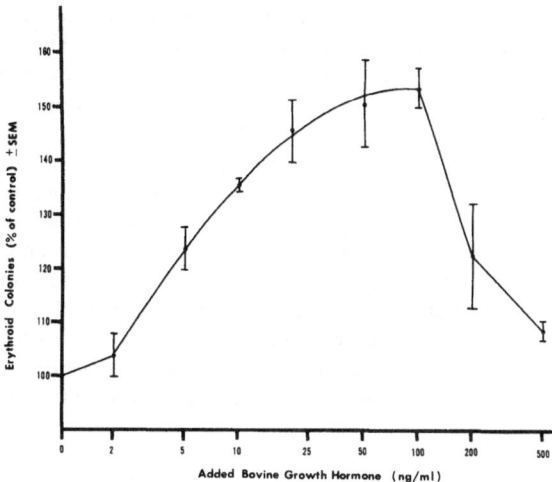

Figure 12-1 Effect of bovine growth hormone on murine erythroid colony formation *in vitro* (0.5 units/ml EP).

formation and was slightly inhibitory at concentrations above 25 ng/ml.[9] The plasmin-cleaved fragment Cys(Cam)[53]-HGH-(1-134) of human growth hormone[21] showed modest but definite potentiation at 100 ng/ml and a human chorionic somatomammotropin also showed some potentiation at this concentration.

We tested the effect of various growth hormone preparations on the cloning of murine Friend erythroleukemia cells *in vitro*. Human growth hormone potently enhanced erythroleukemia colony formation in the serum-free system with peak activity at 200 ng/ml (Table 12-3). As little as 0.5 ng/ml of growth hormone was regularly detectable in this system. An oxidized and biological inert human growth hormone produced by performic acid oxidation was found to have no effect on erythroleukemia colony formation *in*

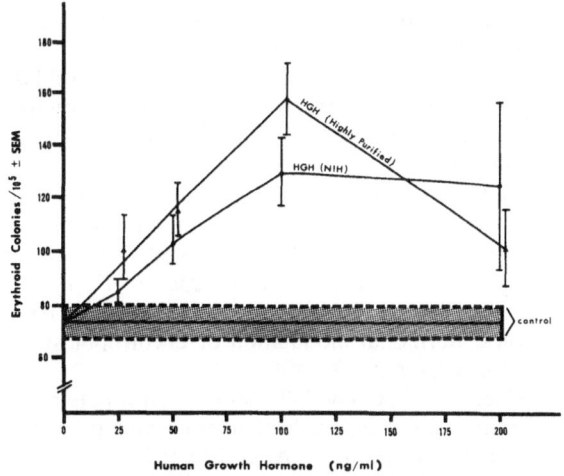

Figure 12-2 Effect of human growth hormone on human erythroid colony formation *in vitro* (1.0 units/ml EP).

Table 12-3 Effect of human growth hormone on erythroleukemia cell colony formation *in vitro*.

Growth Hormone Concentration (ng/ml)	Erythroid Colonies (% of control)
0.5	130
1.0	144
5.0	165
10.0	190
100.0	200
200.0	220
500.0	100

vitro. Human chorionic somatomammotropin and the HGH fragment Cys(Cam)[53]-HGH-(1-134) caused definite enhancement in colony formation but substantially less than that produced by human growth hormone. Thus, the effect of growth hormone on murine erythroleukemia cell proliferation roughly paralleled the effect on normal erythroid progenitors. The serum-free system, however, permits a more precise quantitation of these hormonal effects.

The Friend erythroleukemia cells showed a stimulatory response to prolactin with a maximal increase in cloning of 175% of control at 100 ng/ml. The erythroleukemia cell response therefore differed from that of the normal erythroid progenitors in that the latter was not stimulated by prolactin.

Summary and Conclusions

The results from our laboratory and those of other investigators clearly indicate that a variety of non-hematopoietic hormones influence erythroid cell proliferation *in vitro*. A number of these hormones have clinical utility and have effects on erythropoietin generation as well as direct effects on the erythroid precursor cell. Studies on normal erythropoiesis *in vitro* permit certain inferences regarding the hormonal receptors present on erythroid precursor cells. One can postulate that differentation, as determined by response to a specific hormone such as erythropoietin, may be defined as the expression of a receptor. Thus, at some time in the development of the pluripotent stem cell, signals are given leading to the expression of the postulated erythropoietin receptor. At this point, an erythropoietin-responsive cell is generated. The factors regulating feed-in from the pluripotent stem cell compartment to the committed erythroid compartment are little understood although the hematopoietic inductive microenvironment is thought to play an important role. The sensitivity of response to erythropoietin may relate to the number of receptors expressed, their affinity for the hormone, and factors identified in other hormone receptor systems, such as negative cooperativity.

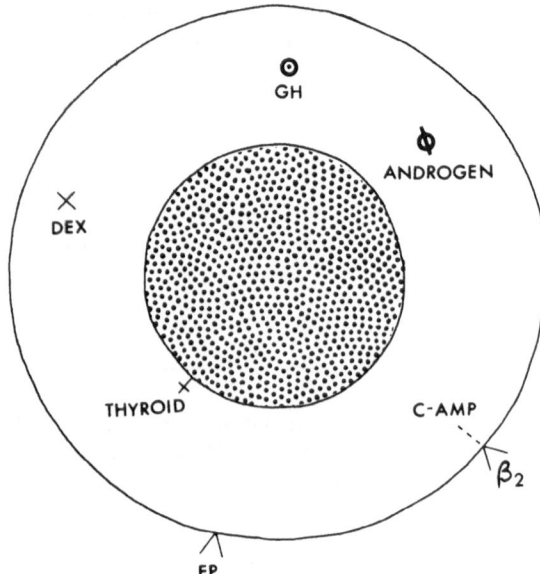

Figure 12-3 Schematic representation of an erythropoietin-responsive cell showing postulated receptors for other hormones.

The presence of other hormonal receptors may be inferred from the *in vitro* responses of the erythroid colony-forming cell. Figure 12-3 shows a hypothetical view of an erythropoietin-responsive cell indicating the presence of other hormone receptors. This is a theoretical conceptualization because the presence of these receptors has never been directly demonstrated. Nonetheless, it is useful to think of erythropoietin-responsive cells as possessing receptor mechanisms for other hormones. During malignant transformation in erythroleukemia, hormone responsiveness may be retained or altered in a manner similar to that found in other tissue such as breast and prostate. Studies of the hormonal responses of normal and neoplastic erythroid precursor cells *in vitro* will lead to a clearer understanding of the physiology of these cells and perhaps lead to important therapeutic insights.

Acknowledgments

The author acknowledges the major contributions of Noelle Bersch and Dr. C. H. Li in this work.

This work was supported by grants USPHS CA–15619, CA–15688, and RR–00865.

Human urinary erythropoietin was supplied by the Heart, Lung, and Blood Institute, National Institutes of Health. Purified growth hormones were obtained from Dr. C. H. Li of the Hormone Research Laboratory of the University of California, San Francisco.

References

1. Krantz SB, Jacobson LO: Erythropoietin and the Regulation of Erythropoiesis. Chicago, University of Chicago Press, 1970.
2. Peschle C, Marone G, Genovese A, Sacchetti L, Condorelli M: The hormonal influences on red cell production: physiological significance and mechanism of action, in Nakao K, Fisher JW, Takaku F (eds): Erythropoiesis. Proceedings of the Fourth International Conference on Erythropoiesis. Tokyo, University of Tokyo Press, pp. 99–117, 1975.
3. Golde DW, Bersch N, Cline MJ: Potentiation of erythropoiesis *in vitro* by dexamethasone. *J Clin Invest* 57:57–62, 1976.
4. Golde DW, Bersch N, Li CH: Growth hormone: species-specific stimulation of erythropoiesis *in vitro*. *Science* 196:1112–1113, 1977.
5. Iscove NN, Sieber F, Winterhalter KH: Erythroid colony formation in cultures of mouse and human bone marrow: analysis of the requirement for erythropoietin by gel filtration and affinity chromatography on agarose-concanavalin A. *J Cell Physiol.* 83:309–320, 1974.
6. Moriyama Y, Fisher JW: Effects of testosterone and erythropoietin on erythroid colony formation in human bone marrow cultures. *Blood* 45:665–670, 1975.
7. Singer JW, Samuels AI, Adamson JW: Steroids and hematopoiesis. I. The effect of steroids on *in vitro* erythroid colony growth: structure/activity relationships. *J Cell Physiol* 88:127–134, 1976.
8. Gidari AS, Levere RD: Glucocorticoid-mediated suppression of erythroid colony formation. *Clin Res* (Abs) 24:631A, 1976.
9. Gospodarowicz D: Localisation of a fibroblast growth factor and its effect alone and with hydrocortisone on 3T3 cell growth. *Nature* 249:123–127, 1974.
10. Kletzien RF, Pariza MW, Becker JE, Potter VR: A "permissive" effect of dexamethasone on the glucagon induction of amino acid transport in cultured hepatocytes. *Nature* 256:46–47, 1975.
11. Golde DW, Bersch N, Cline MJ: Polycythemia vera: hormonal modulation of erythropoiesis *in vitro*. *Blood* 49:399–405, 1977.
12. Koeffler HP, Golde DW: Vitro erythropoietin responsiveness in erythroleukemia and preleukemia. *Exp Hematol* 5 (Abs) (Suppl 2):61, 1977.
13. Brown JE, Adamson JW: Modulation of *in vitro* erythropoiesis. The influence of β-adrenergic

agonists on erythroid colony formation. *J Clin Invest 60:*70–77, 1977.
14. Przala F, Gross DM, Dargon PA, Fisher JW: Effects of *in vitro* beta-adrenergic activation on rabbit bone marrow erythroid colony forming cells. *Proc Soc Exp Biol Med 155:*334–338, 1977.
15. Brown JE, Adamson JW: Modulation of *in vitro* erythropoiesis: enhancement of erythroid colony growth by cyclic nucleotides. *Cell Tissue Kinet 10:*289–298, 1977.
16. Dukes PP: Potentiation of erythropoietin effects in marrow cell cultures by prostaglandin E_1 or cyclic 3',5'-AMP. *Intra Sci Chem Rept 6:*73–75, 1972.
17. Golde DW, Bersch N, Chopra IJ, Cline MJ: Thyroid hormones stimulate erythropoiesis *in vitro*. *Br J Haematol* (in press).
18. Fuhr JE, Gengozian N, Overton M: *In vitro* stimulation of primate hemoglobin synthesis by L-thyroxine. *Blood 49:*407–413, 1977.
19. Malgor LA, Blanc CC, Klainer E, Irizar SE, Torales PR, Barrios L: Direct effects of thyroid hormones on bone marrow erythroid cells of rats. *Blood 45:*671–679, 1975.
20. Popovic WJ, Brown JE, Adamson JW: The influence of thyroid hormones on *in vitro* erythropoiesis. Mediation by a receptor with beta adrenergic properties. *J Clin Invest 60:*907–913, 1977.
21. Li CH, Gráf L: Human pituitary growth hormone: isolation and properties of two biologically active fragments from plasmin digests. *Proc Natl Acad Sci USA 71:*1197–1201, 1974.

13 Regulatory Mechanisms of Erythroid Stem Cell Kinetics

C. Peschle, M. C. Magli, C. Cillo, F. Lettieri, F. Pizzella, G. Migliaccio, and G. F. Sasso

Erythropoietin (Ep) Influences on Erythroid Stem Cells Kinetics

Reliable methods have recently been developed to assay murine erythroid precursors in either plasma clot or methylcellulose cultures.[1-3] At least two populations of erythroid stem cells have been identified, i.e., the erythroid burst- (BFU-e) and colony-forming unit (CFU-e), which give rise to large or small erythroid colonies respectively, colonies peaking at 8 to 12 days or 36 to 48 hr after Ep addition.[3-5] Furthermore, evidence has been presented indicating that in the erythropoietic pathway the BFU-e and the CFU-e represent, respectively, an early and a late erythroid precursor.[3-7] Recently, Gregory[8] has identified a third population of erythroid stem cells, which is apparently intermediate between BFU-e and CFU-e pools.

Mechanisms regulating CFU-e kinetics have been partially elucidated. In this regard, the CFU-e compartment is increased in face of enhanced Ep activity,[9] but depleted in both transfusion-induced[6,9] and posthypoxia polycythemia[10]; the size of this pool is thus clearly correlated with Ep activity. Additionally, erythroid colony formation is strictly Ep-dependent, thereby indicating that Ep fully controls the differentiation of CFU-e to the recognizable erythroid precursors.

On the other hand, mechanisms controlling BFU-e kinetics are still under scrutiny. The present results suggest that Ep plays a significant although not exclusive role in the regulation of BFU-e kinetics. Transfusion of RBC in normal mice causes an early, although temporary depletion of the BFU-e pool, versus a sustained decline of the CFU-e number. Studies from our laboratory indicate a similar pattern for BFU-e kinetics after administration of anti-Ep serum.[9] Vice versa, injection of purified Ep in polycythemic mice induces an early amplification of the BFU-e pool, which is followed by a gradual depletion to lower-than-normal levels. It is of further interest that the expansion of the CFU-e pool induced by Ep is more delayed and prolonged than that of BFU-e.

Materials and Methods
Polycythemic Mice
In these studies CD_1 female mice weighing 20 to 25 gm were maintained on a diet of lab pellets and tap water *ad libitum*. The animals, pretreated intramuscularly with 1 mg of iron dextran, were rendered polycythemic by exposure to hypoxia (0.42 atm air for 19 hr/day) up to a total of 209 hr.

Erythropoietin
The urine from a pure red cell aplasia patient, collected at $-20°C$ and concentrated 100:1 against glycolpolyethylene at $+4°C$, was finally lyophilized (specific activity, 20 to 27 IU/mg of protein). This "crude" Ep preparation was either employed for rabbit immunization or further purified by means of the three-step chromatographic procedure reported by Iscove[5,6]; the final specific activity exceeded 500 IU/mg of protein. This "purified" Ep was virtually free of both activity-stimulating myeloid-macrophage colonies and inhibitor(s) of erythroid colony formation.

Number of BFU-e, CFU-e, and Myeloid-Macrophage Colony-Forming Unit (CFU-c)
The number of BFU-e, CFU-e, and CFU-c in tibial marrow or spleen was evaluated at sequential time intervals after administration of "purified" Ep (2.4 IU/mouse subcutaneously) or homologous RBC (1 ml/mouse of a 75% solution of RBC in saline intraperitoneally) in normal mice subjected to light ether anesthesia. Control animals received an equivalent volume of physiological saline buffered with 5% of mouse normal serum; in some studies, a 5% albumin solution was administered. Each group was comprised of three mice. The assay of BFU-e, CFU-e, and CFU-c was performed by means of methylcellulose cultures, according to a slight modification of the method reported by Iscove et al.[2-4] The animals were killed by cervical dislocation under light ether anesthesia. The number of nucleated cells in tibial marrow and splenic pulp, flushed out into Dulbecco's modified Eagle's medium, was evaluated by means of a ZB 1 coulter counter. Each 1-ml plate contained the following components in Dulbecco's modified Eagle's medium*: methylcellulose (0.8%, final con-

---

* Dulbecco's medium was modified here to contain L-alanine (25 $\mu g/ml$), L-asparagine · H_2O (50 $\mu g/ml$), L-aspartic acid (30 $\mu g/ml$), L-cysteine (70 $\mu g/ml$), L-glutaminic acid (75 $\mu g/ml$), L-proline (40 $\mu g/ml$), sodium pyruvate (110 $\mu g/ml$), vitamin B_{12} (0.025 $\mu g/ml$), and biotin.

Table 13-1 Number of BFU-e, CFU-e, and CFU-c in marrow from mice subjected to RBC transfusion at 0 hr.

Treatment	BFU-e/10^{-3}/tibia (mean ± SEM)	CFU-e/10^{-3}/tibia (mean ± SEM)	CFU-c/10^{-3}/tibia[a] (mean ± SEM)
Controls	3.0 ± 0.2	49.6 ± 2.5	19.5 ± 0.6
Transfusion + 12 hr	1.4 ± 0.3[b]	30.8 ± 0.8[b]	23.05 ± 0.5[a]
Transfusion + 24 hr	1.6 ± 0.2[b]	28.2 ± 0.1[b]	16.8 ± 0.6[a]
Transfusion + 48 hr	2.8 ± 0.1	15.0 ± 1.1[b]	17.4 ± 0.4
Transfusion + 3 days	3.2 ± 0.2	11.3 ± 0.6	18.6 ± 0.5
Transfusion + 5 days	3.3 ± 0.4	12.4 ± 0.9	19.3 ± 0.2

[a] $P < 0.05$ when compared with control group.
[b] $P < 0.01$ when compared with control group.

centration), α-thioglycerol (10^{-4} M), FCS (30 to 45%), in some studies horse serum (3%), 2 or 3 × 10^5 nucleated cells from marrow and spleen, respectively, and either Ep* or lung-conditioned medium†. Preliminary experiments indicated that these amounts of Ep or lung-conditioned medium induced maximal growth of, respectively, either BFU-e–CFU-e or CFU-c colonies in the various conditions evaluated here.

The plates were incubated in a humidified 7.5% CO_2/92.5% air atmosphere at 37°C. CFU-c colonies containing a minimum of 50 cells were scored on days 7 to 8. CFU-e and BFU-e colonies, containing a minimum of either 8 or 200 cells, were scored, respectively, at 36 to 48 hr or 10 to 14 days. The identification was performed in situ on the basis of the morphological criteria described by Iscove et al.[2,3] The validity of these criteria had been previously demonstrated by control studies involving benzidine staining of colonies smeared on glass slides.

Results

As indicated in Table 13-1, the number of BFU-e in marrow is significantly depleted at 12 to 24 hr after transfusion. Thereafter, it rebounds up to and over control values, starting from day 2 through day 5 after transfusion. This secondary rebound has already been described by Axelrad et al.[1] and Iscove.[6] However, the early depletion has not been previously reported. On the other hand, RBC transfusion causes a progressive decline of the CFU-e number, starting from 12 hr through days 2 to 3 after transfusion, which is thereafter maintained through day 5. Finally, it is noteworthy that the kinetic patterns observed after transfusion for both the BFU-e and CFU-e pools closely resemble those described after injection of anti-Ep serum.[9]

* Either Sheep Ep Step III (2 IU/plate) (4.3 to 4.6 IU/mg of protein, Connaught Medical Research Laboratories, Toronto) or "purified," human urinary Ep (3 to 4 IU/plate) were employed.
† The lung-conditioned medium (0.1 ml/plate) was prepared according to Sheridan IW, Metcalf D, Stanley ER; J Cell Physiol, 84:147, 1974.

The sustained depletion of the CFU-e compartment after transfusion has been interpreted in terms of total control by Ep of the CFU-e pool size. In line with this concept, the initial, transitory decrease of the BFU-e number after transfusion may indicate that the size of this population is partially Ep-dependent. In this last regard, an Ep-independent mechanism apparently counteracts the depleting action exerted initially by transfusion.

As indicated in Figure 13-1, administration of "purified" Ep in polycythemic mice induces an early wave of amplification of the BFU-e compartment in marrow, reaching peak values at 4 hr. Starting from 12 through 48 hr after Ep, the BFU-e number declines gradually to lower-than-normal values. Previous studies have described this secondary depletion.[10] Additionally, Ep injection induces a marked amplification of the CFU-e compartment in marrow, which initiates at approximately 8 hr and gradually progresses up to peak values at 18 to 24 hr (Figure 13-2). Thereafter, the CFU-e number declines gradually toward control values. A similar pattern for both

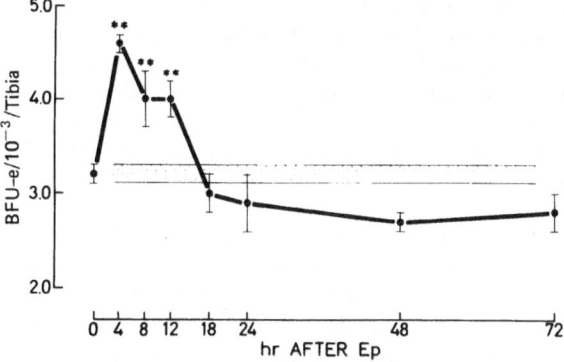

Figure 13-1 Number of BFU-e in tibial marrow from polycythemic mice after administration of purified Ep (2.4 IU/mouse). Each point represents mean ± SEM of mean values of a minimum of five groups, each comprising at least three mice. The shaded area indicates mean ± SEM of 15 control groups. ** $P < 0.01$ when compared with control values on the basis of the analysis of variance (i.e., Dunnett's test).

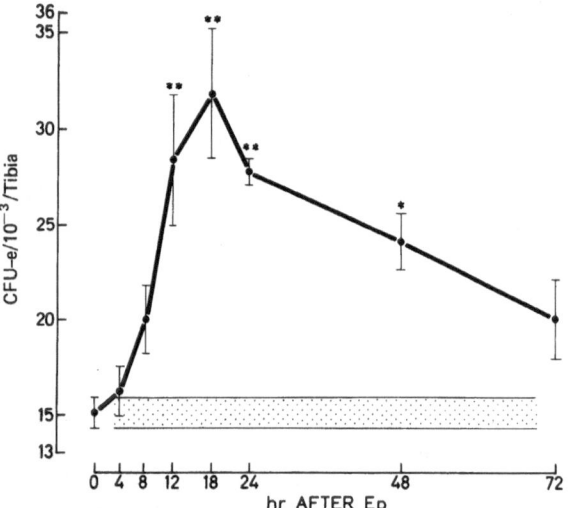

Figure 13-2 Number of CFU-e in tibial marrow from polycythemic mice after administration of purified Ep (2.4 IU/mouse). Each point represents mean ± SEM on mean values of a minimum of five groups, each comprising at least three mice. The shaded area indicates m ± SEM of 15 control groups. ** P < 0.01 when compared with control values on the basis of the analysis of variance (i.e., Dunnett's test).

BFU-e and CFU-e kinetics has been observed at the spleen level (unpublished observations).

It is thus apparent that the wave of erythropoiesis evoked by Ep in polycythemic mice is preceded by expansion of the erythroid stem cell populations. At the marrow level, the early amplification of the BFU-e pool, peaking at 4 hr, is followed by a larger expansion of the CFU-e compartment, up to 18 hr. The latter phenomenon is followed by the even larger amplification of the erythroblastic population, peaking at 48 hr, and finally by the 72-hr reticulocytosis in peripheral blood.[11,12] Therefore, the Ep stimulus induces sequential waves of amplification of progressive magnitude in the erythroid differentiation pathway, starting from the earlier erythroid precursor through circulating reticulocytes. In particular, the present studies suggest that the initial target of the Ep stimulus may be represented by the BFU-e population.

On the other hand, in preliminary studies not reported here, injection of purified Ep in polycythemic mice induced an increase of the BFU-e proliferative rate at 4 hr, as evaluated on the basis of in vitro ³H-TdR killing index.[6] These findings suggest that the 4-hr amplification of the BFU-e pool induced by Ep is at least partially mediated by enhanced proliferation of this compartment. On the other hand, it is suggested that the subsequent decrease of the BFU-e number is mediated by both a wave of BFU-e–CFU-e differentiation and marrow–spleen BFU-e migration.[10]

Conclusion

It is emphasized that the kinetics of BFU-e after Ep administration shows a pattern opposite to that after transfusion. Thus, the significance of the former studies is strengthened by the latter ones and vice versa. In this regard, evidence is presented indicating that, under physiological conditions, the amplification of the BFU-e compartment is partially Ep-dependent. Furthermore, these studies suggest that, in face of stress erythropoiesis, the BFU-e may be the primary target cell responding to the enhanced Ep stimulus.

Mechanisms Underlying Testosterone Action on Erythropoiesis

Although extensively investigated, the mechanism(s) underlying the stimulatory action exerted by testosterone (T) on erythropoiesis have not been elucidated yet. In both rodents and humans, the T action is at least partially mediated by a rise of Ep production.[13,14] On the other hand, a direct stimulatory influence of T at the stem cell level is indicated by enhanced cycling of the spleen colony-forming unit (CFU-s) at 6 to 24 hr after injection of T propionate (TP).[15]

Moriyama and Fisher[16,17] recently indicated that T causes an increase of the number of CFU-e, both in vitro[16] (i.e., on addition of androgen in human marrow cultures) and in vivo[17] (i.e., in marrow from T-treated animals). Furthermore, T has been shown to enhance CFU-e formation in culture,[18] thereby suggesting an interaction between Ep and androgen.

In the present studies, BFU-e and CFU-e kinetics in both marrow and spleen have been evaluated in polycythemic mice at sequential time intervals after a single injection of TP. Other parameters evaluated here included the erythropoietic activity, the serum Ep level, and the number of CFU-c in marrow and spleen. A second series of similar experiments involved simultaneous injection of both TP and anti-Ep serum (A-Ep).

It is suggested that the 24-hr expansion of the BFU-e pool after TP administration may be mediated by a direct stimulatory effect of androgen on stem cells, which requires, however, the "permissive" action of background Ep activity. On the other hand, the 60-hr expansion of the CFU-e compartment is apparently mediated by enhanced production of Ep, which initiates at 36 to 48 hr.

Materials and Methods

Polycythemic Mice

TP (5 mg/mouse) or its vehicle were injected under light ether anesthesia on day 6 or 7 posthypoxia. In some experiments, NRS or A-Ep were administered intraperitoneally under light ether anesthesia 30 min prior to the TP injection. The parameters considered here (erythropoietic rate; Ep level in serum; number of BFU-e, CFU-e, and CFU-c in marrow and spleen) were assessed sequentially starting from 6 through 96 hr after TP administration.

Ep and A-Ep (I and II)

A-Ep I was raised in rabbits according to a slight modification of the method reported by Schooley and Garcia.[19] In this regard, 500 IU of "crude" Ep, diluted in H_2O and emulsified in complete Freund adjuvant (1:1, v/v), were administered subcutaneously into rabbits at weekly intervals. Seven days after the fourth injection, an equivalent dosage of nonemulsified Ep was injected subcutaneously. The rabbits were bled 7 days later. Both A-Ep I and normal rabbit serum were subjected to preliminary adsorption with murine cells from both polycythemic marrow and spleen. Thus, 10^7 nucleated cells from marrow/1 ml of A-Ep I or NRS were incubated at 37°C for 30 min in a water bath with constant shaking. The cellular suspension was then centrifuged and the supernatant collected. The sera were then similarly adsorbed with the splenic cells. The final supernatant was lyophilized and stored at $-20°C$. The neutralizing titer of respectively adsorbed and nonadsorbed A-Ep I, evaluated in exhypoxic polycythemic mice,[20,21] was approximately 60 and 65 anti-human Ep IU/ml.[21]

In some studies, A-Ep II was employed: this serum had a potency of 25 anti-human Ep IU/ml.

Erythropoietic Activity in TP-Treated Polycythemic Mice

Twenty-four-hour % RBC-^{59}Fe incorporation values were employed as the basic index for red cell production. Thus, 0.5 μCi ^{59}Fe citrate, diluted in 0.2 ml sterile physiological saline, were administered intravenously starting from 12 through 96 hr after TP. Each group comprised a minimum of five to six mice. The animals were bled by cardiac puncture 24 hr after radioiron administration. Blood volume was assumed to be 7% of body weight. Mice with a final hematocrit of less than 56% were discarded.

Ep Levels in Serum of TP-Treated Polycythemic Mice

Ep levels in serum were assessed starting from 12 through 72 hr after TP on the basis of the exhypoxic polycythemic mouse assay.[20,21] Each group comprised 16 to 20 donor and 6 to 7 recipient animals. The donor mice were bled by cardiac puncture under light ether anesthesia. The blood samples from each group were pooled and centrifuged at $+4°C$; the serum was dialyzed at $+4°C$ against H_2O and finally stored at $-20°C$. Recipient animals received test samples subcutaneously on day 3 and 4 posthypoxia.

Radioiron (0.5 μCi ^{59}Fe citrate in sterile physiological saline) was injected intravenously on day 5. The 48-hr % RBC-^{59}Fe incorporation values were thereafter determined. Blood volume was assumed to be 7% of body weight. Mice with final hematocrit values of less than 56% were discarded. The radioiron uptake values were converted to IU of Ep by comparison to a log dose/response regression line obtained by injecting standard Ep (International Reference Preparation II, National Institute for Medical Research, London) at the 0.05, 0.20, and 0.80 IU dose levels.

Number of BFU-e, CFU-e and CFU-c in Tibial Marrow or Spleen from Polycythemic Mice

Results

As indicated in Figure 13-3, TP injection causes a gradual rise of the erythropoietic rate, as evaluated on the basis of RBC radioiron uptake, which initiates at 48 hr. Ep levels in serum also show a progressive increase starting from 36 hr, thus indicating that the subsequent wave of erythroid differentiation is at least partially mediated by enhanced Ep production. In line with this concept, simultaneous administration of TP and A-Ep (I or II) causes a total suppression of the erythroid response to androgen (Figure 13-4).

Figure 13-5 shows that administration of TP, associated or not with NRS, induces an early wave of amplification of the BFU-e pool in marrow, reaching peak values at 24 hr. Subsequently, the BFU-e number, although gradually decreased at 30 to 36 hr, rises toward control values at 48 to 60 hr. It is relevant that simultaneous administration of TP and A-Ep (I or II) leads to a total suppression of the early rise of the BFU-e number (Figure 13-5).

Figure 13-6 indicates that TP injection causes initially a progressive depletion of the CFU-e pool in marrow, which becomes significant starting from 30

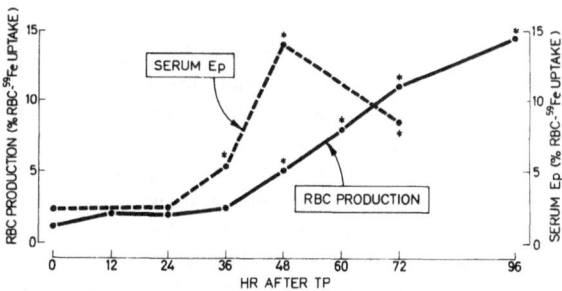

Figure 13-3 Erythropoietic activity (mean 24-hr % RBC-^{59}Fe incorporation values) and Ep levels in donor serum (mean 48-hr % RBC-^{59}Fe incorporation values in assay mice) in polycythemic animals at different time intervals after a single injection of TP. * $P < 0.01$ when compared with control values.

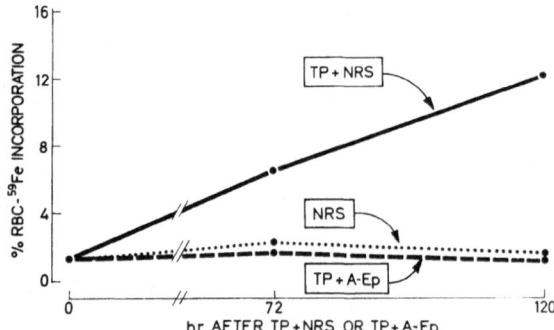

Figure 13-4 Erythropoietic activity, evaluated on the basis of mean 24-hr % RBC-^{59}Fe incorporation values, in polycythemic mice bled at 72 or 120 hr after injection of TP, associated with either NRS (I) or A-Ep (I). Some animals received NRS alone. There was a minimum of six mice per group.

For further details, see text.

through 48 hr. This is followed by a 60 to 72 hr amplification over control values. Administration of A-Ep, after enhancing the early decline of the CFU-e number, fully abolishes its subsequent increase.

As indicated in Figure 13-7, the CFU-c pool size in marrow is not significantly modified by administration of TP.

These observations confirm that the stimulatory action exerted by TP on erythropoiesis is at least partially mediated by enhanced production of Ep. In line with this postulate, the erythroid response to TP injection is preceded by a rise of Ep activity. Furthermore, the wave of erythropoiesis induced by TP is fully abolished by A-Ep.

The kinetics of erythroid stem cells in marrow of polycythemic mice given TP is characterized by: (a) the 24-hr rise of the BFU-e number and its subsequent decline to lower-than-normal levels; (b) the initial depletion of the CFU-e pool, which is followed by its 60-hr amplification.

Mechanisms underlying the 24-hr amplification of the BFU-e pool are apparently of considerable complexity.

Two lines of evidence indicate that this phenomenon is not mediated by enhanced Ep production. Thus, Ep levels in serum are not elevated at 12 to 24 hr after TP. This finding, however, does not exclude the possibility of a nonmonitored enhancement of Ep activity in the initial 24 hr after TP injection. On the other hand, it must be emphasized that the CFU-e pool in marrow is not amplified at 12 to 24 hr and is depleted at 30 to 48 hr. In this last regard, administration in polycythemic mice of as little as 0.05 to 0.1 IU of Ep, although not leading to a significant rise of bioassayed Ep activity in serum, constantly induces an increase of the CFU-e number in marrow, reaching peak values at 18 to 24 hr (unpublished observations). Lack of CFU-e amplification in the 12 to 48 hr period therefore represents strong although indirect evidence against an enhancement of Ep production in the BFU-e peak period or earlier.

On the other hand, simultaneous treatment of polycythemic mice with both TP and A-Ep causes a total suppression of the 24-hr BFU-e peak. Since enhancement of Ep production in the first 24 hr after TP injection represents only a remote possibility, this suppression indicates that background Ep activity is required to allow the 24-hr expansion of the BFU-e pool. Therefore, this amplification is apparently mediated by a direct influence of TP on stem cells, which requires, however, the "permissive" action of background Ep activity. In line with this concept, administration of A-Ep leads to abolition of the BFU-e peak via neutralization of background Ep.

In this regard Byron[15] observed that TP injection induces a 6 to 24-hr enhancement of the proliferative rate at the level of the CFU-s compartment. In view of the close relationship between CFU-s and BFU-e pools in the erythropoietic pathway, the possibility may be envisioned that TP injection enhances BFU-e proliferation, thus leading to the 24-hr rise of their number. Preliminary studies performed in our laboratory on the BFU-e cycling after TP injection, evaluated on the basis of % ^{3}H-TdR *in vitro* killing index,[4]

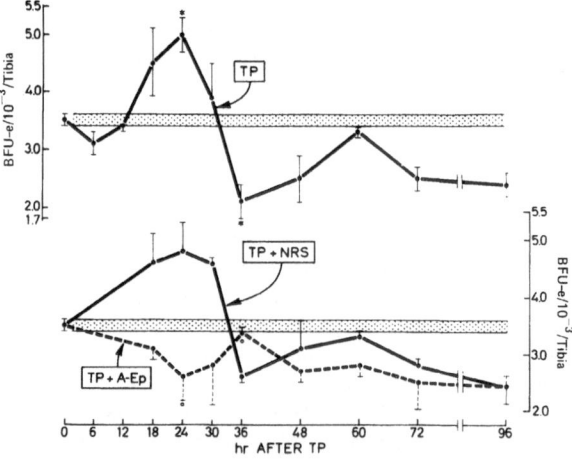

Figure 13-5 Upper panel: BFU-e number in tibial marrow from polycythemic mice after TP administration. Each point represents mean ± SEM of mean values of a minimum of five groups, each including at least three mice. The shaded area indicates mean ± SEM of 14 control groups. * $P < 0.01$ when compared with control values on the basis of the analysis of variance (i.e., Dunnett's test).

Lower panel: BFU-e number after administration of TP, associated with either NRS (I or II) or A-Ep (I or II). Each point represents mean ± SEM of mean values of a minimum of three groups, each including at least three mice (only one group/point received NRS I or A-Ep I). The shaded area represents mean ± SEM of six control groups. ° $P < 0.02$ when compared with the corresponding TP + NRS group on the basis of Student's *t* test.

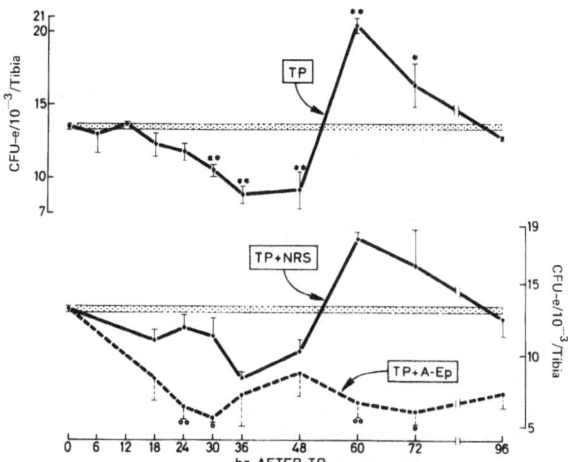

Figure 13-6 Upper panel: CFU-e number in tibial marrow from polycythemic mice after TP administration. Each point represents mean ± SEM of mean values of a minimum of five groups, each comprising at least three mice. The shaded area indicates mean ± SEM of 16 control groups. ** $P < 0.01$ and * 0.05 when compared with the control group on the basis of the analysis of variance (i.e., Dunnett's test).

Lower panel: CFU-e number after administration of TP, associated with either NRS (I or II) or A-Ep (I or II). Each point represents mean ± SEM of mean values of a minimum of three groups, each comprising at least three mice (only one group/point received NRS I or A-Ep I). The shaded area represents mean ± SEM of seven control groups. °° $P < 0.01$ and ° 0.05 when compared with the corresponding TP + NRS group on the basis of Student's t test.

Conclusion

The present results indicate that TP exerts a stimulatory action on erythropoiesis via a twofold mechanism. The early amplification of the BFU-e pool is apparently mediated by a direct influence of TP on stem cells, apparently via enhancement of CFU-s and BFU-e proliferative rates. This phenomenon, however, requires the "permissive" action of background Ep activity. An alternative, but less likely possibility envisions that TP induces an early elevation of Ep production, not monitored by the bioassay, which in turn leads to the BFU-e peak. On the other hand, the secondary amplification of the CFU-e pool is apparently mediated by the rise of Ep production, which also induces the CFU-e pool to differentiate irreversibly into the erythroblastic population.

Mechanisms Underlying Estradiol Influences on Erythropoiesis

Although it is well-established that in rodents administration of estradiol exerts an inhibitory influence on the rate of erythropoiesis,[22] the mechanism(s) underlying this phenomenon is not elucidated yet.

After priming with estradiol benzoate (EB, 5 to 10 µg/day or 150 µg/week), the Ep response to hypoxia is diminished in rats but enhanced in mice.[23,24]

On the other hand, polycythemic mice primed with EB (0.4 µg) show a reduction of the erythroid response evoked by Ep, thus suggesting an EB-

support this contention. Additionally, an elevation of the CFU-s–BFU-e input following the increased CFU-s cycling after TP injection may contribute to the expansion of the BFU-e pool.

The CFU-e peak at 60 hr is apparently mediated by an enhancement of Ep levels in serum at 36 to 48 hr. In this regard, it is well-established that elevation of either exogenous or endogenous Ep activity leads to the amplification of the CFU-e pool, peaking approximately at 18 to 24 hr.[7] On the other hand, the possibility cannot be excluded that the 60-hr CFU-e peak is partially dependent upon the 24-hr expansion of the BFU-e pool: the latter phenomenon is presumably followed by a wave of BFU-e–CFU-e differentiation, which may finally contribute to the 60-hr amplification of the CFU-e compartment. Indirect evidence supporting this mechanism may derive from the 36 to 48 hr depletion of the BFU-e pool, i.e., the enhanced BFU-e–CFU-e would certainly flux lead to a depletion of the BFU-e compartment. Further analysis of this hypothesis is hampered by present uncertainties on the time period required *in vivo* for the BFU-e–CFU-e differentiation.

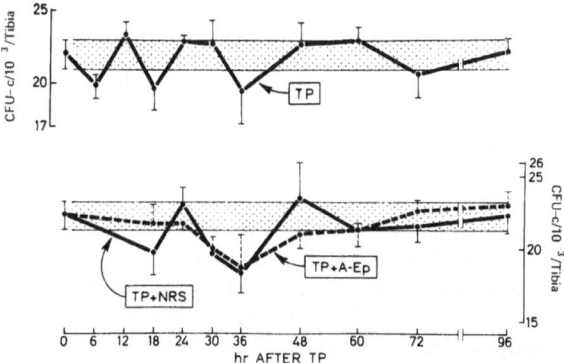

Figure 13-7 Upper panel: CFU-c number in tibial marrow from polycythemic mice after TP administration. Each point represents mean ± SEM of mean values of a minimum of five groups, each comprising at least three mice. The shaded area indicates mean ± SEM of nine control groups.

Lower panel: CFU-c number after administration of TP, associated with either NRS (I or II) or A-Ep (I or II). Each point represents mean ± SEM of mean values of a minimum of three groups, each comprising at least three mice (only one group/point received NRS I or A-Ep I). The shaded area represents mean ± SEM of six control groups.

induced inhibition of the stimulatory action of Ep at the stem cell level.[25] In this regard Fried et al.[26] observed a depletion of the spleen colony-forming unit (CFU-s) pool in marrow and spleen from EB-treated mice (150 μg/week), thereby postulating that estrogen exerts a damaging influence on the pluripotent stem cell. Finally, the possibility of an estradiol-induced migration of erythroid precursors from marrow to spleen has been considered.[24]

The present studies have been undertaken in an attempt to evaluate the CFU-e kinetics in marrow and spleen from normal mice injected with either small or relatively large amounts of EB (0.2 or 5 μg). A further investigation was focused on Ep and the erythroid response to hypoxia in polycythemic animals primed with EB (0.2, 1, 5 or 25 μg).

Materials and Methods
Normal Mice
EB, diluted in sesame oil, was injected subcutaneously under light ether anesthesia. Control mice received similarly a corresponding volume of the vehicle.

Evaluation of Erythropoietic Rate and Serum Ep Activity in Hypoxia-Exposed Polycythemic Mice Primed with EB
The mice received EB (0.2, 1, 5, or 25 μg/day) or the vehicle starting from day 3 to 7 posthypoxia. Immediately after the last injection, the animals were exposed to hypoxia (0.42 atm air for 18 hr). The erythropoietic rate was evaluated on the basis of 24-hr % RBC-^{59}Fe incorporation values. Ep levels in serum were assessed on the basis of the exhypoxic polycythemic mouse assay.[20,21] Each group comprised 15 to 20 donor and 6 to 7 recipient animals. The donor serum was dialyzed for 24 hr at +4°C against H$_2$O (volume ratio: 1/1,000) to remove EB, and then stored at −20°C.

Results
As indicated in Figure 13-8, injection of EB in normal mice at both physiological (0.2 μg) and larger dose levels (5 μg) leads to a progressive depletion of the CFU-e pool in tibial marrow, starting from 12 through at least 48 hr after estrogen. Although results are not presented here, similar depletion of the CFU-e number in marrow has been observed in exhypoxic polycythemic mice primed with 0.2, 1, 5, or 25 μg/day × 5 of EB. Of further interest is that, in normal mice injected with 5 μg of EB, the CFU-e pool in spleen, although apparently depleted at 12 hr, shows a wave of amplification peaking at 48 hr after estrogen administration (Table 13-1).

Figure 13-9 indicates that in polycythemic mice the erythroid response to hypoxia, although markedly dampened by small dosages of EB (0.2

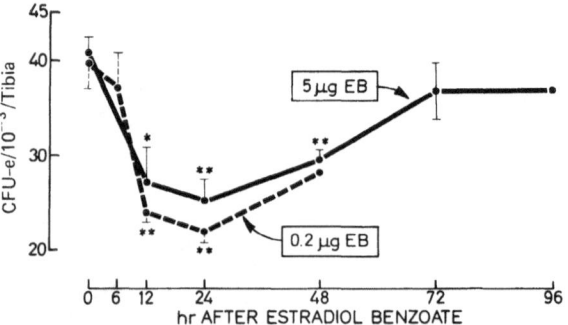

Figure 13-8 Number of CFU-e in tibial marrow from normal mice after administration of 0.2 (---) or 5 μg EB (—). Each point represents mean + SEM of mean values of a minimum of four groups, each comprising at least three mice. * $P < 0.05$ when compared with control values. **$P < 0.01$ when compared with control values.

μg/day × 5), is either mildly or not significantly decreased by larger amounts of estrogen (1, 5, or 25 μg/day × 5). In this regard, hypoxia-induced serum Ep levels in the former versus the latter groups respectively show either a reduction or a dose-related enhancement, when compared with the vehicle-treated controls (Figure 13-10).

The present results indicate that EB, injected in mice at either small (0.2 μg) or large dose levels (5 μg), induces a progressive depletion of the CFU-e pool in marrow, starting 12 hr after estrogen: this depletion apparently underlies the decrease of the erythropoietic response evoked by Ep in EB-primed animals.[24]

The possibility cannot be excluded that this EB-induced decrease of the CFU-e number is mediated by a diminished *in vitro* response of CFU-e to the Ep

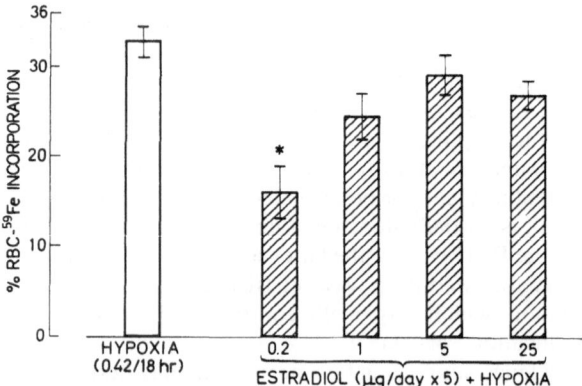

Figure 13-9 Erythropoietic activity (mean 24-hr % RBC-^{59}Fe incorporation values ± SEM) in polycythemic mice primed with estradiol benzoate (0.2, 1, 5 or 25 μg/day × 5) or its vehicle and subjected to hypoxia (0.42 atm air/18 hr) starting immediately after the last injection. *$P < 0.01$ when compared with the control group.

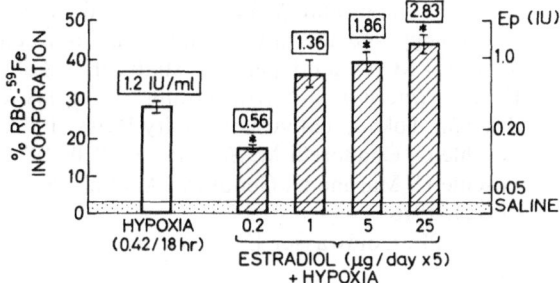

Figure 13-10 Ep serum levels (IU/ml) as evaluated on the basis of mean ± SEM radioiron uptake values in polycythemic assay mice, in polycythemic mice primed with estradiol benzoate (0.2, 1, 5 or 25 μg/day × 5) or its vehicle and subjected to hypoxia (0.42 atm air/18 hr) starting immediately after the last injection. * P 0.01 when compared with the control groups.

stimulus. Further studies are currently in progress in an attempt to substantiate this hypothetical mechanism.

It is of interest that in polycythemic mice exposed to hypoxia, administration of 5, and 25 μg/day of EB causes a sharp enhancement of Ep serum activity. In terms of erythropoietic rate, this increase apparently compensates the depleting action exerted simultaneously by EB on the size of the CFU-e pool. Thus, the erythropoietic rate resulting from these opposing influences (i.e., potentiation of the Ep stimulus and depletion of CFU-e target cells) is either comparable or slightly lower than in vehicle-treated controls.

On the other hand, in mice treated with smaller amounts of EB (0.2 μg/day), hypoxia-induced Ep levels are lower than in vehicle-treated controls. Accordingly, these animals show a marked decrease of the erythroid response to the hypoxic stimulus, i.e., the depletion of the erythroid stem cell pool is not compensated here by enhancement of the hypoxia-induced Ep stimulus.

It must be emphasized that both species specificity and estrogen dose levels exert a marked influence on hypoxic Ep production. In this last regard, administration in rats of either small (5 to 20 μg/day) or large (1 mg/day) amounts of EB causes a significant decrease of the Ep activity evoked by hypoxia.[23,27] On the other hand, the present results indicate that, in female mice, small amounts of EB (0.2 μg) induce a drop of hypoxic Ep levels, whereas larger dosages (5 to 25 μg/day) enhance Ep activity. The latter finding is in line with previous observations.[24]

A further variable may be represented by sex differences. Thus, in male mice, large amounts of EB (250 μg/week) induce a decrease rather than a rise of hypoxic Ep levels.[27]

It is of interest that normal mice administered with 5 μg of EB show an initial depletion and a subsequent wave of amplification of the splenic CFU-e pool, respectively, at 12 and 48 hr. The latter phenomenon might depend on the enhanced Ep activity induced by EB at the 5 μg dose level. Alternatively, the possibility may be envisioned that EB induces a transitory migration of erythroid stem cells from marrow to spleen; this speculative concept, however, requires experimental substantiation. Further studies are therefore in progress to evaluate the number of erythroid precursors in peripheral blood after EB.

Conclusion

Evidence is presented indicating that the inhibitory action induced by EB on the erythroid response to Ep is mediated by a depletion of the CFU-e pool. In mice injected with relatively large dose levels of EB (5 to 25 μg), this phenomenon is partially compensated, in terms of erythropoietic rate, by enhanced Ep production.

Acknowledgments

This work was supported by grants from EURATOM, Bruxelles (No. 159–76–7–B101); Volkswagen Foundation, Hannover; and CNR, Rome (No. 75.01009.65 and 76.01467.04).

We express our appreciation to Mr. P. Ciaglia and P. Barba for their excellent technical help.

A-Ep II was kindly provided by Dr. J. C. Schooley, Berkeley, California.

References

1. Axelrad AA, McLeod DL, Shreeve MM, et al.: Properties of cells that produce erythrocytic colonies *in vitro*, in Robinson WA (ed): *Hemopoiesis in culture*. Washington, DC, U.S. Government Printing Office, 1974.
2. Iscove NN, Sieber F, Winterhalter KH: *J Cell Physiol* 65:760, 1974.
3. Iscove NN, Sieber F: *Exp Hematol* 3:32, 1975.
4. Iscove NN: *Cell Tiss Kin* 10:323, 1977.
5. Heath DS, Axelrad AA, McLeod DL, et al.: *Blood* 47:777, 1976.
6. Gregory CJ: *J Cell Physiol* 89:289, 1976.
7. Gregory CJ, Tepperman AD, McCulloch EA, et al.: *J Cell Physiol* 84:1, 1974.
8. Peschle C, Magli MC, Cillo C, et al.: *Br J Haemat* (in press)
9. Peschle C, Magli MC, Cillo C, et al.: Erythroid stem cell kinetics: experimental and clinical as-

pects, in Brecher G, Bessis M (eds): *Stem Cell Differentiation*. New York, Springer-Verlag, (in press)
10. Hara H, Ogawa M: *Exp Hematol 5:*141, 1977.
11. Stohlman Jr. F: Control mechanism in erythropoiesis in Gordon AS, Condorelli M, Peschle C (eds): *Regulation of erythropoiesis*. Il Ponte, Milan, 1972.
12. Gurney CW, DeGowin R, Hofstra D, et al.: Application of erythropoietin to biological investigation, in Jacob LO, Doyle M, (eds): *Erythropoiesis*. New York, Grune & Stratton, 1962.
13. Mirand EA, Gordon AS, Wenig J: *Nature 206:*270, 1965.
14. Fried W, Gurney CW: *Nature 206:*1160, 1965.
15. Byron JW: *Blood 40:*198, 1972.
16. Moriyama Y, Fisher JW: *Blood 45:*665, 1975.
17. Moriyama Y, Fisher JW: *Proc Soc Exp Biol Med 149:*178, 1975.
18. Prchal J. Adamson JW, Steinmann L, et al.: *J Cell Physiol 89:*489, 1976.
19. Peschle C, Marmont A, Perugini S, et al.: Physiopathology and therapy of adult pure red cell aplasia (PRCA): a cooperative study, in Takaku F (ed): *International Symposium on Aplastic Anemia*. Tokyo, Tokyo University Press, 1977.
20. Peschle C, Condorelli M: *Science 190:*910, 1975.
21. Peschle C, Marone G, Genovese A, et al.: *Blood 47:*325, 1976.
22. Dukes PP, Goldwasser E: *Endocrinology 69:*21, 1961.
23. Peschle C, Rappaport IA, Sasso GF, et al.: *Endocrinology 92:*358, 1973.
24. Anagnostou A, Zander A, Barone J, et al.: *J Lab Clin Med 88:*700, 1976.
25. Jepson JH, Lowenstein L: *Proc Soc Exp Biol Med 123:*457, 1966.
26. Fried W, Tichier T, Dennerberg I, et al.: *J Lab Clin Med 83:*807, 1974.
27. Mirand EA, Gordon AS: *Endocrinology 78:*325, 1966.

14 Hormonal Influences on Erythroid Colony Growth in Culture

John W. Adamson, William J. Popovic, and James E. Brown

Introduction

The growth and function of a number of cellular systems are influenced by interactions of different hormones and small molecules.[1,2,3] Several studies have shown that a variety of hormones, incapable of initiating a metabolic event or differentiative function by themselves, have the capacity to enhance the effect of primary regulatory hormones. As an example, dibutyryl cyclic adenosine monophosphate (db-cAMP) enhances the induction of tyrosine aminotransferase in cultured hepatoma cells, but only in the presence of or following preconditioning of the cells by a glucocorticoid such as dexamethasone.[4]

Recent work from our laboratory, employing the *in vitro* growth of erythroid colony-forming cells, has shown that cAMP and related compounds enhance the growth of erythroid colonies from marrow cell cultures from a variety of mammalian species.[5] Similar to other interactions, however, the cyclic nucleotides were incapable of initiating colony growth by themselves, erythropoietin (ESF) being a necessary constituent of the culture medium. The response to cyclic AMP (cAMP) was specific and enhancement was not seen with cyclic guanosine, inosine, or cytosine nucleotides.

These results suggest that other hormones, thought to influence cellular function by raising intracellular levels of cAMP, might also influence erythroid colony growth. To determine if this were, in fact, the case, studies were carried out employing hormones whose action is mediated by cell surface receptors linked to adenyl cyclase. Examples of such a class of hormones are catecholamines.[6] The results of these studies demonstrate that a variety of adrenergic agonists enhance erythroid colony formation by interacting with receptors having β_2 subspecificity. In addition, other naturally occurring hormones, such as thyroid hormone and its analogues, also enhance erythroid colony growth. Because thyroid hormones are believed to regulate the number of adrenergic receptors in selected mammalian cells, thyroid hormone–catecholamine interactions have been investigated in detail. The results indicate that both catecholamines and thyroid hormones influence the colony-forming cell through receptors having similar characteristics and suggest strategies for examining the significance of thyroid hormone modulation of erythropoiesis in the intact animal.

Materials and Methods
Preparation of Marrow Cells
Bone marrow cells were aspirated under sterile conditions from lightly anesthetized animals as described previously.[5] Precise details of the preparation and culture technique can be found in Appendix 7.[7]

Effect of Other Hormones and Small Molecules on Colony Numbers
To examine their influence on erythroid colony numbers the following hormones were dissolved in Hank's balanced salt solution (BSS) and added in a wide range of concentrations to appropriate cultures: L-phenylephrine (Schwartz-Mann Division, Becton-Dickinson and Co. Orangeburg, New York); L-nor epinephrine (Sigma Chemical Co. St. Louis, Missouri); L-epinephrine (Schwartz-Mann); L-isoproterenol (Sigma Chemical Co.); albuterol (Schering Corp.); metaproterenol (Boehringer–Ingelheim, Ltd. Indianapolis, Indiana); db-cAMP (Sigma Chemical Co.); the nonspecific adenyl cyclase stimulator, cholera enterotoxin; and the cholinergic compound carbamylcholine (Sigma Chemical Co.). The phosphodiesterase inhibitor, RO-20-1724, (Roche Diagnostics Division, Hoffman-LaRoche, Inc. Nutley, New Jersey) was dissolved in ethanol and also tested.

Thyroid hormones tested included L-thyroxine (L-T_4), D-thyroxine (D-T_4), L-triiodothyronine (L-T_3), D-triiodothyronine (D-T_3), diiodo-L-thyronine, diiodo-L-tyrosine, iodo-L-tyrosine (all from Sigma Chemical Co.); triiodothyroacetic acid (Triac), and tetraiodothyroacetic acid (Tetrac; K & K Laboratories, Inc.). L-T_4, D-T_4, Triac, and Tetrac were dissolved in 70% ethanol with 1N NaOH and L-T_3 and D-T_3 were dissolved in 95% ethanol with 2N HCl. These compounds were initially dissolved in a concentration of 10^{-5} M. The remaining compounds were dissolved in and all compounds were subsequently diluted with BSS. Control marrow cultures contained either ethanol alone or with NaOH or HCl present in the concentrations found with the added thyroid hormones. At the concentrations employed no adverse effect on erythroid colony growth was observed.

Effect of Adrenergic Antagonists
The catecholamine antagonists tested included phentolamine (CIBA-Geigy Corp.); D,L-propranolol (Schwartz-Mann); L-propranolol and D-propranolol (Ayerst Laboratories); practolol (Ayerst Labora-

tories); and butoxamine (Burroughs-Wellcome Co.). All were diluted in BSS.

Characterization of Responsive Cells by Velocity Sedimentation

Physical properties of the colony-forming cell populations responding to either ESF or ESF plus enhancing agents were examined by velocity sedimentation at unit gravity. The technique employed was that of Miller and Phillips[8] as modified in this laboratory.[9] The cell concentration and the procedure for loading and unloading the chamber have been detailed previously.

Results

When dog marrow cells are cultured in the presence of ESF under the described conditions, colonies of hemoglobin synthesizing cells appear within the first 24 hr and reach peak numbers between days 2 and 3. Colony size ranges from 8 to 64 cells and both colony numbers and size are dependent on the concentration of ESF. Virtually no colonies (less than $10/10^5$ nucleated cells plated) form in the absence of added ESF. Plateau colony numbers are achieved routinely at approximately 1 unit of ESF/ml.[5] For studies of the effects of hormone interactions on colony growth, 0.5 unit ESF/ml was employed routinely.

Effect of Other Hormones and Small Molecules on Colony Numbers

When agents known to participate in the adenyl cyclase–cAMP system are added to culture, enhanced colony numbers observed, as shown in the following results of a study using canine erythroid colonies.*

Control colonies	460 (± 20)
+ db–cAMP (10^{-5} M)	783 (± 32)
+ RO–20–20–1724 (10^{-5} M)	988 (± 19)
+ Cholera enterotoxin (100 ngn/ml)	1,011 (± 41)
+ Isoproterenol (10^{-7} M)	67 (± 14)

This enhancement is specific for cyclic adenosine nucleotides and is greatest for the dibutyryl derivative of cAMP.[5] Enhancement was also seen with the phosphodiesterase inhibitor, RO–20–1724. Isoproterenol and cholera enterotoxin, two compounds believed to interact with surface receptors to activate adenyl cyclase and thereby raise intracellular levels of cAMP, also enhanced colony numbers.

Distinction between the mechanisms of action for these various compounds was demonstrated by coculturing them in the presence of the β-adrenergic antagonist, propranolol (10^{-7} M). The addition of propranolol completely blocked isoproterenol-enhanced colony growth but had no significant effect on the stimulation of colony numbers induced by db-cAMP, RO-20-1724 or cholera enterotoxin (data not shown). This would be expected since db-cAMP and RO-20-1724 presumably act intracellularly while cholera enterotoxin is believed to interact with specific gangliosides on the cell membrane to achieve adenyl cyclase activation.[10]

When naturally occurring and synthetic catecholamine hormones, including a variety of adrenergic compounds, were added to culture, enhancement of colony growth was seen only with selected agents†:

Control colonies	345 (± 9)
Norepinephrine ($\alpha\beta_1$)†	330 (± 17)
Phenylephrine (α)	366 (± 11)
Epinephrine ($\alpha\beta_1\beta_2$)	621 (± 32)
Isoproterenol ($\beta_1\beta_2$)	683 (± 12)
Albuterol (β_2)	458 (± 23)
Metaproterenol (β_2)	483 (± 18)

Thus, phenylephrine and norepinephrine were ineffective in stimulating colony growth over a wide concentration range from 10^{-10} to 10^{-5} M, whereas epinephrine, isoproterenol, albuterol, and metaproterenol were all active. These agents enhanced colony growth by as much as 100% in the presence of a suboptimal concentration of ESF. Analysis of the receptor specificities of the various compounds revealed that those agents that were active have in common β_2 receptor specificity as defined in other physiological systems. Cholinergic agents were inactive.

Effect of Adrenergic Antagonists

To confirm the specificity of catecholamine enhancement of colony growth, experiments were carried out to determine the effects of various adrenergic blocking agents. Thus, when propranolol and butoxamine (10^{-7} M) were added to cultures containing isoproterenol, agonist-enhanced colony growth was inhibited completely, as shown below‡:

Control colonies	260 (± 17)
+ Isoproterenol ($\beta_1\beta_2$)	363 (± 22)
Isoproterenol + Phentolamine (α)	397 (± 16)
Isoproterenol + Practolol (β_1)	416 (± 30)
Isoproterenol + Propranolol ($\beta_1\beta_2$)	285 (± 13)
Isoproterenol + Butoxamine (β_2)	265 (± 10)

* Mean ($\pm$ SEM) colony numbers are shown in this and all subsequent lists. Control colony numbers represent those obtained with 0.5 unit/ml ESF.

† Receptor specificity. In this experiment, all agonists were tested at 10^{-7} M concentration.

‡ Isoproterenol and all antagonists were tested at 10^{-7} M concentrations.

Table 14-1 The influence of thyroid hormone analogues on erythroid colony growth.

Control Colonies 426(±17)	Concentration (M)						
	10^{-11}	10^{-10}	10^{-9}	10^{-8}	10^{-7}	10^{-6}	10^{-5}
+L-T$_4$	—	469(±11)	515(±23)	790(±40)	1107(±28)	686(±25)	445(±13)
+D-T$_4$	—	420(±17)	415(±21)	533(±8)	869(±31)	528(±14)	432(±11)
+L-T$_3$	431(±30)	595(±13)	924(±43)	781(±38)	600(±19)	481(±6)	—
+D-T$_3$	—	442(±18)	577(±37)	778(±20)	491(±19)	424(±19)	—
+Tetrac	—	436(±27)	418(±16)	507(±13)	661(±41)	984(±36)	463(±15)
+Triac	479(±20)	1116(±63)	878(±25)	681(±19)	478(±27)	422(±30)	—

In contrast, practolol and phentolamine had no inhibitory effect. These results confirm the specificity of the receptors involved in the response to adrenergic agonists in that the active blockers are known to react with receptors having β_2 specificity. Those antagonists which were inactive are not known to recognize such receptors. Further specificity of the antagonist effect was documented by demonstrating that the L-isomer of propranolol was biologically active whereas the D-isomer was ineffective:

Control colonies	205 (±15)
+ Isoproterenol (10^{-7} M)	410 (±13)
Isoproterenol + D-propranolol (10^{-9} M)	405 (±9)
Isoproterenol + L-propranolol (10^{-9} M)	215 (±24)

Even at very high concentrations (10^{-4} M), none of the blocking agents inhibited ESF-stimulated colonies, confirming that the blocking effect is not only specific but indicating the non-identity of the receptors for catechols and for ESF.

Thyroid Hormone Effects on Colony Growth

When thyroid hormones were examined for their effects in culture, enhancement of colony formation was seen with all hormone analogues tested (Table 14-1) with the exception of diiodo-L-thyronine, diiodo-L-tyrosine and iodo-L-tyrosine. This enhancement of colony growth depended on the concentration of the thyroid hormone analogue. The peak activity and results in order of potency for the various hormones were as follows: Triac (10^{-10} M); L-T$_3$ (10^{-9} M); D-T$_3$ (10^{-8} M); L-T$_4$ (10^{-7} M); D-T$_4$ (10^{-7} M); Tetrac (10^{-6} M). This order of potency is the same as the order of calorigenic potency established by in vivo studies and there is the expected stereospecificity in that the L-form of T$_3$ was tenfold more active than D-T$_3$.

Because thyroid hormone function has been linked to adrenergic activity in the intact organism,[11] the effect of various blocking agents on thyroid hormone-enhanced colony growth was examined. Again, inhibition of colony growth was seen only with adrenergic blockers having β_2 specificity*:

Control colonies	390 (±14)
+L-T$_4$	810 (±21)
+Phentolamine (α)	803 (±29)
+Practolol (β_1)	817 (±30)
+Propranolol ($\beta_1\beta_2$)	395 (±7)
+Butoxamine (β_2)	405 (±16)

However, there was a 10^2 to 10^3 greater concentration requirement for blockade (Table 14-2). Inhibition was only observed with L-propranolol, meeting the requirement for stereospecificity of the antagonist effect (data not shown).

Characterization of Responsive Cells by Velocity Sedimentation

When dog bone marrow cells were separated at unit gravity by velocity sedimentation, those cells responding to ESF formed a broad peak at the most

* L-T$_4$ and all adrenergic antagonists were tested at 10^{-7} M concentrations.

Table 14-2 The influence of various concentrations of propranolol on L-thyroxine- and isoproterenol-enhanced erythroid colony growth.

Control Colonies 1160(±90)	No Blocker	+ Propranolol (M)			
		10^{-10}	10^{-9}	10^{-8}	10^{-7}
+Isoproterenol (10^{-7} M)	2216(±143)	1318(±214)	1204(±91)	1273(±149)	1149(±209)
+L-T$_4$ (10^{-7} M)	2341(±89)	2407(±174)	2475(±216)	2398(±192)	1295(±202)

rapidly sedimenting portion of the gradient. These cells had a mean sedimentation rate of approximately 8.7 mm/hr. Colony-forming unis responding to isoproterenol or thyroid hormone sedimented somewhat more slowly, at approximately 7.2 mm/hr. The peak of these cells fell onto a shoulder of the ESF-generated profile. The colony-forming units responding to thyroid hormone or isoproterenol had virtually overlapping profiles. Thus, by this criterion, the populations of responsive cells were identical.

Discussion

Hormonal interactions have been suggested to play important roles in determining the expression of differentiated cell function in a number of organs and cell systems. In mammals, an important interaction between thyroid hormones and catecholamines has been postulated.[11] Results of several studies suggest that thyroid hormones regulate the number of adrenergic receptors on catechol-responsive target cells, potentially dramatically affecting end-organ function.[12,13]

We have examined possible hormonal interactions in cultures of mammalian cells, using as end point erythroid colony growth. Enhancement was observed with a number of agents related to the cAMP–adenyl cyclase system. Certain predictions arising from these observations have been fulfilled. Thus, agents known to react widely with cell surface receptors to activate adenyl cyclase, are also found to enhance erythroid colony growth. Such agents include adrenergic agonists and cholera enterotoxin. The specificity and identity of the receptors responsible for such interactions have been defined by a variety of pharmacological techniques, including the employment of agonists of known receptor affinity as well as blocking agents of comparable receptor definition. These results have demonstrated that receptors on colony forming units have β_2 specificity and are distinct from proposed receptors for ESF.

Golde and co-workers were the first to show that thyroid hormones enhance erythroid colony formation directly.[14] Our results confirm and extend those observations and demonstrate that such enhancement occurs in a population of cells that is physically indistinct from those responding to adrenergic agonists. In addition, the combination of optimally effective concentrations of thyroid hormones and β-adrenergic agonists has no further enhancing effect on colony numbers above that seen with either agent alone (data not shown). This suggests that the hormones are influencing identical populations of cells, perhaps through a similar mechanism.

One possible explanation for these effects is that thyroid hormones increase the number of receptors for adrenergic agonists, a proposal suggested from *in vivo* studies.[13] That this is not the case is indicated by the fact that one does not see a synergistic effect on colony growth when suboptimal concentrations of thyroid hormones and catecholamines are employed together. In addition, much larger concentrations of antagonist are required to block the effect of thyroid hormones.

Consequently, the most likely explanation of these results is a direct hormone effect on erythroid colony-forming units. Other experiments fail to support the contention that these enhancing hormones are influencing a "helper cell" population. Thus, enhancement is still observed even at limiting cell dilutions (10,000 cells/ml), when conditioned medium prepared from isoproterenol-exposed cells is used to support the subsequent growth of normal marrow cells, and following removal of adherent cells.[15]

The mechanisms, however, by which these agents influence colony growth at the intracellular level are unknown. One possibility is that cells in a specific state of cycle are made responsive to ESF by having their cell cycle characteristics altered. In keeping with this would be the recent observation that prolongation of the G_1 phase of the cell cycle in Friend virus-infected cells may be an important prerequisite for the initiation of the differentiation program.[16] Consistent with this hypothesis is the fact that cAMP has as one of its influences on the cell cycle the prolongation of G_1. The fact that colony-forming units affected by thyroid hormones and catechols have a slightly slower sedimentation rate than the majority of erythroid colony-forming units is consistent with that portion of a population in G_1 phase of the cell cycle.[17] Since ESF is required for erythroid colony growth and only enhancement is seen with the addition of other hormones, it is likely that their effect is to modulate the influence of the primary regulatory hormone on cell growth. This might be reflected *in vivo* in the form of "fine tuning" of terminal erythroid maturation. However, these results with naturally occurring hormones suggest strategies for examining the clinical relevance of the observations in animals whose endocrine state has been manipulated, e.g., the hypothyroid dog. Such studies should provide the opportunity to determine how accurately results in culture reflect the regulation of selected aspects of erythropoiesis *in vivo*.

Acknowledgments

The authors wish to thank Mrs. Christine Reichgott and Ms. Faith Shiota for excellent technical assistance. The following compounds were gifts: RO–20–1724 from Roche Diagnostics Division, Hoffman-LaRoche Incorporated; albuterol, from Schering

Corporation; metaproterenol, from Boehringer-Ingelheim; practolol and the optical isomers of propranolol from Ayerst Laboratories. These studies were supported by designated research funds of the Veterans Administration and National Institutes of Health grant AM-19410. Dr. Popovic was an advanced specialty resident of the Veterans Administration and Dr. Brown was the recipient of a Research Associate award of the Veterans Administration. Acknowledgment is also given to the National Institute of Allergy and Infectious Disease, Bethesda, Maryland, for the supply of pure cholera enterotoxin, Lot 0172, prepared by Dr. Richard Finkelstein.

References

1. DeAsua LJ, O'Farrell M, Bennett D, et al: Interaction of two hormones and their effect on observed rate of initiation of DNA synthesis in 3T3 cells. *Science* 265:151–153, 1977.
2. Holley RW: Control of growth of mammalian cells in cell culture. *Nature* 258:487–490, 1975.
3. Tomkins GM: The metabolic code. *Science* 189:760–763, 1975.
4. Granner DK, Lee A, Thompson EB: Interaction of glucocorticoid hormones and cyclic nucleotides in induction of tyrosine aminotransferase in cultured hepatoma cells. *Proc Natl Acad Sci USA* 252:3891–3897, 1977.
5. Brown JE, Adamson JW: Modulation of *in vitro* erythropoiesis: enhancement of erythroid colony growth by cyclic nucleotides. *Cell Tissue Kinet* 10:289–298, 1977.
6. Sutherland EW: Studies on the mechanism of hormone action. *Science* 177:401–408, 1972.
7. Adamson JW, Shiota F, Wise C: Erythroid colony growth from dog marrow cells: practical aspects, in Murphy MJ: *In Vitro* Aspects of Erythropoiesis. New York, Springer-Verlag, 1978, p. 254.
8. Miller RG, Phillips RA: Separation of cells by velocity sedimentation. *J Cell Physiol* 73:191–202, 1969.
9. Singer JW, Adamson JW: Steroids and hematopoiesis. II. The effect of steroids on *in vitro* erythroid colony growth: evidence for different target cells for different classes of steroids. *J Cell Physiol* 88:135–144, 1976.
10. Fishman PH, Brady RO: Biosynthesis and function of gangliosides. *Science* 194:906–915, 1976.
11. Waldstein SS: Thyroid–catecholamine interrelationship. *Ann Rev Med* 17:123–132, 1966.
12. Williams LT, Lefkowitz RJ, Watanabe AM et al: Thyroid hormone regulation of beta-adrenergic receptor number: possible biochemical basis for the hyperadrenergic state in hyperthyroidism. *Clin Res* 25:458A, 1977.
13. Williams LT, Lefkowitz RJ, Watanabe AM, et al: Thyroid hormone regulation of β adrenergic receptor number. *J Biol Chem* 252:2787–2789, 1977.
14. Golde D, Bersch N, Chopra IJ et al: Potentiation of erythropoiesis *in vitro* by thyroid hormones. *Clin Res* 24:309A, 1976.
15. Adamson JW: Unpublished observations, 1977.
16. Terada M, Fried J, Nudel U et al: Transient inhibition of S-phase associated with dimethyl sulfoxide induction of murine erythroleukemia cells to erythroid differentiation. *Proc Natl Acad Sci USA* 74:248–252, 1977.
17. Omine M, Perry S: Use of cell separation at Ig for cytokinetic studies in spontaneous AKR leukemia. *J Natl Cancer Inst* 48:697–704, 1972.

15 A Lack of Burst-Forming Unit (BFU-e) and Lymphocyte-Mediated Suppression of Erythropoiesis in Patients with Aplastic Anemia

Yoshiaki Moriyama, Masatsugu Sato, and Yasutami Kinoshita

Introduction

The pathogenesis of idiopathic aplastic anemia remains obscure.[1] Since all cellular components of the blood are depressed in aplastic anemia, it is reasonable to suggest that the defect resides at the level of the pluripotential hematopoietic stem cells, as a result of a defect in the matrix of the hematopoietic system or of humoral regulation. However, several recent reports[2,3] suggest that immune processes may be involved in the pathogenesis of aplastic anemia in some patients. In work reported in this chapter, using a plasma clot assay for the growth of burst-forming unit (BFU-e) including more mature (CFU-e) erythroid precursors *in vitro*, we examined changes in CFU-e and BFU-e in the marrow and blood of patients with aplastic anemia. Both marrow and circulating CFU-e and BFU-e were found to be markedly reduced in aplastic anemia. In addition, whether or not such a reduction of colony forming capacity *in vitro* may be related to cell-mediated suppression of erythropoiesis[3] was investigated in co-cultures with peripheral blood lymphocytes and normal human bone marrow cells.

Patients and Methods

All eight patients (except one, PNH) with aplastic anemia studied here were idiopathic and had severe pancytopenia, no reticulocytosis and aplastic marrows (nucleated cell count below $3.6 \times 10^4/mm^3$ and marked lymphocytosis). They had no lymphadenopathy or hepatosplenomegaly except one patient who had slight hepatomegaly. Cultures were done before conventional treatments were given.

Bone marrow obtained from the patients and normal donors was transferred to 5 ml of supplemented Eagle's minimum essential medium (HMEM) containing 2% fetal calf serum and 10 units/ml of heparin. After centrifugation, buffy coat was removed and nucleated cells were counted.

Peripheral blood lymphocytes were separated from freshly drawn heparinized venous blood by Ficoll–Hypaque density-gradient sedimentation. The cells were then suspended in HMEM and counted for plating. This procedure contained less than 3.0% monocytes except lymphocytes.

Cultures of erythroid colonies were carried out according to a modification[4] of the method of McLeod et al.[5] Bone marrow cells, in a final concentration of 2×10^4 cells per 0.1 ml, were cultured in the presence of erythropoietin. In the lymphocyte studies, 10^4 lymphocytes per well were co-cultured with the normal bone marrow cells. Cultures were maintained in a humidified atmosphere of 5% CO_2 in air at 37°C. After 7 to 10 days of incubation, the plasma clots were removed and transferred to glass slides, fixed in glutaraldehyde, and stained with benzidine and hematoxylin. Under $\times 100$ magnification, each clot was scanned, and erythroid colonies consisting of eight or more benzidine-positive cells were counted. In the present study, bursts (BFU-e) were counted depending on the spatial orientation of erythroid colonies (Figure 15-1).

Results

As shown in Table 15-1, the erythroid colony-forming capacity in six normal subjects varied, ranging from

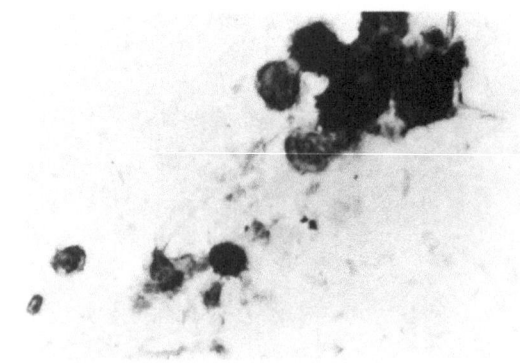

Figure 15-1 Photomicrograph of a burst-forming unit (BFU-e) that developed in the plasma clot culture system containing erythropoietin 10 days after seeding with normal human bone marrow cells. The cytoplasma of most cells in the colonies is stained brown (benzidine positive). $\times 200$.

Table 15-1 The growth of CFU-e and BFU-e in normal subjects and patients with aplastic anemia using a plasma clot culture system.

Case	No. of Experiments	CFU-e per 10⁵ Nucleated Cells	BFU-e/10⁵ Nucleated Cells	
			Erythropoietin units/wall	
			0.2 units	1.0 unit
Aplastic	6	8.6 ± 8.4^a	0	0
Normal	6	156 ± 76^b	3.8 ± 2.7	6.1 ± 4.1^c

[a] Mean ± SEM.
[b] Significantly ($p < 0.001$) greater than the number of colonies in aplastic anemia.
[c] Not significantly different from normal marrow at the dose of erythropoietin (0.2 units).

58 to 272 colonies with an average of 156 colonies per 10⁵ nucleated cells. On the other hand, the growth of CFU-e was significantly ($p < 0.001$) reduced in all patients with aplastic anemia, although all their marrow cells, except in one patient, retained a slight ability to grow colonies *in vitro*. However, there was no ability of the marrow cells from aplastic anemia to grow burst (BFU-e) *in vitro* even in the presence of a large amount of erythropoietin. In addition, the erythroid colony-forming capacity of peripheral mononuclear cells from patients with aplastic anemia (except PNH) was significantly ($p < 0.005$) less than that of normal donors (Figure 15-2).

The effect of peripheral mononuclear cells (lymphocytes) on erythroid colony formation by normal bone marrow is shown in Figure 15-3. Peripheral blood lymphocytes from patients with aplastic anemia, when co-cultured with normal human marrow, significantly ($p < 0.01$) suppressed erythroid colony formation *in vitro*, while lymphocytes from normal donors markedly enhanced the numbers of erythroid colonies formed *in vitro*.

Discussion

In vitro bone marrow culture systems have come into increasing use as a means of assessing human erythropoiesis. In the present studies of the growth

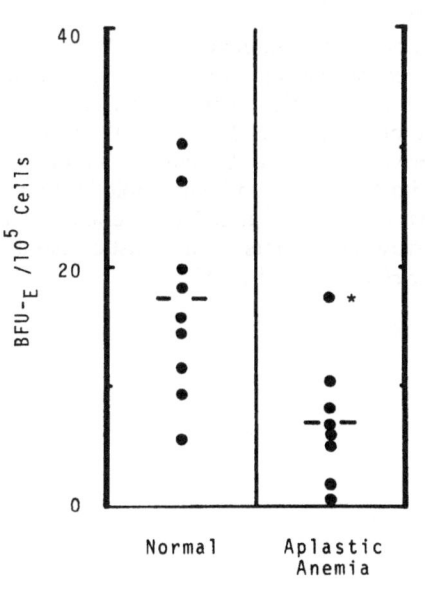

Figure 15-2 Circulating BFU-e in normal subjects and patients with aplastic anemia. Peripheral mononuclear cells (lymphocytes) obtained here were cultured with erythropoietin (1.0 unit/ml).

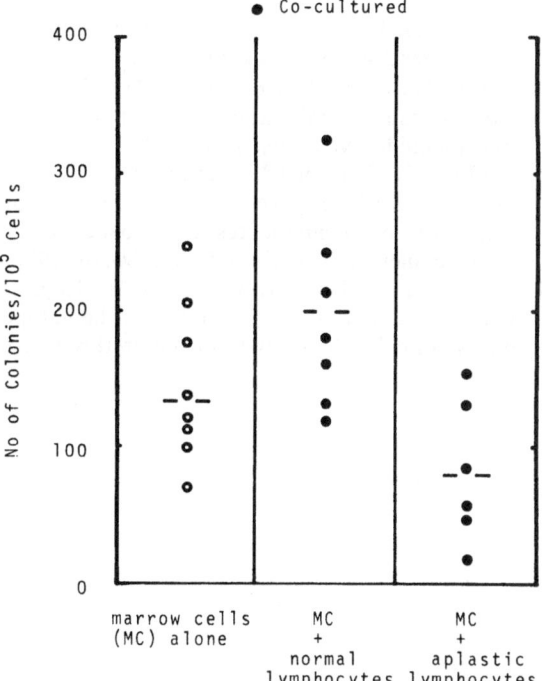

Figure 15-3 Effects of peripheral blood lymphocytes from normal donors and patients with aplastic anemia on erythroid colony formation by normal human bone marrow cells in response to erythropoietin (1.0 unit/ml) *in vitro*.

of erythroid colonies *in vitro* using a plasma clot culture system, both circulating and marrow CFU-e in patients with aplastic anemia were significantly less than those in normal subjects. In addition, the marrow cells from these patients were found to produce no growth of BFU-e *in vitro* that are believed to presumably serve as the progenitor cell for the CFU-e. We have also demonstrated previously, using a double layer agar system, that the numbers of granulocyte colonies formed are markedly reduced in patients with aplastic anemia.[6] These findings suggest that the most possible mechanism of aplastic anemia may be due to a reduction in number or an intrinsic defect in the committed stem cells of these lines. However, the mechanism by which such defects occur in aplastic anemia is unknown.

In the present co-culture studies, lymphocytes from patients with aplastic anemia were found to significantly suppress erythroid colony formation *in vitro*, while normal lymphocytes did stimulate erythroid colony formation. This and the similar data reported by other investigators[3] strongly suggest that lymphocyte-mediated suppression of erythropoiesis may be involved in the pathogenesis of aplastic anemia. Before a definite conclusion is reached, however, further studies of a role of lymphocytes in the control of erythropoiesis are necessary.

Summary

In order to clarify a possible pathogenesis of aplastic anemia, we cultured bone marrow cells as well as peripheral blood lymphocytes from patients with idiopathic aplastic anemia using a plasma clot culture system for the growth of bursts (BFU-e) including more mature (CFU-e) erythroid precursor cells *in vitro*. Both marrow and circulating CFU-e in aplastic anemia were significantly less than those in normal subjects. There was no growth of BFU-e *in vitro* in these patients. In addition, lymphocytes from aplastic anemia, when co-cultured with normal human bone marrow cells, significantly ($p < 0.01$) suppressed erythroid colony formation *in vitro*, while normal lymphocytes stimulated erythroid colony formation. These results suggest that patients with aplastic anemia may have a population of lymphocytes capable of suppressing erythropoiesis *in vitro*.

Acknowledgment

Erythropoietin used in the work described in this chapter was kindly provided by Prof. James W. Fisher, Tulane University School of Medicine, New Orleans, Louisiana.

References

1. International Symposium on Aplastic Anemia, September 3–6, 1976, Kyoto, Japan.
2. Baran DT, Griner PE, Klemperer MR: Recovery from aplastic anemia after treatment with cyclophosphamide. *New Engl J Med 295:*1522, 1976.
3. Hoffman R, Zanjani ED, Lutton JD, Zalusky P, Wasserman LR: Suppression of erythroid-colony formation by lymphocytes from patients with aplastic anemia. *New Engl J Med 296:*10, 1977.
4. Moriyama Y, Kurokawa I, Sato M, Itoga Y, Hayashi N, Kinoshita Y: Studies on hematopoietic stem cells. X. Growth of the erythropoietin-responsive progenitors BFU-e and CFU-e in human bone marrow cultures. *Acta Haemat Jap* (in press).
5. McLeod DL, Shreeve MM, Axelrad AA: Improved plasma culture system for production of erythrocytic colonies in vitro: Quantitative assay method for CFU-e. *Blood 44:*517, 1974.
6. Moriyama Y, Ohno Y, Sato M, Kurokawa I, Kinoshita Y: Studies on hematopoietic stem cells. II. *In vitro* growth of granulocytic colonies from bone marrow of patients with aplastic anemia. *Jap J Clin Haemat 16:*481, 1975.

16 Pharmacological Agents and Erythroid Colony Formation: Effects of Beta-Adrenergic Agonists and Steroids

James W. Fisher, Yasuchico Ohno, Bruno Modder, Franciszek Przala, Gregory D. Fink, and Dennis M. Gross

Introduction

Erythropoietin (Ep) is well known to be the primary regulator of erythropoiesis; its major effect is on the erythroid committed stem cells, causing their differentiation into the heme-synthesizing nucleated erythroid cell compartment.[1-4] As outlined by Stohlman[4] the differentiated erythroid cell compartment is not self-sustaining and must be supported by an influx of cells from a stem cell precursor compartment. Stohlman[4] has suggested that erythropoietin not only initiates but also governs the rate of hemoglobin synthesis and the rate of maturation of erythroid cells. Bottomley[5] found that Ep stimulated the rate-limiting enzyme δ-aminolevulinic acid synthetase, but not heme synthetase, in rabbit bone marrow erythroid cells. Early work has demonstrated that high sustained dosages of erythropoietin are capable of triggering the early release of reticulocytes from the bone marrow.[6,7] It is the primary purpose of this chapter to consider pharmacological agents, some of which are naturally occurring in mammalian species, in modulating erythropoiesis in the stem cell compartment either alone or in concert with erythropoietin. The two classes of agents that have been most extensively studied have been the androgenic steroids[8-18] and the beta-adrenergic (β-adrenergic) agonists.[19-26] Although the sites of action of the steroids and the β_2-adrenergic agonists within the stem cell compartment may be different, there are close similarities in their erythropoietic actions. First, both the androgenic steroids[8-10,17] and β_2-adrenergic agonists[23,24] have been demonstrated to enhance kidney production of erythropoietin and an increase in erythroid colony-forming cells (CFU-e) in the bone marrow.[15,18,19,26-29] β-Adrenergic agonists and androgenic steroids such as testosterone, 5α-dihydrotestosterone, and 5β-dihydrotestosterone are naturally occurring and could be of physiological and/or pathophysiological significance in modulating erythropoiesis. It seems quite likely that β_2-adrenergic agonists could act on specific receptors in the kidney to trigger production of erythropoietin and this mechanism may involve the activation of renal adenylate cyclase to generate cyclic AMP (cAMP).[25,30,31] On the other hand, the mechanism of the effects of steroids in modulating kidney production of erythropoietin is less well understood. It has been postulated that an adenylate cyclase mechanism is also involved in regulating the erythroid committed stem cell compartment.[31,32] This chapter will attempt to clarify the mechanism by which androgenic steroids and β_2-adrenergic agonists stimulate erythropoiesis. The effects of these agents on production of erythropoietin by the kidney and in stimulating the erythroid stem cell compartment will be considered.

Materials and Methods
Erythropoietin Bioassay

Erythropoietin (Ep) was assayed via a modification of the method of Cotes and Bangham[33] using radioactive iron (^{59}Fe) incorporation into newly formed red blood cells (RBC) of exhypoxic polycythemic mice. HAM/ICR strain virgin female mice (25 to 28 gm) were made polycythemic by exposure for 22 hr/day for 2 weeks to 0.42 atm pressure. The mice were removed from the hypobaric chamber and on posthypoxic days 6 and 7 injected (s.c.) with one-half of the total dose of saline (0.5 ml, 2×), human urinary Ep standardized against the International Reference Preparation (IRP-B) or the test substance (0.5 ml, 2×). On the eighth posthypoxic day, each mouse received 0.5 μCi of radioactive iron (^{59}Fe citrate) i.v. via the tail vein. Two days later, posthypoxic day 10, the mice were exsanguinated by cardiac puncture and 0.5 ml of blood was counted per mouse with a Packard Auto-Gamma spectrometer and 48-hr % ^{59}Fe incorporation into newly formed RBC was determined. Five mice were injected with each sample and the mean 48-hr % ^{59}Fe incorporation rate into RBC was used as an indication of the erythropoietic activity of the sample when compared with the response to the IRP.

To determine the erythropoietic effects of the adrenergic agents in the exhypoxic polycythemic mice, they were also prepared according to the above schedule. After their removal from the hypobaric chamber, they were injected on days 6 and 7 with either saline or the appropriate dose of adrenergic agonist drug. Studies involving β-adrenergic blockade of the agonist effect in polycythemic mice required the utilization of DL-propranolol (8 mg/kg). DL-Propranolol was administered (s.c.: 0.25 mg/day) to the polycythemic mice 30 min prior to the injection of the adrenergic agent at a different site. The dose of DL-propranolol utilized was based on the results of a previous study on the effects of β-blockers on Ep production during hypobaric hypoxia.[23] All drug solutions were prepared daily and administered within 15 min following their being dissolved in saline. The remainder of the protocol is the same as outlined for the polycythemic mouse assay. The steroids were injected subcutaneously in propylene glycol on the first, third, and fifth days after removal of the mice from 2 weeks in the hypobaric (0.42 atm) chamber. The mice were injected with radioactive iron (^{59}Fe citrate) on the seventh posthypoxic day for determination of % iron incorporation in newly formed red cells.

Plasma Erythropoietin Assay

To assess the effects of the adrenergic agonists on plasma Ep titers, fully conscious restrained female albino New Zealand rabbits were used. All vascular catheterizations were carried out following subcutaneous infiltration with a 1% solution of xylocaine. The central artery of the right ear was catheterized with a heparin filled PE-50 polyethylene tube that was connected to a Statham pressure transducer for the measurement of mean arterial blood pressure recorded continuously on a Grass polygraph. A marginal vein of the left ear was catheterized with polyethylene tubing (PE-10 or 50) through a small skin incision. This tubing was attached to a Harvard constant rate infusion pump for drug administration.

After the completion of the catheterization procedures, the rabbits were heparinized with 800 to 1,000 units/kg i.v. Ten min following heparinization, arterial blood samples were taken for Ep assay and initial blood gases/pH determinations. Plasma volume lost by the blood sampling was replaced with one-half the plasma volume of dextran 40 in saline. After 40 min equilibration, either saline (0.9% NaCl) or the adrenergic agent dissolved in saline was infused into the marginal ear vein at a rate of 47 to 49 μl/min for 7 hr. Arterial blood pressure though recorded continuously is reported at hourly intervals. After 3 and 7 hr of drug or saline infusion, additional blood samples were taken for the determination of blood gases/pH. The drug infusion was terminated after 7 hr but the rabbits remained in the head restraint for an additional 30 min at which time a final blood sample was withdrawn for Ep determination and final blood Po_2, Pco_2 and pH.

Plasma Clot Techniques for Erythroid Colony Formation (CFU-e and BFU-e)

A brief summary of a modification of the technique of McLeod et al.[34] for erythroid colony formation (CFU-e) and a modification of the technique of Heath et al.[35] for the burst-forming unit (BFU-e) are demonstrated in Appendices 1 and 2. Figure 16-1 is a photomicrograph of a benzidine-positive erythroid colony from a normal rabbit bone marrow utilizing the technique of McLeod et al.[34] Figure 16-2 is a photomicrograph of a "burst type" of erythroid colony that sometimes has more than 1,500 cells utilizing a modification of the technique of Heath et al.[35] Appendix 1 describes the type of CFU-e dose-response curve to

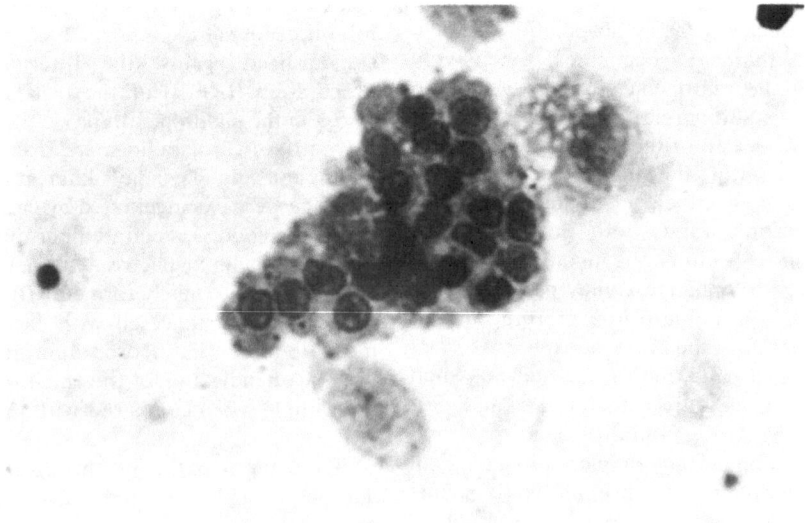

Figure 16-1 Photomicrograph of 4-day erythroid colony (CFU-e) from a normal rabbit bone marrow culture stained with benzidine and counterstained with Giemsa. × 100.

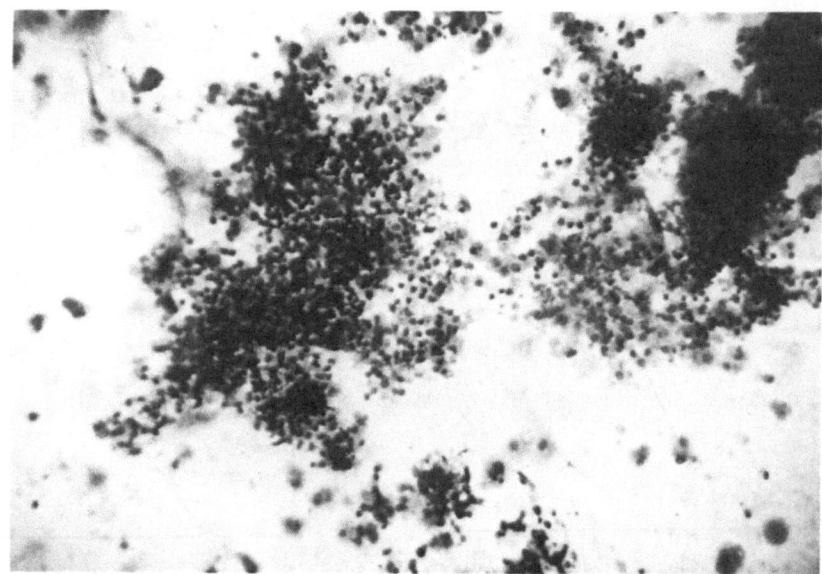

Figure 16-2 Photomicrograph of an 8-day burst-forming unit type of erythroid colony (BFU-e) from a normal rabbit bone marrow culture stained with benzidine and counterstained with Giemsa. × 100.

erythropoietin (Figure 16-3); erythroid colonies formed versus time in culture (Figure 16-4); and the number of colonies formed versus the number of cells plated (Figure 16-5) in normal rabbit bone marrow cultures. Even though the number of colonies are reported per 5×10^4 nucleated bone marrow cells, an attempt was made to quantitatively recover as much of the marrow in the total rabbit femur as possible. However, we felt that the variable recovery from one rabbit femur to another made it difficult, at least in our laboratory, to report erythroid colonies per total femur.

Statistical Analysis Technique

The effect of a single drug at several different time intervals as compared with a control interval was analyzed by the multiple comparison method of Dunnett.[36] All other statistical evaluations were made utilizing the Student's t test for paired and unpaired observations or the ANOVA test for paired comparisons.[37]

β-Adrenergic Activation of Erythropoiesis

Over the past few years Fink et al.,[23–25] Brown and Adamson,[19,32] Byron,[20] and Przala[26] have demonstrated activation of erythropoiesis by β_2-adrenergic drugs. These drugs are well known to activate adenylate cyclase[31] and to trigger kidney production of erythropoietin.[38] In addition, there appears to be an adenylate cyclase activated erythropoiesis in erythroid cells in bone marrow.[32,39–41]

Structure Activity Relationships of the β-Adrenergic Drugs

Figure 16-6 illustrates examples of specific drugs that activate β-adrenergic receptors and are classified as β_1 (heart and blood vessels) and β_2 (metabolic, bronchiolar musculature). A typical agent that will activate both β_1 and β_2 receptors is isoproterenol. An agent that will activate specifically β_1 receptors is the drug dobutamine and the drug salbutamol is a very potent and specific activator of β_2 receptors. A new drug recently introduced in Europe for bronchial asthma because of its potent effect in relaxing bronchiolar smooth muscle is terbutaline, which is a potent β_2 activator. As noted in Figure 16-6 there are

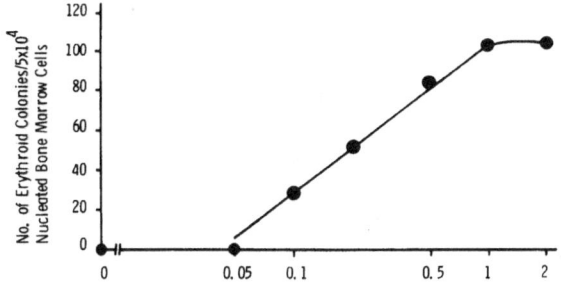

Figure 16-3 Dose-response curve for CFU-e erythroid colonies with increasing concentrations of human urinary erythropoietin in normal rabbit bone marrow cultures.

106 Pharmacological Agents and Erythroid Colony Formation: Effects of Beta-Adrenergic Agonists and Steroids

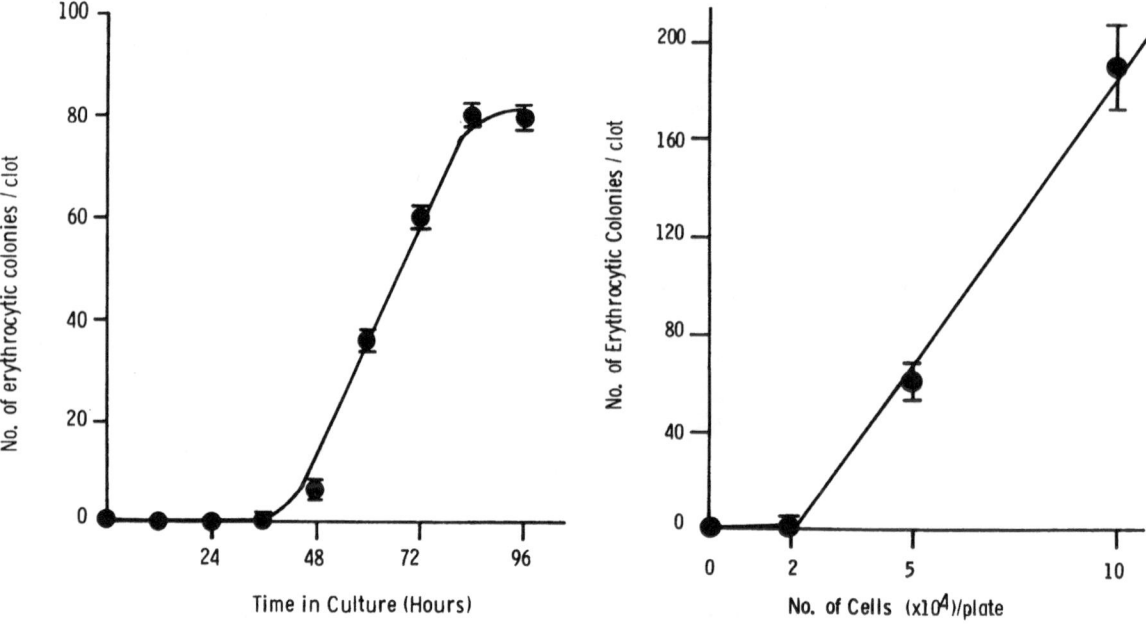

Figure 16-4 Erythroid colony (CFU-e) formation with increasing time in culture with normal rabbit bone marrow containing human urinary erythropoietin (0.2 unit/well). n = 3; I = SEM.

Figure 16-5 Erythroid colony formation (CFU-e) with increasing numbers of rabbit bone marrow nucleated cells/well (0.2 unit Ep/well). n = 3; I = SEM.

specific antagonists of β_1 receptors such as the drug practolol, the β_2 receptor blocker butoxamine and the antagonist propranolol that will inhibit both β_1 and β_2 receptors. Note from Figure 16-6 that there is an interesting structure activity relationship between the β-adrenergic agonists and their respective antagonists. Substitutions have been made on the aromatic ring, the α and β carbon atoms and the terminal amino group to yield a great variety of compounds with selective β_1 and β_2 agonist and antagonist activities. For example, salbutamol, a selective β_2 receptor stimulant useful in the treatment of asthma, has a

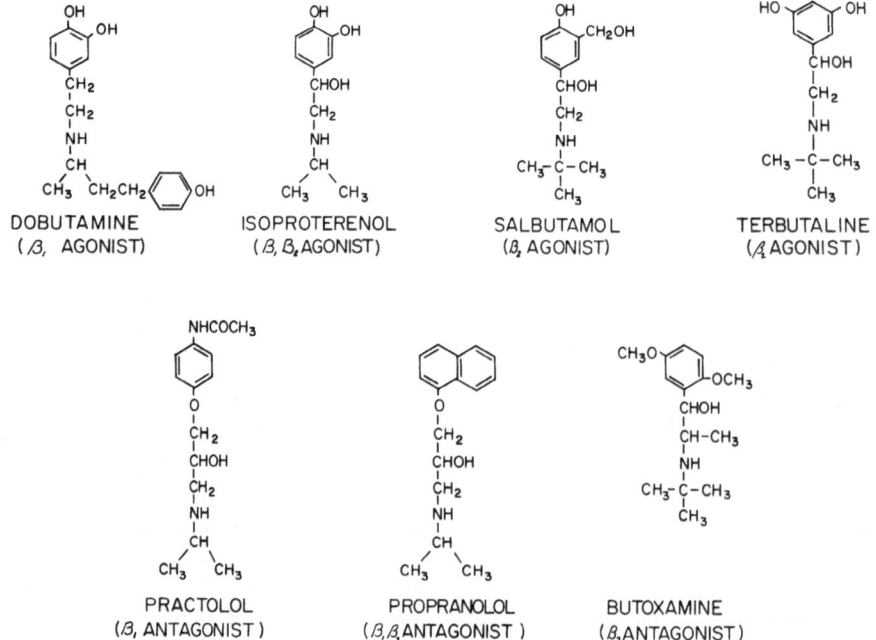

Figure 16-6 Chemical structures of some β-adrenergic agonist and antagonist drugs.

CH$_2$OH substituent in position 3 and a hydroxy group at position 4 of the aromatic ring and is an important exception to the general rule that hydroxy groups must be present in the 3 and 5 positions to confer β_2 receptor selectivity on compounds with large amino substituents. Terbutaline has hydroxy groups at the 3 and 5 positions and a large amino substituent, thus causing relaxation of the bronchiolar musculature in asthma without causing significant tachycardia. On the other hand, practolol does not contain a hydroxy group in either the 3 or 5 position and therefore lacks β_2 receptor selectivity but acts primarily as a β_1 antagonist.

β-Adrenergic Activation of Erythropoiesis in Polycythemic Mice

Note in Figure 16-7 that when the drug isoproterenol was injected on days 6 and 7 out of the tank as described under the Materials and Methods section in dosages of 50 up to 50,000 µgm/kg a very significant increase in radioactive incorporation in newly formed red cells was seen. When the specific β_2 adrenergic agonist salbutamol was injected in dosages of 50 to 5,000 µgm/kg subcutaneously (Figure 16-8), note the significant increase in iron incorporation of red cells with the dosage of 500 micrograms, making it a more potent stimulus than isoproterenol. Note in Figure 16-9 that DL-propranolol, a drug blocking both β_1 and β_2 receptors, completely inhibited the effects of salbutamol on β_2-adrenergic stimulated erythropoiesis in polycythemic mice. On the other hand, when the β_1 agonist dobutamine was used in a dosage of 500 to 50,000 µgm/kg subcutaneously no significant increase in radioactive iron incorporated in newly formed red cells of polycythemic mice was seen (Figure 16-10). It was of interest to learn whether the effects of a newly

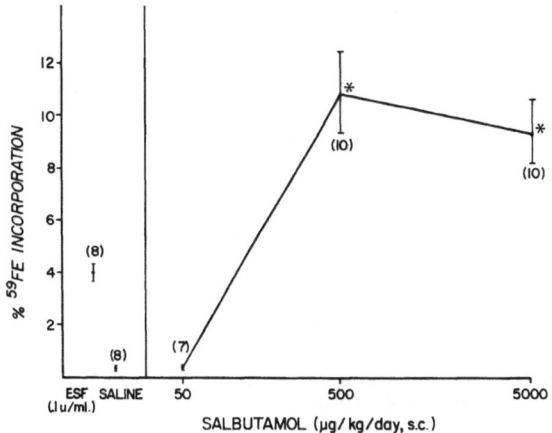

Figure 16-8 The effect of salbutamol on radioactive iron incorporation in red cells of exhypoxic polycythemic mice.

developed specific β_2 adrenergic agonist terbutaline was effective in stimulating iron incorporation in polycythemic mice. Note in Figure 16-11 the effects of a dosage of 500 µgm/kg terbutaline, which produced a more significant enhancement in iron incorporation than was seen with the same dosage of salbutamol. Interestingly, the effects of this potent β_2 agonist was completely blocked by DL-propranolol.

β_2-Adrenergic Activation of Erythropoietin Production

In that the polycythemic mouse responds to erythropoietic stimuli to increase iron incorporation in newly formed red cells by either activating kidney production of erythropoietin or stimulating erythropoiesis directly in the marrow, we decided to determine whether β_2-adrenergic activation of erythropoiesis

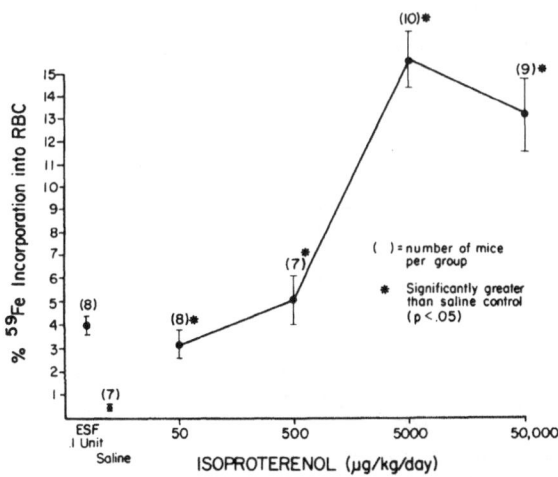

Figure 16-7 The effect of isoproterenol on radioactive iron incorporation in red cells of exhypoxic polycythemic mice. (From Fink and Fisher: *J Pharmacol Exp Ther* 202:192, 1977).

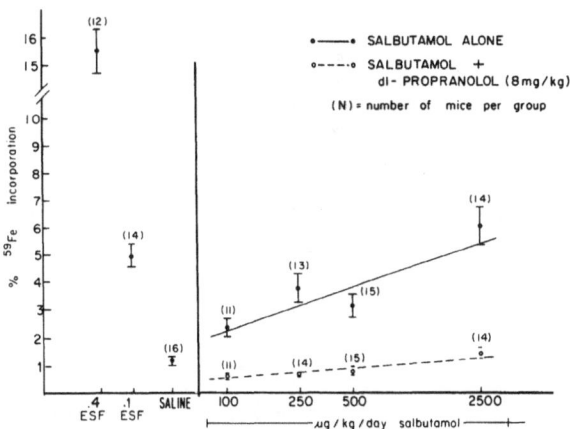

Figure 16-9 The effect of propranolol on salbutamol-induced erythropoiesis in polycythemic mice. Bars represent the SEM. Radioactive iron incorporation values in salbutamol-treated mice were significantly higher than those of mice treated with propranolol plus salbutamol at all doses of salbutamol studied. ESF = erythropoietin.

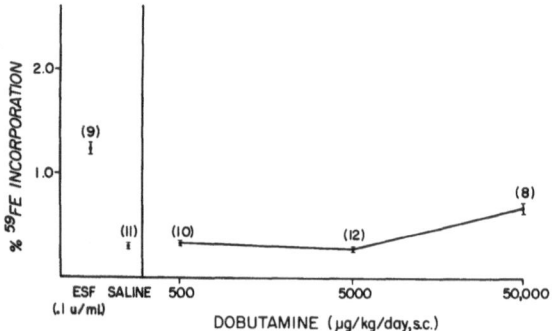

Figure 16-10 The effect of dobutamine on radioactive iron incorporation in red cells of exhypoxic polycythemic mice. n = number of mice per dose.

was related to enhancement of production of erythropoietin. These studies were carried out in the rabbit in which albuterol (salbutamol) was infused over a 7-hr period and plasma levels of erythropoietin determined. Note in Figure 16-12 that albuterol produced a significant elevation in plasma levels of erythropoietin after 7 hr infusion. The mean ^{59}Fe iron incorporation values on the 7 hr plasma sample when assayed in polycythemic mice was equivalent to 0.1 unit of erythropoietin per ml plasma. Butoxamine, the specific antagonist of β_2 receptors, completely blocked the rise in plasma levels of erythropoietin produced by terbutaline. These data clearly demonstrate that β_2-adrenergic agonist drugs are capable of enhancing erythropoietin production as a part of their mechanism of stimulation of erythropoiesis. Fink et al.[25] have found that these β_2-adrenergic agonists do not increase plasma levels of erythropoietin following bi-

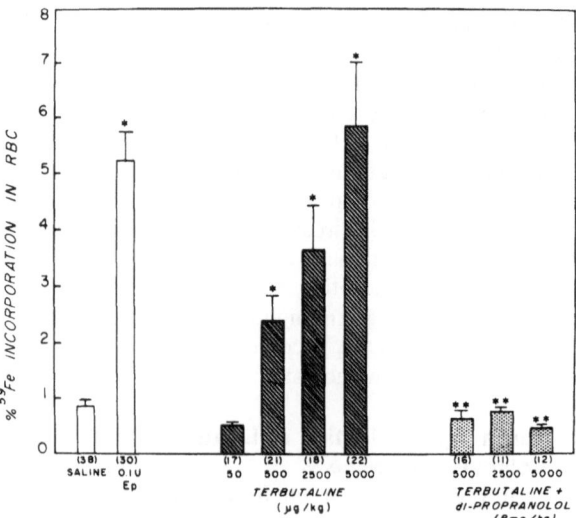

Figure 16-11 The effect of terbutaline on radioactive iron incorporation in red cells of exhypoxic polycythemic mice.

lateral nephrectomy, indicating that the kidney is the target site for the action of these agents.

Effects of β_2-Adrenergic Activation of Erythroid Colony Forming Cells (CFU-e)

Brown and Adamson[19,39] and Przala et al.[26] have previously demonstrated an increase in CFU-e in bone marrows of rats[19,39] and rabbits[26] following β_2-adrenergic activation. This β-adrenergic activation of erythropoiesis may be linked to the activation of a membrane-bound adenylate cyclase leading to an increased level of intracellular cAMP as postulated by

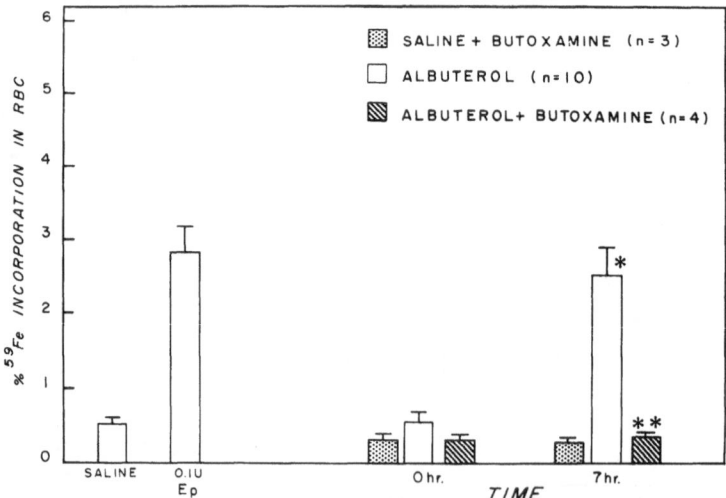

Figure 16-12 The effect of butoxamine pretreatment on albuterol-stimulated erythropoietin production. Each bar is the mean ± SE. * Significantly different from 0 hr and saline control at some interval ($p < .05$). ** Significantly different ($p < .05$ from albuterol infusion group at same interval.

Sutherland.[31] Neither albuterol nor terbutaline were demonstrated to produce an increase in ^{59}Fe incorporation in the extractable heme when studied alone *in vitro* in rabbit bone marrow cultures.[25,26] Byron[20] has reported that β-adrenergic agonists initiated DNA synthesis in mouse bone marrow and may effect the cycling of stem cells. The present studies were undertaken to determine the mechanism of β-adrenergic activation of erythropoiesis using the specific β$_2$-adrenergic agonist albuterol. Albuterol was used *in vitro* with erythropoietin to study the effects of β$_2$-adrenergic agonists on the erythroid colony-forming cells in the rabbit bone marrow culture system. Thus far our studies on β$_2$-adrenergic agonists have been limited to studies of the CFU-e. Note in Figure 16-13 that when albuterol (salbutamol) was added to the normal rabbit bone marrow culture system in dosages of 10^{-14} up to 10^{-4} M concentrations that a significant increase in erythroid colonies was seen at 10^{-10} and 10^{-8} M. Higher dosages of albuterol in the range of 10^{-6} to 10^{-4} M were less effective in enhancing erythroid colony formation than dosages of 10^{-10} to 10^{-8} M. DL-Propranolol (10^{-8} M) completely blocked the increase in erythroid colony formation induced by albuterol. Note in Figure 16-14 when the β$_2$ agonist drug terbutaline was used that a significant increase in erythroid colony formation was seen at concentrations of 10^{-10}, 10^{-8}, and 10^{-6} M. Higher dosages in the range of 10^6 to 10^{-4} M terbutaline were less effective in increasing erythroid colony formation than that of the lower dosages. Note again that DL-propranolol significantly inhibited

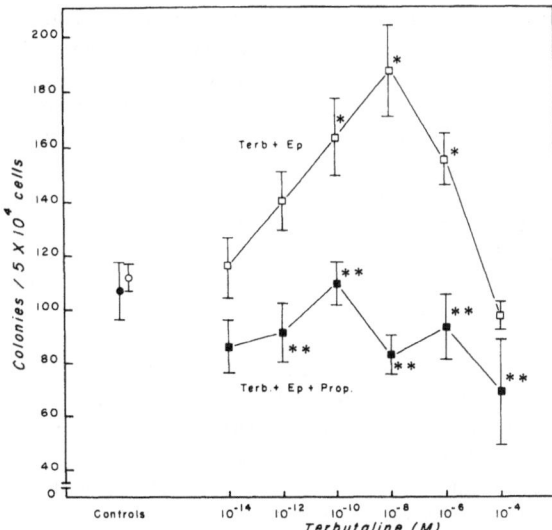

Figure 16-14 *In vitro* effects of terbutaline sulfate on the formation of erythroid colonies (CFU-e) in a rabbit bone marrow plasma clot culture system. Values are the means ± SE of quadruplicate determinations from five to seven separate experiments. Controls: open circles, Ep (0.02 unit/well); closed circles, Ep (0.02 unit/well) + DL-propranolol (10^{-8} M). Experimental: open squares, Ep (0.02 unit/well) + terbutaline sulfate; closed squares, Ep (0.02 unit/well) + terbutaline sulfate + DL-propranolol (10^{-8} M). *Significantly different from controls ($p < 0.05$). **Significantly different from terbutaline + Ep ($p < 0.05$).

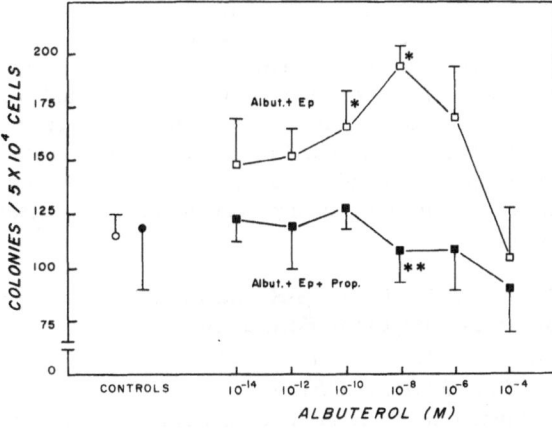

Figure 16-13 Effect of albuterol, erythropoietin and DL-propranolol on erythroid colony (CFU-e) formation in rabbit bone marrow. Each point is the mean ± SE of five separate experiments. Controls: open circles, Ep; closed circles, Ep + DL-propranolol. Experimental: open squares, Albuterol + Ep; closed squares, Albuterol + Ep + DL-propranolol. * Significantly different from control ($p < .05$). ** Significantly different ($p < .05$) from albuterol + Ep at same concentration.

this increase in erythroid colony formation induced by the β$_2$-adrenergic agonist drug terbutaline. Preliminary studies currently underway in our laboratory indicate that this increase in erythroid colony formation induced by terbutaline can be significantly antagonized by the specific β$_2$-adrenergic antagonist drug butoxamine.

Testosterone and Steroid Metabolites on Erythropoiesis

Androgenic steroids have been demonstrated to stimulate erythropoiesis in several mammalian species.[8–12,15–18] Androgenic steroids were demonstrated several years ago to stimulate erythropoiesis in plethoric mice[9,10,42,43] and erythropoietin production in the isolated perfused kidney.[17] Androgenic steroids have been used in the treatment of several refractory anemias, including aplastic anemia[44,45] and the anemia of renal insufficiency.[46] However, these steroids often produce masculinizing, anabolic and other hormone effects that may be related to their ability to stimulate erythropoiesis. During the past few years, several 5β-H steroid metabolites have been reported to increase heme synthesis *in vitro*.[47,48] When the polycythemic mouse assay has been used to assay these 5β-H steroids variable results have been obtained.[11,12,49-51] For example, some investigators have reported that the 5β-H steroids are more effective than testosterone in increasing iron incorporation in red cells of polycythemic mice.[11,12] On the other hand, other workers have reported that 5β-H steroids do not possess significant erythropoietic activity when administered to either mildly plethoric[51] or exhypoxic polycythemic mice.[17,49,50] In a recent well controlled study where several 5β-H androstanes and pregnanes were compared with testosterone, 5α-dihydrotestosterone, and 5β-dihydrotestosterone, only testosterone and 5α-dihydrotestosterone produced a significant increase in radioactive incorporation in red cells of exhypoxic polycythemic mice.[28] None of the 5β-H androstanes or pregnanes were found to possess significant erythropoietic activity in exhypoxic polycythemic mice in this cooperative assay.[52] On the other hand, Granick and Kappas,[47] Fisher et al.[50] and Singer and Adamson[29] have demonstrated that testosterone, 5β-H androstanes and some 5β-H pregnanes were effective in increasing erythroid colony growth. The purpose of this investigation is to compare the *in vivo* erythropoietic activities in polycythemic mice and *in vitro* effects on erythroid colony growth of testosterone, 5α-dihydrotestosterone, and 5β-dihydrotestosterone.

Structure Activity Relationship

Figure 16-15 shows the chemical structures of testosterone, 5α-dihydrotestosterone (5α-DHT), and 5β-dihydrotestosterone (5β-DHT), the three steroids that were used in these studies. Note that 5β-dihydrotestosterone deviates from the planarity of the steroid in the usual projection and is behind the plane of the steroid. The 5 hydrogen is *cis* to the 19-methyl group. They have been demonstrated to be essentially nonandrogenic. On the other hand, the 5α-DHT steroid configuration is planer and the 5-H is *trans* to the 19-methyl group. The 5α-DHT steroids are markedly an-

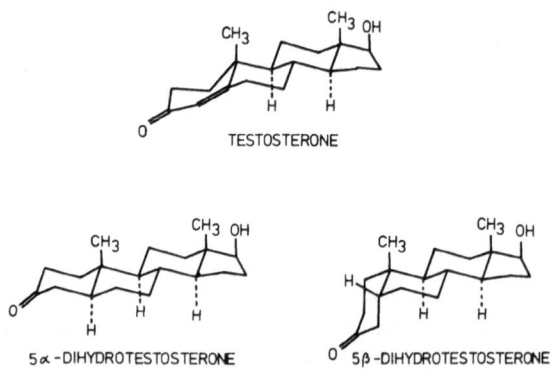

Figure 16-15 Chemical structures of testosterone, 5αdihydrotestosterone (5α-DHT) and 5β-dihydrotestosterone (5β-DHT).

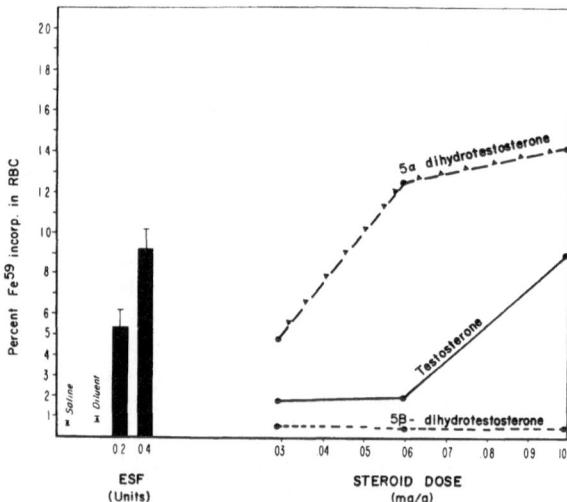

Figure 16-16 Effects of testosterone, 5α-dihydrotestosterone (5α-DHT) and 5β-dihydrotestosterone (5β-DHT) on radioactive iron incorporation in red cells of exhypoxic polycythemic mice. Lines at the top of each bar represent SEM. (From Fisher et al: *Blood Cells* 1:573, Springer-Verlag, 1975).

drogenic. The purpose of the present investigation was to compare the 5α-H and 5β-H androstanes as to their erythropoietic activities both *in vitro* and *in vivo*.

Effects of 5α-H and 5β-H Androstanes in Polycythemic Mice

As indicated in Figure 16-16, when testosterone was compared with 5α-DHT and 5β-DHT in exhypoxic polycythemic mice, 5α-dihydrotestosterone was the most potent steroid in stimulating iron incorporation in red cells of exhypoxic polycythemic mice. Testosterone itself did possess significant erythropoietic activity in all dosages tested. As we have reported previously[49,50] and as reported in a recent cooperative assay of several 5α and 5β steroids[52] 5β-dihydrotestosterone was found to be devoid of erythropoietic activity in our exhypoxic polycythemic mouse assay system.

Effects of 5α-H and 5β-H Androgens on Erythroid Colony Formation

Effects of In Vivo *Steroid Pretreatment on Bone Marrow CFU-e*

Testosterone, 5α-DHT, and 5β-DHT were administered to normal rabbits 18 hr before sacrifice in order to determine the effects of these steroids on bone marrow erythroid colony-forming cells (CFU-e). As shown in Figure 16-17 testosterone pretreatment (5 mg/kg s.c.) produced a significant ($p < .01$) increase in the number of CFU-e per 5×10^4 bone marrow cells in comparison with controls treated with the vehicle alone. In a similar series of experiments 5β-

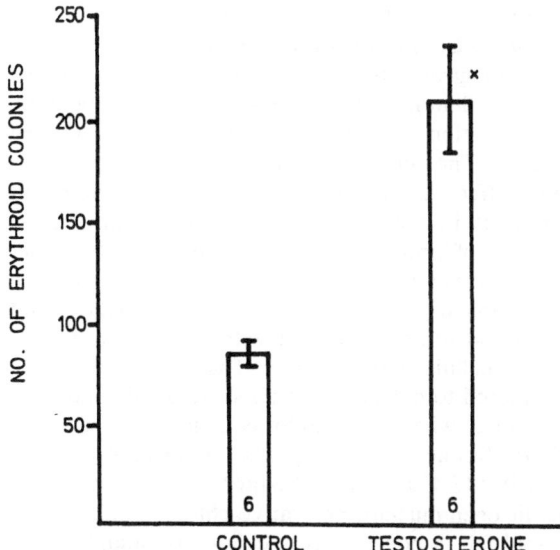

Figure 16-17 Erythroid colony formation in rabbit bone marrow cultures after *in vivo* pretreatment with testosterone. Testosterone —5 mg/kg i.m. and 0.2 unit Ep/ml culture media. Each bar is the mean ± SEM and the number at the bottom of each bar indicates the number of experiments. *Significantly different from control ($p < 0.001$).

themic mice 5α-DHT pretreatment did not produce a significant increase in the numbers of CFU-e in the normal rabbit bone marrow (Figure 16-19). Serum erythropoietin titers, when assayed via a modification of the method of Cotes and Bangham[33] in exhypoxic polycythemic mice, were undetectable in all animals studied. These data indicate that testosterone and the 5β-H steroids have a potent effect in increasing CFU-e numbers when administered *in vivo* to normal rabbits. The reason why 5β-DHT was not effective in the erythroid colony formation but highly effective in polycythemic mice requires further study.

In Vitro Effects of Testosterone, 5α-DHT and 5β-DHT on CFU-e and Normal Rabbit Bone Marrow Cultures

Plasma clot cultures from normal untreated rabbits were utilized to determine the *in vitro* effect of testosterone, 5α-DHT, and 5β-DHT on erythroid colony formation. Each steroid was added to the culture system in a final concentration ranging from 10^{-4} to 10^{-9} M. For all three steroids tested, 10^{-4} M was toxic and the numbers of colonies observed were far below control levels. As shown in Figure 16-20 testosterone in concentrations of 10^{-6} to 10^{-8} M produced a marked increase in CFU-e/5×10^4 bone marrow cells. A significant ($p < .01$) increase in CFU-e was noted in cultures containing 10^6 to 10^{-8} M 5β-DHT. Concentrations of 10^{-5} M and 10^{-9} M 5β-DHT were without effect. None of the concentrations of 5α-DHT employed produced a significant increase in

DHT was found to produce a significant ($p < .01$) increase in CFU-e in rabbit bone marrow cultures (Figure 16-18). This increase was similar to but slightly less than that seen following testosterone pretreatment. On the other hand, in contrast to the very marked erythropoietic effects in exhypoxic polycy-

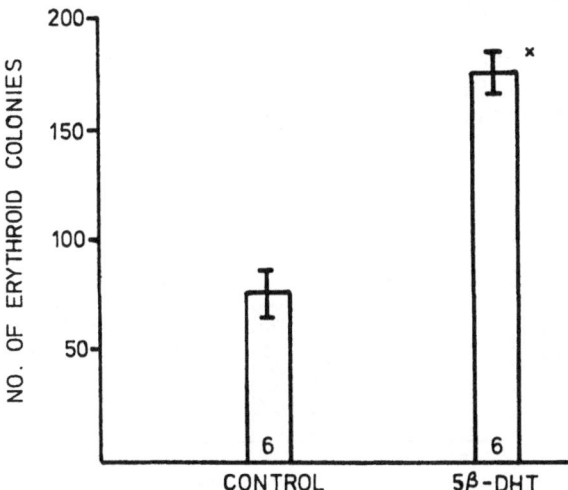

Figure 16-18 Erythroid colony formation in rabbit bone marrow cultures after *in vivo* pretreatment with 5β-dihydrotestosterone. 5β-dihydrotestosterone—5 mg/kg i.m. and 0.2 unit Ep/ml culture media. Each bar is the mean ± SEM and the number at the bottom of each bar indicates the number of experiments. *Significantly different from control ($p < 0.01$).

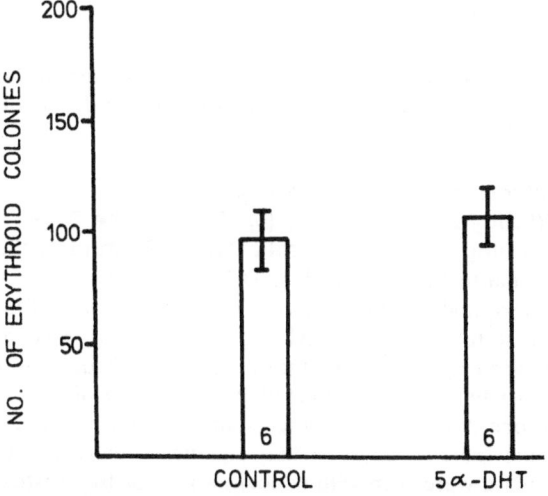

Figure 16-19 Erythroid colony formation in rabbit bone marrow cultures after *in vivo* pretreatment with 5α-dihydrotestosterone (5α-DHT). 5α-dihydrotestosterone—5 mg/kg i.m. and 0.2 unit Ep/ml culture media. Each bar is the mean ± SEM and the number at the bottom of each bar indicates the number of experiments. *Significantly different from control ($p < 0.01$).

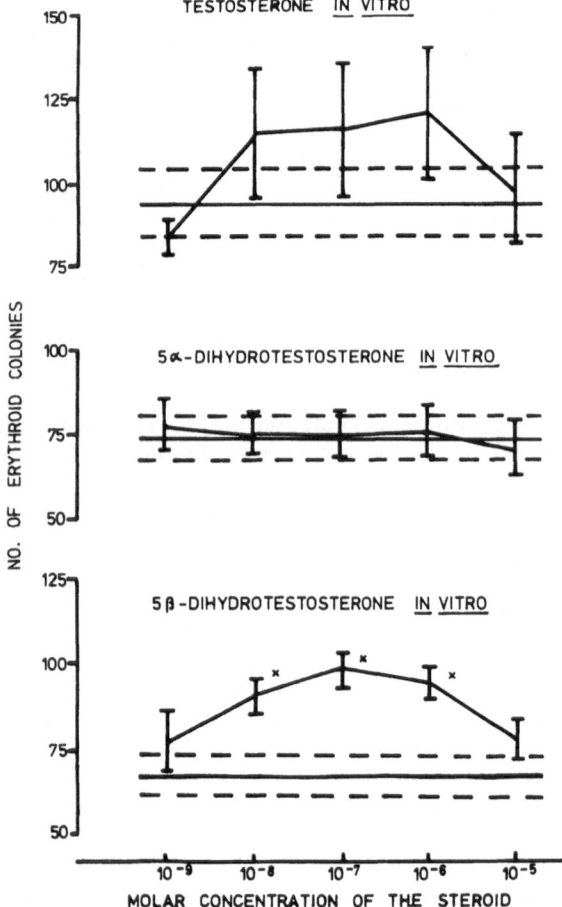

Figure 16-20 Effects of steroids on erythroid colony formation *in vitro*, mean ± SEM of four experiments. Vertical lines indicate mean ± SEM as percent of control. *Significantly different from control ($p < 0.01$)

CFU-e, which correlated well with our finding of the lack of a significant *in vivo* effect of 5α-DHT on CFU-e.

Effects of Testosterone and 5β-Dihydrotestosterone (5β-DHT) on the Burst Forming (BFU-e) Erythroid Colony-Forming Cells

In that testosterone and 5β-DHT were very effective *in vivo* and *in vitro* in increasing CFU-e, it was of interest to learn if the BFU-e compartment was increased by these two steroids. Note in Figure 16-21 that when concentrations of 10^{-6} to 10^{-8} M of testosterone and 5β-DHT were studied in normal rabbit bone marrow cultures that significant increases in BFU-e were seen with concentrations of 10^{-8} testosterone (Test) and 10^{-6} and 10^{-7} of 5β-DHT. It is quite possible that the increase in BFU-e seen with these two steroids is due to the stimulation of the pluripotent stem cell (CFU-s) compartment. However, direct effects of testosterone and 5β-DHT on BFU-e in the presence of Ep cannot be excluded.

Effects of Busulfan on Steroid-Induced Erythroid Colony Formation *In Vivo*

The biological action of busulfan on bone marrow cells was demonstrated several years ago to be done to an action on early, undifferentiated, resting cells.[53,54] The subsequent hematological and antileukemic effects of busulfan were found to be the result of a delay in cellular production rather than the destruction of already formed cells. In contrast, the actions of other nitrogen mustard derivatives such as chlorambucil were very rapid,[55] short-lived,[56] and was found to be associated with a considerable degree of cellular destruction.[53] Thus, chlorambucil is considered to damage cells about to divide and cells in mitosis, whereas busulfan is believed to decrease mitotic frequency but appears not to influence mitosis itself.[55] Csanyi and Elson[53] demonstrated that the nitrogen mustard derivative chlorambucil caused a rapid fall in bone marrow cellularity linked with a decrease in DNA content and DNA synthesis capacity. However, with busulfan the fall in number of nucleated cells in the bone marrow is a much slower process, recovery is more gradual, and myeloid regeneration lags behind erythroid recovery. Busulfan did not produce the initial fall in DNA synthesis capacity seen with chlorambucil but some increase in specific activity and labelling index occurred.[53] Csanyi and Elson[53] concluded from this work that the effect of busulfan can be explained by an effect of this agent on the stem cell population (CFU-s) which, while not resulting in actual killing of these cells, renders them temporarily unable to produce a normal differentiated cell population. Lajtha[56] explained this effect by assuming that a proportion of stem cells are in a resting or G_0 state and, as such, are relatively insensitive to nitrogen mustards such as chlorambucil but yet are very sensitive to the drug busulfan. In addition, Reissmann and Udupa[57] found that a single dose of busulfan suppressed CFU-s in polycythemic

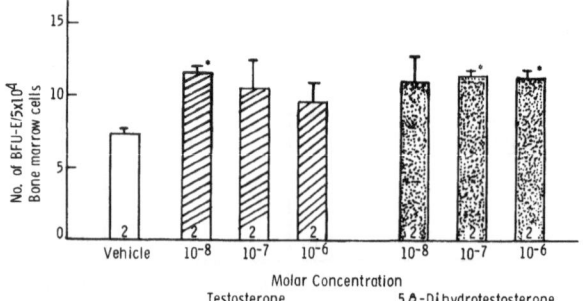

Figure 16-21 Effects of testosterone and 5β-dihydrotestosterone on burst-forming units (BFU-e) in normal rabbit bone marrow cultures. *Significantly ($p < .05$) different from control. The number at the bottom of each bar represents the number of culture experiments (3 plates/culture). I = Standard error of mean.

mice to approximately 1% of normal for a period of 2 weeks. In the absence of erythropoietin, CFU-e (ERC) were not detectable after busulfan for 12 days. Erythropoietin injections had no effect on CFU-s but restored the CFU-e population (as demonstrated indirectly by an increase in radioactive iron incorporation into heme) in proportion to the dose of erythropoietin. In that repeated injections of erythropoietin repopulated the CFU-e compartment without accelerating CFU-s recovery, Reissmann and Udupa[57] interpreted this data to mean that busulfan had an early selective effect on CFU-s without affecting the CFU-e compartment. It is well known that the broad class of alkylating agents affect biological receptors by two distinctly different mechanisms, unimolecular SN1 and biomolecular SN2 processes.[58] Compounds such as chlorambucil apparently react by an SN1 mechanism and are distributed throughout the organism and react at a rate determined by the dielectric constant of the medium, whereas the rate of reaction of compounds reacting by an SN2 mechanism such as busulfan would depend on the concentration of alkylatable centers. It is of interest that busulfan that reacts by an SN2 mechanism has been demonstrated to produce a greater depressant effect on circulating neutrophils, whereas the azo compound such as chlorambucil reacting by an SN1 mechanism produces a greater effect on lymphocytes.[59]

In the current study an attempt was made to use busulfan to suppress CFU-s very early with the hope of blocking any significant inflow into the CFU-e compartment. In this way, it should be possible to study the CFU-e compartment in the absence of any significant inflow from the CFU-s compartment. Thus, an attempt was made to determine the site of action of the steroids on CFU-e in rabbit bone marrow cultures by administering busulfan (20 mg/kg

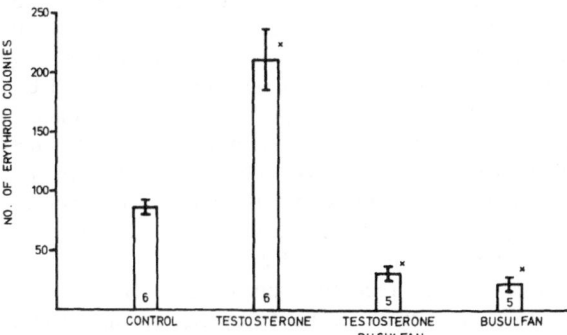

Figure 16-22 Erythroid colony formation in rabbit bone marrow cultures after *in vivo* pretreatment with testosterone and busulfan. Testosterone—5 mg/kg i.m., busulfan—20 mg/kg p.o., 0.2 unit Ep/ml culture media. Each bar is the mean ± SEM and the number at the bottom of each bar indicates the number of experiments. x = significantly different from control ($p < 0.001$).

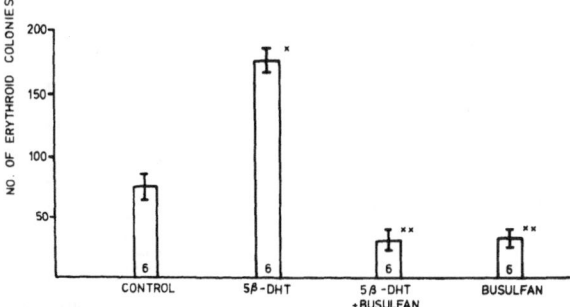

Figure 16-23 5β-dihydrotestosterone *in vivo*. x = significantly different from control ($p < 0.001$). xx = ($p < 0.01$).

p.o.) 24 hr before Test, 5β-DHT, or vehicle. The rabbits were exsanguinated by cardiac puncture and the femurs removed. Studies of the effects of Test and 5β-DHT on erythroid colony-forming cells were carried out on the bone marrow cultures according to the same procedure outlined in the Methods Section by a modification of the method of McLeod et al.[34] When busulfan (20 mg/kg), which is postulated to inhibit differentiation of the pluripotent stem cell (CFU-s), was administered 24 hr prior to testosterone, the increase in CFU-e seen in rabbits receiving testosterone alone was completely blocked (Figure 16-22). Even though in the presence of erythropoietin and busulfan the CFU-e were reduced to approximately one-third of that of the cultures with erythropoietin alone, significant CFU-e formation was seen in the busulfan-Ep treated marrow. The reduction in CFU-e by busulfan probably occurred due to inhibition of entry of BFU-e and subsequently CFU-e from the CFU-e compartment. In a similar manner 5β-DHT treatment while producing a significant ($p < .001$) increase in CFU-e in the presence of erythropoietin *in vitro*, pretreatment with busulfan completely blocked the increase in CFU-e seen with the steroid and Ep (Figure 16-23). These data suggest that the action of Test and 5β-DHT on erythroid colony formation may be due to an effect of the steroid in triggering the CFU-s or perhaps BFU-e to differentiate into the CFU-e compartment. Udupa and Reissmann[60] found that when the spleen colony-forming units (CFU-s) had been eradicated by busulfan in mice that they failed to show a granulopoietic response to androgens.

Analysis of Possible Sites of Action of Erythropoietin, Androgenic Steroids, and β₂-Adrenergic Agonists on Erythropoiesis

Figure 16-24 summarizes our postulated sites of action for erythropoietin, steroids, and β₂-adrenergic agonist drugs on erythropoiesis. It seems clear that a primary site of action of erythropoietin is on the com-

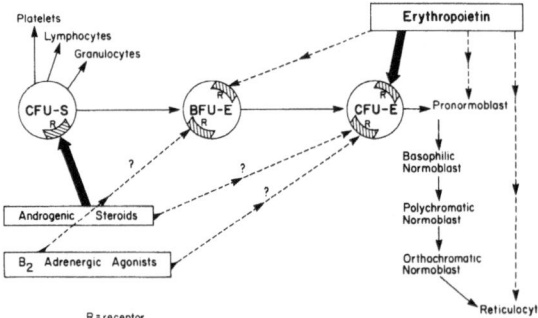

Figure 16-24 Model for sites of action of pharmacological agents on the erythroid stem cell compartment.

mitted erythroid stem cell compartment (CFU-e). In the presence of erythropoietin in bone marrow cultures, colonies of the CFU-e type usually appear within 2 days and consist of small differentiated erythroid colonies that vary from 8 to 60 cells and are presumed to arise from the late committed erythroid stem cells (CFU-e) that have a rather limited proliferative capacity.[61,62] The second type of colony arises usually after 8 to 10 days in culture but is comprised of very large numbers (approximately 10^4) of differentiated erythroid cells.[63,64] This second type of colony is thought to arise by extensive proliferation of stem cells, the so-called BFU-e, which are also committed to the erythroid line. Direct effects of high dosages of erythropoietin on the pronormoblast have also been reported.[4,6,65] After the pronormoblast erythroid precursor cells (80 to 90% pure) from fetal liver cells, isolated by an immune-lysis procedure, are cultured for 10 to 22 hr, the cellular level of globin messenger RNAs increased 5 to 100-fold during which time most of the cells differentiated to basophilic normoblasts and began the synthesis of globin chains.[66] However, hemoglobin was not detected cytochemically until 44 hr of culture.[65] Direct effects of erythropoietin have also been reported on δ-aminolevulinic acid synthetase, the rate-limiting enzyme for heme synthesis, in erythroid cells.[5]

The early release of reticulocytes demonstrated in the isolated perfused hind limbs[6,7] with high dosages of erythropoietin suggests a direct effect of erythropoietin on the late erythroid cell compartment and an increase in the apertures in the walls of bone marrow sinuses with cells in migration at a time when there is no increase in bone marrow cellularity.[67] However, the possibility that the impure erythropoietin used in these studies could have contained contaminants which accounted for the earlier release of reticulocytes cannot be excluded. It seems clear that the primary effect of erythropoietin in the physiological day-to-day control of erythropoiesis is at the level of the CFU-e causing its orderly replication and differentiation.

The effects of androgenic steroids and β_2-adrenergic agonist drugs on erythropoiesis are more complex. It seems apparent that both androgenic steroids[15,18,28,29,68,69] and β_2-andrenergic agonists[19,26,38] when given *in vivo* or *in vitro*, are capable of increasing the number of erythroid colony-forming cells. It also seems clear that androgens are capable of increasing the number of spleen colony-forming units, the CFU-s and granulocytic colonies (CFU-c) both of which were blocked by busulfan.[60] In addition, the effects of androgenic steroids in increasing CFU-e were also found to be blocked by busulfan.[15,28,68] With the assumption that busulfan blocked the proliferation of the pluripotent stem cell compartment (CFU-s) but did not significantly inhibit the CFU-e compartment directly,[57] it would appear that the primary effect of the androgenic steroids is on the CFU-s compartment. On the other hand, in view of the finding that the androgenic steroids increase CFU-s,[60] BFU-e,[70] and CFU-e,[15,18,28,29,68] the possibility cannot be excluded that any of these stem cells which are in the resting phase of the cell cycle can be recruited into the cell cycle by the androgens. If there are fewer CFU-e and BFU-e in a resting phase of the cell cycle, when compared to that of the CFU-s, this may explain the fact that the CFU-s are more sensitive to the effects of busulfan than the BFU-e and CFU-e.[57]

The stem cell compartment is apparently also triggered by β_2-adrenergic agonists leading to an increase in CFU-e.[19,26,38] Brown and Adamson have demonstrated using a velocity sedimentation gradient technique for bone marrow cell separation a slightly different population of cells responding to isoproterenol than that seen with erythropoietin.[19] However, much further work is necessary to determine whether the differences in profiles of the cells responding to isoproterenol and erythropoietin are significant and represent different target cells for Ep and B_2-adrenergic agonist drugs. It would be of interest to study further the effects of busulfan on β_2-adrenergic agonist activation of CFU-e as well as to determine whether β_2-adrenergic agonists significantly increase the CFU-s compartment. It is quite possible that β_2-adrenergic agonists effect all three stem cell compartments. However, further work is necessary to determine the particular stem cell compartment which responds predominantly to the β_2-adrenergic agonist drugs.

Some comment should be made concerning the nature of the receptors on CFU-s, BFU-e, and CFU-e. Recent experiments by Brown and Adamson[19,39] suggest that there may be receptors in the stem cell compartment linked to adenylate cyclase. The work of Sutherland[31] indicates that β-adrenergic agents may act on an adenylate cyclase membrane receptor. In addition, Brown and Adamson[39] have found that marrow erythroid colony growth was increased by dibutyryl cAMP and that this effect of cAMP is enhanced by a phosphodiesterase inhibitor. It is not known whether this type of adenylate cyclase erythroid cell membrane receptor is located on the CFU-s, BFU-e, and/or CFU-e. Further work is necessary to completely clarify the nature of the receptors for erythropoietin, androgens, and β_2-adrenergic agonists.

Summary

β_2-Adrenergic agonist drugs such as terbutaline and albuterol were demonstrated to produce a significant increase in erythroid colony-forming cells in normal rabbit bone marrow cultures. This activation of the β_2 receptors was blocked by DL-propranolol. β_2-Adrenergic agonist drugs were also found to increase plasma levels of erythropoietin that was blocked by the β_2-adrenergic blocker butoxamine. Testosterone and 5β-dihydrotestosterone were found to increase erythroid colony formation in rabbit bone marrow cultures both *in vivo* and *in vitro*. Testosterone and 5β-DHT also produced an increase in BFU-e in bone marrow cultures *in vitro*. 5α-Dihydrotestosterone was not active in increasing erythroid colony formation either *in vivo* or *in vitro*. The effects of testosterone and 5^b-dihydrotestosterone in stimulating erythroid colony formation was completely blocked by the drug busulfan. These data suggest that testosterone acts on the pluripotent stem cell compartments (CFU-s) to cause its orderly differentiation into the erythroid committed stem cell compartment. A model for the sites of action of erythropoietin, androgenic steroids, and β_2-adrenergic agonists is presented.

Acknowledgments

The studies reported in this paper were supported by USPHS Grant AM-13211 and USPHS Hematopharmacology Training Grant HL-05969.

The authors gratefully acknowledge the excellent technical assistance of Mr. Rene Stiaes, Mr. Jesse Brookins and Miss Patricia Dargon. The following companies are also due thanks for their generous supply of drugs: Allen and Hanburys (salbutamol), Ayerst Laboratories (propranolol and practolol), Burroughs Wellcome and Company (butoxamine) and Eli Lilly and Company (dobutamine).

References

1. Alpen EL, Cranmore D, Johnston ME: Early observation of the effects of blood loss, in Jacobson LO, Doyle M (eds): *Erythropoiesis*. New York, Grune & Stratton, p. 184, 1962.
2. Fried W, Plzak L, Jacobson LO, Goldwasser E: Studies on erythropoiesis. III. Factors controlling erythropoietin production. *Proc Soc Exp Biol Med* 94:237–241, 1957.
3. Stohlman F, Jr: Humoral regulation of erythropoiesis. XIV. A model for abnormal erythropoiesis in thalassemia. *Ann NY Acad Sci* 119: 578–585, 1964.
4. Stohlman F, Jr: Erythropoietin and erythroid cell kinetics, in Fisher JW ed: *Kidney Hormones*, New York and London, Academic Press, pp. 331–341, 1971.
5. Bottomley SS: Effect of erythropoietin on bone marrow delta-aminolevulinic acid synthetase and heme synthetase. *J Lab Clin Med* 74:445, 1969.
6. Fisher JW, Lajtha LG, Buttoo AS, Porteous DD: Direct effects of erythropoietin on the bone marrow of the isolated perfused hind limbs of rabbits. *Br J Haematol* 11:342–349, 1965.
7. Kuna S, Gordon AS, Morse BS, Lane FB, III, Charipper HA: Bone marrow function in isolated perfused hind legs of rats. *Am J Physiol* 196:769, 1959.
8. Alexanian R: Erythropoietin excretion in man following androgens. *Blood* 28:1007, 1966.
9. Fisher JW, Roh, BL, Halvorsen S: Inhibition of erythropoietic effects of hormones by erythropoietin antisera in mildly plethoric mice. *Proc Soc Exp Biol Med* 126:97–100, 1967.
10. Fried W, Gurney CW: Erythropoietic effect of plasma from mice receiving testosterone. *Nature* 296:1160, 1965.
11. Gordon AS, Zanjani ED, Levere RD, Kappas A: Stimulation of mammalian erythropoiesis by 5β-H steroid metabolites. *Proc Natl Acad Sci* 65: 919–924, 1970.
12. Gorshein D, Gardner FH: Erythropoietic activity of steroid metabolites in mice. *Proc Natl Acad Sci* 65:564–568, 1970.
13. Gorshein D, Oski FA, Delivoria-Papadopoulos M: Effect of androgens on the red cell 2,3'-diphosphoglycerate hemoglobin oxygen affinity and red cell mass in mammals. *Proc Soc Exp Biol Med* 147:616, 1974.
14. Gorshein D, Reisner EH, Gardner FH: Tissue culture of bone marrow. V. Effect of 5β-H steroids and cyclic AMP on heme synthesis. *Am J Physiol* 228:1024, 1975.
15. Moriyama Y, Fisher JW: Effects of testosterone and erythropoietin on erythroid colony formation in rabbit bone marrow cultures. *Life Sci* 15:1181–1188, 1974.
16. Necheles TF, Rai US: Studies on the control of hemoglobin synthesis: The *in vitro* stimulating ef-

fect of a 5β-H steroid metabolite on heme formation in human bone marrow cells. *Blood* 34:380, 1969.
17. Paulo LG, Fink GD, Roh BL, Fisher, JW: Effects of several androgens and steroid metabolites on erythropoietin production in the isolated perfused dog kidney. *Blood 43*:39–47, 1974.
18. Singer JW, Samuels AI, Adamson JW: Steroid and hematopoiesis. I. Effect of steroids on *in vitro* erythroid colony growth: Structure/activity relationship. *J Cell Phys 88*:127, 1976.
19. Brown James W, Adamson, JW: Modulation of *in vitro* erythropoiesis: The influence of β-adrenergic agonists on erythroid colony formation. *J Clin Invest 60*:70–76, 1977.
20. Byron JW: Evidence for a β-adrenergic receptor initiating DNA synthesis in hemopoietic stem cells. *Exptl Cell Res 71*:228–232, 1972.
21. Byron JW: Manipulation of the cell cycle of the hemopoietic stem cell. *Exptl Hematol 3*:44–53, 1975.
22. Congote LF, Solomon S: Isoproterenol stimulation of heme synthesis in cultures of human fetal liver from early gestation. *Endocrinol 100*:1303–1305, 1977.
23. Fink GD, Paulo LG, Fisher JW: Effect of beta-adrenergic blocking agents on erythropoietin production in rabbits exposed to hypoxia. *J Pharmacol Exptl Therap 193*:176–181, 1975.
24. Fink GD, Fisher JW: Stimulation of erythropoiesis by beta-adrenergic agonists. I. Characterization of activity in polycythemic mice. *J Pharmacol Exp Therap 202*:192–198, 1977.
25. Fink GD, Fisher JW: Stimulation of erythropoiesis by beta-adrenergic agonists. II. Mechanisms of action. *J Pharmacol Exptl Therap 202*:199–208, 1977.
26. Przala F, Gross DM, Dargon, PA and Fisher JW: Effect of in vitro beta-adrenergic activation on rabbit bone marrow erythroid colony forming cells. *Proc Soc Exptl Biol Med 155*:334–338, 1977.
27. Byron JW: Effect of steroids on the cycling of haematopoietic stem cells. *Nature 228*:1204, 1970.
28. Moriyama Y, Fisher JW: Increase in erythroid colony formation in rabbits following the administration of testosterone. *Proc Soc Exp Biol Med 149*:178–180, 1975.
29. Singer JW, Adamson JW: Steroids and hematopoiesis. II. The effect of steroids on *in vitro* erythroid colony growth: Evidence for different target cells for different classes of steroids. *J Cell Physiol 88*:135, 1976.
30. Rodgers GM, Fisher JW, George WJ: The role of renal adenosine 3′,5′-monophosphate in renal control of erythropoietin production. *Am J Med 58*:31–38, 1975.
31. Sutherland EW: Studies on the mechanism of hormone action. *Science 177*:401–408, 1972.
32. Brown JE, Adamson JW: Studies of the influence of cyclic nucleotides on *in vitro* hemoglobin synthesis. *Br J Haematol 35*:193–208, 1977.
33. Cotes PM, Bangham DR: Bioassay of erythropoietin in mice made polycythemic by exposure to air at a reduced pressure. *Nature (Lond) 191*:1065–1067, 1961.
34. McLeod DL, Shreeve, MM, Axelrad AA: Improved plasma culture system for production of erythrocytic colonies *in vitro*: quantitative assay method for CFU-E. *Blood 44*:517–534, 1974.
35. Heath DS, Axelrad AA, McLeod DL, Shreeve MM: Separation of the erythropoietin-responsive progenitors BFU-E and CFU-E in mouse bone marrow by unit gravity sedimentation. *Blood 47*:777–792, 1976.
36. Dunnett CW: Multiple comparison procedure for comparing several treatments with a control. *J Am Stat Assoc 50*:1096–1121, 1955.
37. Sokal RA, Rohlf FJ: Biometry: the Principles and Practice of Statistics in Biological Research. San Francisco, W. H. Freeman, 1969.
38. Przala F, Gross D, Fisher JW: Influence of albuterol on erythropoietin production and erythroid stem cell activation. *Am J Physiol* (submitted for publication).
39. Brown W, Adamson JW: Modulation of *in vitro* erythropoiesis: Enhancement of erythroid colony growth by cyclic nucleotides. *Cell Tissue Kinet 10*:289–298, 1977.
40. Byron JW: Effect of steroids and dibutyryl cyclic AMP on the sensitivity of haematopoietic stem cells to 3 H-thymidine *in vitro*. *Nature 234*:39, 1971.
41. Rodgers GM, Fisher JW, George WJ: Increase in hematocrit, hemoglobin and red cell mass in normal mice after treatment with cyclic AMP. *Proc Soc Exp Biol Med 148*:380–382, 1975.
42. Fried W, Gurney CW: Use of mild plethora to demonstrate an erythropoietic effect from small amounts of androgens. *Proc Soc Exp Biol Med 120*:519, 1965.
43. Gordon AS, Mirand EA, Wenig J, Katz R, Zanjani ED: Androgen action on erythropoiesis. *Ann NY Acad Sci 149*:318–335, 1968.
44. Duarte K, Sandoval RL, Esquivel F, Sanchez-Medal L: Androstane therapy aplastic anaemia. *Acta Haematol 47*:140–145, 1972.
45. Shahidi NT, Diamond LK: Testosterone-induced remission in aplastic anemia of both acquired and congenital types. *New Engl J Med 264*:953, 1961.
46. Hendler D, Coffinet JA, Ross S, Longnecker R, Bakovic E: Controlled study of androgen therapy in anemia of patients on maintenance hemodialysis. *New Engl J Med 291* (20):1046–1051, 1974.
47. Granick S, Kappas A: Steroid control of pro-

phyrin and heme biosynthesis: A new biological function of steroid hormone metabolites. *Proc Natl Acad Sci USA 57:*1463, 1967.
48. Gulinato AM, Salvatorelli G: Preliminary results of the action of some steroid hormones on organotypic cultures of chick (embryo) bone marrow. *Anat Comp 2:*23, 1967.
49. Fisher JW, Samuels AI, Malgor LA: Androgens and erythropoiesis. *Israel J Med Sci 7:*892, 1971.
50. Fisher JW, Moriyama Y, Modder B: Effects of steroids on *in vitro* erythroid colony growth, in *International Symposium on Aplastic Anemia.* Kyoto, Japan, September 3–6, 1976.
51. Schooley JC, Mahlmann LJ: Stimulation of erythropoiesis in the plethoric mouse by cyclic AMP and its inhibition by antierythropoietin. *Proc Soc Exp Biol Med 137:*1289, 1971.
52. Fisher JW, Adamson JW, Fried WA, Gordon AS, Camiscoli JF, Schooley J, Zanjani E: Cooperative erythropoietic assay of several steroid metabolites in polycythemic mice. *Steroids* (in press).
53. Csanyi E, Elson LA: Action of 'Mustard' and 'Myleran' type alkylating agents on the cellular proliferation and DNA metabolism of the haemopoietic organs of the rat. *Ann Rep Br Emp Cancer Campaign 11:*32, 1967.
54. Elson LA: Effect of 'radiomimetic' chemicals on leukocytes, in *Proceedings of the Eighth Congress of the European Society of Haematology.* Karger, Basel, p. 340, 1962.
55. Elson LA, Galton DAG, Till, M: The action of chlorambucil (CB 1958) and busulphan (Myleran) on the haemopoietic organs of the rat. *Br J Haemat 4:*235, 1958.
56. Lajtha LG: Kinetics of a bone marrow stem cell population. *Ann NY Acad Sci 113:*742, 1964.
57. Reissmann KR, Udupa KB: Effect of erythropoietin on proliferation of erythropoietin-responsive cells. *Cell Tissue Kinet 5:*481–498, 1972.
58. Ross WCJ: Biological Alkylating Agents. London, Butterworth, 1972.
59. Elson LA: Radiation and radiomimetic chemicals. London, Buttersworth, 1963.
60. Udupa KB, Reissmann KR: Acceleration of granulopoietic recovery by androgenic steroids in mice made neutropenic by cytotoxic drugs. *Cancer 34:*2517–2520, 1974.
61. Iscove NN, Sieber F, Winterhalter KH: Erythroid colony formation in cultures of mouse and human bone marrow: Analysis of the requirement for erythropoietin by gel filtration and affinity chromatography on agarose-concanavalin A[1]. *J Cell Physiol 83:*309–320, 1974.
62. Stephenson JA, Axelrad AS, McLeod DL, Shreeve M: Induction of colonies of hemoglobin-synthezing cells by erythropoietin *in vitro. Proc Natl Acad Sci 68:*1542–1546, 1971.
63. Axelrad AA, McLeod DL, Shreeve MM, Heath DS: in Robinson WA (ed): *Hemopoiesis in culture.* 226–237 DHEW Publication No. NIH 74-205, 1973.
64. Iscove NN, Sieber F: Erythroid progenitors in mouse bone marrow detected by macroscopic colony formation in culture. *Expl Hemat 3:*32, 1975.
65. Ramirez F, Gambino R, Maniatis GM, Rifkind RA, Marks PA, Bank A: Changes in globin messenger RNA content during erythroid cell differentiation. *J Biol Chem 250:*6054–6058, 1975.
66. Terada M, Cantor L, Metafora S, Rifkind RA, Bank A, Marks PA: Globin messenger RNA activity in erythroid precursor cells and the effect of erythropoietin. *Proc Natl Acad Aci 69:*3575–3579, 1972.
67. Chamberlain JK, LeBlond PF, Weed RI: Reduction of adventitial cell cover: An early direct effect of erythropoietin on bone marrow ultrastructure. *Blood Cells 1:*655–674, 1975.
68. Fisher JW, Moriyama Y, Modder B: Effects of steroids on *in vitro* erythroid colony growth and erythropoietin production. *Symposium Aplastic Anemia* Tokyo, Japan. University of Tokyo Press, pp. 65–79, 1977.
69. Moriyama Y, Fisher JW: Effects of testosterone and erythropoietin on erythroid colony formation in human bone marrow cultures. *Blood 45:*665–670, 1975.
70. Peschle C, Magli MC, Lettieri F, Cillo C, Genoveve A, Pizzella F: Increased erythroid burst formation after treatment with testosterone propionate. *Life Sci 21:*773–778, 1977.

17 *In Vitro* Assessment of Similarities between Erythroid Precursors of Fetal Sheep and Patients with Polycythemia Vera

Esmail D. Zanjani, Rona Singer Weinberg, Benet Nomdedeu, Manuel E. Kaplan, and Louis R. Wasserman

Introduction

Erythrocyte production in normal animals and in man is regulated by the hormone erythropoietin (Ep).[1,2] Alterations in the rate of production of this hormone can lead to changes in the numbers of circulating erythrocytes.[3] It is generally accepted that the increased production of erythrocytes in all cases of secondary polycythemias is associated with increased Ep formation.[4,5] Such an association cannot be demonstrated in polycythemia vera (PV). In this regard, the decreased levels of Ep in PV,[6] and the unusual *in vitro* behavior of erythroid precursors[2,7–10] have provided evidence for the existence of at least two populations of erythroid stem cells in this disorder. *In vitro* studies have shown that while both cell populations are responsive to Ep, one appears to be significantly more sensitive to Ep.[2,9] This population of erythroid precursors is believed to be responsible for the formation of "endogenous erythroid colonies (EEC)",[2] and probably represents the more active of the two populations *in vivo*.[10] Erythropoiesis in the normal mammalian fetus is also regulated by Ep.[11–13] Unlike in the normal adult, however, when hematopoietic precursors of the fetus are cultured in the plasma clot, a significant number of EEC are formed.[14,15] We have investigated the mechanism of EEC formation by fetal erythroid precursors *in vitro* and have found that (a) EEC-forming cells in both the fetus and in patients with PV represent a distinct component of the erythroid precursor compartment, and (b) the proliferation and/or differentiation of the EEC-forming cell is subject to regulation by Ep.

Materials and Methods
Human Subjects

All bone marrow aspirations were performed at the posterior iliac crest. In all instances, informed consent was obtained from both patients and normal donors. Three patients with PV were studied. None had received any prior chemotherapy or radioactive phosphorous. All were diagnosed to have PV based on criteria adopted by the National Polycythemia Vera Study Group, had elevated red cell mass (48.5, 44.6, and 52.7 ml/kg), and exhibited below normal levels of Ep. At the time of the study, two patients were untreated and one had been phlebotomized (3 units) 3 weeks earlier. White blood cell counts were elevated in all three, platelets were increased in one, and spleen was palpable in two. Three hematologically normal donors served as controls.

Fetal Sheep

Bone marrow cells were obtained from three sheep fetuses at 120 days of gestation by aspiration from both femors (removed by surgery). Maternal bone marrow cells were obtained at the time of sacrifice by suction through openings in surgically exposed femors.

Bone Marrow Cultures

Bone marrow cells (human and sheep) were cultured in the plasma clot culture system described by Tepperman et al.[16] Dispersed bone marrow cells were cultured in the presence or absence of either 2 IU human urinary Ep (human bone marrow) or 0.4 IU Step III sheep plasma Ep (sheep bone marrow) at 37°C in a humidified atmosphere of 5% CO_2 in air. The clots were fixed, stained, and examined at either 3 to 4 days (sheep) or 7 days (human).

In studies designed to ascertain the role of Ep in EEC formations, dispersed bone marrow cells were cultured in the presence or absence of Ep in media composed of either untreated fetal calf serum (FCS) and citrated bovine plasma (CBP), normal rabbit serum IgG treated FCS and CBP, or FCS and CBP treated with serum IgG from rabbits immunized against human urinary Ep. Anti-Ep was prepared in rabbits against a preparation of Ep (5 IU/mg protein) obtained from urine of a patient with aplastic anemia after the procedure described by Schooley et al.[17] Another preparation of human urinary Ep with a greater potency (630 IU/mg protein) was used to boost the antibody titers in these animals. One ml of the immune serum after absorption against normal human urinary protein was found to neutralize approximately 90 IU of human urinary Ep. Absorption against human urinary protein was achieved by incubating 20 ml of the antiserum with an amount of urinary protein separated by dialysis and ultrafiltration from 3 liters of normal urine for 1 hr at 37°C followed by an additional incubation for 5 hr at 4°C. The mixture was centrifuged at 20,000 g for 15 min and the supernate was used in subsequent studies. In these studies, the IgG fraction of the immune serum separated by DEAE-Sephadex (Pharmacia Fine Chemicals Inc., Piscataway, New Jersey) chromatography[18] was used. Similarly prepared gammaglobulin fraction from normal rabbit serum was used as control. Treatment of FCS and CBP with IgG was as follows. To 50 ml of each was added either 3 mg of normal rabbit serum IgG or IgG from immunized rabbits, incubated at 37°C for 2 hr with constant shaking and allowed to stand at 4°C for 22 hr. To each mixture was then added 6 mg of IgG separated from a commercially prepared goat anti-rabbit gammaglobulin serum (GARGG) (Antibodies Inc., Davis, California), mixed thoroughly, and then allowed to stand for 2 hr at 4°C. The mixtures were spun at 10,000 g for 20 min, the precipitate discarded, and the supernate was then passed through a Millipore filter (Millipore Corp., Bedford, Massachusetts) and used in cultures. The untreated FCS and CBP were also subjected to a similar procedure but without the addition of the IgG to insure uniform conditions. The use of IgG fraction was dictated by the fact that addition of whole serum to these cultures proved to be toxic to the cells.

Unit Gravity Sedimentation

The procedure employed has been described by Miller and Phillips.[19] Bone marrow cells (1.5×10^8 cells) suspended in 20 ml NCTC–109 with 2% FCS were loaded in the chamber (diameter, 11.5 cm) and allowed to sediment for 4.0 hr at 4°C. One percent and 2% solutions of BSA were used to prepare the gradient. Fractions were collected in 10 ml volume at a rate of 15 ml/min. The first 50 ml of medium was discarded. Each fraction was centrifuged at 1,000 rpm for 10 min and the cells resuspended in NCTC–109 at desired concentrations. These cells were cultured as described above.

Results

The results of culturing normal human bone marrow cells in the presence or absence of Ep are shown in Table 17-1. No erythroid colonies were formed in the absence of exogenous Ep. When Ep was present, a progressive increase in the number of colonies formed was noted as a function of increasing concentrations of cells plated. Table 17-2 demonstrates that significant numbers of colonies (EEC) were formed when bone marrow cells from patients with PV were cultured in the absence of exogenous Ep. Addition of Ep resulted in further enhancement of erythroid colony formation. Erythroid colony formation by maternal (adult) sheep bone marrow cells is shown in Table 17-3. A linear increase in colony formation occurred when increasing numbers of cells were cultured in the presence of Ep. However, no colonies were produced when no exogenous Ep was present. By contrast, as is shown in Table 17-4, bone marrow cells from 120 day-old fetal sheep gave rise to significant numbers of EEC in the absence of exogenous Ep. Significantly greater numbers of colonies were formed when 0.4 IU Ep was added to the culture.

The results presented in Table 17-5 demonstrate that pretreatment of FCS and CBP used in culture with anti-Ep caused a significant inhibition of EEC formation by PV bone marrow cells. Such a treatment had no demonstrable effect on the total number of erythroid colonies formed in the presence of exogenous Ep. Treatment of FCS and CBP with normal rabbit serum IgG exerted no significant influence on

Table 17-1 Erythroid colony (CFU-e) formation by normal human bone marrow cells *in vitro*.

	Total No. CFU-e[a]	
No. of Cells Cultured	No Ep	With Ep
2×10^5	0	247 ± 39
4×10^5	0	392 ± 40
6×10^5	0	587 ± 28
8×10^5	0	947 ± 46

[a] Each value represents the mean ± 1 SEM of three separate studies, each involving a different donor.

Table 17-2 Erythroid colony (CFU-e) formation by bone marrow cells from patients with polycythemia vera (PV) *in vitro*.

	Total No. CFU-e[a]	
No. of Cells Cultured	No Ep	With Ep
2×10^5	82 ± 18	384 ± 52
4×10^5	129 ± 20	532 ± 68
6×10^5	247 ± 28	774 ± 44
8×10^5	297 ± 24	1047 ± 93

[a] Each value represents the mean ± 1 SEM of three separate studies, each involving a separate donor.

Table 17-3 Erythroid colony (CFU-e) formation by bone marrow cells from maternal sheep *in vitro*.

	Total No. CFU-e[a]	
No. of Cells Cultured	No Ep	With Ep
2×10^5	0	279 ± 36
4×10^5	0	642 ± 71
6×10^5	0	814 ± 80
8×10^5	0	1106 ± 112

[a] Each value represents the mean ± 1 SEM of 11 different studies, each involving a separate donor sheep.

Table 17-4 Erythroid colony (CFU-e) formation by bone marrow cells from 120 day-old sheep fetuses *in vitro*.

	Total No. CFU-e[a]	
No. of Cells Cultured	No Ep	With Ep
2×10^5	124 ± 18	408 ± 51
4×10^5	198 ± 36	749 ± 101
6×10^5	262 ± 49	983 ± 82
8×10^5	344 ± 47	1472 ± 177

[a] Each value represents the mean ± 1 SEM of three separate studies, each involving a different fetal sheep.

erythroid colony formation in the presence or absence of Ep (Table 17-5).

The effect of the neutralization of Ep present in FCS and CBP on EEC formation by fetal sheep bone marrow cells is shown in Table 17-6. Formation of EEC was significantly inhibited by the pretreatment of FCS and CBP with anti-Ep, a process that exerted no influence on colony formation when exogenous Ep was present. Table 17-7 demonstrates that cells giving rise to EEC in both PV and fetal sheep possess physical and physiological characteristics that are different from those exhibited by normal CFU-e. However, the proliferative and/or differentiative activity of EEC-forming cells like those of CFU-e requires the presence of Ep (Table 17-7).

Discussion

In vitro culture procedures for the growth and differentiation of erythroid precursors have become increasingly useful as a tool in the study of mechanisms underlying normal and abnormal erythropoiesis in animals and man. In polycythemia vera, a panmyelopathy involving all three formed elements of the blood, the use of these procedures has helped establish the clonal nature of this disorder,[10] as well as the role of Ep in erythropoiesis.[2,9] A regulatory role of Ep in erythropoiesis was suggested by *in vivo* observations that patients with PV respond to anemia[20] and hypoxia[21] by increased red cell production. The finding that this increase occurred only in those patients in whom a rise in Ep production was detected[22] further strengthened this belief. The *in vitro* evidence supporting this view has been derived from the observations that the formation of the so-called "endogenous erythroid colonies," once thought to be Ep-independent, was in fact regulated by Ep.[2,9]

Unlike normal human bone marrow, cells from marrow of patients with PV give rise to significant numbers of erythroid colonies (EEC) in the absence of exogenous Ep. The results presented here, as well as those previously reported,[2,9] indicate that the EEC-forming cell is far more sensitive to Ep than the CFU-e. This is a characteristic that is shared by certain erythroid precursors present in the fetus. Thus, as in PV, bone marrow cells from fetal sheep gave rise to EEC in the absence of exogenous Ep. Adult sheep bone marrow produced colonies only when exogenous Ep was present. Inhibition of EEC formation by fetal marrow was achieved when the cells were deprived of the small amounts of Ep present in culture medium. Similar suppression of EEC formation was reported for PV marrow[2] and was also demonstrated here. This was also true when the EEC-forming cells were separated by sedimentation at unit gravity and examined for their ability to form EEC in the total absence of Ep. The results in Table 17-7 show that when these cells were cultured in media containing FCS-CBP that had been pretreated with anti-Ep, the numbers of EEC formed were reduced by nearly 85%. It appears, therefore, that in the fetus and in patients with PV there exists a population of erythroid precursors with exquisite sensitivity to Ep. In PV, the activity of this population may be responsible for the increased circulating red cell mass.[10]

The EEC-forming cell sedimented with an average

Table 17-5 Effect of pretreatment of fetal calf serum and citrated bovine plasma with anti-erythropoietin (Ep) on erythroid colony (CFU-e) formation by bone marrow cells from patients with polycythemia vera in the presence or absence of Ep *in vitro*.[a]

	Total No. CFU-e					
	No Ep			With Ep[b]		
Donors	Untreated	Normal IgG Treated	Anti-Ep Treated	Untreated	Normal IgG Treated	Anti-Ep Treated
1	192	214	13	693	649	748
2	262	253	21	784	812	732
3	287	346	10	845	797	783

[a] See text for explanation.
[b] 2 IU human urinary Ep was used.

Table 17-6 Effect of pretreatment of fetal calf serum and citrated bovine plasma with anti-erythropoietin (EP) on erythroid colony (CFU-e) formation by bone marrow cells from fetal sheep in the presence or absence of Ep *in vitro*.[a]

	Total No. CFU-e					
	No Ep			With Ep[b]		
Donors	Untreated	Normal IgG Treated	Anti-Ep Treated	Untreated	Normal IgG Treated	Anti-Ep Treated
1	276	243	8	1208	939	1002
2	322	361	0	1399	1512	1174
3	434	359	7	1809	1264	1593

[a] See text for explanation.
[b] 0.4 IU Step III Sheep Plasma erythropoietin was used.

velocity of 8.9 (PV) and 7.8 (fetal sheep) mm/hr, while the sedimentation rates for CFU-e were 5.7 and 6.3 mm/hr respectively. This difference suggests that a distinctly separate cell population is involved in the formation of EEC. These cells appear to be larger than CFU-e (i.e., have a higher modal sedimentation velocity) and do not require large concentrations of Ep for proliferation and differentiation. Moreover, tritiated thymidine suicide studies have shown that these cells exist at a higher proliferative state than CFU-e. These findings are compatible with the EEC-forming cell representing a more differentiated component of the erythroid precursor series. Their relationship to CFU-e is probably similar to that of the latter cell and the BFU-e. Progressively lower concentrations of Ep appear to be needed as the erythroid stem cell becomes more differentiated, with the EEC-forming cell comprising the most differentiated functionally identifiable component of the series.

It is apparent from these studies that at least some of the erythroid precursor cells in patients with PV and mammalian fetuses are far more sensitive to Ep than the normal adult CFU-e. At least in PV, the presence of these cells is compatible with increased erthropoiesis in the face of low levels of circulating Ep in these patients. However, the relationship between fetal erythropoiesis and red cell production in PV is not clear. In patients with PV, hematopoiesis is not confined to the bone marrow but can frequently occur in the spleen and the liver.[23] This feature of the disorder is not shared by the normal mammalian adult; rather, it is a main characteristic of fetal hematopoiesis. Whether these shared characteristics (EEC-forming cells and extramedullary hematopoiesis) are indicative of a reversion to fetal-type hematopoiesis in PV remains to be determined.

Acknowledgments

The studies reported here were supported by grants CA–18755, CA–23021, CA–10728 from the National Cancer Institute of the National Institutes of Health, and Verterans Administration Research Fund.

Ep used in the studies reported in Tables 17-1–7

Table 17-7 Erythroid colony (CFU-e) formation by different sedimentation velocity fractions of bone marrow cells from fetal sheep and patients with polycythemia vera (PV) in the presence or absence of erythropoietin (Ep) *in vitro*.[a]

		Total No. CFU-e[b]		
		No Ep		With Ep
Marrow Donors	Average Sedimentation Velocity	Untreated	Anti-Ep Treated	Untreated
Fetal Sheep	6.3 ± 0.3 mm/hr	<10	0	876 ± 134
Fetal Sheep	7.8 ± 0.4 mm/hr	598 ± 108	84 ± 29	982 ± 170
PV	5.7 ± 0.4 mm/hr	<10	0	912 ± 178
PV	8.9 ± 0.3 mm/hr	388 ± 32	37 ± 13	682 ± 80

[a] Cells were cultured in media containing either untreated fetal calf serum (FCS) and citrated bovine plasma (CBP), FCS and CBP that were treated with normal rabbit serum IgG, or FCS and CBP which were pretreated with anti-Ep.
[b] Values are the mean ± 1 SEM of three separate studies, each involving a different donor.

was collected and concentrated by the Department of Physiology, University of Northeast, Corrientes, Argentina, and further processed and assayed by the Hematology Research Laboratories, Children's Hospital of Los Angeles, Los Angeles, California, under grant HE-10880 from the National Heart, Lung and Blood Institute of the National Institutes of Health.

References

1. Gordon AS: Erythropoietin. *Vit Horm 31:*105, 1973.
2. Zanjani ED, Lutton JD, Hoffman R, et al: Erythroid colony formation by polycythemia vera bone marrow *in vitro:* dependence on erythropoietin. *J Clin Invest 59:*841, 1977.
3. Gurney CW: Pathogenesis of the polycythemias, in Klein H (ed): *Polycythemia: Theory and Management.* Springfield, Ill., Thomas, p. 42, 1973.
4. Krantz SB, Jaconson LO: Erythropoietin and the Regulation of Erythropoiesis. University of Chicago Press, Chicago, Ill. pp. 1-330, 1970.
5. Ossias AL, Zanjani ED, Zalusky R, et al: Studies on the mechanism of erythrocytosis associated with a uterine fibroma. *Br J Hematol 25:*179, 1973.
6. Adamson JW: The erythropoietin/hematocrit relationship in normal and polycythemic man: implications of marrow regulation. *Blood 32:* 597, 1968.
7. Krantz SB: Response of polycythemia vera marrow to erythropoietin *in vitro. J Lab Clin Med 71:*999, 1968.
8. Zucker S, Howe DM, Weintraub LR: Marrow response to erythropoietin in polycythemia vera and chronic granulocytic leukemia. *Blood 39:* 341, 1972.
9. Golde DW, Cline MJ: Erythropoietin responsiveness in polycythemia vera. *Br J Hematol 29:* 567, 1975.
10. Adamson JW, Fialkow PJ, Murphy S, et al: Polycythemia vera: stem cell and probable clonal origin of the disease. *New Engl J Med 295:*913, 1976.
11. Zanjani ED, Horger EO, Gordon AS, et al: Erythropoietin production in fetal lamb. *J Lab Clin Med 74:*782, 1969.
12. Zanjani ED, Poster JP, Mann LI, et al: Erythropoietin production in the fetus: role of the kidney and maternal anemia. *J Lab Clin Med 83:*281, 1974.
13. Zanjani ED, Man LI, Burlington H, et al: Evidence for a physiological role of erythropoietin in fetal erythropoiesis. *Blood 44:*285, 1974.
14. Stephenson JR, Axelrad AA, McLeod DL, et al: Induction of colonies of hemoglobin-synthesizing cells by erythropoietin *in vitro. Proc Natl Acad Sci 68:*1542, 1971.
15. Zanjani ED, Poster J, Mann LI, et al: Regulation of erythropoiesis in the fetus, in Fisher JW (ed): *Kidney Hormones.* New York, Academic Press, 1977.
16. Tepperman AD, Curtis JE, McCulloch EA: Erythropoietic colonies in cultures of human marrow. *Blood 44:*659, 1974.
17. Schooley JC, Garcia JF: Some properties of serum obtained from rabbits immunized with human urinary erythropoietin. *Blood 25:*204, 1965.
18. Sober HA, Gutter FJ, Wyckoff MM, et al: Chromatography of proteins. II. Fractionation of serum proteins on anion exchange cellulose. *J Am Chem Soc 78:*756, 1956.
19. Miller RG, Phillips RA: Separation of cells by velocity sedimentation. *J Cell Physiol 73:*191, 1969.
20. Stohlman F., Jr: Pathogenesis of erythrocytosis. *Sem Hematol 3:*181, 1966.
21. Bomchil G, Carmena AO, Segade A, et al: Studies on the response to hypoxia and relative hyperoxia in two polycythemia vera patients. *Ind J Med Res 55:*543, 1967.
22. Adamson JW: The regulation of erythropoiesis in polycythemia vera and related myeloproliferative disorders, in Clarke WJ, Howard EB, Hackett PL (eds): *Myeloproliferative Disorders of Animals and Man.* Washington, DC, US Atomic Energy Commission, p. 440, 1970.
23. Wasserman LR, Gilbert HS: Complications of polycythemia vera. *Sem Hematol 3:*228, 1966.

Discussion (Chapters 12–17)

Dr. Urabe: Concerning the effect of hormones on Friend virus-transformed cells that you mentioned, did you mean there was an effect on proliferation or on differentiation?

Dr. Golde: I was just referring to proliferation. We found the serum free cloning technique is highly sensitive; in fact, one could say perhaps too sensitive. We've repeated some of this with uridine and thymidine incorporation, which has the advantage of being done rapidly but has the problems in that it may not always reflect proliferation. We looked at the effect of growth alone on differentiation and found none.

Dr. Testa: You mentioned that in some of your experiments you could rule out indirect effects due to cell interaction. Have you any evidence in other experiments that the effects you observed were indirect, for example, through a cell other than the CFU-e?

Dr. Golde: No, I was just suggesting that some of the discrepancies might be explained by cellular interaction. Obviously our interest is to know what is happening with erythroid precursors themselves, but we are not using pure cell populations. There are some controls that one can use. If you are using a hormone whose action can be blocked, that's always helpful, but still there could be another cell with a receptor that is interacting. The Friend cells are, of course, a cloned population. The effects we observe there are at least free of problems with interacting cell populations and serum factors.

Dr. Gordon: Some years ago we were concerned with the effects of hormones on erythropoiesis in the intact and hypophysectomized animal. An important observation that emerged from these studies was that the bone marrow of the hypophysectomized rat was more sensitive to the action of these hormones than marrow in normal rats. There is a rule in endocrinology, namely, that you can generally increase the sensitivity of the target tissue by removing the source of the hormone that is acting on that tissue. I wonder whether or not it might be profitable for you to explore the bone marrow response to these hormones in hypophysectomized animals and to compare the response that is evoked in the normal animal.

We also found that combinations of hormones were more effective in bringing the hypophysectomized picture back to normal. Hypophysectomy in the rat results in hypoplasia of the bone marrow associated with the development of anemia. About 50% of the marrow cellular population disappears and is replaced by fat. Now if you treat the animal with one hormone you will obtain some effect but more potent actions are evoked with combinations like, for example, a mixture of growth hormone, thyroid hormone and androgen. This approach I believe would be profitable since it might reproduce in your *in vitro* system the situation that occurs in the living animal.

Dr. Golde: I agree. We have not begun to look at combinations of these hormones. With *in vivo* work there is always a problem about effects on erythropoietin generation, and many of these very hormones do effect erythropoietin generation.

Dr. Peschle: I should like to emphasize what you've just mentioned. Some of these hormones have been demonstrated to enhance erythropoietin production and now you are investigating the action of these hormones at the level of the erythroid stem cells. We have been pursuing a similar approach in regard to testosterone, and have observed a synergistic interaction between stimulatory action of androgens on erythropoietin production and on erythroid stem cells. Maybe the same interaction mediates the influence of other hormones such as growth hormone, and maybe even adrenal steroids.

Dr. Golde: I did just briefly mention that there are systems where a cell is unresponsive to a primary hormone unless another hormone is present. A good example is the Leydig which will not respond to gonadotropin without glucocorticoids being present.

Dr. Zanjani: Even negatives in some of these systems may not be meaningful until you work out the entire story.

Dr. Murphy: Dr. Golde, do you get any effect with growth hormone or any of the other agents used on normal CFU-e in the absence of exogenously added erythropoietin?

Dr. Golde: Well, I'm going to answer that a little categorically. This problem revolves around the fact that we have background colony formation. I can get rid of that background by plating fewer cells, and maybe I will in the future. Let me state it this way: we believe all the erythropoiesis we observe is erythropoietin-stimulated. None of these hormones in their own right are erythropoietic; they merely potentiate the effect of erythropoietin.

Dr. Zanjani: You mentioned that with some of these agents, you can see the effect much better when you use suboptimal doses of erythropoietin. One would think that the background amount of erythropoietin is really suboptimal. Do you see a differential effect? In other words, when you just add the agent without any exogenous erythropoietin, is there any difference?

Dr. Golde: Well, that depends on the hormone. With thyroid hormone, the amount of potentiation was the same at any amount of erythropoietin. So we didn't have evidence for an interaction there. On the other hand, dexamethazone is most potent at the minimal level of erythropoietin and potentiates least at optimal erythropoietin concentrations. That's why we postulate that dexamethazone may modulate the response to erythropoietin, as it does in many other systems.

Dr. Zanjani: Is that true of growth hormone as well?

Dr. Golde: Well, growth hormone seems not to modulate the erythropoietin response so much as to give a fixed increment.

Dr. Rifkind: Just a comment. And I think the introduction of the Friend cell system into your conceptual package may only lead to more confusion before it finally simplifies it. For example, dexamethazone, which has this growth-promoting effect that you describe under your special conditions exposed to cells grown under conditions that are appropriate for the induction of differentiation, is a very profound inhibitor of differentiation, and promotes cell growth by a unique mechanism, namely by preventing terminal differentiation.

Dr. Golde: Yes. What we've tried to do is to circumvent a lot of this. We are looking just at the proliferative response, and, as you point out, dexamethazone inhibits that, probably by the mechanism of viral induction. It's true that if you add the DMSO and then block differentiation, with steroid, it will look like an increase in proliferation.

Dr. Ogawa: I think it's becoming clear that the intermediate BFU-e, such as 3- or 4-day BFU-e, would respond to erythropoietic stimuli, and contribute to the expansion of erythropoietic system. Studies like yours done on the intermediate BFU-e may have more relevance since CFU-e cannot really be stimulated any more.

Dr. Golde: Yes, we will look at intermediate colonies and bursts also.

Dr. Ogawa: The precursors may be more sensitive to hormones.

Dr. Fisher: I am interested to know if you have any ideas about the mechanism by which the hormones increase the CFU-e and BFU-e in the bone marrow culture?

Dr. Golde: Well, we don't know how the polypeptide hormones work at all. Most of them are potentiators, and we think that dexamethazone may be the one hormone that may regulate erythropoietin sensitivity. In the human system we're looking at intermediate colonies now, because we count on day 8 and there are a lot of different-sized colonies present.

Dr. Adamson: Dr. Peschle, have you had occasion to treat a hypertransfused mouse with anti-Ep to see what happens to the number of burst forming units in the tibia? And, if you find no reduction in the number of BFU-e, in the stable, hypertransfused animal, would that not strongly support your contention of two mechanisms perhaps operating *in vivo* regulating the size of the burst compartment?

Dr. Peschle: This is a good suggestion. Indeed, we were planning to do this experiment. After transfusion, in the very early stage, the BFU-e compartment is depleted, apparently in response to the drop of Ep activity. Thereafter, a very effective Ep-independent mechanism allows the BFU-e pool to "escape" this early, depleting influence. Your suggestion is well taken: we should verify whether in steady-state conditions, a few days after transfusion, the BFU-e kinetics are still partially responsive to Ep or not.

Dr. Zanjani: Aren't you surprised a little that the effect of a single injection of anti-Ep persists for such a long period of time?

Dr. Peschle: We gave a single injection: 0.2 ml of anti-Ep per mouse. The titer against human Ep was 60–65 anti-units per ml, or approximately 12 anti-human Ep units were injected per mouse.

I'm not surprised by this phenomenon, because Schooley demonstrated that, in mice, the half-life of the injected anti-Ep serum is 5 days. This figure is well in line with that phenomenon. Eight days after anti-Ep treatment, approximately 20% of the antibody should be left in the mouse. That is apparently not sufficient to neutralize the endogenous Ep production, which is progressively enhanced by the gradual drop of the hematocrit. Thus, as may be expected, the erythropoietic activity recovers to normality approximately on day 8 after anti-Ep injection.

Dr. Gordon: Dr. Peschle, what is your interpretation of the increase in plasma erythropoietin that occurs with the higher doses of estradiol? Do you think this is mediated through the kidney?

Dr. Peschle: I don't know. In our hands, both small and large amounts of estradiol dampen Ep activity in rats. In mice, however, we confirmed that Ep production is enhanced by large dosages of estradiol. Experiments in nephrectomized animals are now in progress, to elucidate whether this phenomenon is mediated by renal and/or extrarenal Ep.

Dr. Gordon: There is another possibility that would have to be explored relative to the rate of disappearance of the hormone from the circulation. Is estradiol really increasing the erythropoietin production or decreasing its rate of elimination?

Dr. Peschle: We performed the necessary experiments. After estradiol injection, the kinetics of Ep were not modified. Thus, after estrogen, the Ep titer in serum is reflective of Ep production, rather than diffusion and/or inactivation.

Dr. Iscove: One of the observations that Dr. Wagemaker made was that in animals subjected to an erythropoietic stress, the burst-promoting activity in serum increased. A local release of burst-promoting activity could explain the early rise in the BFU-e compartment.

Dr. Peschle: I agree. But if we accept it, we must also accept a series of other hypotheses. Since the BFU-e amplification occurs early after injection of either crude or purified Ep, we are bound to postulate that, in different Ep preparations, the ratio between Ep and the BFU-e enhancing factor is fairly constant. This is only the first hypothesis. We must also postulate that the anti-Ep sera employed here neutralized the burst-promoting factor. The third postulate would be that transfusion induces a decrease of the activity of both Ep and the BFU-e factor. We may still accept this long series of hypotheses, but if so, we may as well conceive the existence of a whole series of factors, paralleling Ep, and regulating selectively the kinetics of either BFU-e, or CFU-e, or erythroblasts, and so on. Dr. Iscove, your hypothesis is attractive but remote in my opinion.

Dr. Murphy: Dr. Wagemaker, would you like to make a comment regarding that?

Dr. Wagemaker: We have evidence that the burst feeder activity, or burst-promoting activity, as Dr. Iscove calls it, is not paralleling erythropoietin. I can't quite tell whether the changes you observe are due to erythropoietin or another factor. For the time being, I agree with you that it might be due to erythropoietin only.

Dr. Peschle: I may add that you have differentiated those two factors *in vitro*. Identification of the BFU-e factor *in vivo*, and all of our experiments are performed *in vivo*, is still difficult.

Dr. Rifkind: How profound is the species limitation or specificity? You said the dog was optimal. What are the limitations imposed by species?

Dr. Adamson: I don't think that there are many limitations imposed. I think Dr. Golde saw a cyclic AMP effect on colony growth with the mouse. We've seen it with rabbit, human, and dog marrow. We chose dog, because it seemed to work best for us. I'm not sure that there is much species specificity.

Dr. Gordon: From what I said after Dr. Golde's presentation, I would have predicted the opposite result. One must bear in mind that you're dealing with only one hormone here.

Dr. Adamson: That's right.

Dr. Gordon: I would suggest that at some time you use purified growth hormone along with the thyroid hormone.

Dr. Adamson: We'd be happy to.

Dr. Gordon: I say this because these two hormones act synergistically in the hypophysectomized animal.

Dr. Adamson: I think that's a reasonable comment. I think that this reflects the fact that endocrine effects of simple thyroid ablation may not be so simple.

Dr. Fisher: In your hypothyroid studies, I think it would be interesting to know whether isoproterenol is truly triggering adenyl cyclase in the bone of these animals. I realize the difficulties in measuring adenylate cyclase activity in bone marrow with such a mixed cell population. Have other tissues been studied in the hypothyroid situation, determining if adenylate cyclase activity is decreased and if this is the reason that the isoproterenol is not effective in the bone marrow of the hypothyroid animal?

Dr. Adamson: I think that's been done with cardiac cells. So I think the answer to the question is "yes," but not in hematopoietic tissue. I would have to resist the temptation to try to measure adenyl cyclase or cyclic AMP changes in such a heterogeneous population of cells in response to an agonist such as isoproterenol.

Dr. Sassa: We have been interested in the permissive effects of hormones in chick embryo liver cells incubated in a serum-free medium. In this system, cells are prepared in the absence of serum and cells can be incubated in the complete absence of serum for a 6-day period. These cells respond to a chemical inducer, such as allylisopropylacetamide (AIA) to increase the level of δ-amino-levulinate synthase (ALAS), provided that insulin is added to the serum-free medium. Insulin itself does not have inductive response but exerts permissive effects on the induction of ALAS synthesis. On the other hand, when insulin is given to the animal *in vivo* together with AIA, insulin blocks the induction of ALAS synthesis. Thus, insulin effect in tissue culture causing permissive effects in tissue culture cannot be extrapolated to its effect *in vivo*. My question is how much do you think one can extrapolate the finding *in vitro* to *in vivo* pathophysiology?

Dr. Adamson: Well, since I'm becoming hypoglycemic, I'll just sidestep the insulin question entirely. I think the thrust of your comment is that one has to be extraordinarily careful in drawing physiological or *in vivo* correlates solely from *in vitro* observations. That's why we have attempted, admittedly in the initial phase, to study these effects in the intact animal that has been manipulated in such a way that certain predictions emerge. I think you raise a precautionary tale that I accept. We're trying to see with what accuracy the culture systems reflect what we're trying to learn about the modulation of terminal erythropoiesis in certain disorders that one encounters in man.

Dr. Golde: As you point out rightly, extrapolations from *in vitro* work to *in vivo* are always dangerous and our experience has been somewhat bitter in both directions. For example, I would have interpreted what you said in terms of the insulin data, that the effect you observed *in vitro* was in fact the valid effect and I think it's the *in vivo* effect that confused. Because you know, if you inject insulin into a whole animal, growth hormone concentrations go up and many may antagonize the insulin-like activities.

Dr. Golde: Dr. Adamson, can I ask you about the thyroid? You've shown the induction of a receptor here that is intriguing. The receptor disappears in the hypothyroid dog and shows up after you add thyroid to it. Is there any interaction or does it change the response to erythropoietin?

Dr. Adamson; Yes, it does. The response to erythropoietin is a little more complex and I really probably should not say too much about it. But it appears, in the hypothyroid state, that cloning efficiency falls, only to increase with incubation of cells with thyroid hormone. Now I'm not sure exactly what means, and, before we say anything else about it, we should finish the studies we are carrying out in the intact animal to quantitate the magnitude of response under certain physiological conditions, and to see what the reflection of this sort of *in vivo* experiment is in culture.

Dr. Sassa: I agree with Dr. Golde and Dr. Adamson in a sense that a serum-free culture provides a very useful system to understand the action of hormones in cells in culture. However, one has to be always careful concerning the limitation pertinent to the tissue culture system. For example, certain *in vitro* findings correlate well with what happens *in vivo*. On the other hand, no such relationships exist for certain other hormones as I mentioned in the case of insulin. Unless the effect of a hormone can be tested or proven *in vivo* as well, *in vitro* findings alone may not be very meaningful and sometimes could even be dangerous.

Dr. Adamson: I'm afraid I have to come back to Dr. Golde's point. And that is, when you inject cyclic AMP into an animal all sorts of things go on. That becomes a very complicated sort of comparison to make. All of these are complicated; I don't know any way around that.

Dr. Gordon: The point I made before I'd like to express a little more succinctly and perhaps more clearly. The pituitary, like most organs in the body, requires thyroid hormone for its maintenance and functionality. Thus, in a hypothyroid animal one must consider not only thyroid lack, but also pituitary insufficiency. I believe your *in vitro* experiments should take this into account. That's why I mentioned that growth hormone addition would be important as well as supplementation with other hormonal factors.

Dr. Adamson: That may be. However, there are other aspects of this study that may reflect uniquely thyroid-dependent phenomena; for instance, the absence of adrenergic receptors and their restitution with thyroid hormone. I think these are limited extensions of the *in vivo* model.

Dr. Murphy: Dr. Moriyama, it's been our experience that 7 day CFU-e in normal human peripheral blood are virtually absent.

Dr. Ogawa: How long did you incubate?

Dr. Moriyama: 14 days.

Dr. Ogawa: About 15% of the growth that you see from peripheral blood mononuclear cells present the morphology of large colonies. Maybe you are counting only these large colonies.

Dr. Zanjani: I just wanted to comment that Prchal reported the presence of CFU-e colonies growing within 7 days in peripheral blood of patients with myeloproliferative disorders.

Dr. Ogawa: I haven't examined many patients, but I have grown peripheral blood mononuclear cells from several patients who had aplastic anemia and I did not find as many colonies or bursts as you reported. I wonder if there are differences in the patients with aplastic anemia between Japan and the United States. I heard from somebody that maybe there are more drug-induced aplastic anemia in Japan. Are all these patients cases of ideopathic aplastic anemia?

Dr. Moriyama: There are more patients with aplastic anemia who showed BFU-e. We get 5×10^4 BFU-e per cubic millimeter.

Dr. Ogawa: But did any of your patients have drug-induced aplastic anemia?

Dr. Moriyama: No.

Dr. Zanjani: Dr. Moriyama showed, like we have, that normal blood peripheral lymphocytes stimulate erythroid colony formation. So that, if you're talking about crossing the genetic barriers, there's a mixture of one individual's lymphocytes with another. There has to be just more than that. I was going to ask Dr. Moriyama, in view of Dr. Adamson's comment that part of the suppressive activity may be related to sensitization with blood transfusion products, whether your aplastic patients included any who had not received transfusion before you studied them, and whose lymphocytes inhibited CFU-e.

Dr. Moriyama: They had never received a transfusion at the time of this study.

Dr. Ogawa: In your co-culture experiments, is it possible that you counted subcolonies of bursts derived from added blood lymphocyte fractions?

Dr. Zanjani: In the co-culture studies, when you added lymphocytes to bone marrow cells, how many days were they cultured for, 7 days or 14 days?

Dr. Moriyama: 10 days.

Dr. Sassa: I would like to show some of the recent findings related to 5 α-H steroids, in cultured chick embryo liver cells. Chick embryo liver cells were prepared in the absence of fetal bovine serum and incubated in a serum-free modified Ham's F12 medium, which was prepared in our laboratory and supplemented with insulin and hydrocortisone. Supplementation of insulin and hydrocortisone to the serum-free medium allows cells to restore the ability to express specific functions of hepatocytes. Mainly hepatocytes are present in these cultures and very few fibroblasts are found, and essentially no red cells are present in this culture. Now these cells are treated with an exact pair of 5 α-H compound and a 5 β-H pregnanolone in the absence of serum. Approximately two-fold increase in porphyrin concentration was observed by the 5 β-H compared with the 5 α compound. When the cells are treated with the steroid and Ca Mg EDTA, the latter inhibiting the terminal enzyme of the heme pathway, greater amounts of porphyrin are accumulated. The difference between the effect of 5 β and 5 α steroids becomes smaller when incubation proceeds, or greater concentrations of steroids are used. Using the same pair of steroids, it was also found that induction of ALA-synthase was twofold greater with the 5 β steroid than with the 5 α compound. One crucial question here is whether the same pair of steroids have a similar differential effect *in vivo*. We found that the same pair of 5 β and 5 α steroids has similar inductive effect *in vivo* on ALA-S in the liver of the eggs teated with these steroids for nine hours. In this case, the *in vitro* findings correlated well with the *in vivo* finding. However, as Dr. Fisher pointed out, one should be cautious about extrapolating *in vitro* findings to *in vivo* physiology. Such *in vitro* findings become valid when they are substantiated by similar *in vivo* results.

Dr. Fisher: Your data is very interesting. I wonder whether the steroid receptors are different in various tissues. For example, your work is in the liver, but we don't know about the receptor proteins in the kidney and bone marrow for 5 β and 5 α steroids. It may very well be that there is a different receptor for the 5 β steroids in the bone marrow than in the liver. The stem cell compartment may have receptors for 5 β but not for 5 α steroids and vice versa for the kidney.

Dr. Murphy: Is it not still the case that one of the problems with regard to steroids is the difficulty of their aqueous insolubility *in vitro*?

Dr. Fisher: Actually, we added our steroids to the bone marrow cultures dissolved in 5 microliters of ethyl alcohol. I think John Adamson used propylene glycol to dissolve his steroids before adding them to his culture. Therefore, our own steroids were in solution in our culture system. There are some hemisuccinate salts of steroids that are soluble in aqueous solution, but there is the problem of not knowing whether the succinate salt may be split off and produce an effect itself in the culture system.

Dr. Erslev: Dr. Fisher, you have enumerated the many supposed effects of erythropoietin, including the effect on sinusoids. It is almost too fantastic to imagine that erythropoietin will do all these things. Isn't it possible that some of these effects are due to endotoxin, which seems to be a contaminant of many of the erythropoietin preparations?

Dr. Fisher: This is very possible. The first time that we perfused the isolated hind limb femur in England, we used a very crude sheep plasma erythropoietin. I don't know if it is endotoxin which may be causing the reticulocyte release, but Bob Weed's findings of increased appertures of the bone marrow sinusoidal lining following stimulation by a rather crude erythropoietin preparation is worthy of note. I worry myself that the erythropoietin preparations that we are using contain so many contaminants that we don't know what we are studying. I think it is very important that we soon have enough purified erythropoietin, at least higher specific activity, to be sure that some of the effects that we see are not due to the contaminants. The reticulocyte release and the change in the appertures of bone marrow sinusoids noted by Weed could be due to the contaminants in the erythropoietin peparations used. I certainly hope erythropoietin, when it is purified, will be found to have a stem cell effect!

Dr. Peschle: We should generalize this point: when investigating the *in vivo* or *in vitro* action of Ep, one should first assay for endotoxin activity in the Ep preparation.

Dr. Fisher: Yes, this may be true.

Dr. Iscove: I'd like to suggest that since human urinary erythropoietin is often derived from the urine of patients with aplastic anemia on androgen therapy, isn't it conceivable that some of the activity of that material might reflect contaminating androgens? Are there androgen-binding proteins in urine?

Dr. Fisher: In answer to your question as to whether the patients with aplastic anemia may have androgenic steroids in their urine because they may have been treated with steroids, the erythropoietin used in our studies was human urinary erythropoietin from patients having hookworm anemia. I don't think the patients would have been receiving steroids. It is true that some people use human urinary erythropoietin from androgen treated patients, but one can't really answer this question unless an analysis for steroids in the urine is made. As to whether there are androgen-binding proteins in the urine, I really can't say. I suspect there probably are, because there are androgen-binding proteins in plasma and some of them could be excreted in the urine.

Dr. Sassa: I would like to briefly comment on the hydrophobic nature of steroids in terms of induction of ALA-synthase in the liver cell culture. Most of the inducing compounds are hydrophobic in nature. Within given certainties, it is possible to plot the inducing capability of these compounds as a function of their solubility in lipids. It appears that liver cells try to get rid of these compounds by inducing microsomal mixed-function oxidases that require an increased heme supply for the formation of cytochrome P-450. Thus, hydrophobicity is a necessary part of an inducer, at least in liver cells.

Dr. Fisher: That's in the liver cells and of course we don't know about bone marrow erythroid cells.

Dr. Sassa: That's in the liver where abundant endoplasmic reticulum is present.

Dr. Fisher: We do not know whether erythroid cells contain cytochrome P-450 and the mixed oxidase enzymes. However, I rather doubt it.

Dr. Seidel: There is an animal model of polycythemia vera that we will talk about this afternoon; and I would like to stress that even in the polycythemic Friend mice, there are different possibilities. One possibility with one virus which we use is that within 2 to 3 weeks after virus infection, 100% Ep-independency develops and stays over a prolonged period. There is another strain of the virus, which we got recently, where 100% Ep independency develops also, but this is not constant and Ep dependent colonies can be seen repeatedly during the disease. I could imagine that in human polycythemia vera it might very well be the same as in our second model; that states exist where most colonies are transformed, and I'm talking of CFU-e, not of BFU-e. That is, CFU-e do not require Ep and then, some days later, there are substantial numbers of normal ones which require it.

Dr. Zanjani: This may very well be. But I believe there are significant differences between polycythemia vera (PV) and Friend-virus induced polycythemia in mice. It may be that there are different populations of patients with different kinds of behavior. A while back, a group in Baltimore reported on a patient with polycythemia vera who developed renal failure and the patient continued to maintain a respectable red cell mass, and did not require transfusions. When you look at the data carefully, however, the patient was not polycythemic. Their conclusion was that erythropoiesis in PV was independent of erythropoietin. The patient then developed urate nephropathy. There is a tremendous difference between 1971 and 1976. For example, the patients hemoglobin fell to 9.3 gram % in 1975, whereas it was 21 gram % in 1971; and you can also see that the red cell mass significantly decreased, indicating anemia. The level of erythropoietin measured in plasma and in the urine of this patient was found to be below normal. We looked at inhibitors and did not find any. In 1975 we looked at this patient again and the erythropoietin levels, despite the severe degree of anemia, were nondetectable; indicating that, having kidney failure, the individual was unable to make erythropoietin. But when we looked at the bone marrow from this individual, endogenous colonies formed; again we could not detect any inhibitors of erythropoiesis. This is very important at this time because of the uremic state of the individual. Our conclusion is that the patient retains the capacity to be described, as far as *in vitro* studies are concerned, as a polycythemia vera patient; but having no erythropoietin, cannot make enough red cells.

Dr. Murphy: Dr. Iscove, using completely serum-free methylcellulose culture media, have you had the opportunity to study polycythemic vera marrow without exogenously added erythropoietin? (Dr. Iscove shakes his head "no.") It would be something that would be worth trying.

18 Murine Erythroleukemia: Cell Surface and Cell Cycle-Related Events in Induced Differentiation

Richard A. Rifkind and Paul A. Marks

Introduction

The murine erythroleukemia cell (MELC) appears to be an erythroid precursor cell (perhaps the erythropoietin-responsive target cell) transformed by the Friend virus complex.[1-11] One component of this virus complex, spleen focus-forming virus, is responsible for uncoupling cellular replication from the physiological regulator of this function, erythropoietin. By some mechanism, perhaps selection of cells with the greatest growth potential, cell lines have been established in which the normal process of erythropoietic differentiation (and concommittant terminal cell division) has become suppressed. In addition, a wide variety of agents have been discovered that can significantly alter the rate at which MELC express the program of erythropoietic differentiation.[12,13] This combination of events, proliferation without erythropoietin, and the availability of agents that modulate differentiation, provides an interesting model system for exploring the cellular events that are of significance in the regulation of the expression of differentiated functions in erythroid cells. Many of the properties of MELC are discussed in considerable detail elsewhere.[14-18] This chapter will focus on three aspects of the problem of regulation of MELC erythropoiesis. First, we will review the evidence which indicates that there are two components to induced differentiation, namely, the commitment of the MELC to erythropoietic differentiation, and the program that coordinates the series of biosynthetic and morphogenetic events that constitute the actual expression of the erythropoietic developmental process. Secondly, we will present and review some evidence that indicates that alterations at the plasma membrane may be implicated in the induction of differentiation by some agents. Finally, we will discuss the role of the cell cycle in induced differentiation.

Materials and Methods
Commitment to Differentiation

MELC (strain DS19 derived from strain 745 of Charlotte Friend) display less than 0.5% of spontaneously differentiating erythroblasts in culture.[19] When MELC are exposed to increasing concentrations of the potent polar-planar inducing reagent, hexamethylene bisacetamide (HMBA), an increasing proportion of cells are induced to differentiate as detected by the benzidine reaction for hemoglobin.[20] The total amount of hemoglobin that can be detected in these cultures is directly proportional to the number of benzidine-reactive differentiating erythroblasts. Furthermore, the mean amount of hemoglobin in each differentiating erythroblast is constant, that is, it is independent of both the total number of erythroblasts triggered to differentiate and of the concentration of HMBA. These observations provided the first substantive clue in our laboratory that two distinct programs were required for erythropoietic differentiation: commitment, which determines the number of differentiating cells, and expression, which achieves the biochemical and morphogenetic changes characteristic of erythropoiesis. The first appears to be responsive to the concentration of inducer, whereas the second is not.

In order to study the kinetics of recruitment of MELC to differentiation induced by HMBA, that is, the kinetics of commitment, experiments were designed to examine the duration of exposure to inducer required to commit MELC to express differentiation when removed from the inducing agent.[21] The kinetics of induced differentiation were compared under two sets of experimental conditions. When MELC are exposed continuously to an optimal inducing concentration (5 mM) of HMBA, benzidine-reactive, hemoglobin-containing cells are detected in culture from about 48 hr and the proportion of benzidine-reactive cells increases to over 90% by 96 hr.

A second set of conditions was designed to quantitate commitment to differentiation at the single cell level. To determine the kinetics of commitment to differentiation and the capacity for replication of single cells following culture with HMBA, cells were transferred, after washing, to inducer-free semisolid media in order to detect cells that will produce differentiated progeny after inducer is removed. Under these conditions, both induced and uninduced cells produce colonies that can be scored for differentiation by staining the culture *in situ* with benzidine after a total of 5 days in culture. The background for spontaneously induced colonies, under these conditions, is consistently less than 3% and the cloning efficiency is over 85%, measured as the proportion of cells inoculated detectable as colonies or small clusters on day 5 of culture. Under these conditions, commitment is first detected in cells exposed to 5 mM HMBA for between 12 and 16 hr and the percentage of colonies containing benzidine-reactive cells increases linearly with duration of exposure to inducer, reaching virtually 100% by 50 hr. Commitment is also proportional to concentration of HMBA, between 0.5 and 5 mM; the optimal concentration for this inducer is 5 mM HMBA, and only at concentrations higher than this is there toxic suppression of cloning efficiency.

Colonies derived from cells exposed in suspension to HMBA are of three types, as assayed by the benzidine reaction[21,22]: (a) colonies containing uniformly benzidine-negative cells, (b) colonies containing uniformly benzidine-reactive cells, and (c) colonies containing a mixture of benzidine-reactive and nonreactive cells (mixed colonies). The contribution of mixed colonies to the total population of differentiated colonies is highest at suboptimal concentrations of HMBA or after short periods of exposure to the inducing agent. These observations suggest that, under these conditions, a committed cell may give rise to both differentiated and undifferentiated progeny; that is, the differentiated state may be unstable, or the decision to differentiate may be expressed, in a statistical fashion, at a time subsequent to the exposure to inducing agent.

The number of cells in the colony is related to the proportion of benzidine-reactive cells in the colony. Colonies without benzidine-reactive cells, as well as mixed colonies, continue to increase in size throughout the period of culture. Uniformly benzidine-reactive colonies are smaller, generally showing no increase in size after day 4 and containing not over 16 to 32 cells. This suggests that induction to differentiation is associated with a limitation in the potential for cell division, consistent with the pattern of terminal differentiation characteristic of normal erythropoiesis.[21-24]

Taken together, these studies strongly implicate two phases in chemically induced differentiation: commitment and biochemical development. Considerable work in this and other laboratories is now dedicated to an attempt to define those cellular and biochemical events that are specifically related to the commitment phase of erythropoietic differentiation in MELC.

Cell Surface-Related Events in Differentiation

Following exposure to inducing agents MELC undergo a variety of alterations at the plasma membrane. Some of these are characteristic of normal erythropoiesis and, presumably, reflect expression of the program or erythropoietic development. These include changes in membrane associated erythrocyte-specific proteins and antigens[14,25-27] and receptors for the iron-binding serum protein, transferin.[28] Other changes, including an early decrease in cell volume[29] and an increase in plant lectin agglutinability,[30] may occur sufficiently early to be implicated in the process of commitment. Evidence implicating cell surface-mediated functions in the induction of differentiation also derives from the differentiation-inducing and differentiation-suppressing activities of agents with known or postulated plasma membrane activities, including the inducer, ouabain[31] and the inhibitors cocaine and tetracaine.[32] These agents, the cardiac glycosides and the local anesthetics, have well documented biological effects at the plasma membrane.

In an attempt to develop further evidence implicating the plasma membrane or plasma membrane-related functions in commitment, we have examined the levels of cyclic nucleotide (cAMP) during chemically mediated commitment. Cells of MELC strain DS19 (a cell line sensitive to induction by all inducing agents which we have tested) and strain DR10 (a cell line[33] selected for resistance to induction by dimethylsulfoxide, Me_2SO), were placed in culture and exposed to each of the following inducing agents: HMBA (5 mM), Me_2SO (280 mM), or sodium butyrate (1.5 mM), and the cellular content of cAMP was determined during the first 18 hr in culture.[34] During the first 3 to 6 hr of culture of DS19 cells there is a distinct rise in cAMP content in cells exposed to any of the three inducing agents; there is a small rise in cAMP content in control cells during the same period. By 12 hr the cAMP content of these cells has returned to the initial value. Cells of strain DR10 fail to show a significant increase in cAMP content, compared to control cells, when exposed to Me_2SO, but do show an increase in cAMP when exposed to HMBA or butyric acid, two inducing agents that can induce this Me_2SO-resistant cell line to differentiate.

Since cyclic nucleotide values are closely influenced by the progression of cells through the cell division cycle,[35] studies on the effects of chemical inducers on cAMP levels were redesigned, using MELC synchronized with respect to the cell cycle by the sequential application of the double thymidine blockade procedure and hydroxyurea (ref. 36 and Schildkraut, personal communication). With this procedure control cells and cells exposed to 4 mM HMBA proceed through the first cell cycle, following release from hydroxyurea, synchronously and with the same kinetics. A brisk, almost tenfold, increase in cAMP level was detected in cells exposed to HMBA, in mid-S-phase of the first cell cycle. Taken together with the data derived from studies on nonsynchronized MELC, these observations suggest that an early plasma membrane related effect of HMBA during commitment involves the accumulation of cAMP.

This effect is not observed in the variant cell line (DR10) resistant to induction by Me_2SO, when exposed to that agent.

As an alternative approach, the effects of a series of tumor promoting plant diterpenes, 12-0-tetradecanoyl-phorbol-13-acetate (TPA) for example, on spontaneous and chemically induced differentiation have been studied. Both spontaneous[37,38] and HMBA-mediated differentiation[38] are inhibited by simultaneous exposure of MELC to TPA. Spontaneous differentiation is more sensitive to inhibition by TPA (0.5 to 1.0 ng/ml) than HMBA-induced differentiation (10 to 100 ng/ml). Other diterpenes such as phorbol-12-,13-didecanoate, mezereine, and ingenol dibenzoate, which are active tumor promoting agents, are likewise effective inhibitors of both spontaneous and HMBA-induced differentiation. Related compounds that are ineffective as tumor promotors are ineffective as differentiation inhibitors. Inhibition of differentiation is not a nonspecific toxic effect of these agents. MELC can be incubated, by repeated passages, for prolonged periods of time in the presence of HMBA and TPA, without displaying either toxicity or differentiation; on removing the TPA these cells display their normal responsiveness to HMBA, attaining over 80% benzidine-reactive cells. The relationship between TPA-mediated inhibition of differentiation and the commitment phase of induced differentiation has also been studied. The present evidence suggests that TPA is effective only prior to commitment. Several lines of evidence suggest an effect of the phorbol esters at the plasma membrane. MELC become adherent to the culture dish when incubated with TPA for several hours.[39] This TPA-mediated change in cell surface characteristics is displayed only by TPA susceptible MELC; two TPA-resistant MELC lines, in which differentiation is minimally influenced by exposure even to high doses of TPA, fail to show this response. The tumor promotors characteristically induce plasminogen activator[40] and TPA induces plasminogen activator in TPA sensitive MELC as well.[39] Taken together, these observations suggest that the cellular target for TPA-mediated inhibition of differentiation may be located at the plasma membrane and that the TPA-mediated effect is directed at a step involved in the commitment phase of erythropoietic differentiation.

Induced Differentiaton and the Cell Cycle

Further investigations were designed to evaluate the relationship of cell cycle events to induced MELC differentiation. Previous studies provided evidence that MELC, synchronized with respect to the cell division cycle by exposure to high levels of thymidine,[36] require the presence of Me_2SO during at least one round of DNA synthesis. McClintock and Papaconstantinou[41] and Harrison[42] have provided evidence, as well, that MELC differentiation is dependent on at least one or more rounds of DNA synthesis.

We have further investigated the relationship between cell cycle events and transition to hemoglobin production in induced MELC by examining the pattern of cell cycle transit of MELC cultured without and with inducers.[43] MELC cultured with Me_2SO or other inducing agents develop a prolongation of G_1 or a transient block in initiation of DNA synthesis. This was demonstrated by measuring the rate of DNA synthesis, proportion of cells in S-phase and pattern of DNA accumulation in MELC grown in nonsynchronous cultures with and without Me_2SO, butyric acid, or dimethylacetamide. In MELC cultured without inducer, there is an initial rise in the rate of [^{3}H]-thymidine incorporation, with a maximum value achieved by about 10 hr. The initial increase in rate of thymidine incorporation probably reflects entry into S-phase of cells partially synchronized in the postlogarithmic growth phase of the previous cell passage. Between 10 and 40 hr this rate remains relatively constant, decreasing thereafter to less than 10% of the peak value by 60 hr. The plateau level observed between 10 and 40 hr reflects a constant proportion of cells in S-phase (loss of the partial synchronization). The fall in DNA synthesis after 60 hr coincides with the onset of stationary growth phase culture. In comparison, in cells cultured with Me_2SO, although there is an initial rise in the rate of [^{3}H]-thymidine incorporation, a difference between cultures with and without inducers is observed as early as 4 to 6 hr after initiation of cultures. In the population of cells cultured with Me_2SO, a decrease in the rate of thymidine incorporation is observed between 10 and 20 hr. At 20 hr, the rate of [^{3}H]-thymidine incorporation is at its lowest, about 25% of the rate in control cultures. The initial rise in rate of thymidine incorporation in cells cultured with inducer indicates some cells are proceeding through S-phase prior to the prolonged G_1. As cells pass into G_1, the proportion of cells whose DNA synthesis is blocked increases. In the induced cultures, the rate of [^{3}H]-thymidine incorporation rises between 20 and 30 hr, to a peak value of about 75% of the highest rate in control cultures, where it remains until 50 to 60 hr and then decreases. This latter decrease coincides with the stationary growth phase, and terminal differentiation. These results indicate a prolongation of G_1 in induced MELC cultures and were confirmed by three other types of studies evaluating the pattern of cell cycle. Thus, determination of the proportion of cells in S-phase by radioautography of cells labeled during 20 min exposure to [^{3}H]-thymidine and determination of the relative DNA content per cell using both the fluorescent Feulgen assay and propidium iodide binding measured with flow microfluorometry, provide evidence for a decrease in the proportion of cells in S or prolongation of G_1 during the early period of culture with inducing agent.

The effects of inducers on transit through the cell division cycle was examined further, employing MELC synchronized by culture through sequential periods with 2 mM thymidine and hydroxyurea as already described. After culture with hydroxyurea, with and without inducer, cells were transferred to fresh medium with or without the same inducer. Cells without or with inducer proceed through the first S, G_2 and M in synchronous manner and with similar transit times. Thereafter, MELC cultured with inducers remained in G_1 for 6.5 to 8 hr, compared with 4 hr for cells cultured without inducer.

In addition to prolongation of G_1, there is evidence that in the course of induced MELC differentiation structural changes in chromatin occur. This evidence, presented elsewhere[18,43,44], demonstrates an increased alkalibility of induced DNA and the inducing properties of actinomycin D and ultraviolet irradiation. The data are consistent with the accumulation of single strand breaks in DNA during induced differentiation.

ducers on cyclic nucleotide metabolism. Associated with the induction process are alterations in chromatin structure, including possible accumulation of single strand breaks in DNA. The precise relationship between DNA synthesis, changes at the plasma membrane, prolongation of G_1, alterations in chromatin structure, and the commitment of MELC to differentiate is not yet known. It may be speculated, however, that some inducers, at least, act by stimulating the accumulation of cyclic AMP by an effect at the plasma membrane. This, in turn, may be responsible for a delay in the subsequent G_1 phase of the cell cycle, permitting some step, crucial in the induction of differentiation, to occur. Induced, the low rate of spontaneous differentiation characteristic of most πELC lines maintained in culture may, perhaps, be the consequence of selection for πELC with an abnormally shortened G_1 period. Precisely which critical events occur during G_1, the possible relationship of these events to the acquisition of an altered configuration of DNA or chromatin (perhaps single strand breaks) and the accumulation of globin mRNA and other erythroid cell-specific products, remain to be determined.

Summary

Evidence has been reviewed which suggests that at least one round of DNA synthesis in MELC culture with inducers occurs prior to expression of differentiated functions. In addition, all agents studied to date with respect to their effects on the cell cycle have been shown to cause a prolongation of G_1. This effect may be mediated by cell-surface effects of in-

Acknowledgments

Work by the authors of this review was supported, in part, by grants and contracts from the National Institutes of Health (GM-14552, CA-13696, CA-18316, NO1-CB-4-4008, NO1-CP-6-1008) and the National Science Foundation (NSF-PCM-75-08696).

References

1. Friend C, Patuleia MC, deHarven E: Erythrocytic maturation *in vitro* of murine (Friend) virus-induced leukemia cells. *Natl Cancer Inst Monogr 22*:505–522, 1966.
2. Mirand EA: Virus-induced erythropoiesis in hypertransfused-polycythemic mice. *Science 156*:832–833, 1967.
3. Steeves RA, Mirand EA, Thomson, S et al: Enhancement of spleen focus formation and virus replication in Friend virus-infected mice. *Cancer Res 29*:1111–1116, 1969.
4. Tambourin P, Wendling F: Malignant transformation and erythroid differentiation by polycythaemia inducing Friend virus. *Nature (New Biol) 234*:230–233, 1971.
5. Lilly F: Mouse leukemia: a model of a multiple-gene disease. *J Natl Canc Inst 49*:927–934, 1972.
6. Lilly F, Pincus T: Genetic control of murine viral leukemogenesis. *Adv Cancer Res 17*:231–277, 1973.
7. McGarry MP, Mirand EA: Incidence of Friend virus-induced polycythemia in splenectomized mice. *Proc Soc Exper Biol Med 142*:538–541, 1973.
8. Fredrickson T, Tambourin P, Wendling F et al: Target cell of the polycythemia-inducing Friend virus: studies with myleran. *J Natl Canc Inst 55*:443–446, 1975.
9. Steeves RA: Spleen focus-forming virus in Friend and Rauscher leukemia virus preparations. *J Natl Cancer Inst 54*:289–297, 1975.
10. Tambourin PE, Wendling F: Target cell for oncogenic action of polycythemia-inducing Friend virus. *Nature 256*:320–322, 1975.
11. Nasrallah AG, McGarry MP: *In vivo* distinction between a target cell for Friend virus (FV^P) and murine hematopoietic stem cells. *J Natl Cancer Inst 57*:443–445, 1976.
12. Friend C, Scher W, Holland JG, et al: Hemoglobin synthesis in murine virus induced leukemic

cells *in vitro:* Stimulation of erythroid differentiation by dimethylsulfoxide. *Proc Natl Acad Sci (USA) 68:*378–382, 1971.
13. Takahashi E, Yamada M, Saito M, et al: Differentiation of cultured Friend leukemia cells induced by short-chain fatty acids. *Gann 66:* 577–580, 1975.
14. Ikawa Y, Ross J, Leder P, et al: Erythrodifferentiation of cultured Friend leukemia cells, in Nakahara W et al. (eds): *Proceedings of the Fourth International Symposium of Princess Takomatsu Cancer Research Fund.* Tokyo, University Tokyo Press, 1973.
15. Harrison PR, Affara N, Conkie D, et al.: Regulation of erythroid differentiation in Friend erythroleukemic cells, in Muller-Berat N et al. (eds): *Progress in Differentiation Research.* Amsterdam, North-Holland Publishing Company, 1976.
16. Pragnell IB, Ostertag W, Harrison PR, et al.: Regulation of erythroid differentiation in normal and leukaemic cells, in Muller-Berat N et al. (eds): *Progress in Differentiation Research.* Amsterdam, North-Holland Publishing Company, 1976.
17. Reuben RC, Marks PA, Rifkind RA, et al.: Induction of erythroid differentiation in Friend cells, in Ikawa Y (eds): *Oji International Seminar on Genetic Aspects of Friend virus and Friend cells.* New York, Academic Press, 1977.
18. Marks PA, Terada M, Fibach E, et al.: Induction of murine erytholeukemia cells to differentiate: Cell cycle related events in expression of erythroid differentiation, in Clarkson B, Marks PA, Till J (eds): *Differentiation of Normal and Neoplastic Hematopoietic Cell.* New York, Cold Spring Harbor Laboratory, 1978.
19. Singer D, Cooper M, Maniatis GM, et al: Erythropoietic differentiation in colonies of Friend virus transformed cells. *Proc Natl Acad Sci (USA) 71:*2668–2670, 1974.
20. Reuben RC, Wife RL, Breslow R, et al: A new group of potent inducers of differentiation in murine erythroleukemia cells. *Proc Natl Acad Sci (USA) 73:*862–866, 1976.
21. Fibach E, Reuben RC, Rifkind RA, et al: Effect of hexamethylene bisacetamide on the commitment to differentiation of murine erythroleukemia cells. *Cancer Res 37:*440–444, 1977.
22. Gusella J, Geller R, Clarke B, et al: Commitment to erythroid differentiation by Friend erythroid leukemia cells: A stochiastic analysis. *Cell 9:* 221–229, 1976.
23. Marks PA, Rifkind RA: Protein synthesis: Its control in erythropoiesis. *Science 175:*955–961, 1972.
24. Rifkind, RA, Bank, A, and Marks, PA: Erythropoiesis, The Red Blood Cell. Edited by D. McN. Surgenor. Academic Press, New York, 1974

25. Furusawa M, Ikawa Y, Sugano H: Development of erythrocyte membrane-specific antigen(s) in clonal cultured cells of Friend virus-induced tumor. *Proc Japan Acad 47:*220–224, 1971.
26. Ikawa Y, Furusawa M, and Sugano H: Erythrocyte membrane-specific antigens in Friend virus-induced leukemia cells. *Bibl Haem 39:*955–967, 1973.
27. Arndt-Jovin DJ, Ostertag W, Eisen H, et al: Studies of cellular differentiation by automated cell separation. Two model systems: Friend-virus transformed cells and hydra attenuata. *J Histochem Cytochem 24:*332–347, 1976.
28. Hu H-Y Y, Gardner J, Aisen P, et al: Inducibility of transferrin receptors on Friend erythroleukemic cells. *Science 197:*559–561, 1977.
29. Loritz F, Bernstein A, Miller RO: Early and late volume changes during erythroid differentiation of cultured Friend leukemic cells. *J Cell Physiol 90:*423–438, 1977.
30. Eisen H, Nasi S, Georgopoulos CP, et al: Surface changes in differentiating Friend erythroleukemic cells in culture. *Cell 10:*689–695, 1977.
31. Bernstein A, Hunt DM, Crickley V, et al: Induction by ouabain of hemoglobin synthesis in cultured Friend erythroleukemic cells. *Cell 9:* 375–381, 1976.
32. Bernstein A, Boyd AS, Crickley V, et al: Induction and inhibition of Friend leukemic cell differentiation: The role of membrane-active compounds, in Cook JS (ed): *Biogenesis and Turnover of Membrane Macromolecules.* New York, Raven Press, 1975.
33. Ohta Y, Tanaka B, Terada M, et al: Erythroid cell differentiation: murine erythroleukemia cell variant with unique pattern of induction by polar compounds. *Proc Natl Acad Sci (USA) 73:*1232–1263, 1976.
34. Rifkind RA, Fibach E, Reuben RC, et al.: Erythroleukemia cells: commitment to differentiate and the role of the cell surface, in Clarkson B, Marks PA, and Till J (eds): *Differentiation of Normal and Neoplastic Hematopoietic Cells.* New York, Cold Spring Harbor Laboratory, 1978.
35. Zeilig CE, Johnson RA, Sutherland EW, et al: Adenosine 3':5'-monophosphate content and actions in the division cycle of synchronized HELA cells. *J Cell Biol 71:*515–534, 1976.
36. Levy J, Terada M, Rifkind RA, et al: Induction of erythroid differentiation by dimethylsulfoxide in cells infected with Friend virus: Relationship to the cell cycle. *Proc Natl Acad Sci (USA) 72:* 28–32, 1975.
37. Rovera G, O'Brien TG, Diamond L: Tumor promoters inhibit spontaneous differentiation of Friend erythroleukemia cells in culture. *Proc Natl Acad Sci (USA) 74:*2894, 1977.

38. Yamasaki H, Fibach E, Nudel U, et al: Tumor promotors inhibit spontaneous and induced differentiation of murine erythroleukemia cells in culture. *Proc Natl Acad Sci (USA) 74*:3451–3455, 1977.
39. Yamasaki H, Fibach E, Weinstien IB, et al: Inhibition of murine erythroleukemia cell differentition by tumor promotors, in Ikawa Y (ed): *Oji International Seminar on Genetics Aspects of Friend Virus and Friend Cells*. New York, Academic Press, 1977.
40. Weinstein IB, Wigler M, Pietropaolo C.: The action of tumor promoting agents in cell culture, in Hiatt HH et al. (eds): *Origins of Human Cancer*. New York, Cold Spring Harbor Laboratories, 1977.
41. McClintock PR and Papaconstantinou J: Regulation of hemoglobin synthesis in a murine erythroblastic leukemic cell: The requirement for replication to induce hemoglobin synthesis. *Proc Natl Acad Sci (USA) 71*:4551–4555, 1974.
42. Harrison PR: Analysis of erythropoiesis at the molecular level. Review article. *Nature 262*:353–356, 1976.
43. Terada M, Fried J, Nudel U et al: Transient inhibition of initiation of S-phase associated with dimethylsulfoxide induction of murine erythroleukemia cells to erythroid differentiation. *Proc Natl Acad Sci (USA) 74*:248–252, 1977.
44. Terada M, Nudel U, Fibach E, et al: Changes in DNA associated with induction of erythroid differentiation by dimethylsulfoxide in murine erythroleukemia cells. *Cancer Res* (in press), 1977.

19 Regulation of Heme Biosynthesis in Mouse Friend Virus-Transformed Cells in Culture

Shigeru Sassa, Joel L. Granick, Harvey Eisen, and Wolfram Ostertag

Introduction

Mouse Friend virus-transformed erythroleukemia cells in culture comprise a homogeneous population of transformed erythroid precursor cells that can continuously divide *in vitro*. They do not normally differentiate along the erythroid pathway in culture. These cells, however, can be induced to undergo terminal erythroid differentiation by addition to the culture of dimethylsulfoxide (DMSO) or a variety of other apparently unrelated compounds. Such erythroid differentiation is characterized by the appearance of erythrocyte-specific proteins, e.g., hemoglobin,[1] spectrin,[2] acetylcholine esterase,[3] erythrocyte membrane-specific antigens,[4] a new histone-like chromatin protein, IP25,[5] and enzymes of the heme biosynthetic pathway.[6] Thus, these cells offer a very useful model in tissue culture for erythroid differentiation and permit biochemical studies of events occurring during the process of differentiation.

Hemoglobin in Friend virus-transformed cells appears 3 to 4 days after addition of DMSO. Since heme is clearly an obligatory component for hemoglobin formation, three questions were posed in this study concerning the biosynthesis of heme in these virus-transformed cells. Firstly, we asked whether enzymes in the heme biosynthetic pathway are present at sufficient activities initially or whether they are induced by treatment with DMSO to make sufficient amounts of heme for hemoglobin production. Secondly, we asked whether heme formation is the rate-limiting step for hemoglobin formation, and if so, whether addition of heme can turn on the synthesis of hemoglobin in these cells. Finally we asked whether there are DMSO-resistant mutant clones of Friend virus-transformed cells that might have a defect in one or more enzymes in heme biosynthesis.

Results

Sequential Induction of Enzymes in the Heme Biosynthetic Pathway

The basal levels of δ-aminolevulinate (ALA) synthase, ALA-dehydratase, URO-synthase, ferrochelatase, and heme were found to be low but were definitely detectable in untreated wild-type cells (745A, T3-Cl-2, F4N).

In striking contrast to the untreated cells, the cells treated with DMSO showed a significant increase in ALA-synthase and ALA-dehydratase activity after 1.5 days (Figure 19-1). The example depicted here is clone T3-Cl-2, and essentially similar changes were observed with other wild-type cells. The activity of URO-synthase showed a significant increase only after the second day, which is approximately 12 hr later than that of ALA-synthase or ALA-dehydratase but is considerably earlier than the appearance of heme (Figure 19-2) or hemoglobin.[7] The differences in the times of appearance of ALA-synthase, ALA-dehydratase, and URO-synthase were consistent and significant since they were observed when assaying these enzyme activities in cell suspensions from the same culture flask. Heme concentration in untreated cells was approximately 10 pmol/10^6 cells, whereas a significant increase of heme concentration occurred in cells grown in the presence of DMSO after 4 days of culture (Figure 19-2) and to levels of approximately 500 pmoles of heme/10^6 cells by the sixth day. This is 50 times more heme per cell than in the untreated state.

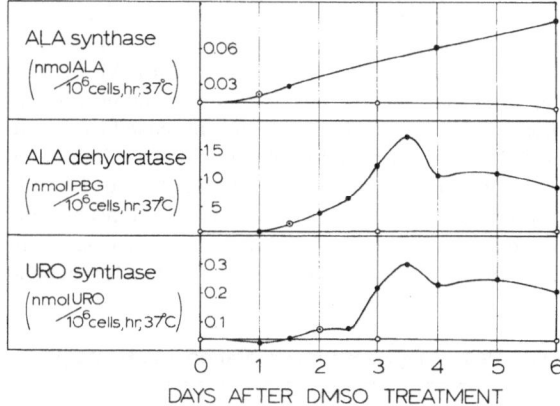

Figure 19-1 Changes in the activity of ALA-synthase, ALA-dehydratase, and URO-synthase in Friend virus-transformed cells after DMSO treatment. Cells of clone T3-Cl-2 were grown in the absence (open circles) or in the presence of 1.5% DMSO (closed circles). There was approximately 20-fold increase in cell number at the end of the experiment. Dotted circles indicate the time when the increase became significant over control. Data are the mean of duplicate ~ triplicate assays. (Reproduced from the Journal of Experimental Medicine, 1976 Vol. 143, pp. 305–315, by copyright permission of The Rockefeller University Press.)

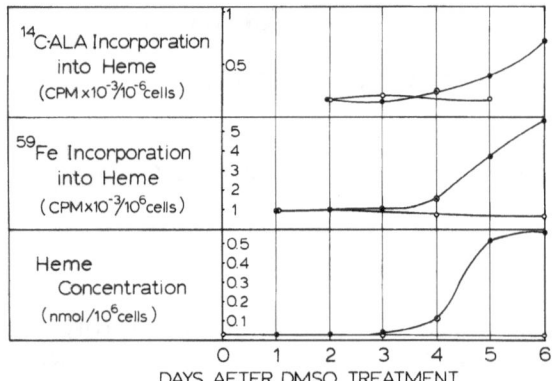

Figure 19-2 Changes in the uptake of ALA and iron into heme, and heme concentration. Cells of clone T3-Cl-2 were grown in the absence (open circles) or in the presence of 1.5% DMSO (closed circles). Data are the mean of duplicate assays. (Reproduced from The Journal of Experimental Medicine, 1976, Vol. 143, pp. 305–315, by copyright permission of The Rockefeller University Press.)

It is clear that cells treated with DMSO for 3 days contain maximally elevated levels of early enzymes of the heme biosynthetic pathway, i.e., ALA-synthase, ALA-dehydratase, URO-synthase, as shown in Figure 19-1. Ross et al.[7] also reported that the maximal formation of globin mRNA occurs on the second day after DMSO addition in the same clone of cells (T3-Cl-2). Thus these cells may be expected to form hemoglobin earlier in greater quantities by incubating them with ALA provided that later enzymes of heme biosynthesis are also present at nonlimiting activities. However, when cells were incubated in the presence of 0.6 mM ALA either with or without DMSO for 3 days, cells accumulated large amounts of protoporphyrin but no increase of heme concentration was observed as the result of incubation with ALA. This fact suggests that the terminal step of heme biosynthesis, i.e., ferrochelatase, is not increased to a sufficient level to form heme. When cells were incubated with ^{14}C-ALA or ^{59}Fe, it was found that DMSO-treated cells showed an increase in ^{14}C-heme or ^{59}Fe-heme formation only after 4 days (Figure 19-2). This fact suggests that the induction of either ferrochelatase or an iron transport system takes place only 4 days after treatment with DMSO, and the utilization of ALA, iron, and presumably protoporphyrin for hemoglobin formation is limited by this delayed step.

The increases in the enzymes of heme biosynthesis were greatly suppressed when actinomycin D was added to DMSO-treated cultures.[6] This finding suggests that continued RNA synthesis is required for induction of heme pathway enzymes to occur after treatment with DMSO. When cells were treated with 5-bromo-2'-deoxyuridine (BrdU), a thymidine analogue, at an early period of induction by DMSO (within 2 days), the induction of heme synthetic enzymes as well as an increase in heme concentration was greatly suppressed. The inhibition by BrdU was completely restored by the simultaneous addition of thymidine but not by uridine. These data suggest that proper transcription is essential for erythroid cell differentiation as well as for the sequential induction of heme synthetic enzymes to take place in these cells.

Hemin Induction of Heme Biosynthesis in Friend Virus-Transformed Cells

When wild-type Friend cells (clone 745A) were incubated with 10^{-4} M of ^{14}C-hemin or ^{59}Fe-hemin, a significant incorporation of the radioactive hemin was observed into hemoglobin 3 days after the treatment in the absence of DMSO (Table 19-1). The fraction of the radioactively labeled heme in the hemoglobin was 60 to 75% of the newly synthesized hemoglobin in these cells with no significant change in the ratio throughout the 5-day incubation period (Table 19-1). These data indicate that exogenously added hemin can be directly incorporated into hemoglobin, but this can only account for approximately two-thirds of the newly synthesized hemoglobin, suggesting that a fraction of heme must be provided by cellular heme formation. To determine the hemin effect on the cellular biosynthesis of heme, cells were incubated with hemin and/or DMSO and aliquots of cells were recovered and assayed for the activities of ALA-dehydratase and URO-synthase. Cells treated with hemin demonstrated considerably higher levels of these enzyme activities at 30 hr posttreatment relative to the

Table 19-1 ^{59}Fe-Hemin incorporation into Hemoglobin.

Treatment	Period (days)	Total Hgb (nmoles)	^{59}Fe-Hgb (nmoles)	^{59}Fe-Hgb/Total Hgb (%)
None	5	<0.10	—	—
^{59}Fe-Hemin 10^{-4} M	1	0.11	0.08	73.6
^{59}Fe-Hemin 10^{-4} M	3	9.11	6.96	76.4
^{59}Fe-Hemin 10^{-4} M	5	12.8	7.97	62.3
^{59}Fe-Hemin 5×10^{-5} M + DMSO 1%	5	40.91	25.49	62.3

Amounts of hemoglobin found are expressed as nmoles/culture. Each culture contained 73 ~ 90 × 10^6 cells. The value of 40.91 nmoles for cells treated with hemin + DMSO corresponds to approximately 8 pg Hgb/cell.

Table 19-2 Levels of ALA-dehydratase and URO-synthase in Friend virus-transformed cells treated with hemin

Treatment	Period (hr)	ALA-Dehydratase (nmol PBG/10^6 Cells, hr)	URO-Synthase (nmol URO/10^6 Cells, hr)
None	19	0.85	0.031
	30	0.64	0.034
	144	0.47	0.018
Hemin, 0.5×10^{-4} M	19	1.1	0.029
	30	1.8	0.076
	144	0.34	0.012
Hemin, 1×10^{-4} M	19	0.54	0.027
	30	2.8	0.084
	144	0.81	0.028
DMSO, 1.5%	19	0.76	0.025
	30	2.4	0.180
	144	1.9	0.058
Hemin, 1×10^{-4} M + DMSO, 1.5%	19	1.3	0.021
	30	2.2	0.056
	144	4.8	0.110

Cells of clone 745A were incubated with hemin and/or DMSO and aliquots of cell suspension were recovered for the analysis of ALA-dehydratase and URO-synthase activities.

control cultures (Table 19-2). It should be pointed out that the induction of these enzyme activities by hemin was observed at 30 hr after DMSO addition but it was shut off after 144 hr of incubation when maximum hemoglobin formation was observed. Although it is difficult to count benzidine positive cells accurately in the presence of large amounts of added hemin by conventional microscopic scoring, flow microphotometric demonstration of benzidine positive cells allowed us to make a distinction between false positive cells and truly benzidine-positive cells (Figure 19-3). Using this technique, it was found that approximately 30% of cells were hemoglobinized after treatment with 1×10^{-4} M hemin alone for 5 days. These findings suggest that hemin, in the absence of DMSO, can cause erythroid cell differentiation as well as the induction of the heme synthetic enzyme activity.

This observation was further confirmed by the following experiments using radioactive tracers for assessing heme formation. In these studies, radioactive tracers were supplied at three different steps in the heme biosynthetic pathway. Firstly cells were incubated with ^{14}C-2-glycine for 5 days to assess heme biosynthetic activity, including ALA-synthase, and the capacity for globin formation. Hemoglobin fractions were partitioned into heme and globin by acid-cyclohexanone and counted for radioactivity. It was found that both DMSO and hemin treatment increased the incorporation of ^{14}C-glycine into heme and globin, suggesting that ALA-synthase activity was increased after treatment with either of these chemicals (Table 19-3).

Although ^{14}C-glycine incorporation into globin was greater after treatment with hemin than after DMSO, the rate of heme synthesis after hemin treatment was less than that with DMSO. This finding also supports the idea that hemin treatment results in an increased synthesis of hemoglobin in Friend virus-transformed cells partly by incorporating the exogenously added hemin, and partly by the induction of cellular heme biosynthetic activity.

Secondly, in cultures parallel to the above, cells were incubated continuously with ^{3}H-leucine and ^{14}C-ALA for 5 days to assess the rate of utilization of exogenously added ^{3}H-leucine compared to the cel-

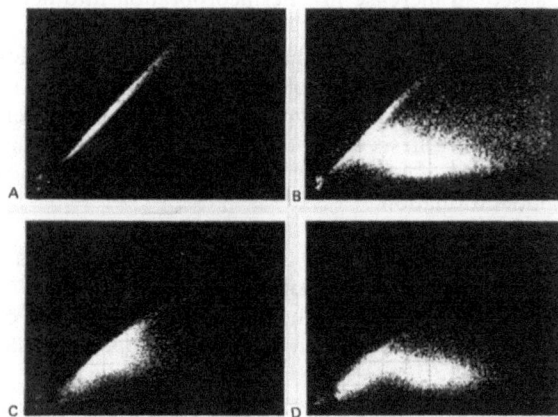

Figure 19-3 Flow microphotometric assay of benzidine-positive cells. Cells of clone 745A were grown in the presence of DMSO and/or hemin for 5 days. Benzidine staining was performed as described in Appendix 12. Benzidine-positive cells were counted in a cytograph model 6300A (Ortho Instruments, Westwood, Massachusetts). Absorption was displayed on the x-axis and scatter was displayed on the y-axis. **A**. Untreated cells: Benzidine-positive cells, 1.8%. **B**. Cells treated with 1.5% DMSO: Benzidine-positive cells, 78%. **C**. Cells treated with 10^{-4} M hemin: Benzidine-positive cells, 34%. **D**. Cells treated with 1% DMSO + 5×10^{-5} M hemin: Benzidine-positive cells, 56%.

Table 19-3 ^{14}C-Glycine incorporation into heme and globin in Friend virus-transformed cells.

Treatment	Heme-^{14}C	Globin-^{14}C
None	100	100
DMSO, 1.5%	3730	630
Hemin, 10^{-4} M	650	1370

Cells of clone 745A were treated with DMSO or hemin for 5 days and cells were recovered for the isolation of hemoglobin. Hemoglobin fraction was purified by CM-Sephadex chromatography. Heme and globin moiety of the hemoglobin was prepared by the acid-cyclohexanone procedure[28]. Rate of incorporation of ^{14}C is expressed as % to the control value.

lular heme biosynthetic activity as assessed by ^{14}C-ALA incorporation into heme. Hemoglobin was fractionated by CM-Sephadex column chromatography, quantitated by 410 nm absorption and counted for ^{3}H labeling globin and ^{14}C labeling heme (Table 19-4). DMSO, hemin and hemin + DMSO treatment all increased heme and globin formation as indicated by increases in the ratios of ^{3}H or ^{14}C uptake over the control value. Hemin increased globin formation to approximately the same level as did DMSO, but it resulted in less of an increase in cellular heme synthesis than did DMSO. Also cells were continuously labeled with ^{59}Fe for 5 days and the hemoglobin fraction was separated into heme and globin. The heme fraction was counted for ^{59}Fe. In addition to DMSO, or DMSO + hemin treatment, hemin treatment alone caused an increase in ^{59}Fe incorporation into heme (data not shown).

These data confirm that hemin-induced hemoglobin formation consists of both the incorporation of added hemin into hemoglobin and the increase in cellular biosynthesis of heme. These results suggest that the supply of heme is the rate-limiting step for hemoglobin formation.

Heme Biosynthesis in DMSO-Resistant Clones of Friend Virus-Transformed Cells

To see whether or not heme determines the rate of synthesis of hemoglobin, we have attempted to analyze heme synthetic enzymes as well as other erythrocyte-specific proteins in DMSO-resistant clones of Friend virus-transformed cells. Such DMSO-resistant clones cannot complete erythroid cell differentiation in the presence of the inducer. All clones were selected for their continued ability to grow in the presence of DMSO.[8]

One DMSO-resistant clone, F4N^{+2}, derived from clone F4N, contains comparable levels of ALA-synthase, ALA-dehydratase, URO-synthase, and heme in the absence of DMSO as the uninduced wild-type F4N cells (Figure 19-4). When F4N^{+2} cells were treated with 1.5% DMSO, they did not increase heme formation although the early enzymes in the heme biosynthetic pathway were induced normally as were other early markers of erythroid cell differentiation such as the synthesis of spectrin, the chromatin protein, IP25, complement-mediated hemolysis, and agglutination by lectins. The induction of ALA-synthase, ALA-dehydratase, and URO-synthase in response to DMSO was completely normal in F4N^{+2} cells, yet the cells did not increase heme content. It should be noted that F4N^{+2} cells increased globin mRNA in the presence of the inducer. These data suggest that F4N^{+2} cells have a defect at a late stage of heme biosynthesis, either at the level of URO-decarboxylase, coproporphyrinogen oxidase, protoporphyrinogen oxidase, or ferrochelatase.

To circumvent the defect of heme formation in F4N^{+2} cells, F4N^{+2} cells were treated with both 10^{-4} M hemin and DMSO. The addition of hemin alone had practically no effect; while hemin addition in the presence of DMSO, caused the differentiation of cells toward the erythroid pathway as judged by the increase in the percentage of benzidine-positive cells from 8.5% to 43% and by the appearance of erythrocyte surface antigens in 80 to 90% of the cells. These cells die off in culture by treatment with DMSO and hemin by completing the terminal erythroid differentiation. This stochastic pattern of differentiation is found in the wild-type cells by DMSO treatment and is an important characteristic of normal erythroid cells.[9]

F4N^{+2} is considered to be a physiological variant of F4N, rather than a mutant for reasons not detailed here but discussed elsewhere.[8] True mutants, however, were selected in the presence of DMSO and/or hemin in methylcellulose containing medium which does not allow physiological variants to survive. The properties of two clones (F4$^+$, F4D5-5) isolated in such a manner are summarized in Figure 19-4 and Table 19-5.

Clone F4$^+$ was isolated after selection in 3% DMSO and was resistant to all inducers examined. Cells of clone F4$^+$ are constitutive for the expression of many, if not all, of the early markers of erythroid differentiation seen in F4N. Namely they are highly agglutinable by lectins and sensitive to complement-

Table 19-4 Incorporation of ^{3}H-leucine and ^{14}C-ALA into hemoglobin of Friend virus-transformed cells.

Treatment	Heme-^{14}C/10^6 Cells	Globin-^{3}H/10^6 Cells
None	100	100
DMSO, 1.5%	940	320
Hemin, 10^{-4} M	460	390

Cells of clone 745A were incubated with DMSO or hemin for 5 days and recovered for the isolation of hemoglobin. Hemoglobin was purified by CM-Sephadex chromatography and counted for ^{3}H labelling globin and ^{14}C labelling heme. Results are expressed as percentage to the control value.

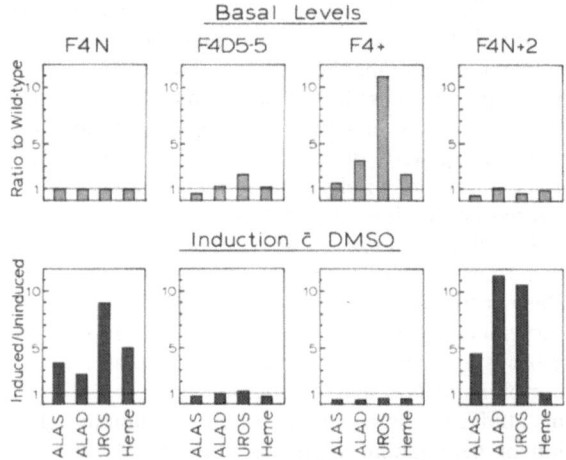

Figure 19-4 Basal levels and induced levels of activities of ALA-synthase, ALA-dehydratase, URO-synthase, and heme concentration in clones derived from F4N. Clone F4N represents the wild-type cells which is inducible with DMSO. Other clones are derived from F4N and DMSO does not induce hemoglobin formation in these cells.

Basal levels of enzyme activities or heme concentration is expressed as a ratio to the wild-type value.

Induction with DMSO was carried out by incubating cells with 1.5% DMSO for 5 days. Induction is expressed as a ratio of the induced level over the uninduced level.

induced lysis whether the inducer is present or not. They also constitutively synthesize spectrin band 3. IP25 is, however, neither present or inducible. Of the heme enzymes examined in these cells before and after DMSO treatment, URO-synthase was synthesized constitutively at a maximum level of the induced wild-type F4N cells. ALA-synthase and ALA-dehydratase levels were two to three times higher in the untreated F4$^+$ cells than the uninduced F4N cells. It is not clear, however, whether increases in ALA-synthase and ALA-dehydratase are due to constitutive synthesis of these enzymes or due to the polyploidy of the cells which may lead to an increase in the basal levels. The presence of the inducer did not increase these enzyme activities in F4$^+$ cells.

Cells of clone F4D5-5 (isolated after selection in 3% DMSO) contained low levels of ALA-synthase, ALA-dehydratase, URO-synthase, and heme. They were completely unresponsive to DMSO treatment. Only small amounts of IP25 was synthesized in response to DMSO. In contrast to F4N^{+2} cells, cells of clones F4$^+$ and F4D5-5 did not make hemoglobin when they were treated with DMSO and hemin simultaneously.

Properties of the other DMSO-resistant clones isolated either from clone 745A (Ma, U91, U99), or from clone T3-Cl-2 (K2) are summarized in Table 19-5. They all behaved differently with regard to levels and inducibility of enzymes in the heme biosynthetic activity. For example, Ma cells contained basal levels of ALA-dehydratase, URO-synthase, and heme similar to the uninduced wild-type cells (745A). The two enzyme levels were increased to the maximally induced level of the wild-type cells after treatment with DMSO; however, little increase in heme synthesis was noted. On the other hand, U91 and K2 cells displayed a considerably higher basal level of ALA-dehydratase and URO-synthase than their corresponding uninduced wild-type cells. Their benzidine-positive cells were equivalent to those of the uninduced wild-type cells. With DMSO treatment their enzyme levels increased to the maximally induced level of the wild-type cells but without an appreciable increase in benzidine positive cells.

Cells of another DMSO-resistant clone U99 did not display benzidine-positive cells after DMSO addition, but approximately 50% of the cells became benzidine-positive after treatment with erythropoietin or with hemin. The properties of this clone are discussed in detail elsewhere in this volume.[10]

Table 19-5 Properties of DMSO-resistant Friend virus-transformed cells.

	Heme Pathway Enzymes				
	Uninduced		Induced		
Clone	Early	Late	Early	Late	Differentiation
F4D-5	−	−	−	−	Primitive
F4$^+$	+++	−	−	−	Early function constitutive
K2, U91	++	−	+++	−	Early function semiconstitutive
Ma, F4N^{+2}, U99	−	−	+++	−	Late function non-inducible c̄ DMSO; inducible c̄ hemin (F4N^{+2}, U99); c̄ erythropoietin (U99); or totally non-inducible (Ma)
745A, T3Cl-2, F4N	−	−	+++	+++	Normal

Early enzymes include ALA-synthase, ALA-dehydratase, and URO-synthase. Late enzymes or events include ferrochelatase, heme and hemoglobin.

Discussion

These studies demonstrate that a sequential induction of enzymes of the heme biosynthetic pathway occurs during erythroid cell differentiation in Friend virus-transformed cells in culture after treatment with DMSO. The induction of ALA-synthase and ALA-dehydratase takes place as early as 36 hr after the addition of DMSO and precedes the increase in concentrations of heme or hemoglobin by approximately 2 days. Increases in these enzyme activities can be regarded as early erythroid functions as they are specific to the early stages of erythroid cell differentiation and, as shown in the studies of DMSO-resistant clones F4N^{+2} and F4$^+$, can take place without terminal erythroid differentiation, which is characterized by the formation of hemoglobin and the synthesis of erythrocyte surface antigens.

At least two possible mechanisms can be considered concerning the sequential induction of the heme synthetic enzymes. One mechanism may be sequential arrangement of genes for heme pathway enzymes in a polycistronic fashion. In fact, genes coding for heme pathway enzymes in *Staphylococcus aureus* are arranged in a polycistronic fashion in the same order in which these enzymes catalyze a series of reactions to form heme.[11] According to this model—termed a polycistronic model—once the inducer (DMSO or DMSO-mediated message) derepresses the operator, all the enzymes in the heme pathway are induced. This model might predict that all the enzymes in the heme biosynthesis would increase simultaneously when the operator is derepressed. Alternatively, genes coding for heme pathway enzymes in the erythroid precursor cells may not be linked to each other, but the product of the first gene may become an inducer for the second gene and the second gene product an inducer for the third gene, and so forth. This model can be called a cascade model. Available evidence favors the cascade model over the polycistronic model: (a) there are considerable lag periods between induction of each enzyme of the heme pathway, and these would be expected to be very small in the polycistronic model; (b) DMSO-treated cells accumulate protoporphyrin, suggesting considerable delay in induction of ferrochelatase activity, or iron supply to this enzyme reaction; and (c) the genetic regulation of ALA-dehydratase in inbred strains of mice has been shown to be distinct from that of URO-synthase.

A similar sequential induction of heme pathway enzymes has been found to occur during erythroid differentiation in the spleen of mice in response to erythropoietin[12] and during the development of fetal mouse liver.[13] Induction of ALA-synthase was found to precede the increase of hemoglobin synthesis by 24 hr in chick blastoderms treated with etiocholanolone.[14] These findings are examples of normal erythroid differentiation occurring with the normal course of development of fetuses or in response to an erythropoietic stimulus. Thus it may be suggested that a sequential induction of heme biosynthetic enzymes is a general phenomenon when the stem cell, either normal or virally transformed, undergoes cell differentiation along the erythroid pathway.

The above finding suggests that the induction of the terminal step of heme biosynthesis, i.e., ferrochelatase or iron supply to the ferrochelatase reaction, is a crucial event for hemoglobin formation. In fact, exogenous hemin has been shown to cause an increase in globin mRNA and hemoglobin synthesis.[15,16] In this study, we have demonstrated that hemin induction of hemoglobin synthesis consists of two processes. Approximately two-thirds of the heme in the newly synthesized hemoglobin is derived from the exogenously added hemin, and the remainder is provided through increased cellular heme synthesis. Evidence has also been presented for the increase by hemin of the enzymes, ALA-synthase, ALA-dehydratase, URO-synthase, and ^{59}Fe incorporation into heme. The stimulation of heme biosynthesis and ALA-synthase by hemin in Friend virus-transformed cells is in contrast to hemin-repression of ALA-synthase in the liver.[17] It has also been shown that hemin does not cause repression of the synthesis of ALA-synthase in prenatal rodent liver, which is predominantly an erythropoietic organ.[18] It appears that ALA-synthase in erythroid cells is under a different regulatory control than the enzyme in liver cells.

A body of evidence indicates that hemin added to an *in vitro* cell-free translational system promotes the rate of globin synthesis by blocking the formation of a soluble inhibitor of globin initiation.[19] More recently it has been shown that hemin stimulates equally the translation of extraneous mRNAs in the cell-free system.[20] The effect of hemin is considered to inhibit the cAMP-independent protein kinase activity associated with the soluble translational inhibitor, leading to an increased formation of an initiation complex, i.e., met-tRNA$_f$-40S ribosomal subunit.[21,22] Thus it is tempting to assume that hemin promotes not only the synthesis of globin, but also enzymes in the heme biosynthetic pathway by interfering with either the formation or the activity of the translational inhibitor.

Regulation of the events occurring during erythroid differentiation is extremely complex as evidenced by studies of other groups and by our studies in the wild-type as well as in the DMSO-resistant clones of Friend virus-transformed cells. It is clear, however, that certain changes such as the synthesis of globin mRNA, sensitivity to lectins and complement, the induction of spectrin and the chromatin protein IP25 and induction of early enzymes in the heme biosynthetic pathway occur early during erythroid differentiation. These changes, termed "early erythroid functions," are thought to lead to normal terminal erythroid differentiation in the induced wild-type Friend virus-transformed cells; this involves the induction of ferrochelatase, and the increased formation of heme and hemoglobin. However, certain DMSO-resistant clones appear to be deficient in one or more of the late processes in erythroid differentiation, such as the formation of heme or hemoglobin. The fact that cells of the DMSO-resistant clone F4N^{+2} display normal induction of certain early erythroid functions and that cells of clone F4$^+$ display constitutive synthesis of certain components in the early erythroid functions suggests that the induction of the late erythroid functions may be regulated independently of the early functions. The complete set of induction of both the early and the late erythroid functions is obviously necessary for erythroid differentiation to be completed. This idea is supported by the findings in cells of clone F4N^{+2}, which can undergo complete erythroid differentiation when treated with both DMSO and hemin. Moreover, another DMSO-resistant clone, U99, also displays terminal erythroid differentiation when treated with hemin or erythropietin. These data suggest that the expression of the late events of erythroid differentiation is caused by hemin or erythropoietin, but not by DMSO.

It is noteworthy that certain aspects of the early erythroid functions that are expressed in the wild-type Friend virus-transformed cells appear to be shut off at the time or shortly after the induction of the late erythroid function (Figure 19-1 and Table 19-2). Certain clonal cells such as F4$^+$, K2 and U91 which cannot complete the terminal erythroid differentiation express the early erythroid function either in a semi-constitutive or constitutive fashion, suggesting that the shut-off mechanism of the biological clock in these cells is defective.

Summary

Mouse Friend virus-transformed cells in culture were treated with dimethylsulfoxide (DMSO) to induce erythroid cell differentiation, and certain enzymes of heme biosynthetic pathway and heme were then determined using sensitive microassays. Incorporation of radioactive tracers into heme or hemoglobin by these cells was also studied to assess in a direct fashion the biosynthetic activity for heme or hemoglobin. We found that the induction of δ-aminolevulinate (ALA)-synthase and ALA-dehydratase occurred within approximately 36 hr of the addition of DMSO to the culture. Induction of uroporphyrinogen-I (URO)-synthase occurred at 48 hr and ^{59}Fe uptake into hemoglobin increased only 96 hr after DMSO addition. These sequential changes in enzymes of the heme biosynthetic pathway occurred in the same order in which these enzymes catalyze a series of intermediary reactions to form heme.

Treatment with hemin, the oxidized form of the end-product of the heme biosynthetic pathway, also caused erythroid cell differentiation in these cells. Approximately two-thirds of the heme in the newly synthesized hemoglobin was derived from this exogenously added hemin; the remainder was provided through increased cellular heme biosynthesis. Evidence was also obtained for the induction of ALA-synthase, ALA-dehydratase, URO-synthase and ^{59}Fe incorporation into heme by hemin treatment. Induction by hemin of cellular heme biosynthesis is in contrast to the finding in the liver cell, in which hemin has been shown to repress the synthesis of ALA-synthase.

Cells of two DMSO-resistant clones of Friend virus-transformed cells, clones F4N^{+2} and Ma, had comparable levels of ALA-synthase, ALA-dehydratase, and URO-synthase in the absence of DMSO as did the wild-type cells. They increased these enzyme activities in a similar manner as did the wild-type cells after DMSO treatment. However, they did not display evidence of increased heme synthesis. Clone F4$^+$ already contained elevated levels of these enzyme activities, but cells did not synthesize increased amounts of heme either in the absence or in the presence of DMSO.

The results of this study suggest that the formation of heme may be the rate-limiting step for hemoglobin formation. It is also clear, as shown by the studies in DMSO-resistant clones, that early events in the biosynthesis of heme can be induced independently from the late processes of heme formation. The complete sequence of induction involving both the early and the late steps of heme biosynthesis is apparently necessary for erythroid differentiation to be completed in these Friend virus-transformed cells.

Acknowledgement

This study was supported in part by grants from the American Cancer Society BC–180A and USPHS ES–01055 to S. Sassa and from the Swiss National Foundation to H. Eisen. We are indebted to Dr. A. Urabe, Sloan-Kettering Institute for Cancer Research, New York, for DMSO-resistant clones U91 and U99. The authors are grateful to Dr. A. Kappas, The Rockefeller University, for his encouragement and support to this study. Technical assistance of Mrs. C. Chang and Mr. S. N. Feltham and secretarial assistance of Mrs. Heidi Robinson are also gratefully acknowledged.

References

1. Friend C, Scher W, Holland JG, et al: Hemoglobin synthesis in murine virus induced leukemic cells *in vitro:* Stimulation of erythroid differentiation by dimethylsulfoxide. *Proc Natl Acad Sci (USA) 68:*378–382, 1971.
2. Eisen H, Bach R, Emery R: Induction of spectrin in Friend virus-transformed erythroleukemic cells. *Proc Natl Acad Sci (USA) 74:*3898–3902, 1977.
3. Conscience JA, Miller RA, Henry J, et al: Acetylcholine esterase, carbonic anhydrase and catalase activity in Friend erythroleukemic cells, non-erythroid mouse cell lines and their somatic hybrids. *Exp Cell Res 105:*401–412, 1977.
4. Furusawa M: Erythroid "differentiation" of Friend cells, in Ito Y, Dutcher RM (eds): *Comparative Leukemia Research, 1973, Leukemogenesis, Biblio Haemat 40;* Tokyo, University of Tokyo Press, pp. 273–274, 1975.
5. Keppel F, Allet B, Eisen H: Appearance of a chromatin protein during the erythroid differentiation of Friend virus-transformed cells. *Proc Natl Acad Sci (USA) 74:*653–656, 1977.
6. Sassa S: Sequential induction of heme pathway enzymes during erythroid differentiation of mouse Friend leukemia virus-infected cells. *J Expt Med 143:*305–315, 1976.
7. Ross J, Gielen, J, Packman S, et al: Globin gene expression in cultured erythroleukemic cells. *J Mol Biol 87:*697–714, 1974.
8. Eisen H, Keppel-Ballivet F, Georgopoulos CP, et al: Biochemical and genetic analysis of erythroid differentiation in Friend virus-transformed murine erythroleukemia cells. In *Cold Spring Harbor Conference on "Differentiation of Hematopoietic Cells,"* 1978 (in press).
9. Gusella J, Geller B, Clarke B, et al: Commitment to erythroid differentiation by Friend erythroleukemia cells: a stochastic analysis. *Cell 9:*221–229, 1976.
10. Urabe A, Murphy M, Jr, Sassa S: Studies on erythroid differentiation in a wild-type and DMSO-resistant clones of Friend-virus transformed cells: Effects of hemin, hemoglobin and erythropoietin, in Murphy MJ Jr (ed): *In Vitro Aspects of Erythropoiesis,* p. 149, New York, Springer-Verlag, 1978.
11. Nakao K, Sassa S, Wada O, et al: Enzymatic studies on erythroid differentiation and proliferation. *Ann NY Acad Sci 149:*224–228, 1968.
12. Freshney RI, Paul J: The activities of three enzymes of haem synthesis during hepatic erythropoiesis in the mouse embryo. *J Embryol Exp Morph 26:*313–322, 1971.
13. Irving RA, Mainwaring WIP, Spooner PM: The regulation of haemoglobin synthesis in cultured chick blastoderms by steroids related to 5 β-androstane. *Biochem J 154:*81–93, 1976.
14. Tien W, White DC: Linear sequential arrangement of genes coding for the biosynthetic pathway of protoheme in *Staphylococcus aureus. Proc Natl Acad Sci (USA) 61:*1392–1398, 1969.
15. Ross J, Sautner D: Induction of globin mRNA accumulation by hemin in cultured erythroleukemic cells. *Cell 8:*513–520, 1976.
16. Dabney BJ, Beaudet AL: Increase in globin chains and globin mRNA in erythroleukemia cells in response to hemin. *Arch Biochem Biophys 179:*106–112, 1977.
17. Granick S, Sassa S: δ-Aminolevulinic acid synthetase and the control of heme and chlorophyll synthesis, in "Metabolic Regulation," (Vol. 5 of Metabolic Pathways), ed. by H. J. Vogel, Academic Press, New York, pp. 77–141, 1971.
18. Woods JS, Murphy VV: δ-Aminolevulinic acid synthetase from fetal rat liver: studies on the partially purified enzyme. *Molec Pharmacol 11:*70–78, 1975.
19. Matthews MB, Hunt T, and Brayley A: Specificity of the control of protein synthesis by haemin. *Nature (New Biol) 243:*230–233, 1973.
20. Beuzard Y, Rodvien R, London IM: Effect of hemin on the synthesis of hemoglobin and other proteins in mammalian cells. *Proc Natl Acad Sci (USA) 70:*1022–1026, 1973.
21. Clemens MJ, Henshaw EC, Rahamimoff H, et al: Met-tRNA$_f^{Met}$ binding to 40S ribosomal subunits: A site for the regulation of initiation of protein synthesis by hemin. *Proc Natl Acad Sci (USA) 71:*2946–2950, 1974.
22. Levin DH, Ranu RS, Ernst V, et al: Regulation of protein synthesis in reticulocyte lysates: Phosphorylation of methionyl tRNA$_f$ binding factor by protein kinase activity of translational inhibitor isolated from heme-deficient lysates. *Proc Natl Acad Sci (USA) 73:*3112–3116, 1976.

20 Comparative Studies on Hemopoietic Stem Cell Pools in Mice after Friend (FV-P) or Rauscher Virus Infection

H. J. Seidel and Uta Opitz

Introduction

In the past years many studies have been published regarding the murine hematopoiesis after infection of mice with the Rauscher or both strains of the Friend virus.[1-24] The results have not always been the same in different laboratories and especially the interpretation of the erythropoietic changes has been subject to discussions in the Rauscher system[8,25] (Nooter and Ghio 1975, Seidel 1976, Opitz et al 1977). Moreover no studies on both diseases, i.e., Rauscher virus induced erythroblastosis and Friend virus induced polycythemia have been performed by the same group.

Our earlier studies on the Rauscher system can be summarized as follows. Mice of different strains develop an anemia with erythroblastosis in the spleen and peripheral blood, an early reticulocytopenia followed by reticulocytosis, accompanied by an increase in total CFU-s content in the enlarged spleen, in some strains (e.g., CBA), also in CFU-s and CFU-c concentration in the spleen.[4-7] Granulocytopoiesis seemed to be affected *in vivo* as bone marrow smears and a neutropenia in the peripheral blood suggested, but a detailed analysis of the differentiation and maturation of granulocytopoietic cells *in vitro* under the influence of CSA, even of different types of CSA, showed a normal pattern.[27] The erythropoietin responsive compartment and the CFU-e's are also above controls in the spleen.[8,26]

Materials and Methods

The Friend leukemia, polycythemic strain, was studied in DBA/2 and NMRI mice. Both mouse strains develop the disease as described originally with reticulocytosis, and increase of hematocrits and an Ep-independent erythropoiesis.[28] Granulocytopoiesis seemed not to be affected.

In our comparative study female NMRI mice were used.

In a first series of experiments the incorporation of ^{59}Fe into bone marrow, spleen, and peripheral blood was determined, in addition to spleen weight changes and the hematocrit (Tables 20-1, 20-2). ^{59}Fe was injected after incubation with normal mouse serum for at least 30 min, 0.25 μCi per mouse. The femora and spleen were taken after 6 hr and after 48 hr the peripheral blood was taken, and counted in a γ-counter. The blood volume was assumed to be 6% of the body weight. The results show common as well as separate findings common to both viruses is the shift of active erythropoiesis as expressed by the 6-hr uptake of radioiron from the bone marrow to the spleen. Within 5 days after infection with Friend (FV-P) as well as with Rauscher virus (RLV) the uptake into the spleen had increased three- to fourfold and increased further. The incorporation into bone marrow was well below controls in both cases 10 days after infection. Within 48 hr after ^{59}Fe injection 55 to 65% of the injected dose had reached the peripheral blood in controls. In FV-P infected mice this was somewhat higher 5 days after infection and remained in the normal range thereafter. The relatively low value 20 days after infection might be underestimated since a normal blood volume was used for the calculations, but there was a considerable increase which, however, was not determined in this experiment. After RLV-infection the radioactivity in the blood was below controls, indicating that not all hemoglobin-synthesizing cells reached the peripheral blood as reticulocytes and erythrocytes. There is—in contrast to the

Table 20-1 Spleen weight and hematocrit of NMRI mice after Friend (FV-P) and Rauscher (RLV) virus infection.

		Contr. I	Days after Infection				Contr. II
			5	10	15	20	
spleen (mg)	FV-P	116 ± 13	379 ± 276	1,960 ± 488	1,197 ± 622	3,635 ± 644	136 ± 19
	RLV		621 ± 195	1,549 ± 1,243	3,062 ± 714	2,853 ± 1,042	
hct. (%)	FV-P	46 ± 2	47 ± 4	42 ± 4	46 ± 9	57 ± 3	48 ± 2
	RLV		43 ± 4	34 ± 6	33 ± 5	30 ± 8	

Table 20-2 FV-P or RLV infection of NMRI mice ^{59}Fe incorporation into spleen, bone marrow, and peripheral blood.

		Contr. I	Days after Infection				Contr. II
			5	10	15	20	
spleen	FV-P	5.6 ± 2.2[a]	19.1 ± 3.0	43.0 ± 4.5	32.7 ± 3.9	29.0 ± 7.0	11.2 ± 4.5
	RLV		18.9 ± 9.2	40.0 ± 5.0	33.1 ± 9.1	38.7 ± 8.7	
b.m.	FV-P	1.3 ± 0.3	1.2 ± 0.3	0.3 ± 0.0	0.2 ± 0.0	0.2 ± 0.0	1.7 ± 0.3
	RLV		0.6 ± 0.2	0.5 ± 0.1	0.1 ± 0.0	0.1 ± 0.0	
p.b.	FV-P	55.4 ± 14.9	71.3 ± 6.2	65.6 ± 9.2	65.0 ± 9.2	54.0 ± 10.4	65.6 ± 9.9
	RLV		25.1 ± 4.7	24.0 ± 16.2	40.3 ± 10.3	49.1 ± 8.6	

[a] Given in % of injected ^{59}Fe dose, spleen and femur were taken 6 hr, peripheral blood 48 hr after ^{59}Fe injection.

Friend disease—ineffective erythropoiesis with the consequence of anemia as seen from the hematocrit. This ineffective cell production was already described by our earlier studies with bone marrow and spleen cell smears, where a maturation arrest on the level of the medium-sized erythroblast was postulated.[4,5,8]

In the first experiment the hemopoietic stem cells CFU-s, CFU-c, BFU-e and CFU-e were studied in Friend or Rauscher virus-infected NMRI mice 12, 24, 36, 48, and 72 hr and 4, 6, and 8 days after virus infection. The results of this experiment are given in Figures 20-1 to 20-8. Controls (shaded areas) were also done each day and in order to reduce the variability of the assays the results of the experimental groups were always expressed as percentages of these daily controls.

CFU-s: During the first 36 hr after infection the CFU-s in the marrow increased in both instances to approximately 140%, and were slightly below thereafter (Figure 20-2). In the FV-P infected mice the CFU-s were reduced to 50 to 60% of control at days 5 through 8. In the spleen a clear fall of CFU-s concentration was present after 24 hr which recovered to control values after 48 hr and later the CFU-s concentration was between 40 and 70% of the controls (Figure 20-3). From day 3 onward, however, the spleen weight was higher (Figure 20-1) and a calculation of the CFU-s content of the organ as a whole again gave approximately normal numbers of CFU-s.

CFU-c: In the bone marrow the CFU-c showed, in contrast to CFU-s, an initial dip with partial recovery to normal concentrations (Figure 20-4). In the spleen there was a marked fall in CFU-c at 24 and 36 hr after infection with overshooting regeneration at day 4 and approximately normal values at days 6 and 8 (Figure 20-5).

BFU-e: This compartment showed several problems in these studies and we must admit that here technical difficulties were also involved. The initial increase in BFU-e in FV-P–infected animals (Figure 20-6) was reproduced many times and the later data, although incomplete, gave approximately normal or slightly increased values. Later after FV-P or RLV infection, however, the BFU-e did not grow at all. This will be shown in the next experimental series.

CFU-e: The CFU-e growth curves confirmed the

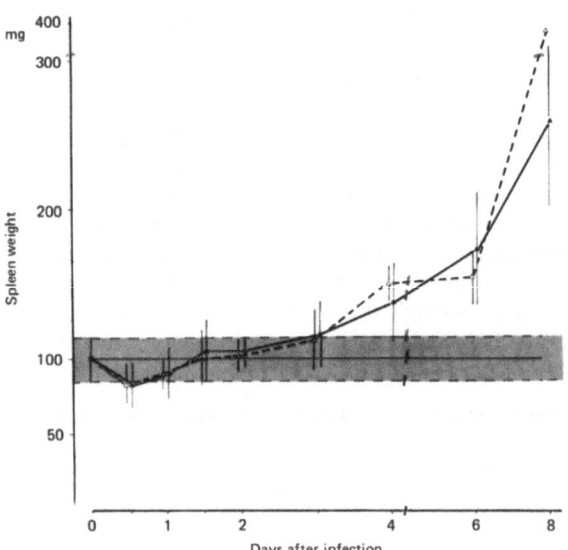

Figure 20-1 Spleen weight versus days after infection. Open triangles, RLV; closed triangles, FVP.

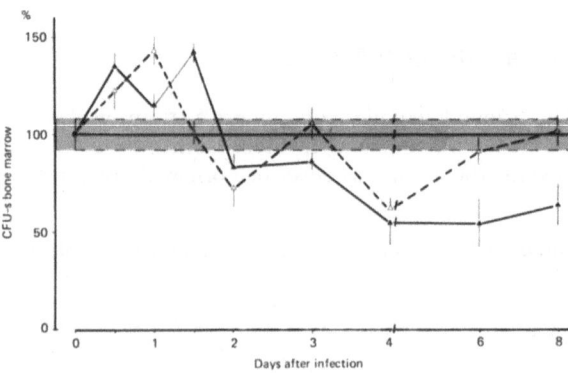

Figure 20-2 Concentration of CFU-s in bone marrow (% versus days after infection. Open triangles, RLV; closed triangles, FVP.

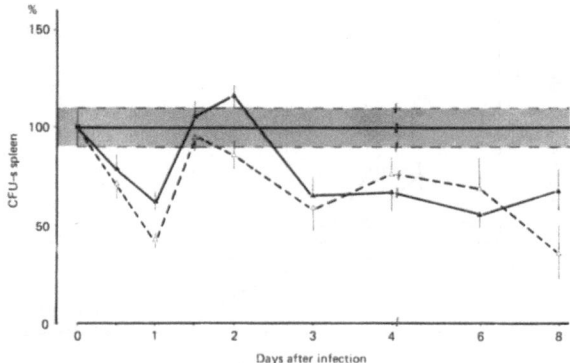

Figure 20-3 Concentration of CFU-s in spleen versus days after infection. Open triangles, RLV; closed triangles, FVP.

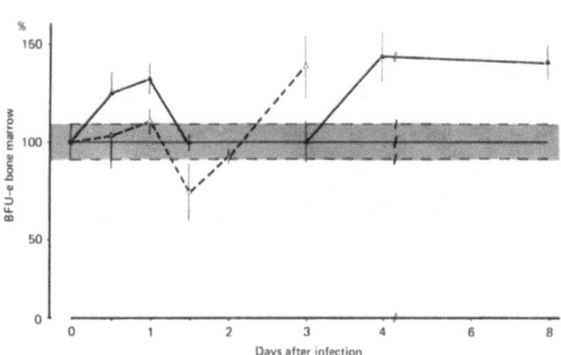

Figure 20-6 Concentration of BFU-e in bone marrow versus days after infection. Open triangles, RLV; closed triangles, FVP.

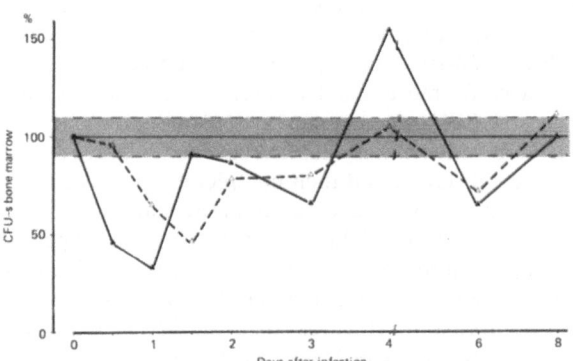

Figure 20-4 Concentration of CFU-c in bone marrow versus days after infection. Open triangles, RLV; closed triangles, FVP.

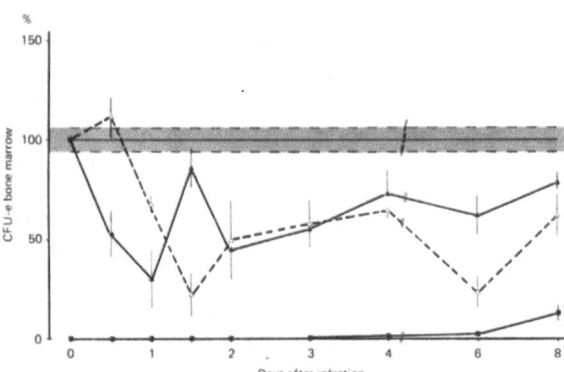

Figure 20-7 Concentration of CFU-e in bone marrow versus days after infection. Open triangles, RLV; closed triangles, FVP + EP; closed squares, FVP − EP.

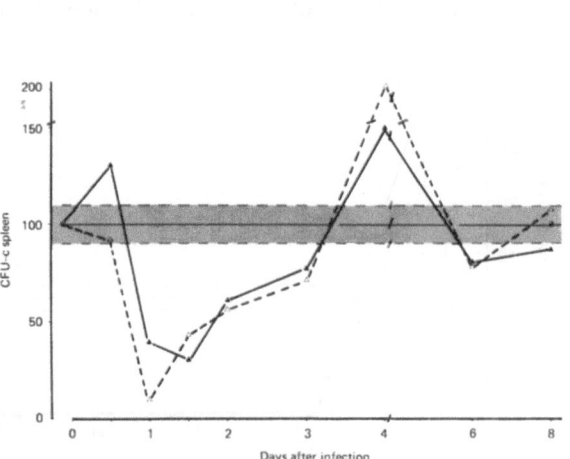

Figure 20-5 Concentration of CFU-c in spleen versus days after infection. Open triangles, RLV; closed triangles, FVP.

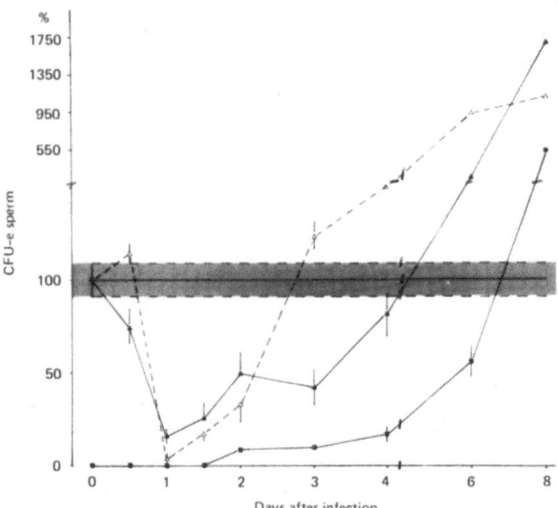

Figure 20-8 Concentration of CFU-e in spleen versus days after infection. Open triangles, RLV; closed triangles, FVP + EP; closed squares, FVP − EP.

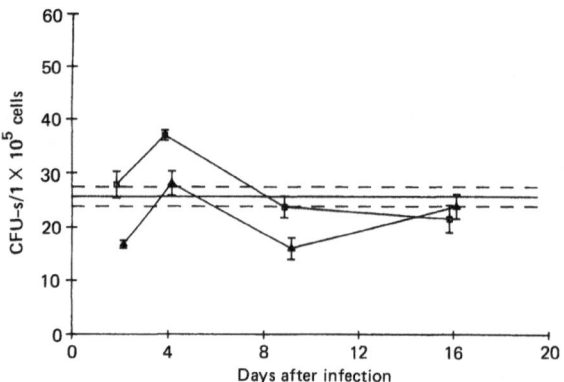

Figure 20-9 Effect of RLV (squares) and FV-P (triangles) infection on concentration of CFU-s in bone marrow.

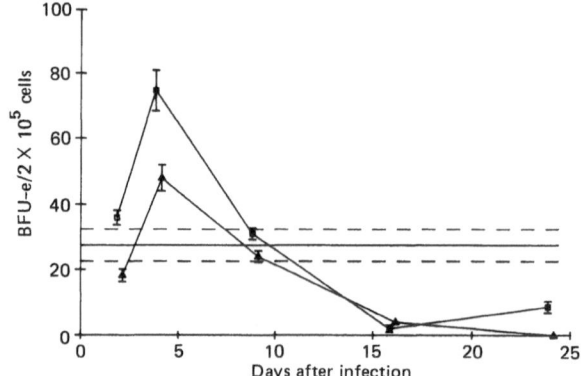

Figure 20-11 Effect of RLV (squares) and FV-P (triangles) infection on concentration of BFU-e cells in bone marrow.

results of the previous experiments.[26,28] In the bone marrow a fall to 20 to 30% of controls was seen between days 1 and 2 after infection and the concentration had not reached control levels at day 8. Colonies without addition of Ep to the plates were first detected at day 4 in FV-P–infected mice but never in RLV-infected ones. In the spleen the initial fall at 24 hr was even more pronounced than in the marrow with rapid recovery after RLV and somewhat slower recovery after FV-P infection. At day 8 the concentration after FV-P infection was approximately fivefold, after RLV infection 17-fold compared to controls. Ep-independent colonies were only detected in FV-P–infected animals, first at day 2 and rapidly increasing at days 6 and 8. Complete Ep independency was not yet achieved at day 8 in this experiment.

CFU-s, BFU-e and CFU-e were also analyzed in a second experiment up to 23 days after RLV or FV-P infection. The results are represented in Figures 20-9 to 20-14. Again the CFU-s in the bone marrow were somewhat below control values (Figure 20-9). In the spleen a rise in RLV-infected mice was observed at day 4 and later the concentration of CFU-s in this organ was approximately one-half the control values (Figure 20-10). The organ weight, however, had increased far more than twofold, so the absolute number of CFU-s per organ had also increased. The BFU-e showed an increase in bone marrow and spleen at day 4, and then completely disappeared at day 9 in the spleen and day 16 in the marrow (Figures 20-11 and 20-12). This loss was rather unexpected and we figured out, as shown later, that this was a pure *in vitro* phenomenon. The CFU-e data gave no new information (Figures 20-13 and 20-14). Complete Ep independency of FV-P infected mice seems to be reached in the marrow at day 14, in the spleen at day 8.

Results

If one tries to correlate all these data the following statements can be made. Both viruses induce quantitative changes in the different stem cell pools. Most

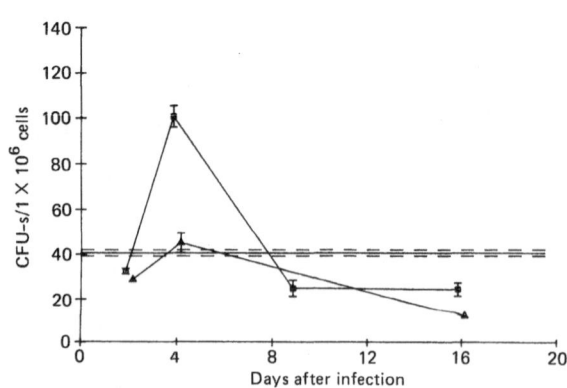

Figure 20-10 Effect of RLV (squares) and FV-P (triangles) infection on concentration of CFU-s in spleen.

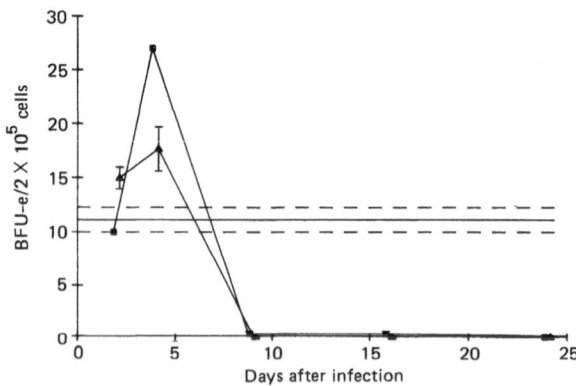

Figure 20-12 Effect of RLV (squares) and FV-P (triangles) infection on concentration of BFU-e cells in spleen.

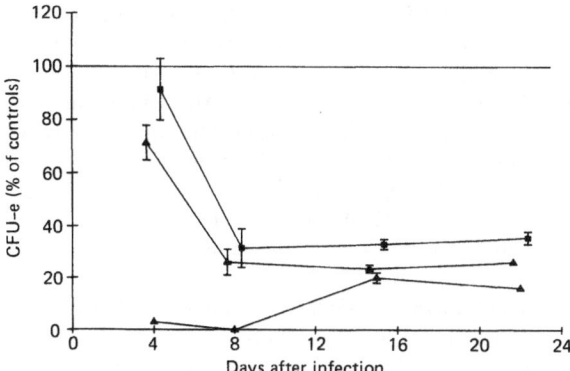

Figure 20-13 Effect of FV-P (dotted triangles), FV-P + EP (closed triangles), and RLV + EP (closed squares) infection on concentration of CFU-e in bone marrow.

striking in nonerythropoietic cells is the decrease in CFU-c in bone marrow and spleen at days 1 through 2, accompanied by an insignificant increase in CFU-s in the marrow. Later, at day 4, a recovery is seen and then CFU-s in the bone marrow and CFU-s concentrations in the spleen stay below control values. These changes have to be seen in relation to a virus infection in general and possibly the induction of interferon production and in relation to changes induced by the specific viruses in the erythroid committed compartments that may be important in the regulation of the compartment sizes of CFU-s and CFU-c, as known, for instance, after injection of high doses of erythropoietin or by bleeding.

The specific effects of the viruses on the BFU-e and CFU-e compartments can also be compared to those after Ep injection, but the decrease in the CFU-e compartment can also be compared to those after endotoxin injection. The shift of erythropoiesis from the marrow to the spleen is well known after different types of stimulation or stress.

One very clear-cut result of these studies was that the Rauscher virus does not induce Ep independency. In a detailed analysis of the CFU-e growth we found that the colonies grew to a larger size after FV-P virus infection and persisted to day 3 before disintegration. The Ep dose-response curve of Rauscher virus-infected marrow and spleen cells and also the curve of the Ep-dependent ones after FV-P infection was completely normal.[28]

The BFU-e loss in both cases was, as already mentioned, an *in vitro* artifact and the studies are in press. In short, we showed that the BFU-e were present in normal concentration after velocity sedimentation by which rather nice BFU-e from CFU-e are separated.[29] By this method a normal profile was seen for BFU-e and CFU-e of FV-P–infected mice in the bone marrow and the spleen.[30] Some more experiments are needed to decide whether the modal sedimentation velocity of normal Ep-dependent CFU-e and -independent ones is identical.

Several other studies have been performed in the laboratory in the past using both viruses but not necessarily in the same protocol. These include infection of polycythemic and bled mice and mice pretreated by Myleran. In all these experiments the same basic phenomena were observed. Polycythemia slightly delayed the CFU-e rise in the spleen and bleeding enhanced it. Myleran, very active in the reduction of CFU-s and CFU-c, had only minor effects on the development of the CFU-e increment or—in the Friend system—the development of Ep-independent colonies, when given 3 or 6 days prior to virus infection.

These results further support the view that the Rauscher and the Friend viruses have targets within the stem cell compartment, most probably on the more immature part of the ERC, perhaps between BFU-e and CFU-e. The report on this study will be a special paper from our laboratory in the near future. The main difference between both diseases remains the induction of Ep-independent effective erythropoiesis by the FV-P and an Ep-dependent erythroblastosis, i.e., ineffective erythropoiesis by the Rauscher virus.

Acknowledgments

This research was supported by the Deutsche Forschungsgemeinschaft, SFB 112.

The technical assistance of E. Barthel and K. Steinhoff is acknowledged.

References

1. Rauscher FJ: A virus-induced disease of mice characterized by erythropoiesis and lymphoid leukemia. *J Natl Cancer Inst* 29:515–543, 1962.
2. Boiron M, Leevy JP, Lasneret J, et al: Pathogenesis of Rauscher leukemia. *J Natl Cancer Inst* 35:865–884, 1965.

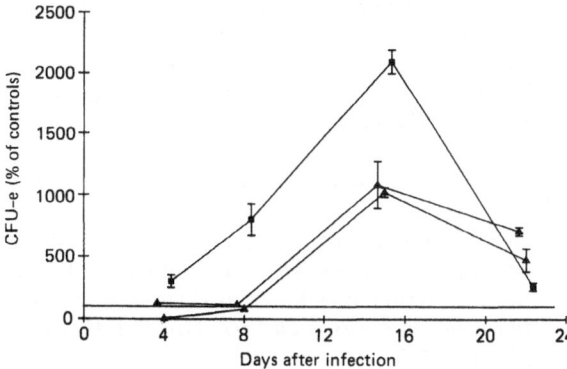

Figure 20-14 Effect of FV-P (dotted triangles), FV-P + EP (closed triangles), and RLV + EP (squares) infection on concentration of CFU-e in spleen.

3. Dunn TB, Green AW: Morphology of Balb/c mice inoculated with Rauscher virus. *J Natl Cancer Inst 36:*987–1001, 1966.
4. Seidel HJ: Die Blutzellbildung bei der Rauscherleukämie (Mäusestamm Balb/c) und ihre beeinflussung durch Hypertransfusion. *Z Krebsforsch 77:*155–165, 1972.
5. Seidel HJ: Pancytopenia in CBA mice after Rauscher virus infection. *J Natl Cancer Inst 48:*959–964, 1972.
6. Seidel HJ: Das Verhalten hämopoetischer Stammzellen bei Mäusen mit Virusleukämie I Milzkolonieversuche am Mäusestamm CBA nach Infektion mit dem Rauscher Virus. *Z Krebsforsch 79:*123–134, 1973.
7. Iturriza RG, Seidel HJ: Stem cell growth and production of colony stimulating factor in Rauscher virus infected CBA/J mice. *J Natl Cancer Inst 53:*487–492, 1974.
8. Seidel HJ: Studies on the erythropoietic cell system in CBA mice after Rauscher virus infection. *Blut 32:*257–268, 1976.
9. Van Beek HJ, Van't Hull E, van Griensven LJLD: Modification of hemopoietic stem cells of Balb/c mice by Rauscher leukemia virus. *Exp Hematol 4:*151–160, 1976.
10. Friend C: Cell-free transmission in adult mice of a disease having the character of a leukemia. *J Exp Med 105:*307–318, 1957.
11. Metcalf D, Furth J, Buffett RA: Pathogenesis of mouse leukemia caused by Friend virus. *Cancer Res 19:*52–58, 1959.
12. Mirand EA: Erythropoietin-like effect of a polycythemic virus. *Proc Soc Exp Biol Med 125:*562–565, 1967.
13. Mirand EA: Virus-induced erythropoiesis in hypertransfused-polycythemic mice. *Science 156:*832–833, 1967.
14. Mirand EA, Steeves RA, Lange RD et al: Polycythemia in mice: erythropiesis without erythropoietin. *Proc Soc Exp Biol Med 128:*844–849, 1968.
15. Wendling F, Tambourin PE, Jullien P: Haematopoietic CFU in mice infected by the polycythemia-inducing Friend virus. I Number of CFU, and differentiation pattern in the spleen colonies. *Int J Cancer 9:*554–566, 1972.
16. Wendling F, Tambourin PE, Jullien P: Hematopoietic CFU in mice infected by the polycythaemia-inducing Friend virus. IV Pattern of blood recovery in irradiated mice grafted with normal or infected bone marrow cells. *Biomedicine 18:*521–529, 1973.
17. Wendling F, Tambourin PE, Gallien-Lartigue O et al: Comparative differentiation and enumeration of CFU-s from mice infected either by the anemia—or polycythemia–inducing strains of Friend virus. *Int J Cancer 13:*454–462, 1974.
18. Tambourin PE, Wendling F: Malignant transformation and erythroid differentiation by polycythaemia-inducing Friend virus. *Nature (New Biol) 234:*230–233, 1971.
19. Tambourin PE, Wendling F: Target cell for oncogenic action of polycythaemia-inducing Friend virus. *Nature 256:*320–322, 1975.
20. Tambourin PE, Gallien-Lartigue O, Wendling F et al: Erythrocyte production in mice infected by the polycythaemia-inducing Friend virus or by the anemia-inducing Friend virus. *Br J Haematol 24:*511–520, 1973.
21. Okunewick JP, Philipps EL: Change in marrow and spleen CFU compartments following leukemia virus infection: comparison of Friend and Rauscher virus. *Blood 42:*885–892, 1973.
22. Liao SK, Axelrad AA: Erythropoietin-independent erythroid colony formation in vitro by hemopoietic cells of mice infected with Friend virus. *Int J Cancer 15:*467–482, 1975.
23. Horoszewicz JS, Leong SS, Carter WA: Friend leukemia: rapid development of erythropoietin-independent hematopoietic precursors. *J Natl Cancer Inst 54:*265–267, 1975.
24. Okunewick JP: The role of committed and uncommitted hematopoietic stem cells as targets for Rauscher and Friend leukemia virus. *Biomedicine 26:*152–157, 1977.
25. Nooter K, Ghio R: Hormone-independent in vitro erythroid colony formation by bone marrow cells from Rauscher virus-infected mice. *J Natl Cancer Inst 55:*59–64, 1975.
26. Opitz U, Seidel HJ, Rich I: Erythroid stem cells in Rauscher virus infected mice. *Blut 35:*35–44, 1977.
27. Seidel HJ, Iturriza RG: Growth of bone marrow cells of normal and Rauscher virus infected mice in suspension cultures stimulated by CSA. *Blut 34:*289–298, 1977.
28. Opitz U, Seidel HJ, Bertoncello I: Erythroid stem cells in Friend virus infected mice. *J Cell Physiol* (submitted for publication).
29. Heath DS, Axelrad AA, McLeod DL et al: Separation of the erythropoietin-responsive progenitors BFU-e and CFU-e in mouse bone marrow by unit gravity sedimentation. *Blood 47:*777–791, 1976.
30. Bertoncello I, Opitz U, Seidel HJ: Analysis of erythroid precursor cells in Friend virus infected mice, in Bentvelzen P, Yohn D (eds): *Comparative Leukemia Research 1977.* Holland, Elsevier-North, 1977.

21 Studies on Erythroid Differentiation in a Wild-Type and Dimethylsulfoxide-Resistant Clones of Friend Virus-Transformed Cells: Effects of Hemin, Hemoglobin, and Erythropoietin

Akio Urabe, Martin J. Murphy, Jr., and Shigeru Sassa

Introduction

Murine Friend virus-transformed cells can be induced to differentiate into erythroid cells by the addition of various agents, for example, dimethylsulfoxide (DMSO),[1] other aprotic solvents,[2] butyric acid,[3] purines and purine analogues,[4] hemin,[5,6] and so forth. These processes that occur after inducer-treatment provide a useful model of erythroid cell differentiation since many erythroid-specific functions are induced such as the production of globin messenger RNA,[7] the synthesis of specific red cell membrane proteins,[8] as well as the *de novo* synthesis of heme pathway enzymes,[9] finally leading to the formation of heme and hemoglobin.

The advantage of the Friend virus-transformed cell system compared with the culture of bone marrow from normal hematopoietic origin is that these cells self-perpetuate in culture. Thus this system offers an opportunity to investigate individual steps of erythroid differentiation in continuously growing cell cultures. Another advantage of this system is that a variety of variant clones can be readily isolated using a semisolid medium containing methylcellulose.[10] Biochemical and genetic characterization of these mutants provides potentially useful information concerning the events that occur during erythroid cell differentiation.

Our studies on erythroid differentiation in a wild-type as well as in two DMSO-resistant clones of Friend virus-transformed cells grown in a semisolid medium constitute this chapter.

Materials and Methods

Cell Culture

A wild-type clone of murine Friend virus-transformed erythroleukemic cells (745A) was maintained in semisolid medium as well as in cell suspension in liquid culture. In liquid culture, as described elsewhere,[9] cells were suspended in a modified Ham's F12 medium[11] with 10% fetal bovine serum (Flow Laboratories, Rockville, Maryland). Cells of clone 745A are inducible with DMSO to undergo erythroid differentiation. In the case of semisolid culture, cells were suspended in the modified Ham's F12 medium containing 0.8% methylcellulose (The Dow Chemical Co., Midland, Michigan), and 10% fetal bovine serum, with or without a variety of inducers, and were incubated in 5% CO_2 and 95% air at 37°C with saturated humidity. Ten thousand cells were suspended in a total volume of 1 ml of the semisolid medium and were placed in a 35 × 10 mm petri dish (Lux Scientific Corp., Newbury Park, California). Cells in the semisolid culture could be incubated for more than 3 weeks, without any change of medium, or addition of reagents. Erythroid differentiation was examined by staining colonies with a benzidine-H_2O_2 reagent as follows: 3% hydrogen peroxide was added to 0.2% (w/v) benzidine dihydrochloride in 0.5 M acetic acid in a proportion of 1:250. Seven to eight drops of benzidine-H_2O_2 solution were then added to each dish. Stained colonies were visualized with an inverted microscope after 1 min. Benzidine-positive colonies were scored between 1 and 5 min after the addition of the reagent. Usually after 5 days of incubation, the hemoglobinization of cells was examined and the colonies containing at least three stained cells were enumerated as positive for benzidine staining.

Selection of DMSO-Resistant Clones

Cells grown in the semisolid medium containing 2% DMSO for 2 weeks were harvested, and were then transferred to a new medium also containing methylcellulose and 2% DMSO. After serial transfers of the cells through the semisolid medium in the continual presence of 2% DMSO, all the cells in the dish were harvested using a rubber policeman and suspended in a liquid medium devoid of DMSO. Cells were then either grown in liquid culture or continuously passaged through the semisolid medium. After repeated passages, these cells were confirmed to have a potential to grow continuously in the DMSO-containing medium but were completely refractory to DMSO since they did not differentiate into erythroblasts.

Preparation of Hemin Solution and Hemolysates

To make 10^{-3} M solution, 3.3 mg of hemin was dissolved in 0.5 ml of N/50 KOH–50% methanol. Then

4.0 ml of modified Ham's F12 medium supplemented with 10% fetal bovine serum was added, and finally 0.5 ml of N/50 HCL was added. The solution was freshly prepared for each experiment and sterilized by Millipore filtration (pore size: 0.45 μm) prior to use. Packed red cells from normal human subjects and normal BDF$_1$ male mice were suspended in modified Ham's F12 medium, and were lysed with freezing and thawing. After centrifugation of the hemolysate at 3000 rpm for 15 min the supernatant was collected, and Millipore filtered (0.45 μm) prior to use. Heme concentration of the hemolysates was determined by the alkaline pyridine hemochromogen method.[12]

Erythropoietin

Human urinary erythropoietin was collected and concentrated. The specific activity of the erythropoietin preparation was 38 IRP units/mg. This lyophilized erythropoietin preparation was reconstituted in α medium (Flow Laboratories) to give a final concentration of 20 units/ml and sterilized by Millipore filtration (0.45 μm) prior to use.

Results

Selection of DMSO-Resistant Clones in the Medium Containing Methylcellulose

Cells of a wild-type clone 745A were maintained in a medium containing methylcellulose. Plating efficiency of such cells when transferred from the liquid culture to the methylcellulose culture is shown in Figure 21-1. Essentially, all inoculated cells made colonies in semisolid medium. Cells present within the

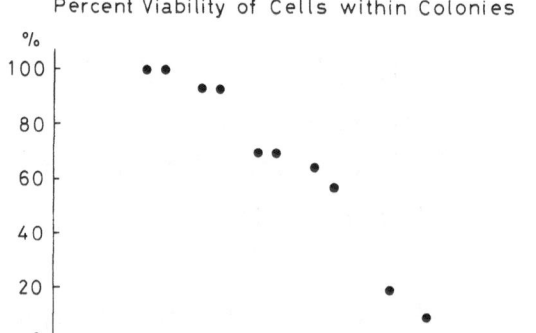

Figure 21-2 Viability of cells within colonies in a semisolid culture was determined by trypan blue exclusion. Cells were recovered and resuspended in 0.75% trypan blue solution in F12 medium. Trypan blue-positive cells were scored within 5 min.

colonies remained alive for at least 2 weeks without any change of medium or addition of reagents as judged by trypan blue exclusion (Figure 21-2).

Erythroid Differentiation of Cells by a Variety of Inducers

A variety of inducers were added at the initiation of incubation in the methylcellulose culture. Erythroid differentiation was examined by the determination of heme concentration, the final product of the differentiation process. Cells that contained heme were stained by benzidine, and were enumerated using an inverted microscope. The percentage of benzidine-positive colonies for 5-day incubation of wild-type cells (745A) with various inducers is given in Table 21-1. DMSO, dimethylformamide (DMF), tetramethylurea (TMU), dimethylacetamide (DMA), and hypoxanthine induced erythroid differentiation as depicted by the increase in benzidine-positive colonies. In addition, hemin clearly induced erythroid differentiation of 745A cells without the presence of other inducers. Erythroid differentiation in the 745A cells by various concentrations of DMSO was also enhanced by coexistence of 10^{-4} M of hemin as depicted in Figure 21-3.

Induction of Erythroid Differentiation by Human and Mouse Hemolysates

Hemolysates from normal human subjects and BDF$_1$ mice also revealed an inductive activity of the erythroid differentiation. To determine whether the effects of hemolysates are caused by a similar mechanism such as by hemin, all preparations of hemolysates and hemin were made to contain an equivalent heme concentration. As shown in Figure 21-4, hemolysates from both human and mouse erythrocytes caused essentially the same effect in a dose-responsive manner as did the purified hemin preparation.

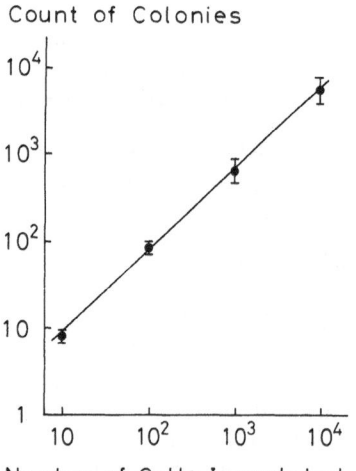

Figure 21-1 Plating efficiency of the wild-type cells (745A) when transferred from the liquid culture to the semisolid medium containing methylcellulose. Numbers of colonies after 5 days of incubation were counted using an inverted microscope.

Table 21-1 Percent benzidine-positive colonies in wild-type cells (745A) treated with inducers[a].

Inducer	Concentration	Benzidine-Positive Colonies/Total Colonies (%)
Dimethylsulfoxide (DMSO)	1%	88
	0.8%	86
	0.4%	56
Dimethylformamide (DMF)	0.8%	63
	0.4%	64
Tetramethylurea (TMU)	0.08%	66
	0.04%	55
Dimethylacetamide (DMA)	0.1%	77
	0.08%	68
	0.04%	65
Hypoxanthine	2×10^{-3} M	72
	1.6×10^{-3} M	67
	0.8×10^{-3} M	27
Hemin	10^{-4} M	44
	10^{-5} M	10
Erythropoietin	2 units/ml	0

[a] Cells were incubated in the modified Ham's F12 medium containing 0.8% methylcellulose and 10% fetal bovine serum. Percentages of benzidine-positive colonies after 5 days of incubation.

These data suggest that induction of erythroid differentiation by human and mouse hemolysates is probably due to the heme in the hemolysates. The data also suggest erythroid differentiation is triggered by hemin rather than by globin, since both human and mouse hemolysates caused erythroid differentiation in a comparable manner.

Heme Synthetic Enzymes in DMSO-Resistant Clones

A DMSO-resistant clone identified as "U 91" was isolated after serial transfer of 745A cells through the methylcellulose culture containing 2% DMSO. This clone was grown in liquid suspension as well as in methylcellulose culture. Response of the cells of clone U 91 to various inducers was also examined. Although U 91 cells were entirely unresponsive to DMSO, cells of clone U 91 showed a slight to moderate response to hemin (10^{-4} M), DMF (0.8%), TMU, (0.08%), DMA (0.1%), and hypoxanthine (2mM) as shown in Table 21-2.

δ-Aminolevulinate dehydratase (ALA-D) and uroporphyrinogen-I synthase (URO-S) were determined by the methods of Sassa et al.,[13] in the wild-type clone (i.e., 745A) and in cells of clone U 91. As

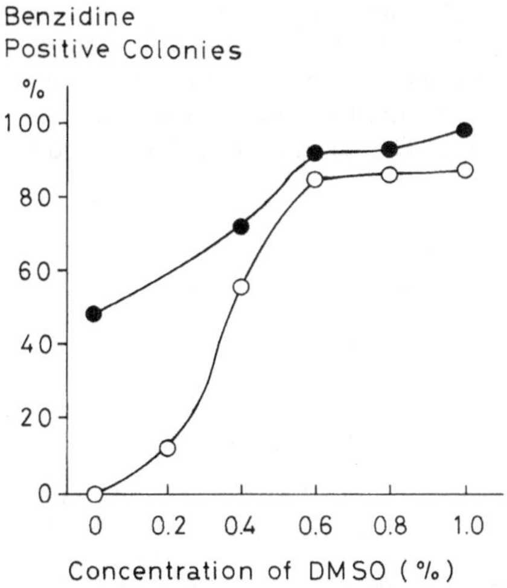

Figure 21-3 Percentages of benzidine-positive colonies in the wild-type 745A cells treated DMSO and/or hemin. Cells were incubated for 5 days with various concentration of DMSO and in the presence or the absence of 10^{-4} M hemin. Closed circles, with hemin (10^{-4} M); open circles, without hemin.

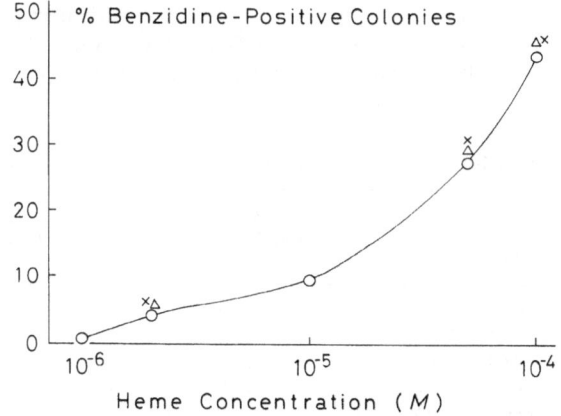

Figure 21-4 Percentages of benzidine-positive colonies in the wild-type 745A cells incubated with hemin, human hemolysate, or mouse hemolysate. Concentrations of heme in the hemolysates are shown on the abscissa. Percentages of benzidine-positive colonies are expressed as a function of heme concentration in each preparation. Circle, hemin solution; triangle, human hemolysate; x, mouse hemolysate.

Table 21-2 Percent benzidine-positive colonies in a DMSO-resistant clone U 91 treated with inducers[a].

Inducer	Concentration	Benzidine-Positive Colonies/Total Colonies (%)
Hemin	1×10^{-4} M	13
	5×10^{-5} M	8
	1×10^{-5} M	3
	5×10^{-6} M	1
	1×10^{-6} M	0
DMSO	2%	0
DMF	0.8%	20
TMU	0.08%	16
DMA	0.1%	17
Hypoxanthine	2×10^{-3} M	14
Erythropoietin	2 units/ml	0
2% DMSO + 10^{-4} M Hemin		10

[a] Conditions of cell culture and scoring benzidine-positive colonies were carried out in the same manner as described in Table 21-1.

shown in Table 21-3, the levels of ALA-D and URO-S, the concentration of heme and the percentage of benzidine-positive cells were increased markedly in the wild-type cells (745A) after treatment with DMSO. Hemin also increased ALA-D and URO-S activities in the wild-type cells.

In striking contrast to those 745A that were not induced, the levels of ALA-D and URO-S were significantly elevated in untreated cells of clone U 91 as shown in Table 21-4. DMSO-treatment of clone U 91 caused an increase in ALA-D and URO-S that approximated the maximally induced level of the wild-type 745A cells. However, heme concentration increased only slightly as determined by fluorometry. This slight increase in heme concentration was not reflected in the number of benzidine-positive cells that remained less than 1%. Treatment with hemin suppressed the levels of ALA-D and URO-S in U 91 cells slightly, but a modest increase of benzidine-positive cells was observed (Table 21-4). The decline in ALA-D and URO-S by hemin in cells of clone U 91 stands in contrast to the hemin-effect in the wild-type cells, but it has also been observed in other DMSO-resistant clones (e.g., K 2[13]).

Another DMSO-resistant clone, designated "U 99" displayed no response to 2% DMSO, but intermediate to moderate increases in benzidine-positive colonies when hemin (10^{-4} M), or erythropoietin (1 to 2 IRP units/ml) were added to the cells after incubation with DMSO. Although clones U 91 and U 99 were selected in a similar fashion, their response to erythropoietin was distinctly different. For example, cells of clone U 91 were totally nonresponsive to erythropoietin as judged by the benzidine-positive cells, whereas approximately 50% of the cells of clone U 99 displayed benzidine-positivity after the addition of erythropoietin (Table 21-5). It is also interesting to note that hemin caused little differentiation of U 91 cells, but considerable erythroid differentiation was observed in U 99 cells after hemin was added. Thus hemin and erythropoietin are evocators of erythropoiesis in U 99 cells.

Discussion

In this study we reported the isolation and the characterization of some properties of two DMSO-resistant clones. One of them (i.e., U 99) demonstrated a unique responsiveness to erythropoietin, although cells of the same clone were totally resistant to DMSO.

Friend et al.[1] have originally reported that DMSO was able to induce the erythroid differentiation of murine erythroleukemic cells in culture. More recently, other substances, such as butyric acid,[3] purine analogues,[4] hemin[5,6] and so forth, were found to also act as inducers of red cell differentiation of Friend cells. Whereas it is clear that in wild-type cells

Table 21-3 Levels of heme pathway enzymes in the wild-type cells (745A)[a].

Treatment	ALA-D[b] [nmol PBG/10^6 cells (hr)]	URO-S[c] [pmol URO/10^6 cells (hr)]	Heme[d] (pmol/10^6 cells)	Benzidine-Positive Cells[e] (%)
None	0.85	35	61	2–3
1.5% DMSO	3.4	225	521	>95
10^{-4} M Hemin	2.2	70	n.d.[f]	n.d.
5×10^{-5} M Hemin plus 1% DMSO	5.0	250	n.d.	>95

[a] Cells were seeded at a density of 5×10^4 cells/ml. After preincubation for 24 hr, DMSO and/or hemin was added. Incubation was continued for 5 days.
[b] δ-Aminolevulinate dehydratase (ALA-D) activity was assayed by a semimicrospectrophotometric method.[13]
[c] Uroporphyrinogen-I synthase (URO-S) activity was assayed by a fluorophotometric method.[13]
[d] Heme concentration was determined by a fluorophotometric method.[13]
[e] Determined by flow microphotometry.[13]
[f] Not determined.

Table 21-4 Levels of heme pathway enzymes in a DMSO-resistant clone (U 91)[a].

Treatment	Cells/mm	ALA-D[b] [nmol PBG/10^6 cells (hr)]	URO-S[c] [pmol URO/10^6 cells (hr)]	Heme[d] (pmol/10^6 cells)	Benzidine-Positive[e] Cells (%)
None	789	2.92	129.2	55.8	0.35
1.5% DMSO	973	4.19	299.2	91.7	0.28
10^{-4} M Hemin	1075	1.75	67.3	n.d.[f]	7.52
1% DMSO + 5 × 10^{-5} M Hemin	901	4.17	202.3	n.d.	0.78

[a-f] See Table 21-3 footnotes.

(745A) DMSO induces the entire process of differentiation into the erythroid cells, such is not the case in a mutant clone (U 91). It is of interest, however, that both ALA-D and URO-S, enzymes early in the heme biosynthetic pathway, were present at high levels before treatment with DMSO in U 91 cells. DMSO also potentiated the levels of these enzymes in U 91 cells, but DMSO did not affect heme synthesis appreciably. Although cells of clone U 91 did not synthesize heme in response to DMSO, several inducers such as DMF, TMU, DMA, hypoxanthine, and hemin had slight to moderate inductive effects on erythroid differentiation in U 91 cells as determined by the frequency of benzidine-positive colonies (i.e., 13% ~ 20%). These results indicate that cells of clone U 91, which were originally selected for their ability to grow yet not to undergo erythroid differentiation in the presence of DMSO, have the capacity to respond to certain other inducers by completing erythroid differentiation.

Preisler et al.[14] have reported that erythropoietin enhanced heme synthesis in the wild-type cells (745A) that had been treated with DMSO. Another clone (U 99) that we have isolated from clone 745A by growing cells in 2% DMSO in a semisolid medium was completely unresponsive to DMSO-treatment, whereas erythropoietin as well as hemin brought about erythroid differentiation in approximately 50% of the cells.

Table 21-5 Percent benzidine-positive colonies in a DSMO-resistant clone U 99 treated with hemin or erythropoietin[a].

Inducer	Concentration	Benzidine-Positive Colonies/Total Colonies (%)
DMSO	2%	0
Hemin	10^{-4} M	59
Erythropoietin	1 unit/ml	26
	2 units/ml	46

[a] Condition of cell culture and scoring benzidine-positive colonies were carried out in the same manner as described in Table 21-1.

It is accepted that erythropoietin is normally the prime regulator of erythropoiesis in mammals.[15] It is still unknown, however, which stage(s) of erythroid differentiation erythropoietin exerts its effect(s). Iscove[16] proposed that physiological levels of erythropoietin did not influence the commitment by multipotential hematopoietic stem cells to become erythropoietin responsive cells (ERC), and that early erythroid committed progenitor cells do not respond to erythropoietin. It is not clear at which stage of differentiation Friend virus-transformed erythroleukemic cells are comparable with normal erythroid cells. It is clear, however, that DMSO induces differentiation of the erythroid series from proerythroblasts to orthochromatic normoblasts in the wild-type cells. This has been demonstrated by the sequential development of heme pathway enzymes finally yielding hemoglobin synthesis,[9] and erythrocyte-specific surface antigens[17] in addition to morphological appearance of normoblasts.

In cells of clone U 91, DMSO did not affect terminal erythroid differentiation, although DMSO increased the early stage enzymes of heme biosynthesis as it did in the wild-type cells. In contrast, hemin increased the number of benzidine-positive colonies from 0 to 13%, but levels of the early stage enzymes, i.e., ALA-D and URO-S, were not increased by the addition of hemin. This result suggests that hemin bypassed the biochemical defect of clone U 91 which is at the terminal stage of heme biosynthesis, thus permitting the final formation of hemoglobin.

Recently, based on studies of mutant clones of Friend erythroleukemic cells, Eisen et al.[18] also proposed that hemin is an inducer of the late-stage program of erythroid differentiation. DMSO, on the other hand, is thought to be an inducer of the early-stage program of erythroid differentiation. In another DMSO-resistant clone (i.e., U 99), hemin, as well as erythropoietin induced erythroid differentiation. These results suggest that erythropoietin, as well as hemin, triggers the late program of erythroid differentiation in these cells in vitro. These data also clearly establish that the mechanisms of induction of erythroid differentiation by DMSO, erythropoietin, or by hemin are independent.

Summary

Two dimethylsulfoxide (DMSO)-resistant clones of Friend virus-transformed erythroleukemic cells were isolated in a medium containing methylcellulose and 2% DMSO. One such clone (U 91), although completely refractory to DMSO, displayed slight responsiveness (13 to 20% benzidine-positive colonies) to certain other inducers as well as hemin.

In addition to hemin, normal human and mouse hemolysates induced erythroid differentiation in these cells. The inductive response caused by the hemolysates was comparable to the effect by hemin on a molar basis of heme concentration.

In these DMSO-resistant clones, DMSO induced early enzymes of the heme biosynthetic pathway, i.e., δ-aminolevulinate dehydratase (ALA-D) and uroporphyrinogen-I synthase (URO-S) as it did in the wild-type cells; however, it did not affect heme synthesis of these cells appreciably.

Hemin and hemolysates as well as erythropoietin caused the terminal erythroid differentiation in another DMSO-resistant clone (i.e., U 99). These results demonstrate that erythropoietin or hemin are required for terminal erythroid differentiation to occur in these cells.

Acknowledgments

This study was supported in part by USPHS grants AM-19741, ES-01055, The Hipple Foundation, and an American Cancer Society grant BC-180A. We are grateful for the secretarial assistance of Ms. Joanne Zisson.

A wild-type clone of murine Friend virus-transformed erythroleukemic cells (745A) was a gift from Drs. C. Friend and W. Scher of the Mt. Sinai School of Medicine, New York, N.Y.

Human urinary erythropoietin used in this study was collected and concentrated by Centro de Estudios Farmacologicos y de Principios Naturales, Buenos Aires, Argentina; further processed and assayed by the Hematology Research Laboratories, Children's Hospital of Los Angeles, under research grant HL-10880 from the National Heart, Lung, and Blood Institute.

References

1. Friend C, Scher W, Holland JG, Sato T: Hemoglobin synthesis in murine virus-induced leukemic cells *in vitro:* Stimulation of erythroid differentiation by dimethyl sulfoxide. *Proc Natl Acad Sci USA* 68:378–382, 1971.
2. Sassa S, Granick S, Chang C, Kappas A: Induction of enzymes of the heme biosynthetic pathway in Friend leukemia cells in culture, in Nakao K, Fisher JW, Takaku F (eds): *Erythropoiesis, Proceedings of the International Conference on Erythropoiesis,* pp. 383–395, Tokyo, University of Tokyo Press, 1975.
3. Leder A, Leder P: Butyric acid, a potent inducer of erythroid differentiation in cultured erythroleukemic cells. *Cell* 5:319–322, 1975.
4. Gusella JF, Housman D: Induction of erythroid differentiation *in vitro* by purines and purine analogues. *Cell* 8:263–269, 1976.
5. Ross J, Sautner D: Induction of globin mRNA accumulation by hemin in cultured erythroleukemic cells. *Cell* 8:513–520, 1976.
6. Dabney BJ, Beaudet AL: Increase in globin chains and globin mRNA in erythroleukemia cells in response to hemin. *Arch Biochem Biophys* 179:106–112, 1977.
7. Kameji R, Obinata M, Natori Y, Ikawa Y: Induction of globin gene expression in cultured erythroleukemia cells by butyric acid. *J Biochem* 81:1901–1910, 1977.
8. Eisen H, Nasi S, Georgopoulos CP, Arndt-Jovin D, Ostertag W: Surface changes in differentiating Friend erythroleukemic cells in culture. *Cell* 10:689–695, 1977.
9. Sassa S: Sequential induction of heme pathway enzymes during erythroid differentiation of mouse Friend leukemia virus-infected cells. *J Exp Med* 143:305–315, 1976.
10. Rovera G, Bonaiuto J: The phenotypes of variant clones of Friend mouse erythroleukemic cells resistant to dimethyl sulfoxide. *Cancer Res* 36:4057–4061, 1976.
11. Sassa S, Kappas A: Induction of δ-aminolevulinate synthase and porphyrins in cultured liver cells maintained in chemically defined medium. *J Biol Chem* 252:2428–2436, 1977.
12. Porra RJ, Jones OTG: Studies on ferrochelatase. 1. assay and properties of ferrochelatase from a pig-liver mitochondrial extract. *Biochem J* 87:181–185, 1963.
13. Sassa S, Granick JL, Eisen H, Ostertag W: Regulation of heme biosynthesis in mouse Friend virus-transformed cells in culture, in Murphy MJ Jr (ed): *In Vitro Aspects of Erythropoiesis,* p. 135, New York, Springer-Verlag, 1978.
14. Preisler HD, Giladi M: Erythropoietin responsiveness of differentiating Friend leukaemia cells. *Nature* 251:645–646, 1974.
15. Gordon AS: Erythropoietin. *Vitam Horm* 31:105–174, 1973.
16. Iscove NN: The role of erythropoietin in regulation of propulation size and cell cycling of early and late erythroid precursors in mouse bone marrow. *Cell Tissue Kinet* 10:323–334, 1977.
17. Sugano H, Furusawa M, Kawaguchi T, Ikawa Y: Differentiation of tumor cells: Induction of

erythrocyte membrane-specific antigens in Friend leukemia cells. *Rec Res Cancer Res* *44*:30–44, 1974.

18. Eisen H, Keppel-Ballivet F, Georgopoulos CP, Sassa, S, Granick J, Pragnell I, Ostertag W: Biochemical and genetic analysis of erythroid differentiation in Friend virus-transformed murine erythroleukemia cells. *Cold Spring Harbor Conference on Differentiation of Hemopoietic Cells, in press*, 1978.

22 Induction of Uroporphyrinogen-I Synthase Activity in Mitogen-Stimulated Lymphocytes: Deficient Induction in Acute Intermittent Porphyria Cells

Shigeru Sassa, Gregory L. Zalar, and Attallah Kappas

Introduction

Acute intermittent porphyria (AIP) is a genetic liver disease that is characterized clinically by a disabling neurological-visceral symptom complex and biochemically by the excessive urinary excretion of the porphyrin precursors, δ-aminolevulinic acid (ALA) and porphobilinogen (PBG). Two enzymic abnormalities of the heme pathway have been described in the livers of clinically manifest AIP patients[1,2,3] as well as an additional defect in the biotransformation of endogenous steroid hormones.[4]

One of the heme pathway enzyme abnormalities is the overproduction of δ-aminolevulinate synthase (ALA-S), the rate-limiting enzyme in the heme biosynthetic sequence in the liver[1,2]; the other is a 50% deficiency of uroporphyrinogen-I synthase (URO-S) activity.[3,5] These two enzyme abnormalities account for the excessive excretion of ALA and PBG in the urine of AIP patients. Decreased URO-S activity has also been found in erythrocytes, cultured skin fibroblasts, and cultured amniotic cells obtained from patients as well as from subjects who are considered to be obligatory carriers of AIP.

Decreased URO-S in AIP subjects may be due to a decreased concentration of the normal enzyme or the production of a catalytically inactive enzyme protein.

We examined this question by studying URO-S activity in lymphocytes undergoing the metabolic activation that is caused by treatment with plant lectins.

Patients and Methods

Nine female patients with clinically manifest AIP and nine latent AIP carriers (five males, four females), as well as thirteen normal subjects (five males, eight females), were studied. All AIP patients and gene carriers were characterized by decreased URO-S activity in their erythrocytes and/or cultured skin fibroblasts.

Preparation of mononuclear cell concentrates and culture conditions were described previously.[6] Contaminating erythrocytes were hemolyzed and removed using $NH_4Cl-KHCO_3$.[7] Phytohemagglutinin (1%) and pokeweed mitogen (1%) were added and then the cultures were incubated for 4 days at 37°C in an atmosphere of 5% CO_2 and 95% air. Enzyme analyses were made on the fourth day. Tubes containing 1 ml cell suspension were treated with ALA (100 µg/ml) on the fourth day and porphyrins were quantitated on the fifth day.

Results

URO-S activity was low until 2 days after treatment with mitogens in both normal and AIP lymphocytes. As incubation proceeded further, URO-S became markedly induced in both lymphocyte preparations (Figure 22-1). In the normal lymphocytes the enzyme activity at 4 days exceeded the basal level by 30-fold; in the AIP lymphocytes, however, URO-S induction reached a level of only approximately one-third that of normal (Figure 22-1). Thus in the mitogen activated state, AIP lymphocytes displayed the biochemical defect of the disease, URO-S deficiency. The extent of deficient induction of URO-S in AIP lymphocytes was similar to the deficient level of this enzyme in cell types such as liver cells and fibroblasts. Both enzymes have the same Km, i.e., 6 μM, for the substrate, but the rate of product formation by the AIP enzyme was considerably less than that by the normal enzyme.

We developed an alternative method for monitoring URO-S activity that proved to be precise and simpler to apply to the study of control and patient populations in larger numbers. In this method, transformed lymphocytes were incubated with the porphyrin precursor, ALA. Lymphocytes synthesize large amounts of protoporphyrin-IX (PROTO) from this precursor over a 24-hr period and the amount of porphyrin formed correlates directly with the level of URO-S as determined enzymatically.

Using this method, PROTO formation from ALA in transformed lymphocytes from groups of normal and AIP subjects was studied. The mean value for 13 normal individuals was 1056 ± SEM of 67 pmol of PROTO formed/mg protein, 24 hr. For nine AIP patients with clinically manifest disease, the mean was

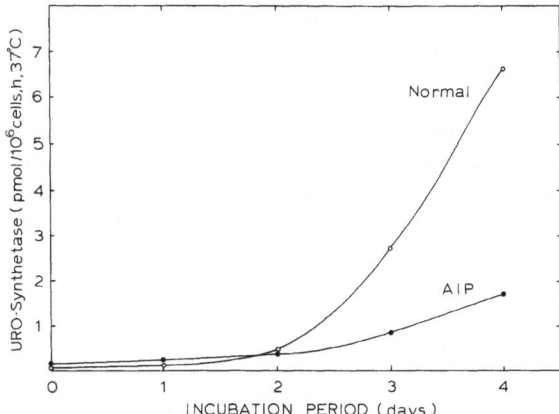

Figure 22-1 Induction of URO-S activity in mitogen-stimulated lymphocytes. Lymphocytes from one normal and one AIP subject were incubated at 37°C in an atmosphere of 5% CO_2 and 95% air in a modified F12 medium supplemented with 10% fetal bovine serum, phytohemagglutinin (1%) and pokeweed mitogen (1%). Cells were harvested on each day and assayed for URO-S activity using PBG as the substrate. Each point represents the mean of triplicate assays.

503 ± 57 pmol of PROTO formed, or 50% less. PROTO formation was 486 ± 54 pmol/mg protein, 24 hr, which was also 50% less than normal, in nine AIP gene carriers in whom the defect had remained entirely latent clinically.

The amounts of PROTO synthesized by lymphocytes in three AIP patients in relapse did not vary significantly between periods of relapse, (498 ± 42 pmol/mg protein, 24 hr) and remission, of their disease (519 ± 154 pmol/mg protein, 24 hr). Thus, as with erythrocytes and cultured skin fibroblasts, the levels of URO-S activity in mitogen-stimulated lymphocytes characterize AIP gene carriers, but do not correlate with clinical activity of the disease.

The rate of DNA formation determined by tritiated thymidine incorporation into isolated DNA was the same in both normal and AIP lymphocyte preparations. RNA synthesis, as determined by tritiated uridine uptake in these lymphocytes, was also equivalent in normal and AIP cells.

Four heme-related cellular parameters were also compared in mitogen-transformed normal and AIP lymphocytes. AIP lymphocytes had the same levels of heme content and activities of catalase and ALA-dehydratase (ALA-D) as did normal lymphocytes. ALA-S activity was also within the normal range. This finding is similar to that in cultured AIP fibroblasts which, as Meyer[8] and Bonkowsky et al.[9] have shown, display normal ALA-S activity despite the concurrent presence of a 50% deficiency in URO-S. Lymphocytes and skin fibroblasts thus differ from liver cells in clinically active AIP patients in whom there exists a deficiency of URO-S as well as a marked overproduction of ALA-synthase. An alternative possibility is that the genetic deficiency of URO-S does not result in a heme deficiency in nonhepatic tissues. It should be reemphasized that AIP gene carriers who develop the overt clinical syndrome of AIP must metabolically differ from gene carriers who never develop the disease since the URO-S deficiency is identical in both groups of subjects.[10]

Discussion

In this study we have demonstrated that the enzyme URO-S is markedly induced by mitogens in human

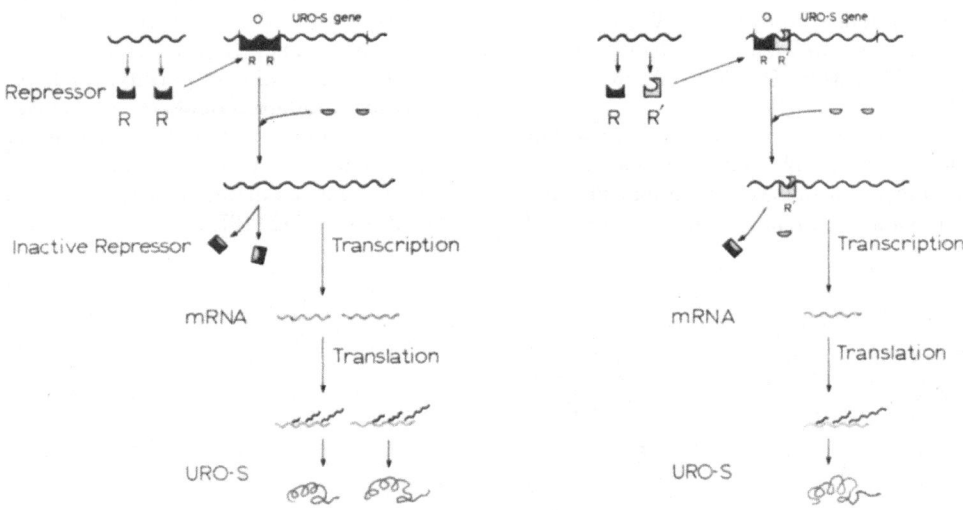

Figure 22-2 Hypotheses concerning the deficient induction of uroporphyrinogen-1 synthase in mitogen-stimulated lymphocytes from patients with acute intermittent porphyria. Left, normal; right, AIP.

lymphocytes and that the gene defect of AIP can be identified in such cells as a result of mitogen-induced transformation. The deficient induction of URO-S in AIP lymphocytes is a specific biochemical lesion since these cells do not differ from normal with respect to heme content or ALA-S, ALA-D, and catalase activities. Moreover, AIP lymphocytes synthesize DNA and RNA at normal rates.

The specific activity of URO-S remains relatively constant in erythrocytes and cultured skin fibroblasts, but it increases markedly in lymphocytes undergoing mitogen-induced transformation. This finding, together with others showing normal physicochemical properties of URO-S enzyme protein in AIP cells,[10] suggests that the deficient induction of URO-S in AIP lymphocytes is related to a defect in the regulation of synthesis of the enzyme (Figure 22-2). An alternative hypothesis to account for the deficient URO-S induction in lymphocytes is that URO-S in AIP cells is a unique structural variant that is catalytically inactive but is normal in its other physicochemical properties (Figure 22-2). A firm distinction between these two hypotheses will require structural analysis of the purified enzyme proteins from normal and AIP cells.

Acknowledgment

This work was supported in part by USPHS grant ES –01055, training grant (5–T32AM07171-02), ACS grant BC–180, and National Foundation grant I–350.

References

1. Tschudy DP, Perlroth MG, Marver HS, Collins A, Hunter G, Rechcigl M: Acute intermittent porphyria: the first "overproduction disease" localized to a specific enzyme. *Proc Natl Acad Sci USA 53:*841, 1965.
2. Nakao K, Wada O, Kitamura T, Uono K, Urata G: Activity of aminolevulinic acid synthetase in normal and porphyric human livers. *Nature 210:*838, 1968.
3. Strand J, Felsher BF, Redeker AG, Marver HS: Enzymatic abnormality in heme biosynthesis in acute intermittent porphyria. Decreased hepatic conversion of porphobilinogen to porphyrins and increased delta aminolevulinic acid synthetase activity. *Proc Natl Acad Sci USA 67:*1315, 1970.
4. Kappas A, Sassa S, Granick S, Bradlow HG: Endocrine-gene interaction in the pathogenesis of acute intermittent porphyria, in Plum F (ed): *Brain Dysfunction in Metabolic Disorders. Res Publ Assoc Nerv Ment Dis,* Vol. 53, New York, Raven Press, pp. 225–237, 1974.
5. Miyagi K, Cardinal R, Bossenmaier I, Watson CJ: The serum porphobilinogen and hepatic porphobilinogen deaminase in normal and porphyric individuals. *J. Lab Clin Med 78:*683, 1971.
6. Sassa S, Zalar GL, Kappas A: Deficient induction of uroporphyrinogen-I synthase activity in mitogen-stimulated lymphocytes from patients with acute intermittent porphyria. *Trans Am Assoc Physics 90:*157, 1977.
7. Sassa S, Kappas A: Induction of δ-aminolevulinate synthase and porphyrin in cultured liver cells maintained in chemically defined medium. Permissive effects of hormones on induction process. *J Biol Chem 252:*2428, 1977.
8. Meyer UA: Intermittent acute porphyria. Clinical and biochemical studies of disordered heme biosynthesis. *Enzyme (Basel) 16:*334, 1973.
9. Bonkowsky HL, Tschudy DP, Weinbach EC, Ebert PS, Doherty JM: Porphyrin synthesis and mitochondrial respiration in acute intermittent porphyria: studies using cultured human fibroblasts. *J Lab Clin Med 85:*93, 1975.
10. Sassa S, Granick S, Bickers DR, Bradlow HL, Kappas A: Studies in porphyria III. A microassay for uroporphyrinogen synthetase, one of three abnormal enzyme activities in acute intermittent porphyria; and its application to the study of the genetics of this disease. *Proc Natl Acad Sci USA 71:*732, 1974.

23 Potential for Differentiation, Virus Production, and Tumorigenicity in Murine Erythroleukemic Cells Treated with Interferon

L. Cioé, A. Dolei, G. B. Rossi, F. Belardelli, E. Affabris, R. Gambari, and A. Fantoni

Introduction

Studies of erythropoiesis suffered in the past from the lack of a suitable *in vitro* system in which the biology of erythropoietic cells could be studied under defined conditions, on one hand, and in which mass cultures would be possible, on the other hand, so that molecular biology could also enter the picture. When Charlotte Friend showed that explants of transplantable subcutaneous tumors, originated from virus-induced murine leukemic spleens and/or livers and adapted to grow *in vitro,* consistently gave rise to tissue culture lines able to differentiate only along the erythroid pathway, the first step had been made toward this goal.[1,2] There were, however, two unfortunate reasons why studies of this system were hindered for some time: (a) the percentage of differentiating cells was too low to allow any molecular biology analysis, and (b) the cultured cells were not sensitive to erythropoietin, which meant that erythropoiesis in this system was not the exact counterpart of the physiological pattern. It was only a few years later that Friend cells became a most popular tool for biologists and molecular biologists engaged in studies of erythropoiesis. This was due to Charlotte Friend's[3] demonstration that the addition of dimethyl sulfoxide (DMSO) to cultures of Friend cells caused a massive shift toward more differentiated stages of erythropoiesis. Although the precise mechanism of this phenomenon is still unknown, the availability of a cell population whereby 70 to 80% of the cells would engage at will in quasiterminal erythroid differentiation immediately prompted an avalanche of studies dealing with several aspects of erythropoiesis. Temin's[4] and Baltimore's[5] discovery of reverse transcriptase made possible, in addition, *in vitro* synthesis of radioactive complementary DNAs (cDNAs) to a given RNA probe. In erythropoiesis this meant the synthesis of the globin cDNA,[6] and thus, the possibility of exact measurements of globin mRNA amounts in the Friend system.

All these achievements did not efface the observation of erythropoietin independence of Friend cells; it is generally agreed by now that the Friend system is a good model for thorough evaluation of "stimulated" later stages of normal erythroid maturation, but not for the process of commitment itself. Nonetheless, the system proved so useful as to be adopted as the *in vitro* tool for understanding the expression of erythroid markers of normal differentiation. Friend et al.[3] originally described the appearance of "normal" orthochromatic cells containing heme and hemoglobin. Later globin mRNAs, intracellular and membrane-associated erythrocyte-specific proteins (for a review see ref. 7) were also shown to be increased on DMSO treatment, so that these cells provide a model for elucidating how a whole set of functions are coordinated during the development of a single differentiation program.

DMSO is no longer the only agent able to promote such changes; an endless list of compounds widely divergent in both chemical structure and functions has been uncovered (see ref. 7). Some of these agents (e.g., purine derivatives, fatty acids, or heme) could be acting physiologically, but this is far from proven. Even less known is the mechanism of action of the other chemical agents. Genetic studies suggest that they may not have a single common mechanism. Recently, however, it has been claimed that an early change of the Na^+/K^+ equilibrium at the plasma membrane level may represent a metabolic step common to all inducers.[8] In this respect some light has also been shed by the observations of Rovera et al.[9] and Yamasaki et al.[10] who found that treatment of Friend cells with tumor promoters such as phorbol esters inhibits spontaneous and stimulated erythroid differentiation, as well as a cAMP burst detected soon after addition of the stimulatory compound. Treatment with these cocarcinogenic drugs also selects for glass-adherent variants of Friend cells, which obviously implies that some as yet undetermined changes of the plasma membrane did occur.

At least in the case of DMSO, sizable evidence supports the claim that one or two cell doublings are required for the stimulation of erythroid differentiation.[11–14] The observation of Leder et al.[15] that treatment of Friend cells with butyric acid may cause an increase of % benzidine-positive (B$^+$) cells in absence of DNA synthesis has been met with some criticism and reinterpreted as possibly due to binucleate cells that are frequently observed in the specific Friend cell clone used in that study.

Under optimal conditions, virtually the entire population of Friend cells can undergo erythroid differentiation in a stochastic fashion, the probability of which depends on the nature and concentration of the stimulant and other environmental factors.[16]

Virus Production

Friend cells are chronically infected with Friend Leukemia Virus (FLV), which is continuously shed into the supernatant fluids and even more so on administration of DMSO to the cultures.[1,17] The role of virus production in the genesis of erythroid differentiation *in vitro* remains to be elucidated. It seems clear that *in vivo* FLV plays a critical role in either transforming previously committed erythroid precursor cells or in rendering its target cells "committed" to erythroid differentiation as a consequence of viral transformation. Circumstantial evidence points to the first alternative as the most likely.[18] *In vitro*, however, the majority of available data seem to exclude any direct involvement of the viral genome in triggering the modulation of erythroid differentiation on stimulation with one or another drug. Although DMSO administration causes an increased production of both hemoglobin (Hb) and viral particles, no etiological relationships have been established between the two phenomena. The analysis of Friend cell variants that have arisen either spontaneously or by selection with DMSO indicates that the potential for Hb production may or may not be lost regardless of virus production. Ostertag et al.,[19] on the other hand, have observed that virus production is an early event during erythropoiesis, as shown by the kinetics of FLV RNA induction, which may therefore play a role in the modulation of globin gene transcription.

Tumorigenicity of Friend Cells

Friend cells obtained from established tissue culture lines remain truly malignant, and will accordingly always cause subcutaneous tumors when inoculated into histocompatible susceptible mice.[1] As few as 500 cells will do it. Treatment with DMSO *in vitro* is known to cause cessation of cell proliferation after 4 to 5 cell doublings. When tested *in vivo* 4 days after DMSO treatment, Friend cells exhibit a 2-log decreased ability to form tumors in DBA/2 mice.[3] The administration of DMSO to mice inoculated with various amounts of Friend cells, instead, was not successful in curing the animals presumably because, in addition to reducing the proliferation potential of the injected cells, the drug was also able to stimulate virus production and hence to cause increased transformation of host cells. (Rossi G.B., unpublished data).

Effects of Interferon on Replication and/or Production of Viral Components of Retroviruses

Since its discovery, interferon (IF) has been known to fully inhibit the replication of the great majority of animal viruses. Interferon is produced by the quasitotality of animal cells on induction with viruses or with nonviral compounds such as Poly I: Poly C, and can induce an antiviral state in both the producing cells and in other cells. IF is species-specific and not virus-specific. Inhibition of viral production is accompanied by elimination of viral products in most systems studied whereby the animal virus used was a conventional cytopathic one. Under these conditions the presence of the virus is no longer demonstrable in the infected cells that become, therefore, virus-free.

When, instead, murine cells chronically infected with an RNA tumor virus are treated with IF, virus release into the supernatant fluids is completely blocked but the synthesis of viral components is not. In several systems studied, such as the AKR Virus,[20] the Rauscher Leukemia Virus,[21] and the Moloney Leukemia Virus,[22] and infected cells, no infectious viral particles banding at 1.16 to 1.18 gm/cm^3 in sucrose gradients were detected in any cases. The intracellular levels of viral antigens, on the other hand, were found either increased[20,21] or unaffected.[22] Taken together, these data have been interpreted as indicating that IF has a dual effect on viruses, namely on the conventional cytopathic ones and on the retroviruses. The mechanism of action of IF is still obscure in both cases. It has been suggested that IF (which does not act directly on viruses) may act by inducing the synthesis of a protein able to specifically suppress the production of virus proteins (the so-called "translation inhibitory protein"). As for retroviruses, it is suggested that IF blocks the release of virus particles from cells either inhibiting the synthesis and/or the assembly of a late viral protein or interfering with the physical release from the plasma membrane of an otherwise unaltered nucleocapsid during the process of budding.[20,22]

Effects of Interferon on Growth, Differentiation and Other Functions of Animal Cells

Due to the perseverance of Ion Gresser,[23] the concept that IF is only an antiviral substance no longer holds true. The list of IF effects on both normal and neoplastic cells is seemingly endless. *In vitro* evidence to date is tentatively summarized in Table 23-1.

Further interest has been derived from the isolation of the so-called Type II interferon or immune IF. This is a cell product that has antiviral properties similar to those of the standard IF (Type I), but differs from it inasmuch as it is produced only by immunologically competent B- and/or T-lymphocytes follow-

Table 23-1 *In vitro* effects of interferons on cell functions not involved in virus replication.

Source of IF	Function Studied	Type of Effect	Reference
Mouse IF Type I	Growth of lymphoid leukemia cells (L1210)	Inhibited	53
Mouse IF Type I	Growth of Friend leukemia cells	Inhibited	40
Human IF Type I	Growth of transformed embryonic cell lines	Inhibited	54
Mouse IF Type I	Interferon production	Increased	55
Mouse IF Type I	PHA-induced DNA synthesis in spleen lymphocytes	Inhibited	56
Human IF Type I	Thymidine uptake in lymphocytes co-coltivated with RDMC cells	Inhibited	25
Human IF Type I	Thymidine uptake and DNA synthesis	Decreased	57
	Uridine uptake and RNA synthesis	Increased	57
Mouse IF Type I	Doubling potential of lymphoid cells (L1210)	Decreased	58
Mouse IF Type I	Cell cycle of mammary tumor cells	Prolonged	59
Mouse IF Type I	Cell cycle of Friend leukemia cells	Prolonged	41
Mouse IF Type I	IgE-mediated hystamine release from basophils	Increased	60
Mouse IF Type I	Expression of histocompatibility antigens in lymphoid cells	Increased	61
Mouse IF Type I	Antibody synthesis in spleen cell cultures	Inhibited	62
Mouse IF Type I	Cytotoxicity by sensitized lymphocytes	Increased	63
Mouse and rat IF Type I	Aryl hydrocarbon hydrolase activity in cell cultures	Increased	64
Rat IF Type I	Tyrosine amino transferase activity and synthesis in cell cultures	Decreased	65
Rat IF Type I	Steroid-induced glycerol-3-phosphate dehydrogenase synthesis in rat cells	Inhibited	66
Mouse IF Type I	Hemoglobin synthesis in Friend leukemia cells	Inhibited	46
Mouse IF Type I	Con A-binding capacity in lymphoid cells (L1210)	Increased	67
Mouse IF Type II	Immune response to sheep erythrocytes	Inhibited	68
Mouse IF Type II	Immune response to sheep erythrocytes also *in vivo*	Inhibited	69

ing mitogen or antigen activation.[24] It also differs from Type I IF antigenically and in respect to several properties physicochemically. When Type II IF is tested on fibroblasts and on lymphocytes of the same species, by endpoint limiting dilutions, it is apparently more active on lymphocytes than on fibroblasts (Sonnenfeld et al., submitted for publication). This would imply that besides being species-specific, IF may have also some tissue specificities. Quite recently, Trinchieri et al.[25] have shown that under certain experimental conditions, unsensitized (natural killers?) lymphocytes produce Type I IF on co-cultivation with given neoplastic cell lines. In this instance, there is no evidence for any viral or immune stimulus and IF is apparently produced as a consequence of cell contact or of a transfer of an unknown mediator.

Results
Phenotypic Changes of Stimulated Friend Cells
Cell Variants and Stimulatory Compounds

When Charlotte Friend described the "DMSO effect," Friend cell lines available in other laboratories, including ours, all originating from Friend's laboratory, were obviously tested for DMSO stimulation. In our hands, DMSO administration did not result in any significant increase of spontaneous (0.5 to 1% B$^+$ cells) erythroid differentiation of Friend cells. The same thing happened in J. Paul's laboratory, where our cells had come from. The apparent discrepancy was resolved when Friend sent to both Paul and us the cells she was actually working with in her laboratory at that time (clone 745A). The DMSO effect was then readily reproduced in our hands too, but our original cell lines were constantly "insensitive" to DMSO. It became apparent that a variant cell line had originated that was not inducible to erythroid differentiation by DMSO. Since that time, the number of Friend cell variants has enlarged as to become almost limitless. Some have a spontaneous origin, most have been selected by continuous growth in DMSO-supplemented medium, and a few originate from cloning of parental cells in semisolid medium or according to the limiting dilutions method of Curtis and Weissman.[26] At this time, we work with three Friend cell lines: the 745A clone ("sensitive" to DMSO), the Fw clone ("insensitive" to DMSO) and the A°1 clone, obtained by E. Affabris from 745A cells seeded according to the method described in ref. 26. A detailed description of the features of this clone will be published elsewhere.[27] Briefly, cells of this clone do not spontaneously differentiate but show a much earlier and more pronounced erythroid differentiation (as measured by the percentage of B$^+$ cells) on exposure to DMSO. In addition, the numbers of globin mRNA molecules (detected by molecular hybridization with ^{3}H-globin cDNA) have been determined in 745A and A°1 cells at various days after DMSO treatment (Table 23-2). Baseline values (day 0) are comparable in the two cell populations, but already on day 1 globin mRNA molecules are increased almost sixfold in A°1 cells as opposed to no significant rise in 745A cells. On day two, globin mRNA mole-

Table 23-2 Comparative time-course study of the expression of erythroid markers in DMSO-stimulated 745A and A°1 cells.

Time of Exposure to 1.5% DMSO	745A		A°1	
	% B+[a] Cells (range)	Globin mRNA[b] Copies/Cell	% B+ Cells (range)	Globin mRNA Copies/Cell
Day 0	<1	221	<1	185
Day 1	<1	345	<1	1,109
Day 2	1.5 (1–2)	862	22.5 (20–25)	3,326
Day 3	30 (25–35)	2,045	70 (60–80)	6,028
Day 4	60 (50–70)	4,090	99 (95–99)	9,535

[a] Determined according to Orkin et al.[28]

[b] Determined by molecular hybridization as described in ref. 46 and assuming 220,000 as the average molecular weight for globin mRNA.

cules continue to accumulate in A°1 cells and a sizable percentage of B+ cells also appear whereas the corresponding values observed in 745A cells are much lower. From day 3 onward, differences in percent of B+ cells and amounts of globin mRNA molecules between the two cell clones are still detectable but to a lesser extent. More globin mRNA molecules and B+ cells are, therefore, produced in A°1 than in 745A cells, but the most significant feature of this clone appears to be its early switch-on of the globin gene transcription, as compared with 745A cells and with most data from the literature.

Another interesting observation is the apparent discrepancy in time-course detectability of globin mRNA molecules versus B+ cells. In both clones levels as high as 862 and 1109 globin mRNA molecules/cell are not matched by any significant percent level of B+ cells. This phenomenon may have several explanations. One is that the benzidine staining technique is a very insensitive one[28] and therefore may not show any positivity when Hb is present in low amounts. A less trivial possibility is as follows: as shown by Chan in chick erythroid cells,[29] hemoglobin becomes detectable only when 75% or more of total cellular globin mRNA is present in the cytoplasm, as if a given mRNA concentration were required for successful competition with other cellular mRNAs and, thus, for initiation of globin synthesis. A third possibility is that globin synthesis commences immediately after the first molecule of globin mRNA appears in the cytoplasm, but cannot be detected as Hb until some time later.

All the clones available in our laboratory have been tested for Hb production following exposure to a battery of stimulatory drugs other than DMSO. They were chosen so that they were as widely divergent as possible for both structure and function. As already described,[30] hexamethylenbisacetamide (HMBA) is the most potent stimulant on both 745A and A°1 cells. Hemin and butyric acid are, at the same time, the less potent stimulants for differentiation and the most physiological compounds for the cells. They stimulate almost as well both Fw and 745A cells. Hypoxanthine, on the other hand, is a potent agent in 745A cells, but does not turn on efficiently the DMSO-resistant Fw clone.

Spectrin Accumulation

Spectrin is a peripheral protein complex of the plasma membrane of red blood cells accounting for about 30% of the total membrane proteins. It apparently consists of two large polypeptide chains of approximately 210,000 to 250,000 molecular weight, and is not antigenically detectable in nonerythrocyte cells.[31]

Spectrin has been detected in untreated 745A and A°1 cells. It accumulates following exposure of the same cells to DMSO.[32,33] In our laboratory both immunofluorescence studies and electrophoretical analysis of immunoprecipitated low-ionic-strength extracts of Friend cells have been carried out using a rabbit monospecific serum against spectrin. The accumulation of the protein occurs essentially during the first day after exposure to DMSO, reaches its peak on day 3 and declines thereafter. The ratios between the amounts of spectrin detected on day 3 in DMSO-treated A°1 and 745A cells and those detected in the untreated controls are approximately 5 and 4, respectively. These data provide evidence for an early and membrane-specific marker for DMSO-stimulated erythroid differentiation of Friend cells.

Pattern of Non-Histone Chromatin Proteins

Non-histone chromatin proteins (NHCP) have been shown to be deeply involved in the regulation of transcription of histone genes.[33] A 150,000 daltons molecular weight NHCP is not apparently detectable any more during cartilage differentiation.[34] A regulatory activity NHCP on cell differentiation has also been postulated.[35]

The pattern of NHCP in differentiating Friend cells has been investigated by several groups. Those obtained by Lunadei et al.[36] are summarized here. Briefly, NHCP extracted from whole purified nuclei of Friend cells have been run on both SDS- and SDS-urea-polyacrylamide gels. One day after DMSO stimulation, a chromatin protein component of approximately 32,000 daltons molecular weight is not visible in extracts from DMSO-treated 745A cells as compared with extracts from untreated 745A and Fw cells. Since extracts from DMSO-treated non-differentiating Fw cells harvested from 6 to 72 hr after treatment do show the above protein component, it seems unlikely that its reported disappearance differentiating 745A cells is due to a mere physiochemical interaction of DMSO with chromatin. Other authors[36-38] have reported, on the other hand, the late (72 hr) appearance of a 23,000 to 25,000 daltons molecular weight protein in differentiating Friend cells.

Effects of Interferon on Growth, Cell Cycle, and Differentiation of Friend Cells

Growth and Cell Cycle of Unstimulated Friend cells

As in most cell systems studied (see also Table 23-2), the administration of IF to Friend cells seeded in absence of DMSO results in a marked and dose-dependent inhibition of cell growth[40] under conditions of unaffected overall protein synthesis (Table 23-3). Data obtained from Friend cell cultures treated with mock-IF (supernatant fluids from uninduced cultures) or heterologous rabbit serum IF were consistently superimposable on those obtained from control untreated cultures and are omitted from all presentations heretofore.

Inhibition of growth rate obviously involves change(s) of cell cycle parameters. A detailed description of such changes has been published.[41] Briefly, autoradiograph analysis of growth parameters of Friend cells during treatment with IF demonstrates that the rate of entry into the S phase, the percent decline of unlabeled mitoses, and the mitotic index are significantly lower in IF-treated cell cultures than in controls when tritiated thymidine was added 12 hr after administration of IF. These data indicate that fractions of IF-treated cell populations are delayed in both G_1 and in G_2 phases of the cell cycle. This was confirmed by exact measurements of the length of the various phases of the cycle. In control cells G_1 and G_2 phases lasted 0.9 and 3.2 hr, respectively, whereas in IF-treated cells the lengths of G_1 and G_2 phases were 1.6 and 6.4 hr, respectively. These changes fully account for the increased total length of the cycle (17.6 hr in IF-treated cells versus 13.6 hr in untreated controls).

The IF-directed inhibition of growth of Friend cells is reversible after removal of the compound. Autoradiograph data obtained from control cultures and from cultures previously treated with IF, washed and reseeded if IF-free medium, demonstrate that during the first 12 hr after IF removal, a large majority of the cells previously treated with IF had a deranged flow into the S phase, a high number of unlabeled mitoses, and a low mitotic index. These data provide further evidence for the above-mentioned prolongations of G_1 and G_2 phases of the cell cycle. All growth parameters tested reverted to normal values within 12 hr after washing out IF.

Globin Gene Expression in Differentiating Friend Cells Treated With IF

The administration of low doses of IF (100 units or less/ml/10^5 cells) to cultures of DMSO-stimulated 745A cells causes a slight (if any) reduction of cell growth rate accompanied by a small but significant increase of Hb production (Table 23-4). This IF-directed enhancement of globin gene expression is dose-dependent in an inverse fashion, i.e., the maximal effect is caused by doses in the range of 25 to 50 units/ml whereas the administration of increasing dosages of IF (up to 150/units/ml) progressively effaces the effect. The data on IF-directed inhibition of erythroid differentiation, to be described later, only apparently contradict these. As a matter of fact, it seems clear that very low doses of IF (25 to 50 units/ml) stimulated erythroid differentiation, intermediate amounts of IF (100 to 300 units/ml) apparently do nothing with respect to differentiation, and large doses of IF (above 500 units/ml) are inhibitory. A similarly small but significant increase in the rate at which globin mRNA and Hb accumulate after

Table 23-3 ^{3}H-Arginine incorporation into TCA-precipitable material in Friend cells treated with IF on day 0.

Treatment	CPM × 10^{-6} Cells	
	15 min	30 min
None	1814	3205
IF (640 units/ml)	2460	3707

Table 23-4 Erythroid differentiation of DMSO-stimulated 745A cells given, on day 0, low dosages of IF.

Treatment	% B$^+$ Cells		
	Day 4	Day 5	Day 7
None	<1	<1	<1
DMSO 1.5%	32	40	72
DMSO + IF (25 units/ml)	48	64	96
DMSO + IF (50 units/ml)	49	69	97
DMSO + IF (75 units/ml)	47	56	87
DMSO + IF (100 units/ml)	46	55	86
DMSO + IF (125 units/ml)	41	53	74
DMSO + IF (150 units/ml)	39	43	70

Table 23-5 Influence of DMSO treatment on inhibition of RNA synthesis by α-amanitin.

Treatment (3 hr)	RNA Synthesis[a] (% of control)		
	Untreated 745A Cells	DMSO-Stimulated 745A Cells	DMSO-Stimulated 745A Cells Washed and reseeded in DMSO-Free Medium
8 μg/ml α-amanitin	100.4	49.7	51.4
15 μg/ml α-amanitin	78.7	36.2	15.6

[a] Determined by 15 min pulse incorporation of ^{3}H-Uridine into acid-insoluble material.

IF treatment (50 units/ml) of DMSO-stimulated 745A cells has also been observed by Luftig et al.[42]

During the last 2 to 3 years our group has been working on the interactions of large doses of IF with cultures of DMSO-stimulated 745A cells. We shall briefly summarize here published evidences and present in some detail recent unpublished data.

The erythroid differentiation of DMSO-stimulated Friend cells is inhibited by the administration of large dosages of IF. Hb and heme amounts are reduced nine- and twofold, respectively; thus heme is not the limiting factor involved. This effect is dose-dependent (500 units/ml is the threshold dosage), is not mediated by cytotoxicity,[40] and takes place under growth conditions that fullfill all the experimental prerequisites for DMSO imprinting of Friend cells, as stated by Gusella et al.[16,43] This is to say that, although an influence of growth inhibition on the reduction of Hb synthesis cannot entirely be ruled out, the effect is likely to be due to action of IF on the expression of globin gene.[40] Moreover, in order to minimize possible cytotoxic effects of the combined addition of both agents, IF dosages were always given 24 hr after seeding the cells with 1.5% DMSO. This protococol was used in all experiments described heretofore. Both the rationale and the experimental data warranting its choice have been described.[40,44,45]

As already described for IF-directed inhibition of growth of Friend cells, the block of Hb synthesis is also fully reversed by thorough washing to remove IF, further confirming the specific character of its effect.[40] The prompt resumption of Hb production in the absence of DMSO after IF removal suggests that IF action had not effaced the DMSO imprinting of Friend cells. The possibility, however, that reversion of the block of Hb synthesis was due to *de novo* transcription of globin mRNA occurring after IF removal had not been ruled out. In order to answer this question, we resorted to the use of α-amanitin that specifically acts on RNA Polymerase II and thus inhibits transcription of messenger RNAs.

The action of α-amanitin may be measured by the incorporation of ^{3}H-uridine in short pulses. Under these conditions a 50% reduction of uridine incorporation is taken to represent full inhibition of the RNA polymerase II activity. Treatment of cells with this drug usually requires their pretreatment with DEAE-dextran or other polications to allow the entry of α-amanitin into the cells. In our system, however, the presence of DMSO makes this step useless, as shown in Table 23-5. Untreated Friend cells given 8 μg/ml of α-amanitin did not show any reduction of RNA synthesis, whereas DMSO-stimulated cells had their levels of RNA synthesis reduced to 50% as a result of the administration of the drug. It is also noteworthy that the DMSO-induced change(s) of plasma membrane apparently persist shortly after its removal. DMSO-stimulated cells, in fact, washed and reseeded in DMSO-free medium were still fully sensitive to the action of α-amanitin.

Five and eight μg/ml of α-amanitin were then administered to DMSO-stimulated 745A cultures 60 hr after cell seeding (designated as time 0). B$^+$ and Hb values (Table 23-6) and globin mRNA amounts (Figure 23-1) were determined after 12 hr of treatment with the drug (see also footnote to Table 23-6). Cells given the lower dosage of the compound had fewer B$^+$ cells and lower amounts of Hb and globin mRNA than untreated control, yet all the values were increased over the time 0 values indicating that the expression of globin gene had been only partially inhibited. Friend cells given 8 μg/ml of α-amanitin, on the other hand, did show values for all parameters tested, practically coinciding with those observed at time 0,

Table 23-6 Globin gene expression in DMSO-stimulated 745A cells treated with α-amanitin.[a]

Treatment	B[a]	Hb[b] (μg × 10^{-7} Cells)
None (Time 0)	15	4.5
None, 12 hr later	25	17.5
5 μg/ml α-amanitin, 12 hr later	20	7.35
8 μg/ml α-amanitin, 12 hr later	15	6.5

[a] DMSO-stimulated Friend cells after 60 hr of culture (Time 0) where treated with α-amanitin for 3 hr; the cultures were then diluted 1:4 with fresh medium without DMSO, and incubated until 12 hr.
[b] As measured with the Benzidine assay technique described by Crosby et al.[70] in 1956.

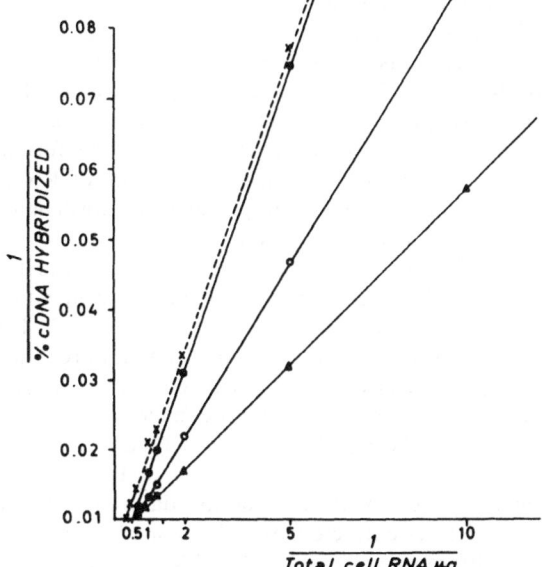

Figure 23-1 Inhibition of globin mRNA synthesis in α-amanitin-treated DMSO-stimulated Friend cells. Double reciprocal plots of saturation curves of hybrids between ³H-globin cDNA and total cell RNAs from DMSO-stimulated Friend cells harvested 60 hr after seeding (closed circles, Time 0). Replicate cultures were washed, reseeded in the same volume of DMSO-free medium with no further treatment (closed triangles), or with 5 μg/ml (open circles) and 8 μg/ml(×) of α-amanitin. All cultures were harvested 12 hr later.

For extraction of RNAs and hybridization and procedures see ref. 46. Amounts of globin mRNA were as follows (ng/μg total cell RNA): Time 0, DMSO-stimulated Friend cells (closed circles) 0.062; a replicate culture, 12 hr later (see above) (closed triangles) 0.116; 5 μg/ml α-amanitin-treated cells (open circles) 0.090; 8 μg/ml α-amanitin-treated cells (×) 0.058.

Total cytoplasmic and nuclear globin mRNA amounts have been determined in DMSO- and DMSO + IF-treated cultures from day 2 onward and were found to be reduced almost threefold on day 2 and one and one-half-fold on day 5,[46] with no evidence of a nuclear storage of globin transcripts. Since overall protein synthesis was also unaffected in these conditions, amounts of globin chains synthesized in IF-treated cells were evaluated by carboxymethylcellulose chromatography to elucidate whether their synthesis or their assembly into Hb was selectively impaired. A drastic reduction of β and α chain synthesis was observed, whereas larger amounts of more cationic materials were detected in DMSO + IF-treated (19.8%) than in DMSO-stimulated cells (7.8%).[46] These data, taken together, indicate that IF may operate on two levels, one involving the transcription of globin mRNA and the other involving its translation.

In view of the observed discrepancy between the globin mRNA contents and the negligible amounts of globin detected, two lines of research are being pursued: (a) the analysis of the structure and the ability of globin mRNA molecules to act as a template in cell-free systems, and (b) a study of the nature of the more cationic material obtained from the carboxymethylcellulose chromatographies.

Poly (A)-containing RNA molecules have been isolated from total cell RNAs extracted from DMSO- and DMSO + IF-treated cells and chromatographed on oligo-(dT)-cellulose columns. Their respective amounts did not vary significantly. Hybrids of the

except for the Hb value, which showed a slight increase (6.5 over 4.5 μg/10⁷ cells) that was probably accounted for by some residual translation of preexisting globin mRNA molecules. Cell mortality at 12 and at 20 hr time intervals was negligible. It thus appears that this dose of the drug completely prevents any *de novo* transcription of globin gene(s) and may therefore be suitably employed to test the mechanism of reversibility of the block of Hb synthesis caused by IF. Data from a preliminary experiment, shown in Table 23-7, suggest that B⁺ values of DMSO + IF-treated Friend cells are comparable with control values 20 hr after removal of IF even if α-amanitin was added to the cultures. Globin mRNA determinations are under way. These data, if confirmed, indicate that the reversion of the block of Hb synthesis caused by IF treatment is apparently not due to *de novo* transcription of globin mRNA. This implies that at least some globin mRNA molecules are transcribed in the presence of IF and are not at all translated.

Table 23-7 Reversal of IF inhibition of differentiation of DMSO-stimulated Friend cells treated with α-amanitin.

Treatment	5 Days of Culture	
	Cells per ml (× 10⁻⁵)	B⁺ (%)
DMSO	23.1	72
DMSO + IF (1.000 units/ml)	4.8	13
Previous Treatment + α-Amanitin	20 hr After Washing and Reseeding in DMSO- and IF-Free Medium at 10⁵ Cells per ml With or Without α-Amanitin[a]	
	Cells per ml (× 10⁻⁵)	B⁺ (%)
DMSO	2.7	50
DMSO + α-amanitin (10 μg/ml)	1.8	44
DMSO + IF	1.6	34
DMSO + IF + α-amanitin (10 μg/ml)	1.2	40

[a] For the α-amanitin treatment see also Table 23-6 footnote.

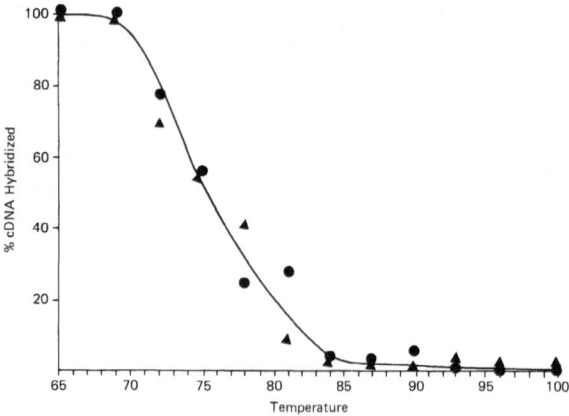

Figure 23-2 Melting profiles of hybrids between ³H-cDNA and Poly-A⁺ RNAs extracted from either DMSO-stimulated or DMSO + IF-treated Friend cells. For RNA extraction and hybridization procedures see ref. 46. Determination of melting points was done in 40 mM NaPh buffer, in 0.2 ml volume. After 72 hr of hybridization, samples were cooled to room temperature, heated at the desired temperature for 10 min, and then assayed for S_1 nuclease digestion.

Poly (A)-containing RNAs with globin cDNA were analyzed for their thermal stability (Figure 23-2). No major alterations of the base sequence between the two RNA populations have been uncovered. The same RNA preparations heated at 90°C for 3 min in 99% formamide have been analyzed in 99% formamide 5% polyacrylamide gels. Preheating of RNAs and gel electrophoresis in presence of formamide provide the most thorough denaturing conditions for RNA molecules.[47] Gel slices were eluted and tested for globin mRNA content by molecular hybridization with globin cDNA, as shown in Figures 23-3, left panel (for RNAs extracted from 2-day cultures) and 23-3, right panel (for RNAs extracted from 5-day cultures). Two observations may be made.

(a) While the profiles shown in Figure 23-3, left panel peak indistinguishably, those of Figure 23-3, right panel peak with a 1-fraction difference in the case of RNAs extracted from IF-treated cells. This slight difference was not attributable to an accidental artifact since materials eluted and rerun on gels did again migrate with the same pattern. Should this find-

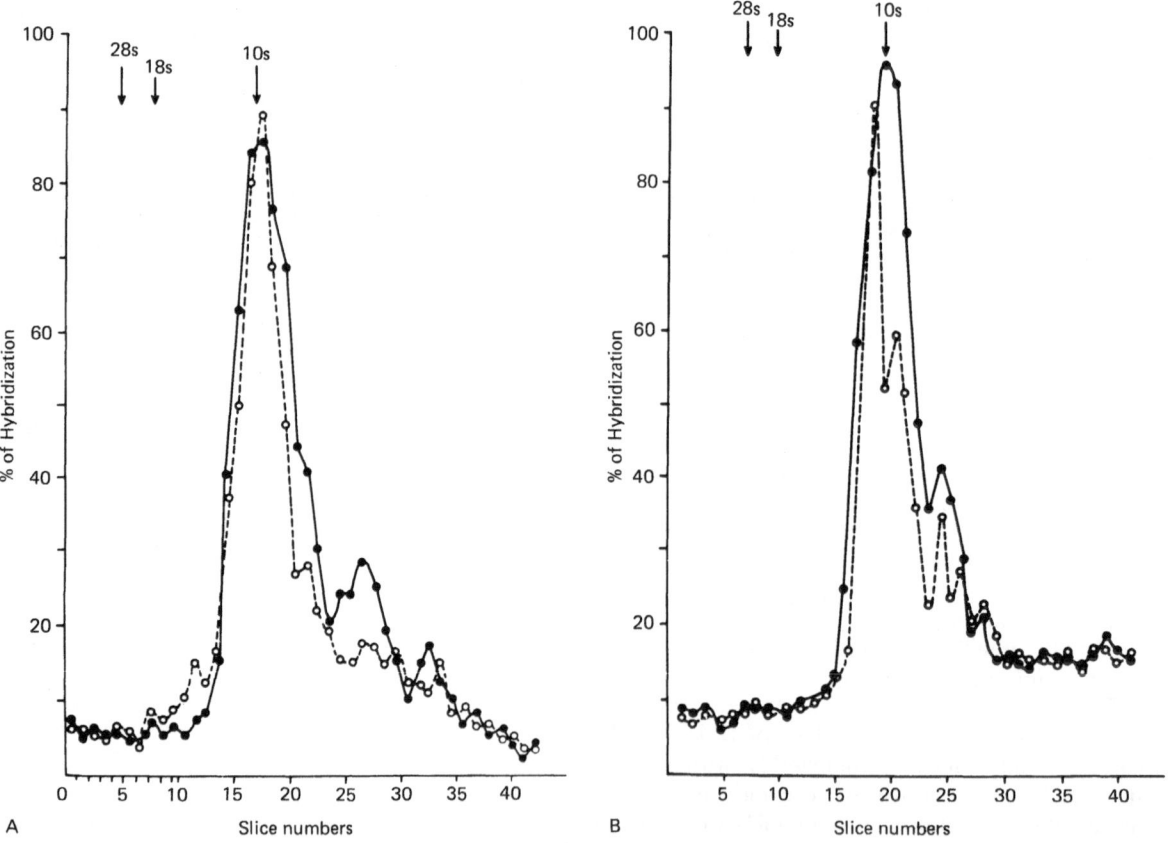

Figure 23-3 Gel electrophoresis of Poly-A⁺ RNAs from DSMO- and DMSO + IF-treated Friend cells. **A**: RNAs from 2-day cultures. **B**: RNAs from 5-day cultures. All samples were preheated at 90°C for 3 min in 99% formamide and run on 99% formamide-5% PAGE. Gel slices were eluted and hybridized with ³H-globin cDNA as described in refs. 46 and 47. The 28S and 18S arrows indicate the mobility of ribosomal RNAs added as internal markers in the gels. The peak of hybridization profile of RNA from DMSO-treated cells is taken to represent a 10S marker. Continuous line: DMSO-stimulated Friend cells. Broken line: DMSO + IF-treated Friend cells.

ing be confirmed by more extensive and thorough analyses, it would mean that the molecular weight of globin mRNA molecules present in IF-treated cells is slightly higher (approximately 10,000 daltons) than that of the molecules present in DMSO-stimulated cells.

(b) In Figure 23-3, on the other hand, the peak profile is expanded on the side corresponding to the lighter molecules for mRNAs derived from DMSO-stimulated cells.

Data illustrated in (a) and (b) may also be interpreted as evidence of a decreased globin mRNA degradation in DMSO + IF-treated Friend cells, which is in keeping with data from the previously reported experiments on reversibility of IF effects with and without the addition of α-amanitin. Definitive proof of the absence of major IF-induced globin mRNA degradation will come from the estimates of half-lives of globin messengers.

Further circumstantial evidence pointing in this direction is provided by experiments done utilizing globin mRNA molecules, present in DMSO + IF-treated Friend cells, as templates for protein synthesis in exogenous and endogenous cell-free systems.

As for the exogenous cell-free system, 9-10S RNA isolated from polyribosomes of Friend cells treated with DMSO and DMSO + IF, respectively, has been added to a wheat germ cell-free protein synthesizing system, prepared according to the method described in refs. 48 and 49, and the products analyzed for size on SDS-acrylamide gels. No detectable differences were observed between the two experimental conditions in terms of ^{3}H-leucine incorporation into SDS-electrophoresis purified globin-sized material (globin synthesis/ng of mRNA was 764 cpm and 860 cpm for DMSO and DMSO + IF-treated cells, respectively).

As for the endogenous capacity to synthesize globin-sized proteins, S_{30} lysates of the the two experimental conditions were tested according to the method described in ref. 50. As shown in the Table 23-8, lysates of cells treated with DMSO synthesize twice as much protein as lysates of the same number of cells treated with DMSO + IF. Addition of tRNA does not result in an increased efficiency of the system, whereas when rabbit globin mRNA is added, lysates from DMSO + IF cells synthesize globin-sized material to the same extent that DMSO-lysates do. These findings suggest that the decreased efficiency of S_{30} lysates from cells treated with DMSO + IF (Table 23-9, line 1) is probably due to a reduced content of endogenous globin mRNA molecules rather than to built-in deficiencies of the protein-synthesizing machinery.

Much work still needs to be done to understand the nature of the more cationic material obtained from the carboxymethylcellulose chromatographies. For the time being, evidence has been obtained that this material, which differs from globin in its charge,

Table 23-8 Incorporation of ^{14}C-leucine into globin size material by S_{30} lysates of DMSO- and DMSO + IF-treated Friend cells. SDS-Page analysis.

Materials	DMSO		DMSO + IF	
	cpm	Ratio	cpm	Ratio
Lysate only	10.110	1.00	6.574	1.00
Lysate + tRNA	10.243	1.01	7.116	1.08
Lysate + rabbit globin mRNA	11.125	1.10	11.571	1.76
Lysate + tRNA + rabbit globin mRNA	13.600	1.34	13.386	2.03

can be immunoprecipitated at a significant extent by a monospecific globin antiserum (Table 23-9). A very preliminary analysis of its size in 10% SDS-polyacrylamide gel electrophoresis has revealed a protein with slower electrophoretic mobility than globin, but one that is still immunoprecipitable with the globin antiserum.

Effects of Interferon on Virus Production and Expression of Viral Antigens in Friend Cells

The effects of IF administration on the biology of Friend Leukemia Virus, which is chronically shed by Friend cell cultures, are essentially similar to those already described with respect to the chronic infections with other murine RNA tumor viruses. Details of data obtained by us have been published.[51] Immunofluorescent staining was used to detect intracellular and membrane-bound viral antigens, and labeled RNA and protein precursors were incorporated into sucrose-banded virions to detect extracellular virus release. As little as 16 units/ml of IF markedly inhibit FLV release into the supernatant fluids. This does not mean, however, that the virus has been eliminated from the cultures, as is customary in cells infected by conventional cytopathic viruses and treated with IF. Removal of IF, in fact, is followed in the Friend system by a burst of virus production and release into the supernatants, and subsequently the viral cycle of these cultures becomes indistinguish-

Table 23-9 Immunoprecipitation of eluted pooled fractions of CM-52 columns with antiglobin serum.

Region	CTR (%)	DMSO (%)	DMSO + IF (%)
Run off	10.7	9.9	11.2
11–13 mM	73	57.5	77.6
15–17 mM	—	58.5	71.4

Fractions from CM-52 chromatographies were pooled as indicated and immunoprecipitated with rabbit immune serum against DBA/2 mouse globin, as previously described.[46] Data are given as percentage of radioactivity recovered in precipitates.

able from that of cultures never treated with IF. These data have been readily explained by the abundant detection of viral antigens both intracellularly and on the outer surface of the plasma membrane. Therefore, it is apparent that, while the production of fully assembled and/or infectious virions is blocked by IF treatment, the synthesis of virus proteins proceeds at a substantial level. It has been demonstrated that exposure of Friend cells to DMSO also causes a marked enhancement of virus production[17,19] and of the accumulation of intracellular virus proteins.[51] Administration of a wide range of IF doses (16 to 1,000 units/ml) to DMSO-stimulated Friend cells results in both the block of virus release and a more marked accumulation of viral antigens in the cytoplasm.[51] When IF was removed, virus production rose and intracellular virus antigens fell in the levels of untreated controls. This raised the question as to whether or not virions released from IF-treated cells after removal of IF were utilizing the virus proteins previously accumulated. In a preliminary experiment, we labeled virus proteins being synthesized in presence of IF, removed IF itself and the labeled precursors, and determined the radioactivity present in virions harvested 24 and 48 hr later. A substantial amount of radioactivity was detected in sucrose-banded virions harvested at both time intervals, suggesting that virus proteins synthesized during IF treatment had indeed been assembled into virions banding at the 1.16 to 1.18 gm/cm^3 density. A profound decrease of the number of virions released into the supernatant fluid and in the number of cytoplasmic vacuolar structures, accompanied by formation of defective particles, has been described in Friend cells treated with IF.[42]

Tumorigenicity of Friend Cells Treated with Interferon

We had shown that the administration of large doses of IF to mice injected with Friend cells caused a marked inhibition of cell growth in the recipient mice.[52] IF *in vitro* treatment also results in the lengthening of the intermitotic time of Friend cells,[41] which may increase cell elimination *in vivo* and possibly account for IF-induced inhibition of the growth of Friend cells. It is also known that DMSO treatment of Friend cells reduces by 2 logs their ability to form subcutaneous tumors in susceptible recipient mice.[3]

We have determined the tumorigenic capacity of Friend cells treated either with IF or DMSO or both. Cells were treated *in vitro* for 48 to 72 hr, and then inoculated in varying amounts in the subcutaneous tissues of histocompatible DBA/2J/Cn mice. The data in Table 23-10 show that the *in vitro* treatment with IF does not affect the tumorigenicity of Friend cells. Exposure of the cells to DMSO, instead, results in a 1.34 log increase of TCID$_{50}$ values, in keeping with the data of Friend et al.[3] The combined treatment with DMSO + IF brings the TCID$_{50}$ value back to control values, as if IF were apparently counteracting the effects of DMSO in this system. Measurements of tumor diameters also afforded the possibility of testing whether tumor growth had been constant or if tumor regressions had taken place. Tumors produced by the inoculation of untreated Friend cells showed no regressions, whereas some of those produced by implanting treated cells did regress during the third to fourth week after cell inoculation.

Table 23-10 Comparative evaluation of tumor takes by Friend cells treated with either DMSO or IF or both.[a]

Treatment	TCID$_{50}$ (According to Reed and Muench Method)
None	$10^{4.5}$
IF[b] (800 units/ml)	$10^{4.5}$
DMSO[b]	$10^{5.84}$
DMSO[b] + IF[c] (800 units/ml)	$10^{4.32}$

Cells were seeded at 10^5/ml, as indicated. On day 3 of culture, cells were counted, resuspended at the desired concentrations into 0.25 ml of saline, and injected subcutaneously into DBA/2/J/C$_1$ mice (kindly provided by Dr. V. Monaco, CNEN-Casaccia, Rome, Italy). Mice were examined for tumor implantation weekly in the first month, and every other week thereafter. Tumor diameters were measured with callipers. The experiment was terminated 90 days after mice injection.

[a] Treatments of Friend cells were done *in vitro*.
[b] Administered on day 0.
[c] Administered one day later (day 1).

Discussion

The most relevant conclusion that can be derived from the data reported and/or reviewed in this chapter stems from the evident widespread and pleiotropic effects caused by the administration of IF to the cell systems studied. Even at first glance, the list of IF-directed effects on cell functions (summarized in Table 23-1) is so impressive as to make the statement, "IF is not only an antiviral substance" almost a trivial one. The time has come, in our opinion, to completely dispel the doubts and disbelief that have so far rewarded the work of those who provided the evidence listed in Table 23-1.

One of the aspects to be considered first is that of IF dosages. With few exceptions, IF doses capable of interacting in one way or another with cell growth, metabolism, and differentiation are consistently higher than those able to induce the antiviral state. Ratios between the former and the latter vary from 10 to 50.

Closely related with the previous problem is the one of the so-called "physiological" doses of IF. It has been taken for granted for a long time that IF dosages in excess of 200 to 300 units/ml and/or able to decrease the cell proliferating capacity were outrightly toxic. All data describing inhibitory effects of IF on cell growth, therefore, were void of any value by definition. In this respect the work published by Dianzani and Baron[71] is of poignant interest. They provided an estimate, based on both experimental and extrapolated data, of IF concentrations actually occurring around producing cells *in vivo;* according to their calculations, these concentrations may be in the range of 3,000 to 30,000,000 units/ml. Even if these figures were wrong by more than one order of magnitude, they would still amply allow for IF doses active on cells to be defined as *physiological*. Cell viability is of course a stringent criteria to be followed in the evaluation of experimental data obtained by treatment with high doses of IF.

Effects dependent on the administration of doses of IF higher than those needed for the establishment of the antiviral state are probably explained by some recent observations about IF interactions with the cell membrane. The cellular uptake of IF is not required for the antiviral effects[72]: sensitivity to IF seems to be determined by receptor(s) located at the plasma membrane level. It has been observed that IF binding involves both cellular gangliosides and cellular glycoproteins.[73-75] If these observations will be found valid also for the cellular effects, then it would be easy to hypothesize that different IF effects on the cells may depend on critical concentrations of IF molecules versus cell receptors.

Interactions of IF with DMSO-stimulated Friend cells appear to be biphasic with reference to the expression of the globin gene. Doses in the range of 25 to 100 units/ml cause a small but significant increase of Hb synthesis.[42,45,76] Slightly higher doses of IF (100 to 300 units/ml) apparently do not influence DMSO-stimulated erythroid differentiation of the cells, whereas doses above 500 units/ml are inhibitory of globin gene expression.[40,44-46,51] Failure to observe the IF-directed inhibition of Hb synthesis by DMSO-stimulated Friend cells is probably best explained by the insufficient IF doses used.[76,77]

As already suggested before, reciprocal concentrations of IF molecules and cell receptors may be critical in determining the type of the ensuing effect. The real issue here is to understand how the impulse originated by IF impact on the plasma membrane is then transferred to the cell nucleus. Interestingly enough, the integrity of the cytoskeleton is needed for the establishment of the antiviral state.[78] Another possibility that remains to be determined is that the antiviral effects of IF are mediated through a second message such as AMP.[79] Biphasic effects are only one reason why the action of IF at the cellular surface resembles in many respects the action of the polypeptide hormones. It is also noteworthy that IF rapid turnover phase in serum is mediated by the same desialylation mechanisms operating for other circulating glycoproteins.[80]

IF-directed inhibition of globin gene expression is apparently characterized by a dual level of action as both the transcription and the translation of globin gene are reduced, although at a variable extent. The magnitude of the effect on translation (tenfold reduction) exceeds that of the inhibition of transcription (roughly twofold). This implies that globin mRNA molecules are accumulated in the cytoplasm while they are not translated. The question arises, therefore, as to whether the structure of these molecules is comparable to that of globin mRNAs present in DMSO-stimulated cells. Our present data concerning this may be summarized as follows: (a) no major alterations in the base sequence of globin mRNAs have been observed; (b) globin mRNA molecules from DMSO + IF-treated cells harvested on day 5 of culture migrate, under the most stringent denaturing conditions, one fraction ahead of those from DMSO-stimulated cells; this finding is not visible when RNAs from 2-day cultures were run on gels. (c) In both cases, instead, the peak profiles are suggestive of less degraded globin sequences present in DMSO + IF-treated as compared to DMSO-stimulated cells. This last observation (a very preliminary data, also the preceding one) is noteworthy as it suggests that globin mRNA molecules prevented from being translated into globin are not degraded any faster than the corresponding molecules from DMSO-treated cells; if anything, the former appear to be slightly less degraded than the latter.

The data derived from the experiments carried out with both cell-free systems indicate that globin mRNA molecules present in DMSO + IF-treated cells are translated into globin-sized products as efficiently as are the corresponding molecules from DMSO-stimulated cells. One has, therefore, to face the fact that apparently "normal" globin sequences present in the cytoplasm of DMSO + IF-treated Friend cells are not translated into globin. This is a novel effect of IF as the synthesis of a specific cellular protein in living cells is inhibited but not that of viral proteins. The mechanism of action of IF in both cases remains to be elucidated. Several groups are working on this topic, but a review of their work done so far would be too far-reaching and beyond the purpose of this chapter.

Globin chains are not produced in DMSO + IF-treated cells, but more cationic material, with a slower electrophoretic mobility than globin and sharing most antigenic determinants of globin, has been observed in larger amounts in these cells than in the controls. Although little is still known about the nature of this material, we venture to offer the following working hypotheses: (a) it could be reminiscent of β globin aggregates observed in some cases of human β-thalassemias; (b) it could represent a different, although related, protein (embryonic globins?); and (c) it could just be an altered protein possibly having nothing to do with erythropoiesis.

Our data on the effect(s) of IF administration on the life cycle of FLV in Friend cells are in good agreement with those already reported for other murine RNA tumor viruses and reviewed in this chapter. The observation that viral proteins produced during the IF-induced block of virus release are indeed assembled into full virions after the removal of IF is noteworthy. It apparently indicates that viral products synthesized in the presence of IF are not defective since they become part of normal virions. This only adds more interest to the fact that IF treatment, while inhibiting virus release in all cases, does eliminate all virus-coded products of conventional cytopathic viruses but does not prevent the synthesis of most virus coded products of tumor viruses. Of course, an obvious difference between the two groups of viruses is that the genome of the latter does integrate into the cell DNA, whereas the genome of the former does not. This puts the genes of tumor viruses in a position similar to that of cell genes. In this respect, models other than the conventional one (overall inhibition of translation or transcription of the virus genome) can be constructed to explain the effect(s) of IF in these systems. For example, IF is known to alter several features of cell membrane.[55,61,63,81] It is tempting to hypothesize that such an altered cell membrane may interfere with the as yet largely unknown process(es) of virus release.

Changes of the plasma membrane may also mediate the findings described in Table 23-10, namely the observation that DMSO + IF-treated Friend cells form subcutaneous tumors in DBA/2J mice much more efficiently than do DMSO-stimulated cells (1.5-log difference in $TCID_{50}$). Since tumor cells pretreated with IF apparently undergo membrane changes and escape the surveillance of natural killer lymphocytes,[25] this mechanism may also be operative in the case of DMSO + IF-treated Friend cells whereby the DMSO may be acting as an amplifier agent. An alternative explanation may be that DMSO-induced Friend cells form tumors less efficiently because a substantial fraction of the cells have a limited capacity for cell division in view of DMSO-stimulated imprinting and differentiation.[16] What IF treatment may be doing in this respect then, is just to decrease the numbers of cells that have already undergone enough cell divisions as to become incapable to form tumors *in vivo*.

Acknowledgment

The skillful assistance of Ms. A. Tamburrini and P. Meo is gratefully acknowledged. This work was partially supported by grants from Consiglio Nazionale delle Ricerche, Progetto Finalizzato Virus, (NN. 77.00300.84 and 77.00304.84) and from N.A.T.O. (N. 1152).

References

1. Friend C, Patuleia MC, De Harven E: *Natl Cancer Inst Monogr 22*:505–522, 1966.
2. Rossi GB, Friend C: *Proc Natl Acad Sci USA 58*:1373–1380, 1967.
3. Friend C, Scher W, Holland JC, et al: *Proc Natl Acad Sci USA 68*:378–382, 1971.
4. Temin HM, Mizutani S: *Nature 226*:1211–1213, 1970.
5. Baltimore D: *Nature 226*:1209–1211, 1970.
6. Ross J, Ikawa Y, Leder P: *Proc Natl Acad Sci USA 69*:3620–3623, 1972.
7. Harrison PR. In Paul J (ed): Biochemistry of Differentiation. London, MTP, in press.
8. Bernstein A, Hunt DM, Crichley V, et al: *Cell 9*:375–381, 1976.
9. Rovera G, O'Brien TG, Diamond L: *Proc Natl Acad Sci USA 74*:2894–2898, 1977.
10. Yamasaki H, Fibach E, Nudel V, et al: *Proc Natl Acad Sci USA 74*:3451–3455, 1977.
11. Mc Clintock PR, Papaconstantinou J: *Proc Natl Acad Sci USA 71*:4551–4555, 1974.
12. Wayne Wiens A, Mc Clintock PR, Papaconstantinou J: *Biochem Biophys Res Comm 70*:824–831, 1976.
13. Levy J, Terada M, Rifkind RA, et al: *Proc Natl Acad Sci USA 72*:28–32, 1975.
14. Terada M, Fried J, Nudel U, et al: *Proc Natl Acad Sci USA 74*:248–252, 1977.
15. Leder A, Leder P: *Cell 5*:319–322, 1975.
16. Gusella J, Geller R, Clarke B, et al: *Cell 9*:221–230, 1976.
17. Sato T, Friend C, De Harven E: *Cancer Res 31*:1402–1417, 1971.
18. Jasmin C, Smadja-Joffe F, Klein B, et al: *Cancer Res 36*:603–607, 1976.
19. Ostertag W, Pragnell IB, Krieg CF, et al: In Oncogenic Viruses and Host Genes. New York, Academic Press, in press.
20. Chang EH, Myers MW, Wong PKY, et al: *Virology 77*:625–635, 1977.
21. Billiau A, Heremans H, Allen PT, et al: *Virology 73*:537–542, 1976.
22. Pitha PM, Rowe WP, Oxman MN: *Virology 70*:324–338, 1976.

23. Gresser I: In Cancer, A Comprehensive Treatise. New York, Plenum Press, 1978.
24. Youngner JS, Salvin SB: *J Immunol 111*:1914–1922, 1973.
25. Trinchieri G, Santoli D, Dee RR, et al: Submitted for publication.
26. Curtis PJ, Weisman C: *J Mol Biol 106*:1061–1075, 1976.
27. Affrabris E, Pulciani S, Rossi GB: Microbiologica, in press, 1978.
28. Orkin SH, Harosi FI, Leder P: *Proc Natl Acad Sci USA 72*:98–102, 1975.
29. Chan LNL: *Nature 261*:157–159, 1976.
30. Reuben RC, Wife RL, Breslow R, et al: *Proc Natl Acad Sci USA 73*:862–867, 1976.
31. Hiller G, Weber K: *Nature 266*:181–183, 1977.
32. Rossi GB, Aducci P, Gambari R, et al: Submitted for publication.
33. Stein G, Stein J, Thrall C, et al: In Stein GG, Kleinsmith LJ (eds): Chromosomal Proteins and their Role in the Regulation of Gene expression. New York, Academic Press, pp. 1–17, 1975.
34. Newman SA, Birbaum J, Yech GCT: *Nature 259*:415–418, 1976.
35. Paul J, Gillmour LS: *J Mol Biol 34*:305–316, 1968.
36. Lunadei M, Matteucci P, Ullu E, et al: Submitted for publication.
37. Peterson JL, Mc Conckey EH: *J Biol Chem 251*:555–558, 1976.
38. Keppel S, Allet B, Eisen H: *Proc Natl Acad Sci USA 74*:653–656, 1977.
39. Lau AF, Ruddon RW: *Exp Cell Res 107*:35–46, 1977.
40. Rossi GB, Matarese GP, Grappelli C, et al: *Nature 267*:50–52, 1977.
41. Matarese GP, Rossi GB: *J Cell Biol, 75*:344–354, 1977.
42. Luftig RB, Conscience JF, Skoultchi A, et al: *J Virol 23*:799–810, 1977.
43. Gusella JF, Housman D: *Cell 8*:263–269, 1976.
44. Rossi GB, Dolei A, Cioé L, et al: Texas Reps on Biol and Med. Interferon Issue. S Baron and F Dianzani, in press, 1977.
45. Rossi GB, Dolei A, Cioé L, et al: Oncogenic Viruses and Host Genes. Academic Press, New York, in press, 1977.
46. Rossi GB, Dolei A, Cioé L, et al: *Proc Natl Acad Sci USA 74*:2036–2040, 1977.
47. Farace MG, Ullu E, Fantoni A, et al: Submitted for publication.
48. Marku K, Dudock B: *Nucleic Acid Res 1*:1385–1389, 1974.
49. Fantoni A, Ullu E, Gambari R, et al: *Ann Immunol* (Ist Pasteur) *127*:881–883, 1976.
50. Bordin S, Farace MG, Fantoni A: *Biochim Biophys Acta 281*:277–288, 1972.
51. Ramoni C, Matarese GP, Rossi GB, et al: *J Gen Virol*, in press, 1977.
52. Rossi GB, Marchegiani M, Matarese GP, et al: *J Natl Cancer Inst 54*:993–996, 1975.
53. Gresser I: In Klein G, Weinhouse S (eds): Advances in Cancer Research. New York, Academic Press, pp. 97–140, 1972.
54. Kuwata T, Fuse A, Morigana N: *J Gen Virol 33*:7–15, 1976.
55. Stewart WE II, Gosser LB, Lockart RZ Jr: *J Virol 7*:792–801, 1971.
56. Lindahl-Magnusson P, Leary P, Gresser I: *Nature (New Biol) 237*:120–121, 1972.
57. Fuse A, Kuwata T: *J Gen Virol 32*:17–24, 1976.
58. Macieira-Coelho A, Brouty-Boyè D, Thomas MT, et al: *J Cell Biol 48*:415–419, 1971.
59. Collyn d'Hooghe M, Brouty-Boyè D, Malaise EP, et al: *Exp Cell Res 105*:73–77, 1977.
60. Ida S, Hooks JJ, Siraganian RP, et al: *J Exp Med 145*:892–899, 1977.
61. Lindahl P, Gresser I, Leary P, et al: *Proc Natl Acad Sci USA 73*:1284–1287, 1976.
62. Gisler RH, Lindahl P, Gresser I: *J Immunol 113*:438–444, 1974.
63. Lindahl P, Leary P, Gresser I: *Proc Natl Acad Sci USA 69*:721–727, 1972.
64. Nebert DW, Friedman RM: *J. Virol 11*:193–197, 1973.
65. Beck G, Poindron P, Illinger D, et al: *FEBS lett 48*:297–300, 1974.
66. Illinger D, Coupin G, Poindron P: *J Gen Virol*, 1978.
67. Huet C, Gresser I, Bandu MT, et al: *Proc Soc Exp Biol Med 147*:52–57, 1974.
68. Johnson HM, Stanton GJ, Baron S: *Proc Soc Exp Biol Med 154*:138–141, 1977.
69. Sonnenfeld G, Mandel AD, Merigan TG: Submitted for publication.
70. Crosby WH, Furth FW: *Blood 11*:380–385, 1956.
71. Dianzani F, Baron S: *Nature 257*:682–683, 1975.
72. Ankel H, Chany C, Galliot B, et al: *Proc Natl Acad Sci USA 70*:2360–2363, 1973.
73. Besançon F, Ankel H: *Nature 252*:478–480, 1974.
74. Kohn LD, Friedman RM, Holmes JM, et al: *Proc Natl Acad Sci USA 73*:3695–3699, 1976.
75. Vengris SH, Reynolds FH Jr, Hollemberg MD, et al: *Virology 72*:486–493, 1976.
76. Lieberman D, Voloch Z, Aviv H, et al: *Mol Biol Rep 1*:477–481, 1975.
77. Swetly P, Ostertag W: *Nature 251*:642–644, 1974.
78. Bourgeade MF, Chany C: *Proc Soc Exp Biol Med 153*:501–504, 1976.
79. Weber JM, Stewart RB: *J Gen Virol 28*:363–372, 1975.
80. Bocci V, Pacini A, Pressina GP, et al: *Experientia 33*:164–165, 1977.
81. Stewart WE, De Clercq E, Billiau A, et al: *Proc Natl Acad Sci USA 69*:1851–1854, 1972.

24 Kinetics of Erythroid and Myeloid CFU Following Rauscher Virus Infection

Suzanne Hasthorpe

Introduction

Rauscher murine leukemia virus was initially described as producing a lymphoid leukemia, with an early phase characterized by erythrocytosis.[1] The primary erythropoietic phase is apparent by 7 days and the secondary phase of lymphocytic leukemia occurs 30 to 45 days post infection. The degree of erythrocytosis has been shown to be an increasing function of virus titer and occurs at a higher frequency in mice of 5 weeks or younger. The mouse strain is also important for the manifestation of splenomegaly and erythrocytosis.[1]

The erythropoietic response to Rauscher virus is characterized by splenomegaly, anemia, and thrombocytopenia[2] and has been shown to be moderated by red blood cell transfusion.[3] Erythropoietin levels have also been reported to be elevated during the course of the disease.[4]

The question of whether or not Rauscher virus-induced disease is an autonomous erythroleukemia still remains open. There are a number of general features that distinguish this disease from the commonly known characteristics of neoplasia. Dunn and Green[5] interpreted their observations as evidence that this disease is nonneoplastic. They observed that the reaction was very rapid *in vivo*, with erythropoiesis occurring in hemopoietic organs and it was not disseminated widely. Although there is a predominance of an erythroblast-like cell, other stages of erythroid maturation were also found, indicating that differentiation was still occurring. Transplantation of Rauscher virus-infected tissues results in a repetition of the erythroblastic reaction in hemopoietic organs, with growth at the site of transplantation absent.[5] In addition, continued responsiveness to the humoral regulator erythropoietin has been demonstrated both *in vivo*[6] and *in vitro*.[7]

This chapter will consider the kinetic changes in the erythroid and myeloid precursor pools during the development of the primary phase of this disease. The responsiveness of erythropoietin will also be discussed for the CFU-e and BFU-e populations in mice infected with Rauscher virus.

Materials and Methods

Female BALB/c mice, 5 weeks old, were inoculated intraperitoneally with 0.4 ml of a cell-free spleen preparation. This aliquot contained approximately 44,000 plaque-forming units and was assayed according to the method of Rowe et al.[8]

Three to four mice were sampled from each group at successive times. Cardiac blood was sampled for hematocrit determinations, spleens were weighed, and cell pools from marrow and spleen were assayed *in vitro*.

Erythroid colony-forming units (CFU-e) were cultured in methylcellulose according to the previously described method of Guilbert and Iscove,[9] except for two modifications, i.e., 5% fetal calf serum and double the concentration of transferrin, with a final concentration of 6.8×10^{-6} M. Colony formation was linear with cell doses over a range from 2×10^5 to 2.5×10^4 cells for normal marrow, and also for marrow from Rauscher virus-injected mice.

Granulocyte/macrophage colony (CFU-c) and erythroid burst- (BFU-e) forming units were cultured in the same way except for different serum concentrations; equal volumes of (10%) fetal calf and horse serum were used with a 1% concentration of bovine serum albumin.

Erythropoietin was obtained from Connaught Laboratories (Willowdale, Ontario, Canada), a step III preparation was used, the batch number was 3015–5 and had a specific activity of 5.9 IU/mg protein. The stimulator that was used for CFU-c cultures was an extract of pregnant mouse uterus and embryo; the preparation of this has been described by Bradley et al.[10] and Stanley et al[11].

Staining for hemoglobin-containing colony cells was carried out in the dish using a 0.2% solution of benzidine containing 3% acetic acid. Individual colonies and cytocentrifuge preparations were stained for hemoglobin according to a previously described method.[12]

Results

Coincident with the development of anemia is a marked increase in spleen weight (Table 24-1). There is also a pronounced increase in CFU-e levels both in spleen and bone marrow, reaching an overall plating efficiency of one CFU-e per 15 to 20 spleen cells in the latter stages of this disease (Table 24-2).

CFU levels in bone marrow are indicated in Figure 24-1. They are expressed as a ratio of corresponding normal femoral marrow values. CFU-e exceed normal femoral bone marrow levels after the fifth day

Table 24-1 Changes in spleen weight, hematocrit, and CFU-e levels in 5-week-old BALB/c female mice following infection with Rauscher virus (RL) compared to normal uninfected mice (N).

Time (days)	Hematocrit (%)	Spleen Weight (gm)	CFU-e/10^5 Cells Spleen		CFU-e/10^5 Cells Bone Marrow	
			RL	N	RL	N
1	39	0.07	82	96	229	236
3	38	0.09	119	146	159	264
5	41	0.13	1,270	66	232	269
7	38	0.25	3,100	49	547	324
10	37	0.47	4,570	45	2,410	543
14	35	1.24	3,860	94	1,400	317
17	30	1.77	2,920	N.D.	520	291
19	23	2.88	5,470	N.D.	624	321
25	22	3.34	5,100	N.D.	568	455
30	19	3.06	5,500	35	793	412
Normal mice (♀♀)	44	0.10				

and persist at higher levels until the third to fourth week. BFU-e fall below normal marrow levels being maintained at approximately 60% of normal. CFU-c fluctuate around normal marrow levels and do not show a consistent deviation from them.

Changes in spleen are more pronounced due to the added effects of concentration of CFU and total organ cellularity. There is an initial dip in all CFU populations, which is followed by an exponential rise (Figure 24-2). It is notable that both types of erythroid CFU are increased as well as the CFU-c. These increases occur in a sequential fashion, which has been observed in normal regenerating spleen following X-irradiation and bone marrow transplantation.[13] The doubling times for CFU-e, CFU-c, and BFU-e are approximately 17, 30, and 42 hr respectively. These doubling times are comparatively longer than in regenerating spleen where Gregory et al.[13] reported 6 and 12 hr for CFU-e and CFU-c doubling.

The question of CFU-e dependence on erythropoietin for colony formation has been subject to some controversy. It has been reported that CFU-e from Rauscher-infected bone marrow cells were not de-

Table 24-2 Bone marrow BFU-e responsiveness to erythropoietin in normal and Rauscher virus-infected mice.[a]

Erythropoietin Concentration (IU/ml)	Normal Marrow (% peak ± SE)	Rauscher Virus Infected Marrow (% peak ± SE)
2	100 ± 4.7	100 ± 3.8
1	93.9 ± 8.9	78.8 ± 3.3
0.5	75.3 ± 5.9	66.8 ± 4.5
0.25	60.8 ± 5.6	69.3 ± 4.9
0.125	27.0 ± 4.0	47.6 ± 3.9
0.063	10.2 ± 3.3	42.8 ± 4.5
0.032	0.6 ± 0.3	30.1 ± 5.5
0.016	0	26.7 ± 4.3
0.008	0	19.7 ± 3.2
0.004	0	14.1 ± 2.4
0.002	0	7.1 ± 3.3
0.001	0	13.4 ± 4.5
0	0	3.8 ± 1.6

[a] Mice were injected with Rauscher virus 2 to 3 weeks prior to sampling. Means represent data from four experiments with normal marrow (100% level is equivalent to 18.9 ± 2.1 BFU-e/10^5 cells). Five experiments were performed with Rauscher bone marrow (100% level is equivalent ot 20.7 ± 3.1 BFU-e/10^5 cells).

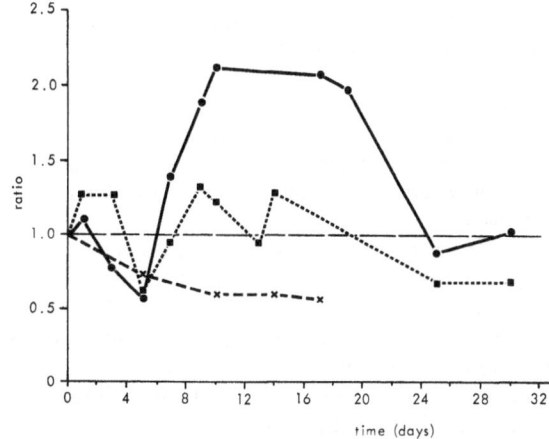

Figure 24-1 The ratios $\left(\frac{\text{Rauscher infected}}{\text{normal}}\right)$ of bone marrow CFU are plotted for various times after infection with Rauscher virus. Femoral marrow CFU-e, CFU-c and BFU-e values are plotted for mice that have been infected with Rauscher virus up to 30 days. The ratio to normal levels was determined from values on corresponding days. The overall normal averages for CFU-e (closed circles), CFU-c (closed squares), and BFU-e (x) were 63,700 ± 7,070; 10,800 ± 830; and 2,900 ± 240 per femur, respectively.

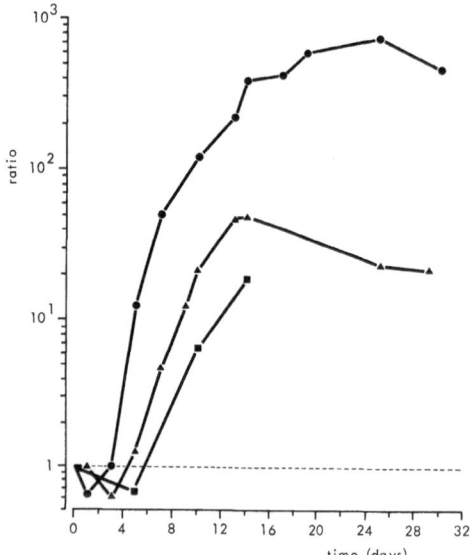

Figure 24-2 Changes in the ratio of *in vitro* CFU are plotted for Rauscher-infected mice at successive days after infection with virus. Splenic ratios of CFU-e (closed circles), CFU-c (closed triangles), and BFU-e (closed squares) are plotted as indicated in Figure 24-1. Normal levels of splenic CFU were determined on corresponding days with the ratio calculated from the overall average of values over the entire experiment. These values for CFU-e, CFU-c, and BFU-e were 119,000 ± 29,000; 3,700 ± 390; and 3,970 ± 690 per spleen, respectively.

pendent on erythropoietin for colony formation.[14,15] However, the present investigation indicates that the dose-response relationship of CFU-e from Rauscher virus-infected bone marrow and normal marrow is comparable over a range from 1 to 0.016 IU of erythropoietin (Figure 24-3). This indicates that responsiveness to erythropoietin is retained for CFU-e from Rauscher virus-infected bone marrow cells. It is also apparent that erythropoietin is necessary for maximal colony formation, as is the case for normal marrow cells.

There appears to be a more critical concentration range for optimal colony formation in the case of Rauscher bone marrow cells. Also, the decline of colony number with decreasing concentration of erythropoietin is more pronounced than that for normal marrow cells. This is in agreement with the finding described by Opitz et al[7].

BFU-e show an altered dose responsiveness to erythropoietin, with colony formation extended to doses at which normal marrow BFU-e are no longer detectable Table 24-2.

Discussion

The kinetic changes of *in vitro* CFU pools indicate that injection of Rauscher virus has a profound effect on hemopoiesis in both spleen and bone marrow. The CFU-s compartment is also affected[16–19] and is presumably instrumental in the expansion of the precursor compartments observed *in vitro*. Although *in vitro* CFU levels become extremely high, the initial exponential phase of growth is characterized by markedly slow doubling times. This phenomenon has also been demonstrated by Okunewick et al.,[19] where CFU-s data shows a doubling time of approximately 46 hr, accompanied by maximal ^{3}H-TdR suicide of only 20.9%. The reason for these kinetic changes remains obscure at present. However, it is apparent that both CFU-s and the *in vitro* CFU progenitor cells are affected; thus one cannot exclude any of these compartments when considering the target for this virus. A detailed review of this subject has been published by OKunewick.[20]

The retention of erythropoietin responsiveness has been well described for the *in vivo* situation and is consistent with that found *in vitro*. CFU-e dose responsiveness to erythropoietin that has been presented here is in agreement with that described by Opitz et al.[7] The discrepancy between these results and those of Nooter and Bentvelzen[15] may be due to different culture conditions and interpretation of benzidine-staining colonies. As discussed elsewhere (pp. 77–80) benzidine stainability is not exclusively a property of erythroid cells.

Erythropoietin dose responsiveness of BFU-e exhibits a markedly lower threshold of sensitivity in Rauscher-infected marrow cells. This may be a consequence of cell cycle redistribution in this situation of heavy and prolonged pressure for erythroid differentiation. Further experiments are being carried

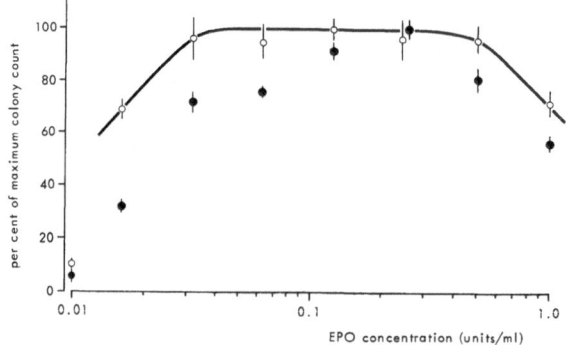

Figure 24-3 Changes in the frequency of CFU-e with varying erythropoietin concentrations are shown, with the maximum colony number equivalent to 100%. CFU-e from bone marrow of normal mice (open circles) and mice infected with Rauscher virus (closed circles) approximately 2 weeks before, are shown for concentrations of 1 to 0.016 IU and no added erythropoietin (Ep). Mean values represent data from three experiments with mean 100% levels equivalent to 320 CFU-e/10^5 nucleated cells for normal marrow and between 325 to 510 CFU-e/10^5 cells for Rauscher virus-infected marrow.

out to investigate this phenomenon in the BFU-e population.

In the following summary, erythropoiesis in Rauscher virus-infected mice and normal mice will be compared.

(a) The *in vitro* progenitor cell kinetics in bone marrow and spleen appear to mimic the events in normal mice following bleeding or erythropoietin injection that has been described by Iscove[21] and Hara and Ogawa.[22] The decrease of BFU-e in bone marrow may reflect BFU-e traffic to splenic tissue where erythropoiesis is favored.

(b) Consistent with our knowledge of normal erythropoiesis are the more pronounced changes of splenic CFU populations as compared to bone marrow. In addition, infection with Rauscher virus causes marked splenomegaly and an accumulation of characteristic proerythroblast-like cells that comprise up to 80% of nucleated cells after the second week. These cells are at a prehemoglobin synthesis stage.

(c) Progressive anemia develops in these mice, although there is apparently a great potential for erythrocyte production. Red cell survival has been demonstrated to be reduced to 50% of normal, following infection with Rauscher virus,[23] which may be due to a direct effect of the virus on erythrocytes.

In accounting for the observed characteristics of this disease it appears to be necessary to postulate the existence of a rate-limiting step or lesion in the maturation pathway of erythropoiesis that leads to an accumulation of immature erythroid cells such that erythrocyte output becomes inadequate in the balance of decreased erythrocyte survival. A scheme of this nature has been proposed by Brommer and Bentvelzen.[16,17]

Work is presently in progress to characterize the level at which these proerythroblast-like cells accumulate. Experimental data suggests that a maturation state nearer the CFU-e level is involved. However, plating efficiencies for splenic CFU-e are low (maximum of 6%) in contrast to the numbers of proerythroblast-like cells (approximately 80% of nucleated spleen cells), which suggests that these two populations are distinct.

Acknowledgments

This work was supported by the Koningin Wilhelmina Fonds, and The Netherlands Organization for the Fight against Cancer.

Virus preparations used in this study were obtained from Dr. C. J. M. Melief, Central Laboratory of the Netherlands Red Cross Blood Transfusion Services, Amsterdam. This virus was originally obtained from Dr. F. J. Rauscher.

References

1. Rauscher FJ: A virus-induced disease of mice characterized by erythrocytopoiesis and lymphoid leukemia. *J Natl Cancer Inst 29:*3, 515, 1962.
2. Brodsky I, Dimitrov NV: Platelet metabolism in Rauscher virus leukemia. *J Natl Cancer Inst 43:*2, 385, 1969.
3. Dunn, TB, Malmgren RA, Carney PG, Green AW: Propylthiouracil and transfusion modifications of the effects of the Rauscher virus in BALB/c mice. *J Natl Cancer Inst 36:*6, 1003, 1966.
4. OKunewick JP, Erhard P: Accelerated clearance of exogenously administered erythropoietin by mice with Rauscher viral leukemia. *Cancer Res 34:*917, 1974.
5. Dunn TB, Green AW: Morphology of BALB/c mice inoculated with Rauscher virus. *J Natl Cancer Inst 36:*987, 1966.
6. Weitz-Hamburger A, Fredrickson TN, LoBue J, Hardy Jr WD, Camiscoli JF, Ferdinand P, Gallicchio V, Gordon AS: Stimulation of erythropoietic differentiation in BALB/c mice infected with Rauscher leukemia virus. *J Natl Cancer Inst 55:*5, 1975.
7. Opitz U, Seidel HJ, Rich IS: Erythroid stem cells in Rauscher virus infected mice. *Blut 35:*35, 1977.
8. Rowe WP, Pugh WE, Hartley, JW: Plaque assay techniques for murine leukemia viruses. *Virology 42:*1136, 1970.
9. Guilbert LJ, Iscove NN: Partial replacement of serum by selenite, transferrin, albumin and lecithin in haemopoietic cell cultures. *Nature 263:*594, 1976.
10. Bradley TR, Stanley ER, Sumner MA: Factors from mouse tissues stimulating colony growth of bone marrow cells *in vitro*. *Aust J Exp Biol Med Sci 49:*595, 1971.
11. Stanley ER, Metcalf D, Maritz JS, Yeo GF: Standardized bio-assay for bone marrow colony stimulating factor in human urine: levels in normal man. *J Lab Clin Med 79:*657, 1972.
12. Hasthorpe S, Hodgson GS: Haemopoietic recovery in spleen and marrow after transplantation of bone marrow from either normal or hydroxyurea treated mice. *Exp Hematol 5:*348, 1977.
13. Gregory CJ, Tepperman AD, McCulloch EA, Till JE: Erythropoietic progenitors capable of colony formation in culture: response of normal and genetically anemic W/W^v mice to manipula-

tions of the Erythron. *J Cell Physiol 84:*1, 1974.
14. Nooter K, Ghio R: Hormone-independent *in vitro* erythroid colony formation by bone marrow cells from Rauscher virus infected mice. *J Natl Cancer Inst 55:*1, 59, 1975.
15. Nooter K, Bentvelzen P: In vitro transformation of murine erythroid cells by Rauscher leukemia virus. *Cancer Lett 1:*155, 1976.
16. Brommer EJP, Bentvelzen P: Interactions between differentiation and leukemogenesis in Rauscher murine leukemia. Unifying concepts of leukemia in Dutcher RM, Chieco-Bianchi L (eds): *Bibl Haematol 39:*929, 1973.
17. Brommer EJP, Bentvelzen P: The haemopoietic stem cell in Rauscher virus-induced erythroblastosis of BALB/c mice. *Europ J Cancer 10:*827, 1974.
18. Seidel H-J: Target-cell characterization for Rauscher leukemia virus *in vivo*. Unifying concepts of leukemia, in Dutcher RM, Chieco-Bianchi L (eds): *Bibl Haematol 39:*935, 1973.
19. OKunewick JP, Phillips EL, Brozovich B: Effect of Rauscher leukemia development on DNA synthesis by hematopoietic CFU-s. *Exp Hematol 4:*143, 1976.
20. OKunewick JP: The role of committed and uncommitted hematopoietic stem cells as targets for Rauscher and Friend leukemia virus. *Biomedicine 26:*152, 1977.
21. Iscove NN: The role of erythropoietin in regulation of population size and cell cycling of early and late erythroid precursors in mouse bone marrow. *Cell Tissue Kinet 10:*323, 1977.
22. Hara H, Ogawa M: Erythropoietic precursors in mice under Erythropoietic Stimulation and Suppression. *Exp Hematol 5:*141, 1977.
23. Brodsky I, Kahn SB: Effect of a leukemia virus on Erythropoiesis. *J Natl Cancer Inst 42:*39, 1969.

Discussion (Chapters 18–24)

Dr. Gallicchio: Have you ever looked at Ep administration in the intact Rauscher animal?

Dr. Hasthorpe: No, I haven't, but Dr. Seidel has looked at that.

Dr. Seidel: We have injected high amounts, 4 units per mouse, in Rauscher virus-infected CBA mice, and we have seen that the reticulocytopenia that usually occurs in CBA mice at day 10, followed by a reticulocytosis, is overcome by Ep injection. The reticulocytopenia was less and the regeneration was quicker, but we haven't done any further characterization.

Dr. Gallicchio: Did it have any effect on life span on the intact animal?

Dr. Seidel: We haven't studied that.

Dr. Gallicchio: We have reported that physiological levels of erythropoietin administration (0.2 units) markedly increased the life span of Rauscher virus-intact animals. These lower Ep doses had a more physiological response than those studies done with higher Ep levels, which seemed to exacerbate the disease process. Therefore, lower Ep doses may have a therapeutic role in Rauscher disease.

Dr. Erslev: I am curious to know the effect of a reduced level of erythropoietin. What happens after hypertransfusion? Do you get remission, or do you prolong the life span?

Dr. Seidel: We did these experiments several years ago. We hypotransfused Balb/c mice after Rauscher virus infection, and got rid of the erythroblastosis. The spleen weight increase was slower. Since these mice die mostly from splenic rupture, the life span is certainly increased by hypertransfusion. In another series, we studied CFU-e and CFU-s in hypertransfused Rauscher virus-infected mice, and the development of CFU-e and CFU-s increase in the spleen was delayed 2 to 4 days, but in principle, at the stem cell level, we did observe the same changes in hypertransfused as well as in normal mice.

Dr. Iscove: How clean was your erythropoietin preparation?

Dr. Urabe: A human urinary erythropoietin preparation with a specific activity of 38 units per milligram was used.

Dr. Rifkind: It seemed to me that the hemin-induced colony that you showed us was very large compared to the colony you showed us as a typical DMSO-induced colony. Do the hemin-induced colonies actually enter terminal differentiation in the sense of acquiring a proliferative capacity limited to five or six generations?

Dr. Urabe: The size of the colonies was different in the culture medium. Some colonies were very large and some colonies were quite small. I don't know exactly how many times the cells divided, but usually during 5 to 7 days of incubation the colony size increased with increased benzidine positive cells.

Dr. Rifkind: Were the hemin-induced colonies mixed colonies regularly?

Dr. Urabe: Usually all the cells were induced into the erythroid series, but sometimes mixed colonies were seen.

Dr. Testa: Did the U 99 cells in the colonies reach the state of nonnucleated cells, after the addition of erythropoietin?

Dr. Urabe: Enucleated cells were not seen.

Dr. Iscove: Can you distinguish between a cell that has synthesized hemoglobin and a cell that has simply taken up hemin from the medium?

Dr. Urabe: I have no technique to distinguish between a cell that has synthesized hemoglobin and the cell that has taken up hemin. But the fully induced cells, such as DMSO-treated wild-type cells or erythropoietin-treated U 99 cells, have similar and characteristic morphology.

Dr. Rossi: Dr. Iscove's question could be answered if one were to add heme to the control replicate culture, say 4 or 5 hours before testing, as opposed to introducing heme a few days before. This could be a way of testing whether there is any specificity or whether you are just staining heme that is present in soluble form in the cell sap. The second comment I wanted to make is that, to the best of my knowledge, nobody has yet reported any kind of enucleation of Friend cells in whatever culture you can think of. Only Yoji Ikawa in Japan has reported formation of erythrocytes in Millipore chambers in animals in which Friend cells were inoculated in Millipore chambers, but they differentiated under *in vivo* conditions, not under *in vitro* conditions.

Dr. Sassa: In my presentation, I should have mentioned that Harvey Eisen, treated $F4N^{+2}$ cells, a variant clone of F4N cells, with hemin and DMSO. These cells completed erythroid differentiation, becoming enucleated erythrocytes. These cells were morphologically normal reticulocytes.

Comment on Dr. Iscove's question about benzidine positivity: We usually find benzidine positive materials as if localized in the nucleus of the cells, if we use the wet benzidine method. We do not understand why it is so, but this particulr localization of the benzidine-positive material in the cells together with the normal morphology of the cells facilitates differentiation between truly positive and false positive cells. We have checked whether hemoglobin is in the nucleus by using a different hemoglobin stain that is based on a different principle, for example, using perchloric acid (PCA) and β-mercaptoethylamine (MA). When cells treated with PCA and MA are exposed to UV light, iron is removed from heme, and a fluorescent product protoporphyrin is visualized. By using this technique, it was found that fluorescence due to heme was also predominantly in the nuclei. So we isolated nuclei from these cells, and then looked for the presence of heme or hemoglobin in the isolated nuclei. We did not find very much heme in the isolated nuclei. The data could be interpreted in two ways: our nuclear isolation method allowed heme to leak out from the nucleus during the isolation procedure, or the staining has a peculiar property so that it shows up better in nuclei. We have also developed a cytographic technique for benzidine staining. If cells are truly benzidine-positive, then they can be completely separated from the benzidine negative cells. When the cells are treated with hemin, there are false positive cells, but their distribution is different from the truly positive cells; thus it is possible to make a distinction between the two cell populations having false and true benzidine-positive materials. Using the cytographic method on benzidine stain, we confirmed Dr. Urabe's finding that cells treated with hemin became truly benzidine-positive.

Dr. Rifkind: Howard Davies, years ago, using Soret band absorption on fixed erythrocytes, demonstrated hemoglobin in the nucleus under conditions in which it was unlikely that it had factitiously migrated there.

Dr. Zanjani: Did your cultures contain any serum?

Dr. Urabe: Yes, 10% serum.

Dr. Murphy: I'd just like to take a moment to iterate something about U 99 that Dr. Urabe has cloned. This has been really a *tour de force* since for the first time we have a cell clone that is capable of proliferation in culture and at least half of which differentiate in the presence of erythropoietin.

Dr. Sassa: Is TPA mitogenic?

Dr. Rifkind: Not to my knowledge.

Dr. Sassa: You mentioned actinomycin D does not affect cyclic AMP level. Is that correct?

Dr. Rifkind: Preliminary data suggest that actinomycin D may behave quite differently from DMSO, and may have very different effects on cell cycle kinetics, for example, and it doesn't appear to have the cyclic nucleotide effect. I think it may be acting at a site distal to the point of induction by polar compounds such as DMSO or HMBA.

Dr. Sassa: If you treat the cells with BUDR together with an inducer, could you prevent the shift of DNA in alkaline density gradient sedimentation?

Dr. Rifkind: That's a very good question, and we haven't tested this. One would expect incorporated BUDR to stabilize some configurations of DNA.

Dr. Rossi: About the ultraviolet experiments, did you measure non-histone chromatin proteins under UV plus or minus DMSO?

Dr. Rifkind: No, we have not examined nuclear proteins under these conditions.

Dr. Hasthorpe: On your last slide, at the commitment level you had "lectin agglutination." Could you just elaborate on that, please?

Dr. Rifkind: A number of years ago we did studies using Con A, looking for a marker for early stages of differentiation. We found, but never published, a very early transient drop in agglutination. This observation has now been established by others. What it means, I have no idea, but I like to think it has something to do with the cell surface effects that lead to a change in cyclic AMP levels.

Dr. Sassa: In collaboration with Harvey Eisen, I looked at the heme pathway enzymes in mutant clones derived from F4N. The appearance of ALA-dehydratase was found to occur just about the time as the induction of spectrin synthesis takes place. Thus, early markers of erythroid differentiation appear to be spectrin synthesis, and certain early enzymes of the heme pathway. Ross et al. also reported that globin mRNA appears early during erythroid differentiation.

Dr. Rifkind: I don't know about that. We get globin messages appearing just about the time we can recognize documented commitment roughly at 24 hours.

Dr. Rossi: We have been doing, concomitantly with Dr. Eisen, some spectrin studies. We see a fourfold increase, in 24 hours, of the amount of spectrin. We didn't study the rate of synthesis; we only studied steady-state conditions by electrophoresis in polyachrylamide gels. We only got onefold increase from the first day to the third day, which is the peak of the accumulation of spectrin. What is also very interesting is the fact that, at variance with globin which keeps going up and being accumulated until the very end of the culture, spectrin decreases from day 3 to day 5. It seems that something is happening in the membrane of these cells, which is different from what happens in red blood cells, where spectrin synthesis takes place early but then it remains as long as the red cell lives.

Dr. Rifkind: We're still in for a lot of trouble interpreting these markers. It is not yet established that there is a transcriptional regulation of globin messenger during induced differentiation. When it comes to something like spectrin, we're going to have a long time before we're ready to determine what is the mechanism for that burst of spectrin accumulation. Is it a transcriptional event, is it a processing event, or is it a translation of existing mRNA?

Dr. Sassa: Do you have any idea how many poly As are present on the 3' end.

Dr. Rossi: You mean the length of the poly-A stretches? No, we don't. We have to check that.

Dr. Sassa: Concerning the larger size of β-globin messenger RNA, you didn't discuss the capping of the messenger RNA at the 5' end. Couldn't it be another possibility that may change the size of mRNA?

Dr. Rossi: Sure, capping could be the other answer.

Dr. Zanjani: Harvey Preisler, in collaboration with us, found that if you expose the Friend leukemic cells to DMSO, and subsequently expose them to erythropoietin, you observe an increase in the amount of heme that is formed as measured by incorporation of radioactive iron into heme. In addition, Ken Goldstein in our laboratory demonstrated that adding erythropoietin to the DMSO-treated cells induced erythroid colony formation *in vitro* so that there was a significant increase over the control cultures. From your last slide, it appears to me that only cells that don't respond to DMSO tend to respond to erythropoietin, and yet in the studies that Harvey Preisler supervised, the DMSO effect was also there, but the treated cells now developed the ability to respond to erythropoietin as well.

Dr. Sassa: The data I have shown were for cells of one of Dr. Urabe's DMSO-resistant clone's, "U 99". These cells undergo erythroid differentiation when they are treated with erythropoietin and do not need DMSO pretreatment for this event to occur. I would interpret Preisler's data to mean that DMSO pretreatment triggers certain cells to complete the early erythroid function, and those cells now express erythropoietin-specific receptors on the surface of the cells. Dr. Urabe's data showed that cells of U 99 display response to erythropoietin without pretreatment by DMSO. These data may be interpreted that this particular clone of cells may have expressed an erythropoietin specific-receptor in a constitutive manner.

Dr. Rifkind: Some of the mutants might be behaving as though they were mitochondrial variants. Do you have any evidence to suggest they they are comparable to some classical mitochondrial variants?

Dr. Sassa: No, we have not yet looked at mitochondria specifically. However, it does not look like any of our DMSO-resistant clones are classical mitochondrial mutants, like "pokey" because in the case of pokey mutants all mitochondrial enzymes should be deficient. Therefore, suppose if all cells are treated with ALA, then ALA should be accumulated as protoporphyrin 9. It partially happens in the case of cells taken from human or bovine protoporphyria, although they are not devoid of mitochondria, but they have a 50% deficiency of ferrochelatase activity. None of Friend cell variants accumulate exclusively protoporphyrin when incubated with ALA. Thus, certain defects at late stages of the heme biosynthetic pathway can be a plain ferrochelatase deficiency. Therefore, it may be a more complex story.

Dr. Murphy: Your demonstration of the radioactive iron-containing hemin directly being incorporated within heme is remarkable. Would you care to speculate on its physiological role? Is this something one would expect to see, and if so, would it be expected to be seen *in vivo* at the level you demonstrated?

Dr. Sassa: I don't know whether this hemin effect shown in Friend virus-transformed cells also occurs in normal erythroid precursor cells incubated *in vitro* with hemin. Induction of heme pathway enzymes by hemin in Friend virus-transformed cells may be explained by the fact that hemin stimulates the synthesis of a variety of proteins in a cell-free system, which was demonstrated by London et al. They reported that hemin stimulates the initiation of the translation of mRNAs added to a cell-free protein synthesizing system by interfering with the action or the formation of a "soluble translational inhibitor" on protein synthesis. The inhibitor is now identified to be cyclic AMP-independent protein kinase.

Dr. Rossi: Since I understand you're working with Dr. Ostertag too, I wonder whether he has tested, or did you make any kind of correlation between the early burst of viral I RNA synthesis, which he describes in the very early hours after DMSO induction, and your early changes, as opposed to globin mRNA, which he interprets as an early change?

Dr. Sassa: No, I can't answer that.

Dr. Rossi: And this is of some importance because he is trying to make a point that maybe some viral genes may be involved in the triggering of the whole system, and that's why it would be important for him to test it.

Dr. Peschle: I wish to comment on your very interesting concept of two populations of CFU-e and one of BFU-e in Friend virus-injected mice: one CFU-e pool is Ep-dependent, the second one Ep-independent, whereas the BFU-e are Ep-dependent. As you pointed out in your presentation, even Ep-independent CFU-e growth occurs in the presence of low levels of Ep in the serum. Would you consider the possibility that the "Ep-independent" CFU-e population is more sensitive to Ep? I also wonder if a similar enhancement of sensitivity to Ep might occur at the BFU-e level. Did you observe BFU-e growth with relatively low dosages of Ep?

Dr. Seidel: We did Ep dose-response curves for the BFU-e's of Friend virus-infected mice, and they were completely normal. With respect to the Ep independency of CFU, it is a complete change in the Ep sensitivity. What we have after Friend virus infection, I called "independent" colonies. It's hard to believe that no intermediate population is ever seen, a population with an intermediate Ep sensitivity. These intermediate sensitivities were not found. This was shown in studies where we used mice with 50% Ep-independent bone marrow cells. The Ep-dose-response curve of those requiring Ep was normal. The other point is that we can't exclude at the moment that there is among the Ep-independent CFU-e a population with a sensitivity to some "logs" less Ep. That we can't exclude from these experiments, and I'm waiting, like many others, for the availability of the serum-free cultures to do this study and for it to really be conclusive. But from the colony numbers and whatever you see in these cultures without addition of Ep, it would be much more sensitive than what has been observed before. It seems to be an all-or-none phenomenon. I think it's a new population, which is induced and which should be compared for instance with fetal liver cells. In our group, Ivan Rich has studied fetal liver cells very precisely and there seems to be no relationship in their Ep sensitivity.

Dr. Gallicchio: Did you do any Ep dose-response curve for "Rauscher" infection? And what effect if any did you see on CFU-e?

Dr. Seidel: We have recently published the dose-response curve after Rauscher virus infection, and I didn't include it in this talk because Susan Hasthorpe will later present those data. In my hands, the dose-response curve for Ep in "Rauscher" CFU-e is completely normal.

Dr. Rossi: Wouldn't Dr. Peschle's question be solved if you were to treat your cells with a vast excess of anti-erythropoietin, so as to make it unlikely that you have any erythropoietin left in your small amount of serum?

Dr. Seidel: Liao and Alexrad did the anti-Ep study, published 2 years ago, and I believe that these colonies grew also with anti-Ep.

Dr. Murphy: Dr. McLeod, would you like to comment regarding Arthur Axelrad's use of anti-Ep in conjunction with that last question?

Dr. McLeod: Dr. Axelrad used Dr. Lange's antisera, and I think the results were quite clear.

Dr. Zanjani: Are you suggesting that the polycythemia that develops in Friend virus-infected mice is erythropoietin-dependent?

Dr. Seidel: No, certainly not. The increase in reticulocytes and hematocrit is certainly Ep-independent. The only thing that I suggested from our target cell studies when we infected mice is that the target cell for the virus might well be in the compartment influenced by the erythropoietic state of the animal.

Dr. Gordon: Dr. Sassa, when are you going to make these cells form hemoglobin?

Dr. Sassa: Let me put it this way. Although these enzymes in the heme pathway are inducible in lymphocytes, the extent of induction is very small compared to the induction occurring in erythroid cells; for example, Friend mutant clones that cannot make hemoglobin have much higher levels of these enzymes. Induction of heme pathway enzymes in the lymphocytes is not in a comparable range as in erythroid cells and certainly I did not find hemoglobin in them. Induction of heme pathway enzymes in mitogen-treated lymphocytes may be required to meet with an increased demand for cytochrome formation that probably takes place in these metabolically activated cells.

25 The Role of Serum Inhibitors of Erythroid Colony-Forming Cells in the Mechanism of the Anemia of Renal Insufficiency

James W. Fisher, Yasuhico Ohno, J. Barona, Maria Martinez, and Arvind B. Rege

Introduction

Uremic toxins have been the subject of investigation for several years to determine their role in the signs and symptoms of uremia. Substances that are known to accumulate in body fluids to produce symptoms of uremia in patients with renal insufficiency were outlined by Schreiner and Maher in 1961[1] and Bergstrom and Bittar in 1969.[2] The success of regular hemodialysis in relieving the symptoms of uremia indicates that toxic substances below 5,000 daltons molecular weight must be removed. Jebsen et al.[3] found that prolonged dialysis can reverse the peripheral neuropathy independent of changes in BUN and creatinine. Thus, there appears to be very little correlation between the toxic manifestations of uremia and plasma levels of creatinine, urea, and uric acid. Babb et al.[4] postulated that uremic products in the middle molecular range of 1,000 to 2,000 daltons may be the cause of the toxic manifestations of neuropathy. It is the purpose of this chapter to examine plasma fractions below 1,000 daltons to attempt to characterize the uremic toxins responsible for the anemia of renal insufficiency. In addition, assay of sera from several predialysis patients with anemia of renal insufficiency for erythropoietin (Ep) using a sensitive radioimmunoassay for erythropoietin was carried out. The anemia of renal insufficiency is a hypoproliferative state of the bone marrow associated with a normocytic normochromic anemia. Anemia is one of the most common presenting symptoms in patients with end-stage renal disease and is usually alleviated to some degree by hemodialysis. However, a large number of patients require frequent blood transfusions even when being chronically dialyzed. Improvements in ferrokinetics and erythropoiesis have been reported in patients after dialysis [5-7] without an increase in plasma erythropoietin levels, indicating that dialysis may be removing toxic substances that are involved in the mechanism of the suppressed erythropoiesis in renal disease. Androgen therapy[8-13] and infusions of erythropoietin-rich plasma[14,15] have been reported during the past several years to relieve the anemia of renal insufficiency.

The several factors postulated to play a role in the anemia of renal insufficiency are listed:

1. Inadequate production of erythropoietin (Ep) relative to the increased demand for Ep created by the anemia.
2. Inhibitors of heme synthesis and/or erythroid colony-forming cells (CFU-e).
3. Shortened red cell life span that may be due to hemolytic factors.

Erythropoietin deficiency is probably the most important factor in the anemia of renal insufficiency that is related to the failure of the kidney to produce a sufficient amount of erythropoietin to meet the increased demands for new red cells created by the shortened red cell life span and inhibitors of erythropoiesis. Inhibitors of heme synthesis[16-20] and/or erythroid colony-forming cells (CFU-e)[18,21,22] have been reported by several investigators to be involved in the mechanism of the anemia of renal disease. Fisher et al.[16-18] and Wallner et al.[19,20] have demonstrated an inhibitor of heme synthesis in the plasma of patients with anemia of renal insufficiency utilizing human and canine bone marrow cultures. Ohno et al. have recently reported an inhibitor of CFU-e in the sera of rabbits using a 5/6th nephrectomy model[22] and in the sera of patients with anemia of renal disease.[23] This inhibitor was partially removed by hemodialysis.[23] In addition, Freedman et al.[21] have reported an inhibitor of CFU-e in plasma of anemia-uremic children. Hemolysis and shortened red cell life span play a variable role in the anemia of renal disease.[24,25] We have reported over the past few years inhibitors of heme synthesis in the acutely nephrectomized rabbit[26,27] and in some but not all patients with anemia of renal disease.[18,28] This inhibitor appears to suppress heme synthesis directly in the rabbit and human bone marrow cultures but does not inhibit the effects of erythropoietin in exhypoxic polycythemic mice.[18,28] Because of the variation in the results obtained with the heme synthesis model we have turned to studies of the effects of the uremic toxins on the erythroid stem cell compartment, (CFU-e).

Materials and Methods
Models for Studies of the Effects of Uremic Toxins on Erythropoiesis

The two models of anemia of renal insufficiency we have utilized in our studies are the 5/6th nephrectomized rabbit 7, 14, 21 and 35 days postnephrectomy

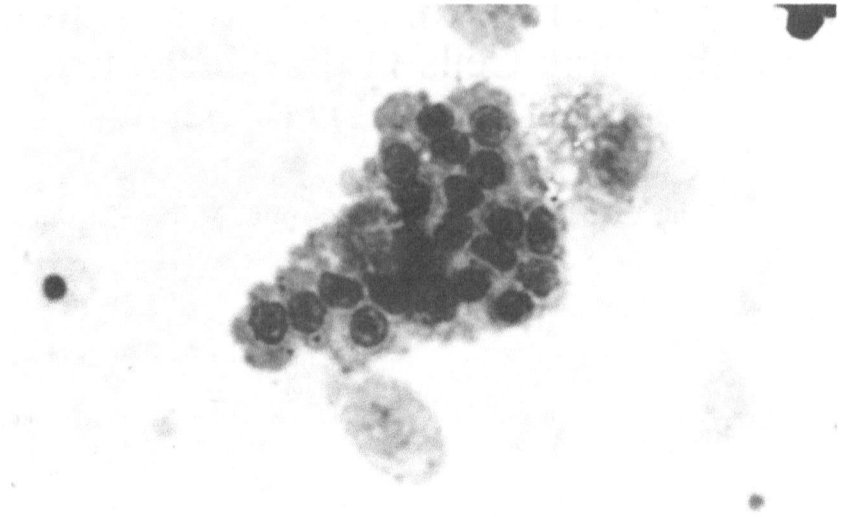

Figure 25-1 Photomicrograph of a 4-day erythroid colony (CFU-e) from a normal rabbit bone marrow culture stained with benzidine and counterstained with Giemsa. ×100.

and patients with end-stage renal disease both pre- and postdialysis.

The chronic uremic rabbit was prepared according to a modification of the technique of Chanutin and Ferris.[29] Using this technique two-thirds of the right kidney was removed through a flank incision in female New Zealand albino rabbits anesthetized with pentobarbital. The left kidney was removed 10 to 14 days later. Sham-operated controls were studied in parallel. Blood was collected 7, 14, 21 and 35 days after the second operation via cardiac puncture and serum separated for studies of uremic toxins.

Erythropoietin Bioassay

Erythropoietin (Ep) was assayed via a modification of the method of Cotes and Bangham[30] using radioactive iron (^{59}Fe) incorporation into newly formed red blood cells (RBC) of exhypoxic polycythemic mice. HAM/ICR strain virgin female mice (25 to 28 gm) were made polycythemic by exposure for 22 hr/day for 2 weeks to 0.42 atm pressure. The mice were removed from the hypobaric chamber and on posthypoxic days 6 and 7 injected (subcutaneously) with one-half of the total dose of saline (0.5 ml, 2×), human urinary Ep standardized against the International Reference Preparation (IRP-B) or the serum sample (0.5 ml, 2×). On the eighth posthypoxic day each mouse received 0.5 µCi of radioactive iron (^{59}Fe citrate) i.v. via the tail vein. Two days later, posthypoxic day 10, the mice were exsanguinated by cardiac puncture and 0.5 ml of blood was counted per mouse with a Packard Auto-Gamma spectrometer and the 48-hr % ^{59}Fe incorporation into newly formed RBC determined. Five mice were injected with each sample and the mean 48-hr % ^{59}Fe incorporation rate into RBC was used as an indication of the erythropoietic activity of the sample when compared with the response to the IRP.

Plasma Clot Techniques for Erythroid Colony Formation (CFU-e and BFU-e)

The details of the technique for CFU-e and BFU-e are shown in Appendices 1 and 2.

Figure 25-1 shows a photomicrograph of an erythroid colony (CFU-e) produced by CFU-e in the normal rabbit bone marrow according to the method of McLeod et al.[31] Figure 25-2 shows a photomicrograph of a "burst" type of erythroid colony produced by the BFU-e according to the method of Heath et al.[32] Figure 25-3 shows the dose-response curve for CFU-e to increasing dosages of erythropoietin.

Statistical Analysis Techniques

The effect of a single treatment at several different time intervals as compared with a control interval was analyzed by the multiple comparison method of Dunnett.[33] All other statistical evaluations were made utilizing the Student's t test for paired and unpaired observations or the ANOVA test for paired comparisons.

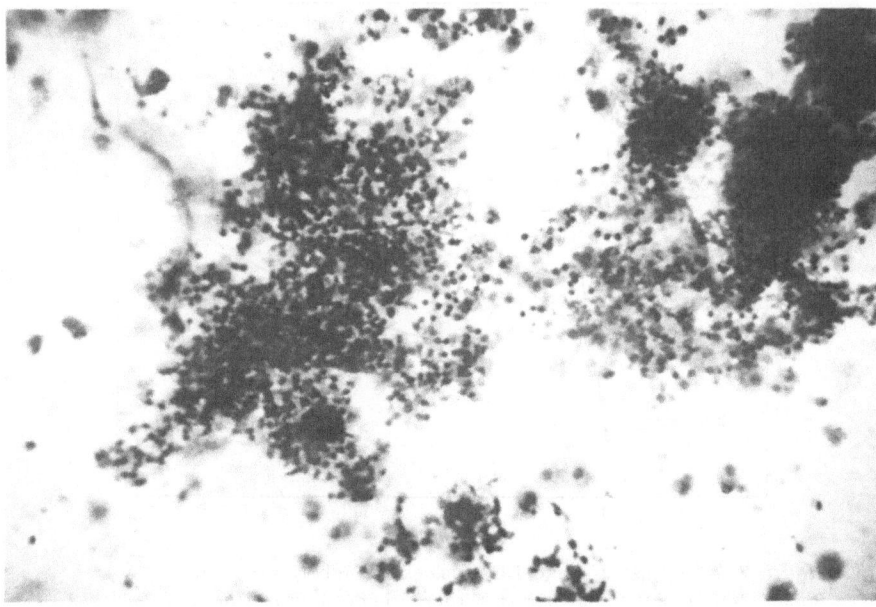

Figure 25-2 Photomicrograph of an 8-day burst-forming unit type of erythroid colony (BFU-e) from a normal rabbit bone marrow culture stained with benzidine and counterstained with Giemsa. ×100.

Studies on Erythroid Colony-Forming Cells in 5/6th Nephrectomized Rabbit Model

Using our 5/6th nephrectomy model, rabbits were made chronically uremic by removing two-thirds of one kidney in the first operation followed by complete removal of the contralateral kidney in the second operation after 14 days. Note in Figure 25-4 the rapid increase in BUN and the slow decline in hematocrit in two representative rabbits over a 50-day period following 5/6th nephrectomy. The animals appear to regenerate their renal mass rather completely after 50 days. However, we selected a 35-day 5/6th nephrectomy to carry out our studies of inhibitors of CFU-e because uremia and anemia were sustained during this period. Note in Figure 25-5 that over the 35-day period of 5/6th nephrectomy the creatinine levels gradually increased and the hematocrit values gradually decreased. On the other hand, the creatinine levels and hematocrit in the sham-operated animals remained relatively constant during the 35-day period. The mean hematocrit of the 5/6th nephrectomized rabbits was significantly ($p < .01$) less than that of the sham-operated controls. The mean creatinine levels in the 5/6th nephrectomized rabbits was significantly higher than that of the sham-operated controls during the 35-day period. Erythropoietic activity in 1.0 ml of sera from the nephrectomized and the sham-operated rabbits at all times were not detectable when assayed in the exhypoxic polycythemic mouse assay.[30] Note in Figure 25-6 the effect of normal rabbit sera and sera from 5/6th nephrectomized uremic rabbits on the numbers of erythroid colony-forming cells in bone marrows 7, 14, 21, and 35 days post-5/6th nephrectomy. Note the significant ($p < .01$) decrease in CFU-e in bone marrows in the 5/6th nephrectomized rabbit at 7 and 35 days. The renal mass gradually increased in size over the 35-day period yet the inhibitor of CFU-e was found to be significantly ($p < .01$) increased after 35 days. Figure 25-7 shows the number of CFU-e in the bone marrows from sham-operated and 35-day chronic uremic rabbits. The mean number of CFU-e in the chronically anemic uremic bone marrows was significantly

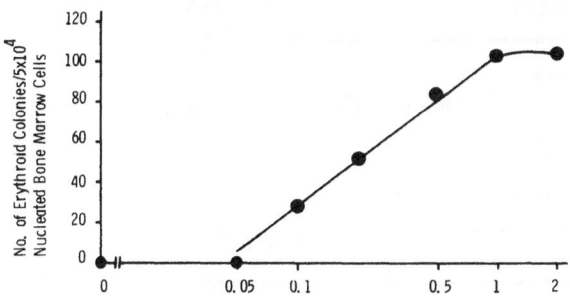

Figure 25-3 Dose-response curve for CFU-e erythroid colonies with increasing concentrations of human urinary erythropoietin in normal rabbit bone marrow cultures.

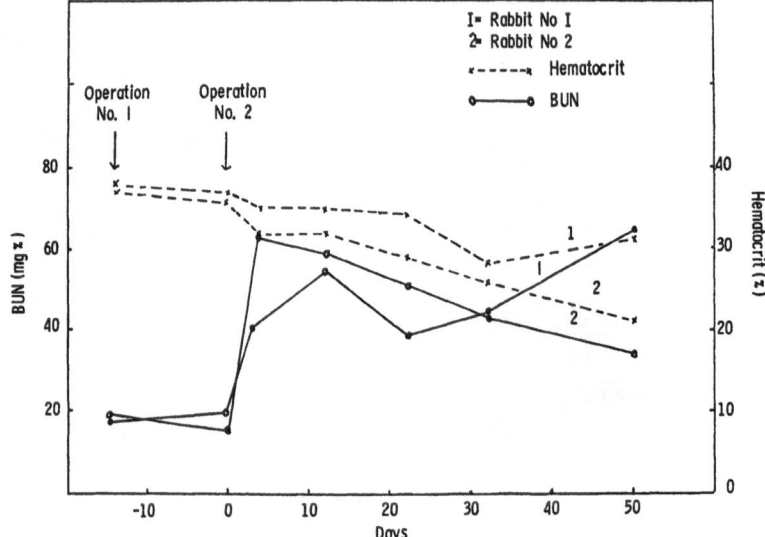

Figure 25-4 Hematocrit and BUN values in 5/6th Nephrectomized Uremic Rabbit Model over a 50-day period. (From Fisher et al. *Proc Clin Dialysis Trans Forum* Vol. VI, p. 42–49, 1976).

($p < .01$) less than that of the sham-operated rabbits 35 days after the operation.

Studies on Erythroid Colony-Forming Cells (CFU-e and BFU-e) Using Sera from Normal Human Subjects and Patients with Anemia of Renal Insufficiency

Patients with anemia associated with renal insufficiency admitted to the New Orleans Charity Hospital Renal Dialysis Center were studied before and following 16 weeks hemodialysis for the presence of inhibitors of CFU-e and BFU-e. Note that the hematocrits ranged from 14 to 28%, creatinine 6.5 to 25.0 mg %, BUN 97 to 258 mg %, and creatinine clearances were 0 to 5.0 ml/min (Table 25-1). These patients were undialyzed and the plasmas were removed the first time that they came in with anemia and uremia associated with renal disease and prepared for hemodialysis. Table 25-1 shows the clinical data on the nine patients selected for study. Figure 25-8 shows the effects of sera from seven hematologically normal

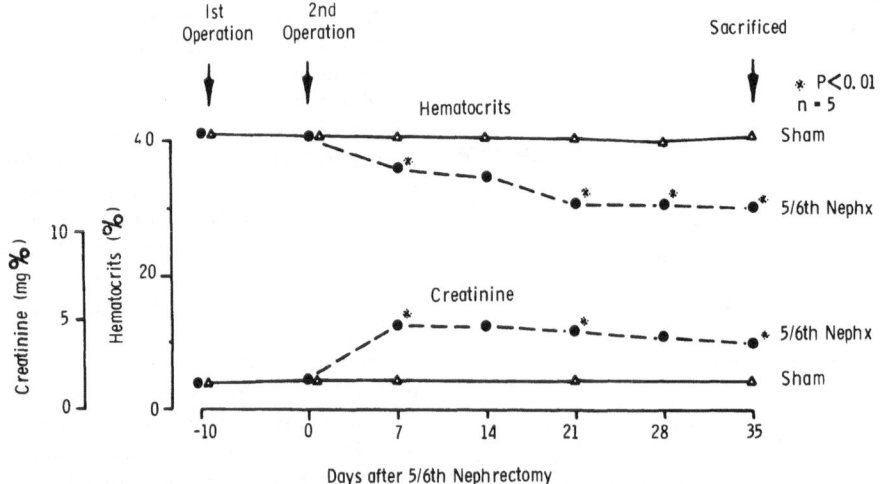

Figure 25-5 Hematocrit and creatinine values in sham-operated and 5/6th nephrectomized uremic rabbit model over a 35-day period.

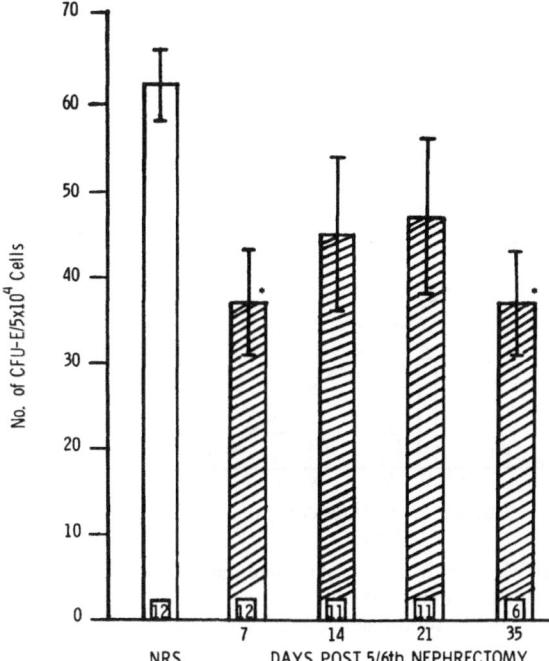

Figure 25-6 Erythroid colony-forming cells (CFU-e) in normal rabbit bone marrow. Cultures incubated with sera from sham-operated rabbits and rabbits 7, 14, 21, and 35 days following 5/6th nephrectomy. Clear bar = NRS: Sham-operated rabbit sera. Striped bar = URS: 5/6th nephrectomized uremic rabbit sera. I = SEM. The number at the bottom of each bar represents the number of rabbits in each group. $p < 0.001$.

human subjects (hematocrits 43 to 48) and sera from nine predialysis patients with anemia and uremia associated with chronic renal disease. The sera were heat-inactivated at 56°C for one-half hr to remove any cytotoxic agents. Note the variation in CFU-e with normal human sera as compared to the significantly ($p < .001$) reduced response to the sera from the anemic uremic patients. The mean CFU-e in the rabbit marrow indicates that the sera from the anemic uremic patients was significantly less ($p < .001$) than

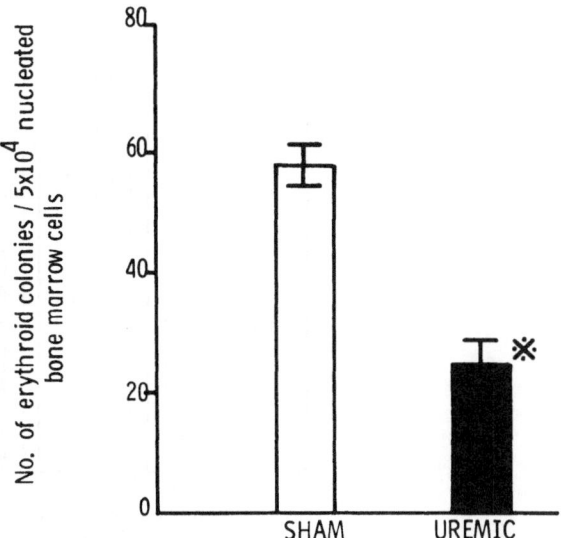

Figure 25-7 Erythroid colony-forming cells in bone marrow cultures from sham-operated and uremic (35 days post-5/6th nephrectomy) rabbits. n = 5; I = SEM; dotted x = $p < 0.01$.

that of the CFU-e in rabbit marrow treated with sera from the seven normal human subjects. These data indicate the presence of an inhibitor of CFU-e in the sera of patients with anemia associated with uremia. However, the possibility cannot be ruled out that the patients with anemia of uremia are deficient in some essential substance that supports erythroid colony growth.

It was of interest to us to next study the effects of normal human sera and sera from predialysis patients with anemia of renal disease on burst-forming units (BFU-e). Note in Figure 25-9 that the mean BFU-e in sera from nine predialysis patients with anemia of renal disease were significantly ($p < .01$) less than that of eight normal human sera. These data support the hypothesis that the inhibitor of erythropoiesis is acting on the stem cell compartment either at the level of the BFU-e or on the pluripotent stem cell

Table 25-1 Clinical data on predialysis anemic-uremic patients used in inhibitor and radioimmunoassay studies.

Patient	Diagnosis	Hematocrit (%)	BUN (mg %)	Creatinine (mg %)	Cr/cl (ml/min)
G.M.[a]	Chronic Pyelonephritis	19	160	16.1	—
E.S.[a]	Nephrosclerosis	24	168	25.0	—
J.V.	Collagen Disease	26	186	25.0	—
A.R.[a]	Nephrosclerosis	22	—	6.5	—
D.G.[a]	Acute Renal Failure	14	>100	>15.0	<5.0
W.S.	Nephrosclerosis	28	131	10.8	4.3
G.S.[a]	Chronic Pyelonephritis	26	97	7.3	4.5
G.R.	Glomerulonephritis	24	80	24.8	—
E.N.	Nephrosclerosis	26	258	20.6	—

[a] Serum assayed for erythropoietin using radioimmunoassay.

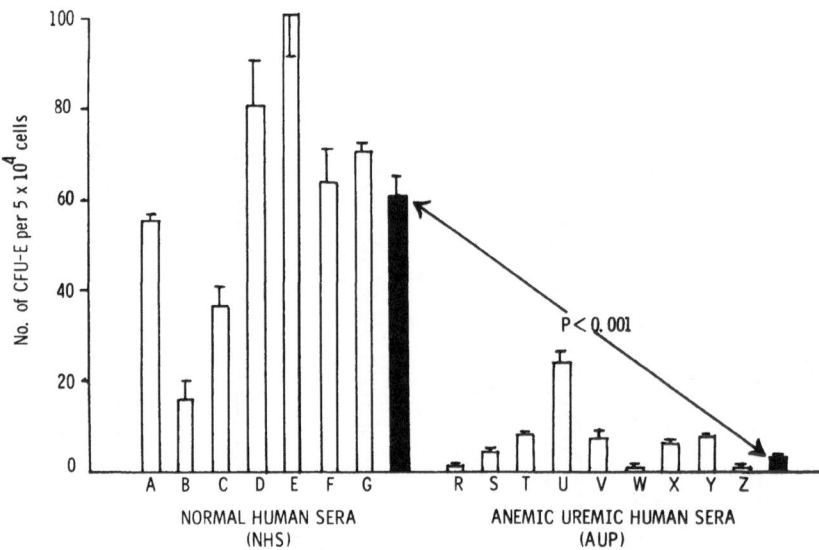

Figure 25-8 Erythroid colony-forming cells (CFU-e) in normal rabbit bone marrow cultures treated with sera from hematologically normal human subjects and predialysis patients with anemia of renal insufficiency. T = SEM. Solid bar = mean number of CFU-e in NHS and AUP.

compartment (CFU-s) to deplete the CFU-e compartment. This depletion of the CFU-e compartment probably subsequently leads to a hypoproliferation of the heme-synthesizing nucleated erythroid cell compartment.

We were next interested to learn if hemodialysis is effective in removing the uremic toxins that suppress the bone marrow stem cell compartment. One of the patients demonstrated to possess significant CFU-e inhibitory activity in his predialysis serum was placed on hemodialysis (cupraphane membrane) 3 days per week (6 hr hemodialysis each day) for a period of 16 weeks. In that we have demonstrated previously that the inhibitor of heme synthesis was localized in a

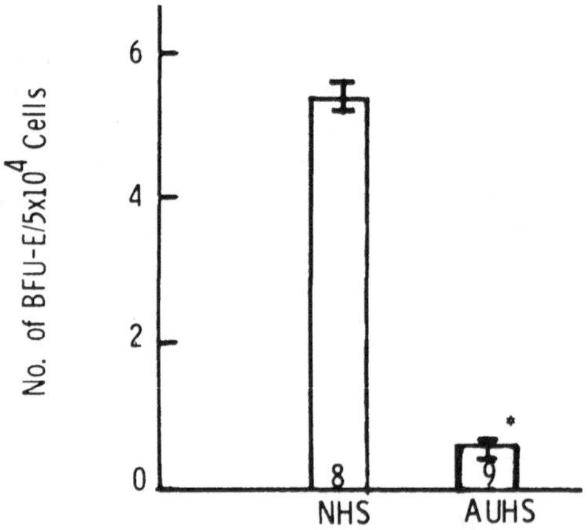

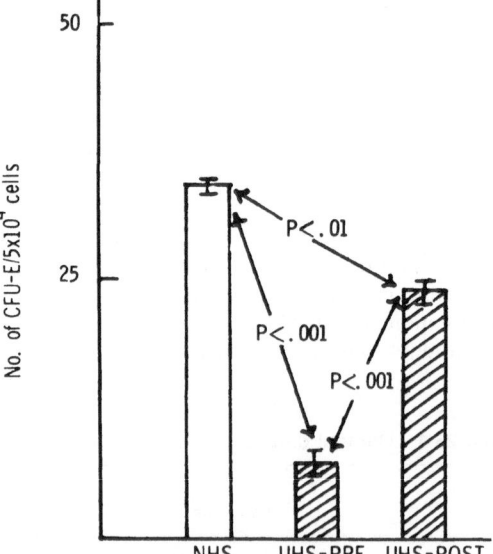

Figure 25-9 Burst-forming unit (BFU-e) colonies in normal rabbit bone marrow cultures treated with sera from normal human subjects and predialysis patients with anemia of renal insufficiency. NHS = Normal human sera. UHS = Uremic human sera (predialysis). I = SEM. The number at the bottom of each bar represents the number of anemic uremic patients or normal human volunteers. $p < 0.001$.

Figure 25-10 Erythroid colony-forming cells (CFU-e) in normal rabbit bone marrow cultures with serum fractions from a normal human subject and a patient before and after 16 weeks of hemodialysis. NHS = Low M.W. [<1,000 daltons] serum fraction from a normal human subject. UHS-PRE = low mol wt [<1,000 daltons] serum fraction from an anemic uremic patient before dialysis. UHS-POST = low mol wt [<1,000 daltons] serum fraction from a uremic patient after 16 weeks hemodialysis (3×/week). I = SEM.

fraction separated by membrane fractionation less than 1,000 daltons,[12,18,28] we fractionated the sera prior to studies on CFU-e in rabbit bone marrow culture. As indicated in Figure 25-10 the less than 1,000 dalton fractions of normal human sera, predialysis human sera, and 16 weeks postdialysis uremic sera were studied as to their effects on CFU-e in normal rabbit bone marrow cultures. Significant ($p < .01$) inhibition of CFU-e was seen with this low molecular weight fraction of predialysis uremic sera when compared with that of normal human sera. However, after 16 weeks on hemodialysis the postdialysis sera had significantly ($p < .01$) less inhibitory effect on CFU-e than that of the predialysis uremic sera (Figure 25-10). Even though significant inhibition of CFU-e was seen with the postdialysis serum fraction, it seems clear that hemodialysis is capable of removing, at least in part, the inhibitor of CFU-e and that this substance is in a low molecular weight fraction less than 1,000 daltons.

Summary

Inhibitors of erythroid colony-forming cells (CFU-e and BFU-e) have been demonstrated in the sera of predialysis patients with anemia of renal insufficiency. The inhibitor of CFU-e was removed in part by hemodialysis and appears to be a low molecular weight substance of less than 1,000 daltons. Inhibitors of CFU-e were found to gradually increase in the sera of rabbits during the first 35 days following 5/6th nephrectomy. It is postulated from these studies that relative erythropoietin deficiency and inhibitors of the erythroid stem cell compartment play a major role in the mechanism of the anemia of renal insufficiency.

Acknowledgment

This study was supported by USPHS grant AM-13211.

References

1. Schreiner GE, Maher J: *Uremia*. Springfield, Illinois, Charles C. Thomas, 1961.
2. Bergstrom J, Bittar EE: The basis of uremic toxicity, in Bittar EE, Bittar N (eds): *The Biological Basis of Medicine* Vol. 6:495. New York, Academic Press, 1969.
3. Jebsen RH, Tenckhoff H, Hoult, JC: Natural history of uremic polyneuropathy and the effects of dialysis. *New Engl J Med* 277–327, 1967.
4. Babb AL, Farrell PD, Uvelli DA, Scribner BH: Hemodialyzer evaluation by examination of solute molecular spectra. *Trans Am Soc Artif Intern Org* 18:98, 1972.
5. Berry ER, Rambach WA, Alt HL, Del Greco F: Effect of peritoneal dialysis on erythrokinetics and ferrokinetics of azotemic anemia. *Soc Artif Intern Org* 10:415–421, 1964.
6. Eschbach JW, Funk D, Adamson J, Kihn I, Scribner BH, Finch CA: Erythropoiesis in patients with renal failure undergoing chronic dialysis. *New Engl J Med* 276:653–658, 1967.
7. Mann DL, Donati RM, Gallagher, NI: Erythropoietin assay and ferrokinetic measurements in anemic uremic patients. *JAMA* 194:1321–1322, 1970.
8. DeGowin RL, Lavender AR, Forland M, Charleston D, Gottschalk A: Erythropoiesis and erythropoietin in patients with chronic renal failure treated with hemodialysis and testosterone. *Ann Intern Med* 72:913–918, 1970.
9. Eschbach JW, Adamson JW: Improvement in the anemia of chronic renal failure with fluoxymesterone. *Ann Intern Med* 78:527–532, 1973.
10. Richardson JR Jr, Weinstein MB: Erythropoietic response of dialyzed patients to testosterone administration. *Ann Intern Med* 73:403–407, 1970.
11. Shaldon S, Koch K, Oppermann F, Patyna WD, Schoppe W: Testosterone therapy for anaemia in maintenance dialysis. *Br Med J* 3:212–215, 1971.
12. Shaldon S, Patyna WD, Kaltwasser P, Werner E, Koch KM, Schoppe WE: The use of testosterone in bilateral nephrectomized dialysis patients. *Trans AM Soc Art Intern Org* 17:104–107, 1971.
13. Stuckey WJ, Fisher JW, Lindholm D, Beltran G, Lertora JJL: *The 5th Annual Contractor's Conference, Artificial Kidney Progress* USPHS-NIAMDD, Bethesda, Maryland pp. 157–158, 1972.
14. Esser U, Muller W, Brunner E: *Klin Wschr* 51:1005–1009, 1973.
15. Larsen O, Andree, JP, Lassen NA: *Ugeskr Laeger* 125:435–441, 1963.
16. Fisher JW, Lertora JJL, Lindholm DD, Tornyos K, Moriyama Y: Erythropoietin production and inhibitors in serum in the anemia of uremia, in *Proceedings of the Clinical Dialysis and Transplant Forum.* 3:22–32, 1973.
17. Fisher JW, Moriyama Y, Rege AB, Tornyos K, Lertora JJL: The role of inhibitors of heme synthesis and bone marrow erythroid colony forming cells in the mechanism of anemia of renal insufficiency, in *Proceedings of the Clinical Dialysis and Transplant Forum.* 4:141–151, 1974.
18. Fisher JW, Foley JE, Moriyama Y, Ohno Y, Modder B, Lertora JJL: Studies on the mechanism of the anemia of renal insufficiency, in *Pro-*

ceedings of the Clinical Diaalysis and Transplant Forum. 6:42–49, 1976.
19. Wallner SF, Ward HP, Vautrin R, Alfrey AC, Mishell J: The anemia of chronic renal failure: *In vitro* response of bone marrow to erythropoietin. *Proc Soc Exp Biol Med 149:*939–944, 1975.
20. Wallner SF, Kurnick JE, Ward HP, Vautrin R, Alfrey, AC: The anemia of chronic renal failure and chronic diseases: *In vitro* studies of erythropoiesis. *Blood 47:*561–569, 1976.
21. Freedman MH, Saunders EF: Erythroid stem cell studies in chronic renal failure, in *The 16th International Congress of Hematology,* Kyoto, Japan. (abs) p. 6, 1976.
22. Ohno Y, Fisher JW: Inhibition of bone marrow erythroid colony forming cells (CFU-e) by serum from chronic anemic uremic rabbits. *Proc Soc Exp Biol Med 156:*56–59, 1977.
23. Ohno Y, Fisher JW, Barona J: Inhibitors of erythroid colony forming cells in sera of chronic renal failure patients. *J Lab Clin Med* (submitted for publication).
24. Castaldi PA, Rozenberg MC, Stewart JH: The bleeding disorder of uremia. A qualitative platelet defect. *Lancet 2:*66, 1966.
25. Shaw AB: Haemolysis in chronic renal failure. *Br Med J 2:*213, 1967.
26. Moriyama Y, Fisher JW: Effects of erythropoietin on erythroid colony formation in uremic rabbit bone marrow cultures. *Blood 45:*659–664, 1975.
27. Moriyama Y, Rege A, Fisher JW: Studies on an inhibitor of erythropoiesis. II. Inhibitory effects of serum from uremic rabbits on heme synthesis in rabbit bone marrow cultures. *Proc Soc Exp Biol Med 148:*94–97, 1975.
28. Fisher, JW, Moriyama Y, Lertora JJL: Erythroid colony forming cells in anemia associated with renal disease, in *The 16th International Congress of Hematology,* Kyoto, Japan. *Excerpta Medica* pp. 1013–1018, 1977.
29. Chanutin A, Ferris EB: Experimental renal insufficiency produced by partial nephrectomy. *Arch Intern Med 49:*767, 1932.
30. Cotes PM, Bangham DR: Bio-assay of erythropoietin in mice made polycythemic by exposure to air at a reduced pressure. *Nature (Lond) 191:*1065, 1961.
31. McLeod DL, Shreeve MM, Axelrad A: Improved plasma clot system for production of erythrocytic colonies in vitro: Quantitative bioassay method for CFU-e. *Blood 44:*517–534, 1974.
32. Heath DS, Axelrad AA, McLeod DL, Shreeve MM: Separation of the erythropoietin-responsive progenitors BFU-e and CFU-e in mouse bone marrow by unit gravity sedimentation. *Blood 47:*777–792, 1976.
33. Dunnett CW: A multiple comparison's procedure for comparing several treatments with a single control. *J Am Stat Assoc 50:*1096–1121, 1955.

26 Modulatory Effect of Macrophages on Erythropoiesis

Martin J. Murphy, Jr. and Akio Urabe

Introduction

The existence of "erythroblastic islands" in the bone marrow has been known for many decades. With the electron microscope, the erythroblastic islands are clearly seen as a distinct anatomical feature of the marrow and mouse spleen.[1] An island consists of one or two central reticular cells (i.e., macrophages) surrounded by a wreath of erythroblasts.

Recently, we have observed spontaneous growth of erythroid colonies in soft agar of marrow from patients with myeloproliferative diseases (e.g., polycythemia vera).[2] In serially sectioned colonies, single or multiple macrophages central within the erythroid colony were frequently observed. The morphology of these erythroid colonies by light and electron microscopy was characteristic of erythroblastic islands containing central iron-containing macrophages, surrounded by erythrocytic precursors, with the more mature erythrocytes found at the periphery of the colony.

These findings suggest that cellular interaction between macrophages and erythroid stem cells may at least be facultative for erythroid colony growth. In this chapter, we describe the stimulatory effect that mouse peritoneal macrophages exert on erythroid colony formation by mouse bone marrow cells.

Materials and Methods

Ten- to twelve- week-old BDF_1 female mice (Cumberland View Farms) were used.

Mouse Adherent Peritoneal Exudate Cells

Mice were killed by cervical dislocation. Approximately 5 ml of α medium (Flow Laboratories, Rockville, Maryland) was infused into the peritoneal cavity of each mouse. The peritoneal lavage was sterilely collected 1 to 2 min later and placed in 35 × 10 mm Petri dishes (Lux Scientific Corp., Newbury Park, California), and then incubated in 5% CO_2 and 95% air at 37°C with saturated humidity for 2 hr. After 2 hr of incubation, nonadherent cells in suspension were discarded, and fresh α medium was added to the dishes. The adherent cells were harvested using a rubber policeman. Cells were washed once with α medium (2000 rpm, 5 min).

Mouse Nonadherent Bone Marrow Cells

Bone marrow cells were flushed from femora of mice into α medium using 1 ml syringes. Cell suspensions in 35 × 10 mm petri dishes were incubated in 5% CO_2 and 95% air at 37°C with saturated humidity for 2 hr. After 2 hr of incubation, nonadherent marrow cell suspensions were collected and washed once with α medium (2000 rpm, 5 min).

Mixing Cell Culture

Mouse nonadherent bone marrow cells (2×10^5/ml) were cultured with various concentrations of mouse adherent peritoneal exudate cells (0.5 to 2.0×10^4/ml) in semisolid medium containing methylcellulose in the presence of 2 units of human urinary erythropoietin (38 IRP units/mg) per ml (Figure 26-1). Details about cell culture are described in Ap-

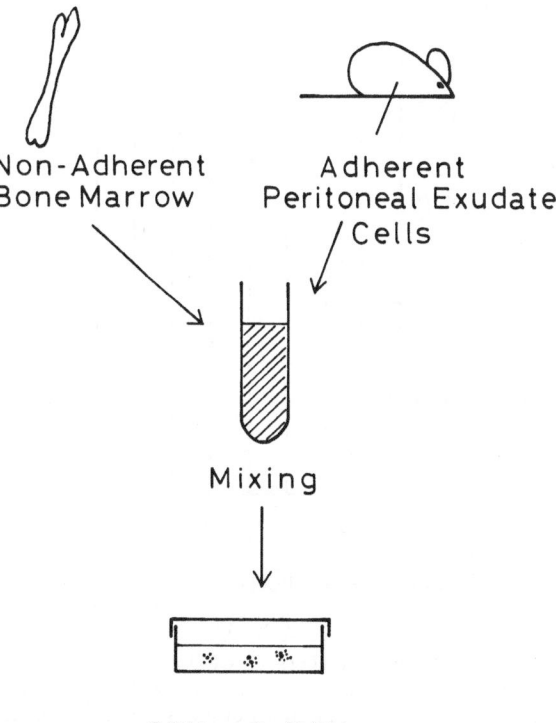

Figure 26-1 Mouse nonadherent bone marrow cells were cultured with mouse adherent peritoneal exudate cells in semisolid medium containing methylcellulose.

pendix 10. The numbers of colony-forming units-erythroid (CFU-e) were counted on day 3, and the numbers of burst-forming units-erythroid (BFU-e) were counted on day 9.[3]

Results and Discussion

As shown in Figures 26-2 and 26-3, the co-incubation of adherent peritoneal exudate cells increased the number of both CFU-e and BFU-e from bone marrow nonadherent cells in a dose-dependent fashion. Adherent peritoneal exudate cells alone did not produce any colonies in response to erythropoietin. Since adherent peritoneal exudate cells are mainly macrophages, this supports the proposition that peritoneal macrophages may enhance the erythroid colony formation of bone marrow cells *in vitro*.

Although erythroid colonies *in vitro* and erythroblastic islands *in vivo* are not necessarily the same, there is a strong suggestion that macrophages play an active role in the microenvironment of the maturing erythroblasts and/or by supplying them with as yet undefined substances. It was this reasoning that led Bessis to refer to the reticulum macrophage as a "nurse cell."[4]

The markedly elevated production of E-type prostaglandins by activated macrophages may be of considerable importance in the regulation of hematopoiesis.[5] The E-type prostaglandins have already been reported to enhance erythropoiesis.[6,7] The stim-

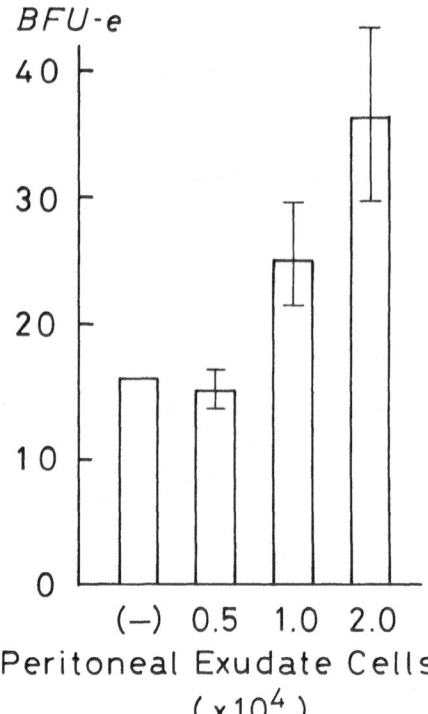

Figure 26-3 The numbers of BFU-e (mean ± SE) in 2×10^5 mouse nonadherent bone marrow cells as a function of different concentrations of mouse adherent peritoneal exudate cells. Erythropoietin concentration was 2 units/ml.

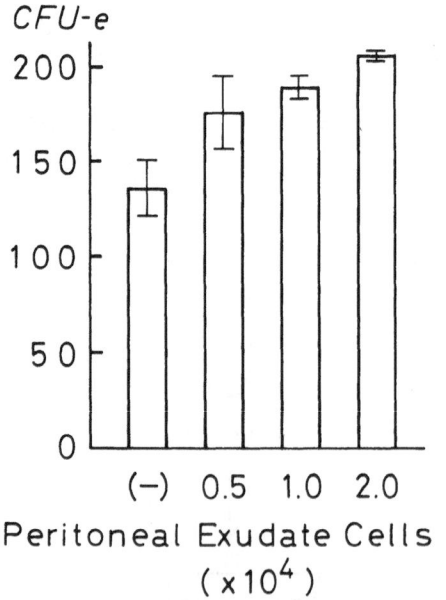

Figure 26-2 The numbers of CFU-e (mean ± SE) in 2×10^5 mouse nonadherent bone marrow cells as a function of different concentrations of mouse adherent peritoneal exudate cells. Erythropoietin concentration was 2 units/ml.

ulatory effect of peritoneal macrophages on erythroid colony formation reported here might be interpreted in this regard.

Summary

Mouse nonadherent bone marrow cells were cultured in a semisolid medium with various concentrations of mouse adherent peritoneal exudate cells in the presence of erythropoietin. The co-incubation of mouse peritoneal exudate cells enhanced erythroid colony formation of both colony-forming units-erythroid (CFU-e) and burst-forming units-erythroid (BFU-e) derived from mouse nonadherent bone marrow cells. Because adherent peritoneal exudate cells are mainly macrophages, these results suggest that macrophages may stimulate erythroid colony formation *in vitro*. These data support the proposition that macrophages may play a significant rôle in erythropoiesis *in vivo*.

Acknowledgments

This work was supported by grants AM–19741 and AM–07266 from the USPHS and The Hipple Foundation.

The erythropoietin preparation used in this study was purified by the Hematology Research Laboratories, Children's Hospital of Los Angeles, under grant HL-10880 from the National Heart, Lung and Blood Institute.

References

1. Bessis M: Living Blood Cells and their Ultrastructure. New York, Heidelberg, Berlin, Springer-Verlag, p. 87, 1973.
2. Horland AA, Wolman SR, Murphy MJ, Jr, Moore MAS: Proliferation of erythroid colonies in semisolid agar. *Br J Haematol 36:*495, 1977.
3. Iscove NN, Sieber F: Erythroid progenitors in mouse bone marrow detected by macroscopic colony formation in culture. *Exp Hematol 3:*32, 1975.
4. Bessis M: L'islot érythroblastique, unité fonctionelle de la moelle osseuse. *Rev Hémat 13:*8, 1958.
5. Kurland J, Moore MAS: Modulation of hemopoiesis by prostaglandins. *Exp Hematol 5:*357, 1977.
6. Schooley JC, Mahlmann LJ: Stimulation of erythropoiesis in plethoric mice by prostaglandins and its inhibition by anti-erythropoietin. *Proc Soc Exp Biol Med 138:*523, 1971.
7. Dukes PP, Shore NA, Hammond D, Ortega JA, Datta, MC: Enhancement of erythropoiesis by prostaglandins. *J Lab Clin Med 82:*704, 1973.

Discussion (Chapters 25–26)

Dr. Murphy: Dr. Urabe would like to make a comment pertinent to Dr. Fisher's presentation.

Dr. Urabe: Bone marrow cells from uremic patients respond to erythropoietin in suspension culture without any significant difference from that of normal bone marrow cells. Freedman and Saunders reported normal to increased numbers of erythroid precursor cells (CFU-e) by bone marrow cells of a patient with chronic renal failure. It is naturally assumed that, in the uremic state, erythroid stem cells are inhibited by factor(s) in the plasma of uremic patients that may play a role in the pathogenesis of the anemia of chronic renal failure. We have reported that the plasma of patients with chronic renal failure suppressed the responsiveness to erythropoietin *in vitro* of normal human bone marrow cells in a dose-dependent fashion (Figure 1).

It is quite probable, therefore, that the inhibitory factor(s) in plasma or serum is of significance with regard to the pathogenesis of the anemia of chronic renal failure.

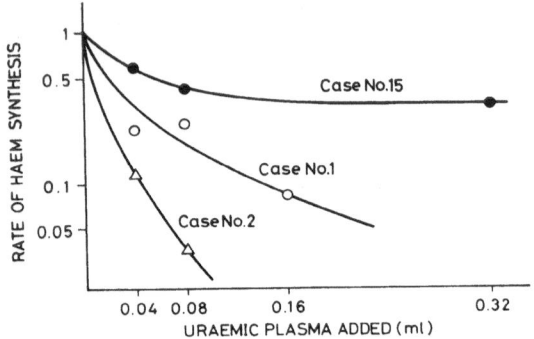

Figure 1 Inhibitory effect of the plasma from patients with chronic renal failure on the response of normal human bone marrow cells to erythropoietin. Normal human bone marrow cells (1.5×10^6 per dish) were suspended in 35×10 mm petri dishes containing 0.48 ml of NCTC 109 medium, and various amounts of plasma from patients (blood group AB) with chronic renal failure, and were incubated in the presence of 0.1 unit of erythropoietin for 72 hr under 5% CO_2 and 95% air with saturated humidity. Total volume of plasma in each dish was adjusted to 0.32 ml by adding plasma obtained from a normal individual (blood group AB). Iron levels of the plasma samples were 95 μg/100 ml (Case 1), 69 μg/100 ml (Case 2), 81 μg/100 ml (Case 15), and 98 μg/100 ml (normal individual). After the incubation, heme was extracted from cultured cells. Radioactivity of the heme was measured by a gas flow counter. (From Urabe A, et al: *Scand J Haematol 17*, 335–340, 1976, with permission.)

Dr. Erslev: The search for inhibitors in transfusion polycythemia and in uremia have been going on for a long time now; and although the hypothesis that they exist is very attractive, *in vivo* studies have never really established their presence. Since they have been shown in *in vitro* studies, the question now is: how well can we translate these *in vitro* results, which are quite striking, into relevant physiological and pathological data? In the case here, I have great trouble with the *in vitro* data. We have a test in which maybe one-tenth or one-twentieth of the testing mixture contains uremic serum, with a dilution of the hypothetical uremic toxin five to ten times. However, even in that dilution, it has a remarkable suppressive effect on CFU-e, as you have just seen in rabbit bone marrow. What would undiluted serum, as we have *in vivo*, do? It should completely knock out the bone marrow, but we don't see that. At the time when these sera were obtained, the red cell production was still very appreciable. Although this is an *in vitro* conference, I still believe that the *in vitro* demonstration of inhibitors does not necessarily mean that they are acting *in vivo*.

Dr. Fisher: May I try to answer Dr. Erslev? I think you are right that we need a good *in vivo* model but, for example, if we use the polycythemic mouse assay for this system, the inhibitor does not seem to affect erythropoiesis *in vivo*. However, you must remember that the polycythemic mouse has a perfectly normal excretory mechanism and can excrete the inhibitor. The mouse is not like the uremic patient, nor is the uremic animal, who also does not have the ability to remove inhibitors. We have the same problem with trying to correlate the *in vitro* and *in vivo* systems, but I think we require a renal insufficiency model to try to study the inhibitor *in vivo*.

Dr. Gordon: I have growing suspicion that the reticuloendothelial system can do practically anything you want it to do, if you provide it with the proper environment and the proper stimuli.

Dr. Murphy: We have had suggestions that macrophages, in certain organs of the body, at least, may produce erythropoietin. So even that seems a possibility.

Dr. Peschle: Did you check if there was an increased titer of erythropoietin in the medium?

Dr. Murphy: There are very few cells in these cultures, and I have not measured Ep. But that certainly is something that could readily be done.

Dr. Ogawa: We have not seen macrophages within *in vitro* derived colonies.

Dr. Murphy: I'd like to comment on that. I don't mean to give the impression that this is the only anatomical feature of erythropoiesis. I believe that there is a position for the macrophage in erythropoiesis, within an erythroblastic island, and that it may be facultative or facilitative, but by no means obligatory.

Dr. Rossi: I'm sure you're aware of at least two compounds that paralyze macrophages; and I'm referring to silicon particles and Carrageenans, which have been mostly used by immunologists; and I think it would be very interesting if you were to use these before mixing in your cultures.

Dr. Murphy: Thank you, that's a very good suggestion.

Dr. Erslev: No comment, just a question. Have you been able so far to show any qualitative or quantitative difference between a colony having a central macrophage and a colony without macrophages?

Dr. Murphy: These currently reported studies are preliminary and investigations designed to answer your questions are currently in progress.

27 Extrarenal Erythropoietin

Brian A. Naughton, Albert S. Gordon, Sam J. Piliero, and Philip Liu

Introduction

Although the kidney is clearly the major site of erythropoietin (Ep) production in the intact adult animal, evidence has steadily accumulated to indicate the existence of potent extrarenal loci of Ep biosynthesis.[1,2] Thus the classic experiments of Jacobson and his group[3] and Mirand et al.[4] indicated that nephrectomy resulted in lowered production of Ep in rodents subjected to certain forms of hypoxia. The reduced elaboration of Ep following renal ablation cannot be attributed to the accumulation of unexcreted wastes in the renoprival state since bilateral ureteral ligation in the rat[3] or implantation of the ureter into the iliac vein of the dog[5] did not appreciably alter their ability to produce Ep in response to different types of hypoxia. Fried and co-workers[6] and Schooley and Mahlmann[7,8] noted that nephrectomized rats exposed to hypoxia produced approximately 10 to 15% of the Ep evoked in intact animals. These workers have demonstrated that a basic chemical similarity exists between exterarenal and renal Ep since anti-Ep neutralized the activity of the plasma from both sources (see refs. 2 and 9 for a more comprehensive treatment of this subject).

Studies performed in the fetus are of considerable interest. Subjection of nephrectomized fetal goats (108 to 115 days of gestation) to bleeding did not prevent their capacity to produce Ep.[10] Thus the kidneys are not necessary for the ability of the fetal animal (i.e., sheep, goat) to produce Ep in response to anemic hypoxia. A similar situation appears to prevail in the neonatal and weanling rat.[11–14] Thus, for example, bilateral nephrectomy performed in neonatal rats does not impair their capacity to produce Ep in response to hypoxia. Therefore, it would appear that more potent extrarenal sites of Ep production exist in the neonatal than in the adult rat.

In seeking the source(s) of extrarenal Ep, Kaplan et al.[15] exposed neonatal rats to hypobaric hypoxia and found significantly elevated levels of erythrogenin in the livers and spleens but not in the kidneys of these animals. This is in contrast to findings in the young and adult rat, in which the primary erythrogenin response to hypoxia is renal. This work served to emphasize that, at the neonatal stage in development, the rodent kidney lacks erythrogenin and probably produces little or no Ep whereas the liver and to a lesser degree the spleen (probably because of its smaller mass) constitute the major sites of Ep production at this period of life. In this regard it has been shown that although nephrectomy does not alter the Ep response of sheep fetuses (95 to 115 days of gestation) to bleeding, a superimposed subtotal hepatectomy almost completely abolishes the production of Ep in these animals.[16] Experiments such as these tend to establish the liver as the chief source of Ep in the fetus and probably the neonatal animal as well.

The Liver as a Site of Extrarenal Ep Production in the Adult

The possibility that the liver may also function as a site of Ep production in the adult was suggested from hepatic perfusion experiments[17,18] and from liver extirpation studies.[19,20] In this regard, erythrogenin has been extracted from the livers and spleens of nephrectomized baboons.[21] Furthermore, Kaplan et al.[22] demonstrated high levels of erythrogenin in the livers and spleens of young nephrectomized male rats subjected to severe hypoxia (0.35 atm of air). The amounts determined were equivalent to those extracted from the kidneys of intact young rats exposed to the same intensity and duration of hypoxia.

With regard to the precise extrarenal cellular site of origin of Ep, the reticuloendothelial system (RES) becomes suspect because both the liver and the spleen possess large numbers of these elements. Thus an increase in the plasma Ep levels is noted in the anephric rat following induction of RES hyperplasia by colloidal carbon[23–25] or after RES blockade by thorotrast.[26] Carbon overload also results in a considerable increase in hepatic erythrogenin activity in adolescent anephric rats.[27] RES elements can produce hematopoietically active substances. Colony-stimulating factor (CSF) has been localized in Kupffer cells[28] as well as in blood monocytes.[29] Immunofluorescent labeling for Ep in liver sections of carbon-injected neonatal rats further implicates the Kupffer cell as the site of Ep synthesis and/or storage.[30]

Studies on the Regenerating Liver

Liver regeneration is a well documented process.[31–33] Surgical removal of a substantial amount of liver tissue induces major ultrastructural and biochemical alterations in the remaining liver stump.[34,35] The following is an account of our recent experiments relating to the effects of subtotal hepatectomy on liver structure during regeneration and on the production of Ep following nephrectomy and exposure to hypoxia.[36-39]

Materials and Methods

Male Long-Evans adolescent (130 to 150 gm body wt) and adult (350 to 400 gm) rats were employed. Hematocrits were performed routinely before and after all surgical procedures. Each experimental group consisted of 12 to 15 rats. A brief description of the techniques used follows.

Surgical Procedures
Hepatectomy
The technique used is a modification of the method of Higgins and Anderson[31] in which the left and median hepatic lobes are ligated and removed (approximately 66% of the liver). We perform an 80 to 90% hepatectomy under light ether anesthesia. A 1-inch midline incision is made posteriorly from the sternum, exposing the abdominal cavity. A ligature is tied distally around the right lobe of the liver and manipulated so that both the right and the median hepatic lobes are drawn out of the body cavity. The left lobe is exposed in a similar manner and a single 4-0 ligature is used to tie off all three lobes simultaneously. The ligated liver tissue is then excised and weighed. The 30 to 40% hepatectomy involves the removal of the right hepatic lobe. With the exception of the blood removed with the liver, the procedure is relatively bloodless. The survival rate is greater than 95%.

Bilateral Nephrectomy
Two incisions are made on the dorsolateral flanks; the kidneys are drawn out of the body cavity, the renal vessels ligated and the kidneys excised.

Splenectomy
After an abdominal incision, all afferent and efferent vessels are ligated and the spleen is extirpated.

Bilateral Ureter Ligation
The ureters are approximated after an abdominal incision and are ligated as close to the kidneys as possible.

Sham Operation
An incision equal in size to that performed for a hepatectomy or nephrectomy is made and the body cavity is closed 5 min later without having removed any tissue.

Histological Procedures
Light Microscopy Studies
To determine the Kupffer cell concentrations in the liver, experimental and control rats are injected i.v. with a marker dose of colloidal carbon consisting of 0.1 ml of a 1:28 dilution of Pelikan ink (Günther Wagner, West Germany) to physiological saline 4 hr prior to sacrifice. The liver tissue is excised, fixed in formalin, dehydrated in ethanol, and cleared in xylene prior to paraffin embedding. Hematoxylin-eosin staining is employed. A Kellner counting ocular is used to determine at 400× the relative numbers of the various hepatic cell types. Photographs of random fields showing the hepatic central vein are taken and a series of concentric circles are drawn around the central vein at 125μ intervals. The relative numbers of the various cell types found at 0 to 125μ, 125 to 250μ, and 250 to 375μ from the central vein are then determined.

Photographs of random fields in all experimental and control groups are taken at 400× and 970×. After a 2× photographic enlargement, individual cells are measured with a Gelman planimeter. The planimeter is a mechanical area integrator that is accurate and easy to use. To measure an area, the tracer point is placed on a marked starting point, the reading of the measuring wheel is noted, the area is circumscribed in a clockwise direction by means of a tracer point, and the measuring wheel is read again. The planimeter is usually used on camera lucida drawings or photographs but may also be employed on tissue images projected from below on ground glass.[40,41] Relative areas of fixed numbers of experimental to normal cells are noted in the results.

Electron Microscopy Studies
After diabutal anesthesia, the liver is perfused with a cold 2.5% glutaraldehyde-Millonig buffer solution[42] and then is excised and placed in a container with the perfusion solution for 2 hr. After Millonig buffer rinsing, the tissue is placed in 1% osmium tetroxide-Millonig buffer for 90 min. The specimens are washed in distilled water and immersed in an aqueous 0.4% uranyl acetate solution for 15 hr. The tissue is dehydrated in ethanol and propylene oxide prior to Dow Epoxy Resin (DER) 332/732 embedding. Sections are stained with uranyl acetate and poststained with citrated lead hydroxide prior to viewing. Planimetric measurements are made of various hepatic cell types in experimental and control groups from photographs taken at 2500×.

Radioisotopic Procedures
Autoradiography
Experimental and control animals are injected i.p. with tritiated thymidine (1 μCi/gm body wt) and after 1 hr are killed. The livers are removed, placed in formalin, and embedded using the procedure previously described. The tissues are sectioned, layered with Kodak Nuclear Track Emulsion NTB-3, and developed for 2 weeks before staining and counting.

Gamma Counting and Scanning-radioactive Technetium Sulfur Colloid

(TSC) A Technetium Colloid Kit (Mallinckrodt Co., New York) was used. This is standardized with a Radx meletron dose calibrator and administered i.v. in dosages varying from 420 to 440 µCi/150 gm rat. Five min after injection of the TSC, imaging is performed for 4 min with the Elscint gamma camera CE–1 (with pinhole collimator) and the Elscint Minim computer. A 25% window is employed in all scans with a constant 64 cm² region of interest. Counts are recorded along with the scintigraphs on polaroid film. Whole organ liver and spleen counts are performed with a Picker Well Counter (Spectroscaler IIIA) 15 min after the original injection of TSC. After counting for 30 sec, the organs are weighed and the activity is expressed as counts per gm of tissue.

Experimental Protocols

Each experimental group is analyzed according to the methods previously described. Serum from each group is assayed for Ep using the exhypoxic polycythemic mouse.[43]

Either 80 to 90% hepatectomy or 30 to 40% hepatectomy is performed on adolescent (130 to 150 gm) or adult (350 to 400 gm) rats. Bilateral nephrectomy is carried out at intervals of 0, 3, 6, 12, 24, 48, 72, 96, 120, 144, and 288 hr after hepatectomy and the animals are subjected to 6 hr of hypoxia at 0.4 atm of air. Three groups of control animals are examined following the same time intervals as for the experimental animals. Group 1 included animals that were sham-hepatectomized, bilaterally nephrectomized, and rendered hypoxic for 6 hr. Group 2 included rats that were hepatectomized, bilaterally nephrectomized, and allowed to remain for 7 hr at ambient pressure (approximately 1 atm of air). Group 3 included animals that were sham-hepatectomized sham-nephrectomized, and subjected to 6 hr of hypoxia. In addition, two other test groups were examined. In the first group, a 30 to 40% hepatectomy was followed by bilateral nephrectomy and hypoxia. The second group consisted of rats subjected to 80 to 90% hepatectomy followed by combined bilateral nephrectomy, splenectomy, and hypoxia for 6 hr.

Multiple hepatectomies were performed only in 130 to 150 gm rats. The procedure for this experiment is described below:

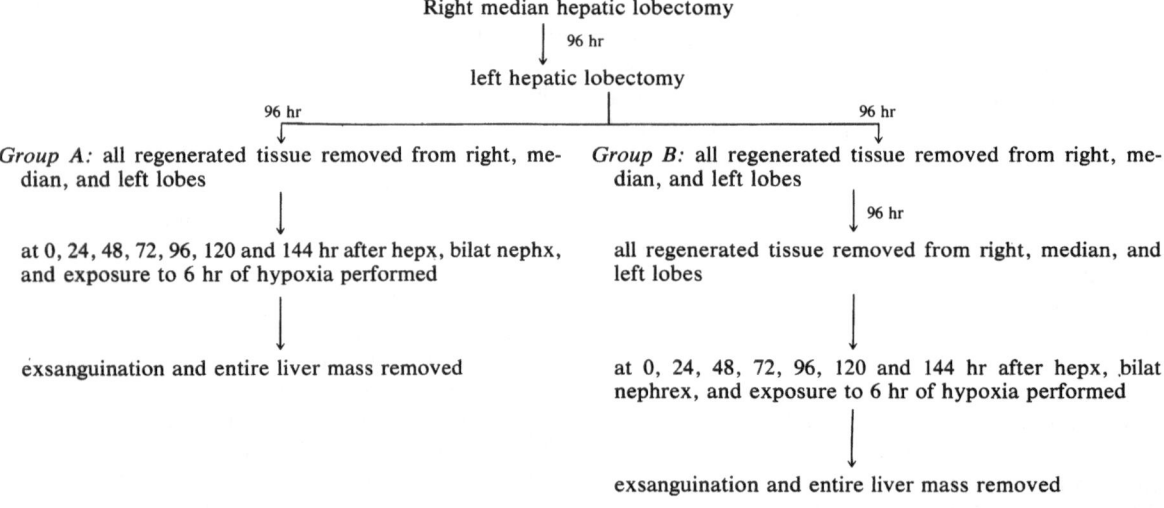

To test for a possible hepatic growth factor in the serum of hepatectomized rats, the following protocols were established:

recipient rats (normal 130 to 150 gm) $\xrightarrow[\text{hepx + nephrex or}\atop\text{hepx + bilat. ureter-ligated donors}]{\text{2 ml serum (i.v.)}\atop\text{from hepatic vein}\atop\text{or dorsal aorta of}}$ $\xrightarrow{18\ hr}$ $\xrightarrow[\text{nephrex}]{\text{bilat.}}$ $\xrightarrow[\text{6 hr}]{\text{0.4 atm}}$ *

c. donor rats (normal 130 to 150 gm) $\xrightarrow{\text{80 to 90\% hepx}}$ at 0, 24, 48, 72, 96, and 120 hr after hepx, either bilat nephrex or bilat ureter ligation; 24 hr later, exsanguination from either the hepatic vein or the dorsal aorta.

recipient rats $\xrightarrow[\text{bilat ureter}\atop\text{ligation}]{\text{bilat nephrex or}}$ $\xrightarrow[\text{hepx + nephrex or}\atop\text{hepx + bilat}\atop\text{ureter-ligated}\atop\text{donors}]{\text{2 ml serum (i.v.}\atop\text{from hepatic vein}\atop\text{or dorsal aorta of}}$ $\xrightarrow{18\ hr}$ $\xrightarrow[\text{6 hr}]{\text{0.4 atm}}$ *

* Serum Ep levels were assayed in all recipients after subjection to the hypoxic stimulus.

Results and Discussion

At the outset, it should be stated that of the 1,300 rats examined in our studies, the livers in 68% of the rats regained their normal gross appearance. Liver mass was restored by hyperplasia of the left lobe in 28% of the rats and 4% possessed livers that did not regenerate or which developed hepatomas. Serum from the latter 4% was not assayed for Ep.

Studies in Rats after Subtotal Hepatectomy
Serum Ep Levels
When compared to the sham-hepatectomized, bilaterally nephrectomized, hypoxic groups, rats subjected to a combination of 80 to 90% hepatectomy, bilateral nephrectomy, and hypoxia displayed decreased serum concentrations of Ep from 0 to 26 hr after liver removal; Ep production then markedly increased, reaching a peak at 72 hr (Figure 27-1 and Table 27-1). The serum Ep level then dropped to that of the sham-operated groups by 96 hr and remained there. In the adolescent Long-Evans rat, liver mass returns to near normal values at 72 to 96 hr post hepatectomy (Figure 27-2)[37,38,44] and the period of most rapid tissue proliferation, as indicated by the point-slopes of a percent regeneration versus time post-hepatectomy graph, is 48 to 72 hr after hepatectomy (Figure 27-2). This period corresponds to the time of

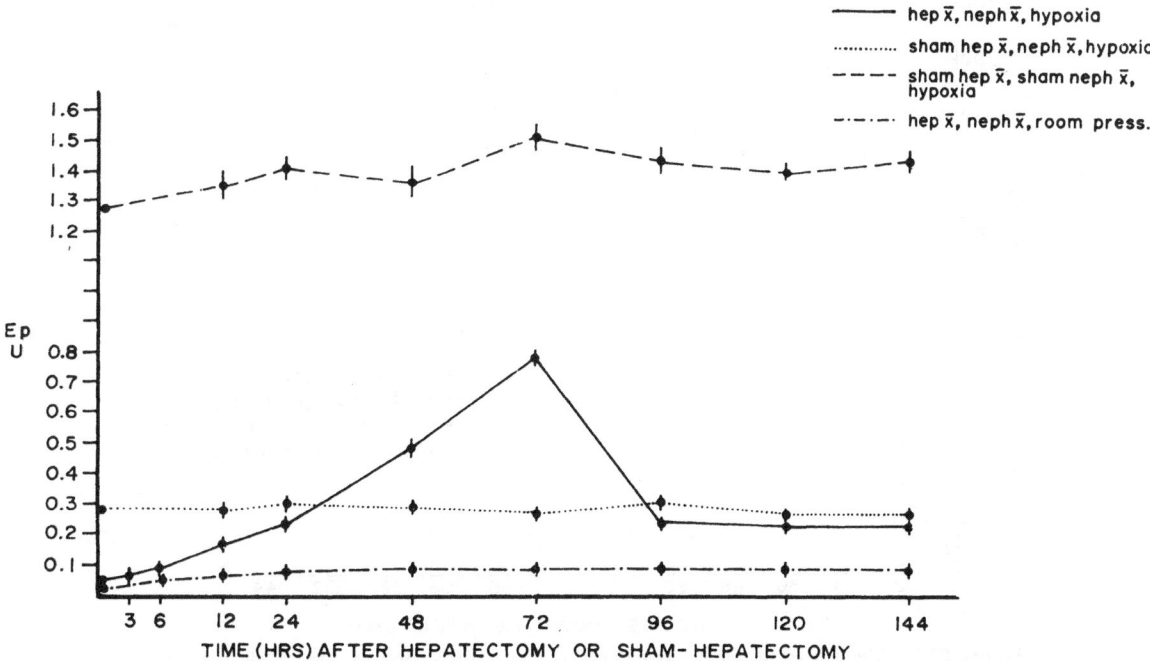

Figure 27-1 Mean quantities of Ep in the serum at various times during hepatic regeneration. Vertical bars through the means indicate ±1 SEM. Hepx = hepatectomy, nephx = nephrectomy.

Table 27-1 Serum erythropoietin (Ep) concentrations and hematocrit values in rats during hepatic regeneration.[a]

Time after Hepatectomy (hr)	Hepatectomy (70–80%), Nephrectomy, and Hypoxia			Hepatectomy (30–40%), Nephrectomy and Hypoxia		Hepatectomy (70–80%), Nephrectomy with Splenectomy and Hypoxia	
	Ep	N	Hematocrit value[b]	Ep	N	Ep	N
0	0.055 ± 0.0049	4	38.3 ± 1.29	0.056 ± 0.0072	3	0.049 ± 0.0036	4
3	0.059 ± 0.0050	3	39.2 ± 1.60			0.051 ± 0.0025	4
6	0.086 ± 0.0097	3		0.058 ± 0.0029	3	0.054 ± 0.0025	4
12	0.160 ± 0.0260	3	38.0 ± 1.45	0.070 ± 0.0005	3	0.148 ± 0.0099	4
24	0.245 ± 0.0071	3	39.1 ± 1.23	0.186 ± 0.0085	3	0.237 ± 0.0120	3
48	0.442 ± 0.0433	3	38.6 ± 1.09	0.359 ± 0.0100	3	0.451 ± 0.0173	3
72	0.780 ± 0.0100	4	39.4 ± 1.52	0.590 ± 0.0150	3	0.783 ± 0.0093	3
96	0.264 ± 0.0099	3	38.1 ± 1.42	0.325 ± 0.0179	3	0.276 ± 0.0123	3
120	0.251 ± 0.0133	3	37.3 ± 2.10	0.268 ± 0.0013	3	0.262 ± 0.0096	3
144	0.253 ± 0.0192	3	38.4 ± 1.26	0.259 ± 0.0017	3	0.263 ± 0.0133	3

[a] The results are expressed as the mean ± standard error of the mean for each experimental group, with the number (N) of experimental trials given in separate columns. The data for Ep represent equivalent units of Ep derived from the standard dose response curves for the international reference preparation (IRP) of human urinary Ep, assayed in the exhypoxic polycythemic mouse.

[b] The hematocrit value in normal rats is 38.7 ± 1.46.

maximum Ep production. In the adult rat, regenerative potential decreases when compared to the young rat (Figure 27-2) as does the ability to produce Ep (Figure 27-3). In the adult Long-Evans rat, functional liver mass as measured by circulating Ep titers does not return to near normal levels until 120 to 144 hr post hepatectomy with maximum tissue proliferation occurring 72 to 96 hr after liver removal. This age-related variation in hepatic regeneration agrees with previous reports.[45,46]

Histology: Light and Electron Microscopic Analysis

Kupffer cell to parenchymal cell ratio is also highest at 72 hr post-hepatectomy in the adolescent animal (Figure 27-4; ref. 47). The initial decline in parenchy-

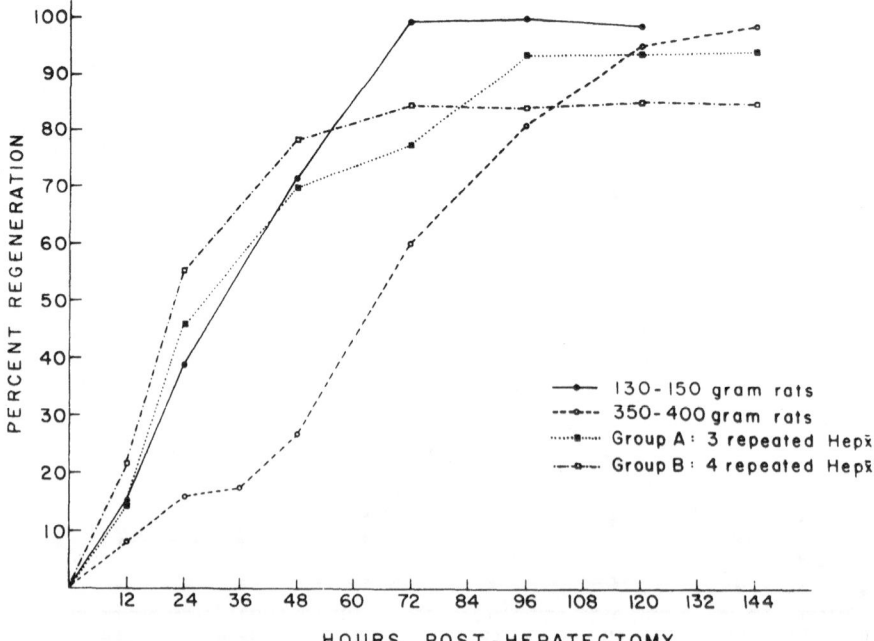

Figure 27-2 Mean percent hepatic regeneration values for various groups as determined by comparing the mass of tissue extirpated with the mass of regenerated tissue (five experimental groups per point). Solid line, venous system; broken line, arterial serum.

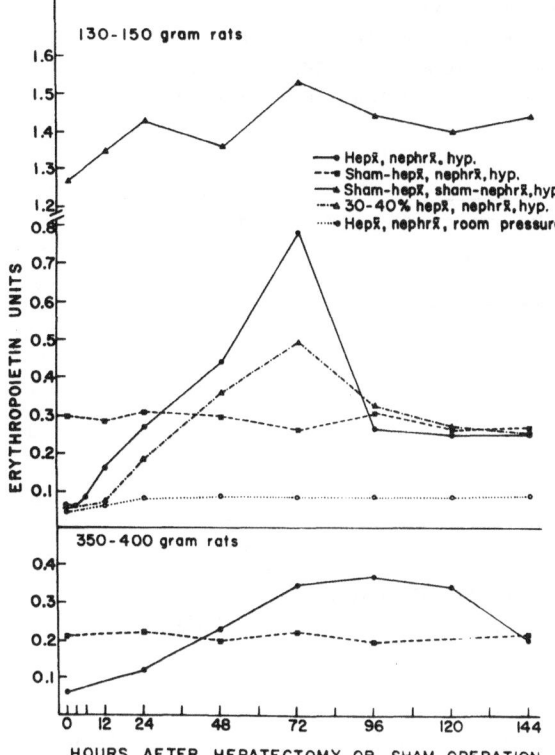

Figure 27-3 The effect of various surgical procedures on Ep production in rats of different ages. The results are expressed as the mean values for experimental groups (three trials apiece). Represented are equivalent units of Ep as compared to the standard dose-response curves for the International Reference Preparation (IRP) of human urinary Ep in the exhypoxic, polycythemic mouse. Normal basal Ep in the non-hypoxic adolescent and adult ranges from non-detectable levels to 0.05 IRP units per ml of plasma. Hepx = hepatectomy, nephrx = nephrectomy, hyp = hypoxia.

(Figure 27-6). This phenomenon could be attributed to either a humoral mechanism governing the growth of these cells or possibly to an increased availability of nutrients.[55] A similar pattern is observed when relative areas of the various cell types are determined by planimetry.[40,41] All of the results are indicated as a ratio of the areas of equal numbers of experimental cells to normal cells of the same cell type. A ratio significantly greater than one is taken as indicative of hypertrophy. Tracings are made from photographs of sections taken at 400×, 970×, and 2500× after a 2× photographic enlargement. In the 970× evaluation, Kupffer cells reveal the greatest degree of hypertrophy at 72 hr after hepatectomy (Figure 27-7A) whereas parenchymal cells are maximum in volume 24 to 48 hr post-hepatectomy (Figure 27-7B). Cellular activity as indicated by hypertrophy is greatest near the vascular supply. Spatial distribution of this nature is not observed in either the relative areas or in tritiated thymidine uptake of sham operated control groups. Presumably, if the spatial distribution, which is noted in these determinations, is due to the availability of nutrients, it would also occur in the sham-hepatectomized controls (Figure 27-8). Since this pattern is not observed in the controls, it would appear that the cells are reacting to a hormonally controlled growth-stimulating mechanism. This contention has been shared by others.[56,57] Further experimental evidence for the existence of humoral factors in liver regeneration is described later in this chapter (see Radionuclide Studies of Hepatic Regeneration).

Ultrastructural analysis of the events occurring after partial hepatectomy has been performed by several groups.[34,58–60] Parenchymal cell elements are the major area of interest in these studies. Cells of all mal cell numbers per unit area at 0 to 72 hr after hepatectomy is not seen in the Kupffer cell compartment (Figure 27-5). In fact, Kupffer cell numbers per unit area are elevated from 24 to 120 hr post hepatectomy when compared to sham-operated controls. Autoradiographic evaluation with tritiated thymidine shows that parenchymal cell labeling is highest at 48 hr post-hepatectomy whereas the Kupffer cell peak labeling does not occur until 72 hr after hepatectomy. This 24 hr lag in Kupffer cell labeling has previously been reported. Some investigators had reported an earlier peak[48,49] in labeling.[50,51] Others report a later initial labeling peak.[52–54] Labeling in these cell types appears to be dependent on their distance from the central vein. Cells located near the vein (0 to 125μ) show consistently higher labeling when compared to elements further away (250 to 375μ), regardless of whether they are Kupffer or parenchymal in nature

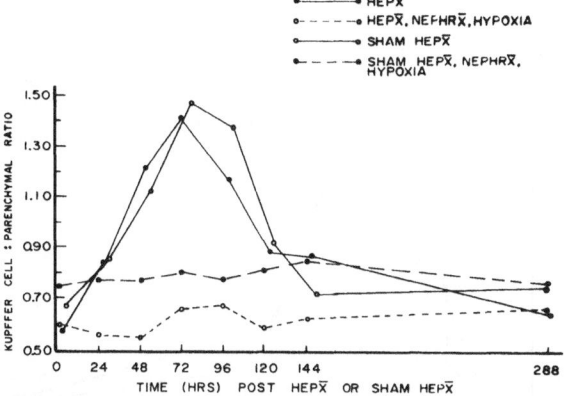

Figure 27-4 The results are expressed as the mean ratio of Kupffer cell numbers to parenchymal cell numbers at various intervals after partial hepatectomy. Sham hepatectomy; hepatectomy, nephrectomy, and hypoxia; or sham hepatectomy, nephrectomy, and hypoxia. Cell numbers in random fields are counted under a constant area of interest. A minimum of 3,000 cells are determined per point.

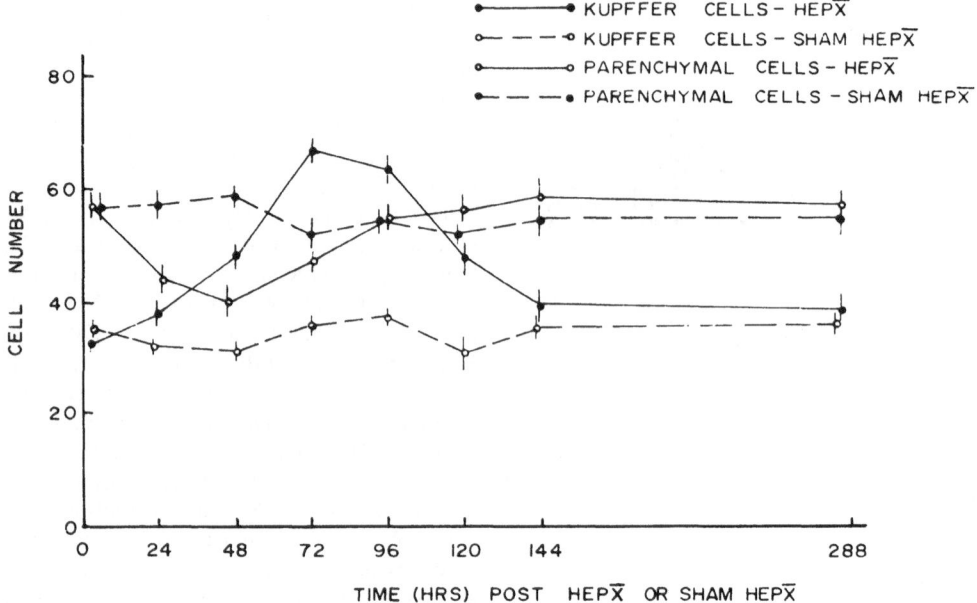

Figure 27-5 Mean Kupffer and parenchymal cell numbers ±1 SEM at various times after hepatectomy or sham hepatectomy. A minimum of 3,000 cells were counted per point.

types appear swollen 12 hr after hepatectomy.[34] The endoplasmic reticulum of parenchymal cells, which is sparse 6 to 12 hr after hepatectomy, begins to reform at 18 hr and the number of mitochondria markedly increases.[58] Glycogen, which is abundant in parenchymal cells of normal liver, decreases considerably after hepatectomy and fails to return to normal levels in the 0 to 144 hr post-hepatectomy period studied.[34,61] Lipid bodies are noted in parenchymal cells 24 to 48 hr after partial hepatectomy. Although smooth endoplasmic reticulum is reformed first, the ribosomes are restored to give the endoplasmic reticulum a near normal appearance by 36 to 48 hr post-hepatectomy.[59,60] The numbers of nucleoli per nucleus steadily increase in parenchymal cells as do the number of nuclei per cell. Kupffer cells on the other hand exhibit few of the ultrastructural changes noted in the parenchymal cells during the course of regeneration. By 24 to 48 hr after hepatectomy, the amount of rough endoplasmic reticulum and the number of polysomes are considerably augmented.[61] Mitotic activity is noted in many of these cells although the density of the ultrastructural components is maintained. Conspicuous, single membrane bound lipid-like drop-

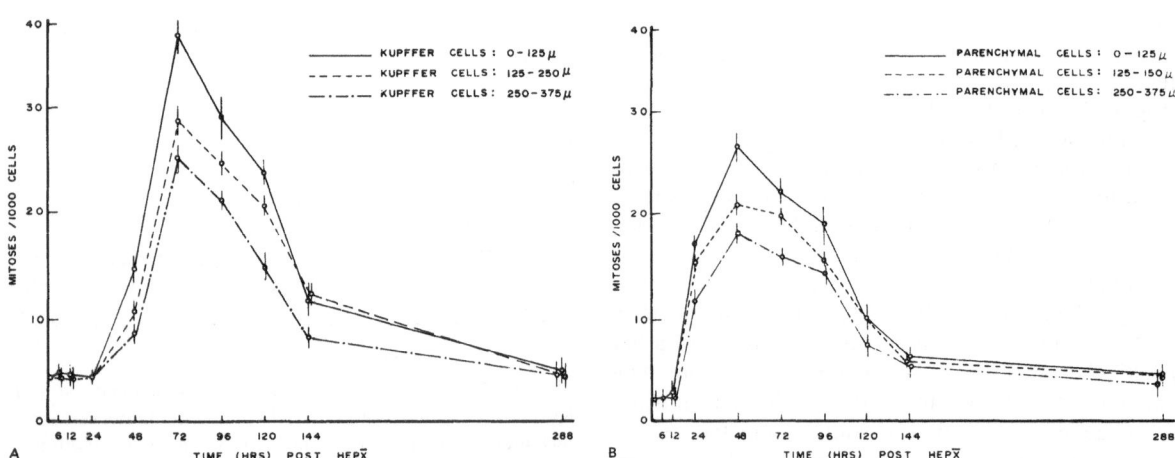

Figures 27-6A and B Mean number ±1 SEM of dividing Kupffer (A) and parenchymal (B) cells per 1,000 total cells of the same type at different distances from the hepatic central vein and at various times after partial hepatectomy as evaluated by tritiated thymidine autoradiography. A minimum of 3,000 cells are counted per point.

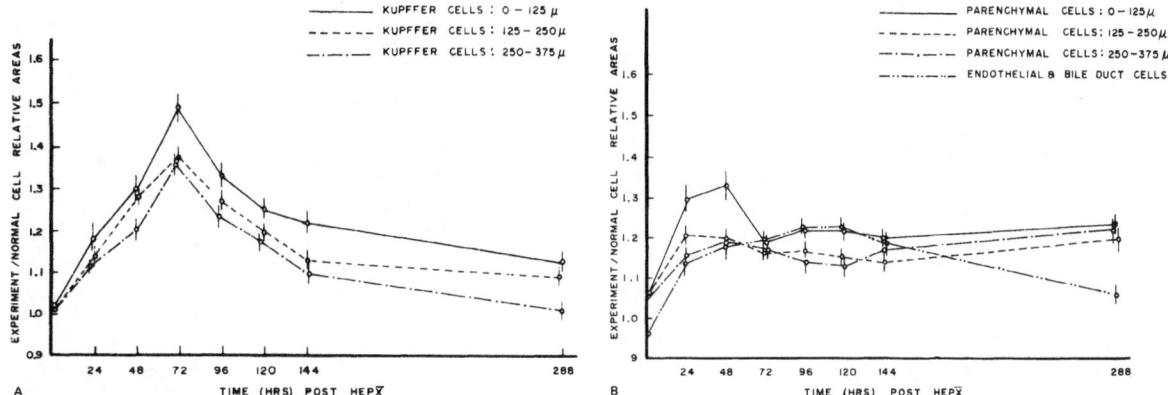

Figures 27-7A and B Mean relative areas of hepatic cell types at different distances from the hepatic central vein and at various times after partial hepatectomy. A minimum of 1,000 cells are traced per point from photographs enlarged 2× after photomicrography at 970×.

lets are found near the highly developed Golgi regions of the Kupffer cells from 24 to 120 hr after hepatectomy. Increased lysosomal activity is also noted during this time. Endothelial and bile duct cells, with the exception of the nuclear changes due to cycling, are ultrastructurally unremarkable.[61]

Numerous erythroblastic islands are found in the regenerating liver 24 to 36 hr after hepatectomy (Figure 27-9). These have previously been observed in fetal liver, which shows substantial erythropoietic activity (Figure 27-9).[62] Erythroblastic nests have been reported in the livers of patients with chronic myeloid leukemia[63] as well as other pathological conditions,[64] although they are normally found in the bone marrow.[65] The islands usually consist of six to eight erythroblasts surrounding a central reticular cell or macrophage.[66] Since *in vitro* growth of erythroblasts requires macrophages as well as Ep, it is possible that stem cell differentiation into the erythroid line is influenced by the presence of special macrophages in cultures.[66,67] It is clear that in the early stages of hepatic regeneration, some erythropoiesis takes place in the liver (Figure 27-9). This appearance is augmented by the imposition of bilateral nephrectomy on hepatectomy although the number of nests found is only 10 to 20% of those found in the 16- to 18-day rat

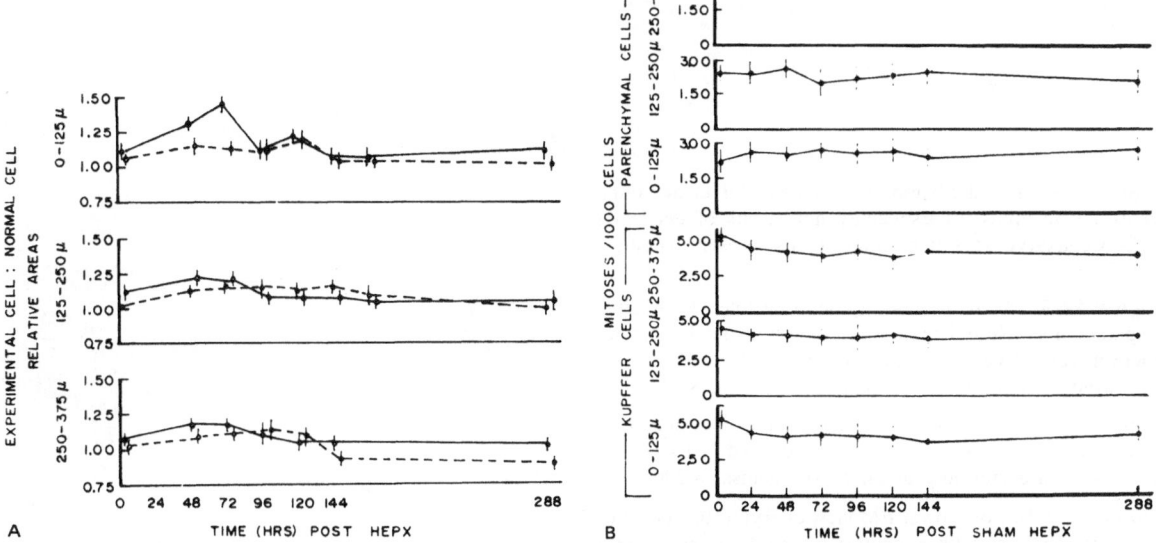

Figure 27-8 A, Mean relative areas ±1 SEM for hepatic cell types at different distances from the hepatic central vein and at various times after sham hepatectomy (970×) as determined by planimetry. Solid line, Kupffer cells; broken line, Parenchymal cells. **B,** Mean numbers of dividing cells ±1 SEM for hepatic cells of the same type at different distances from the hepatic central vein and at various time intervals after sham hepatectomy (970×) as evaluated by tritiated thymidine autoradiography.

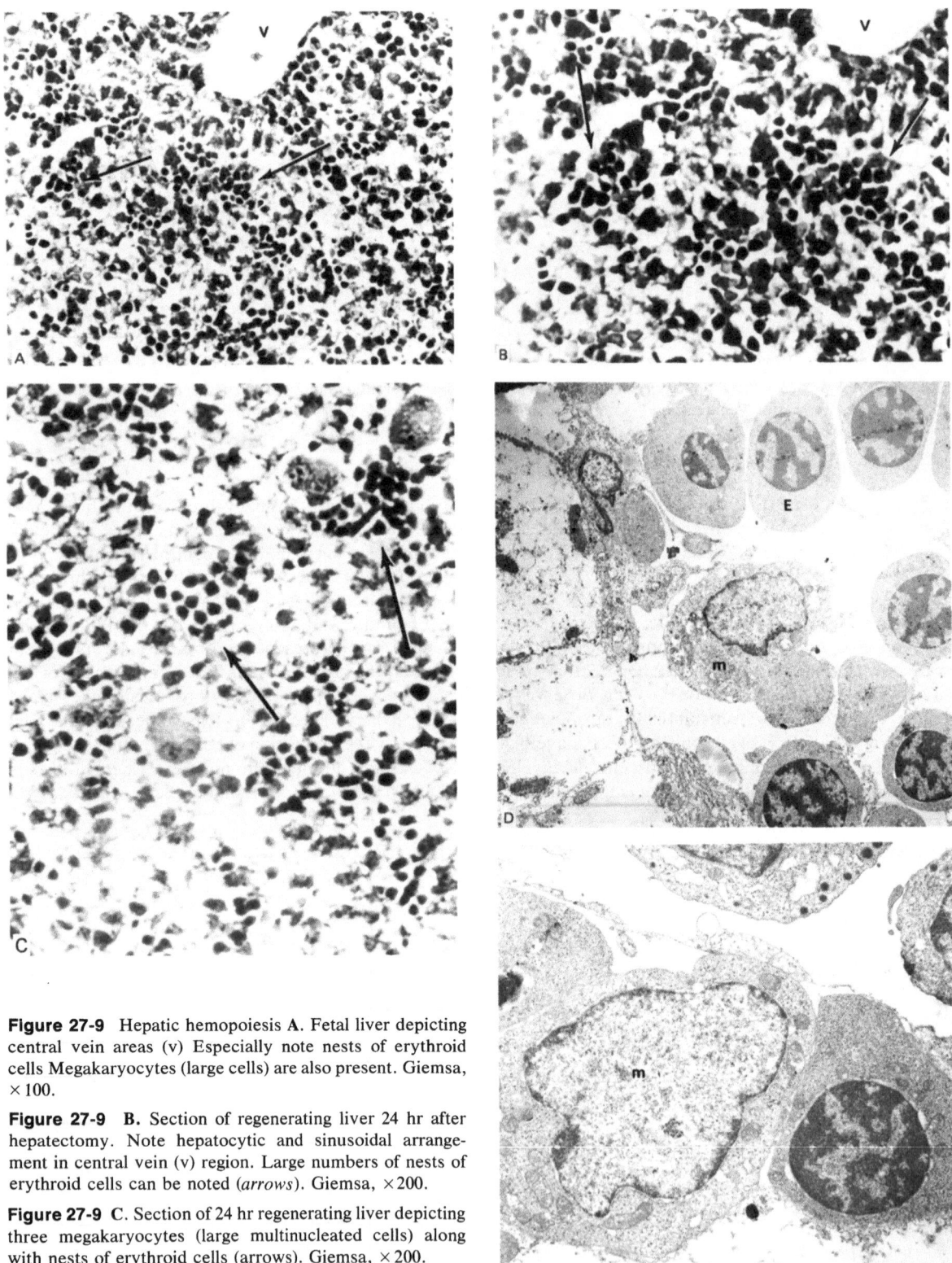

Figure 27-9 Hepatic hemopoiesis **A.** Fetal liver depicting central vein areas (v) Especially note nests of erythroid cells Megakaryocytes (large cells) are also present. Giemsa, ×100.

Figure 27-9 B. Section of regenerating liver 24 hr after hepatectomy. Note hepatocytic and sinusoidal arrangement in central vein (v) region. Large numbers of nests of erythroid cells can be noted (*arrows*). Giemsa, ×200.

Figure 27-9 C. Section of 24 hr regenerating liver depicting three megakaryocytes (large multinucleated cells) along with nests of erythroid cells (arrows). Giemsa, ×200.

Figure 27-9 D. Low power EM view of erythroblasts (E) in regenerating liver with a central macrophagic-monocytic (m) cell. ×1,300.

Figure 27-9 E. Higher power EM view of an area showing continuity of erythroblasts (e) and macrophagic-monocytic (m) cell ×2,600.

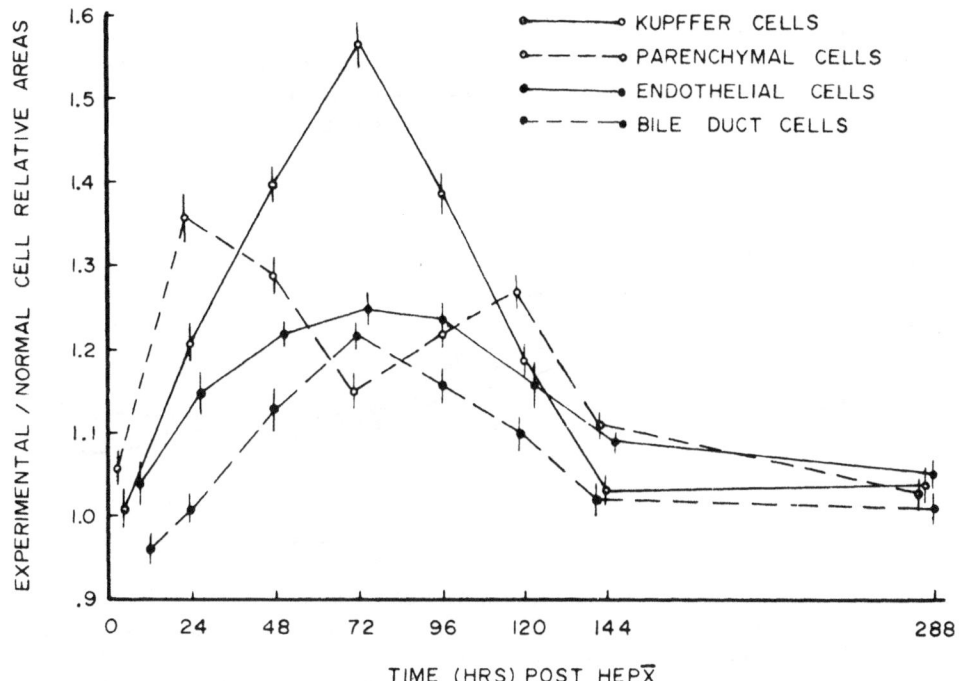

Figure 27-10 Mean relative areas of hepatic cells ±1 SEM at various times after partial hepatectomy, as determined by planimetry. Tracings are made from photomicrographs taken at 2500×. Experimental to normal cell ratios greater than 1 are indicative of hypertrophy.

fetus. The development of these erythroblastic elements may be due to a combination of factors: the hyperplasia and hypertrophy of the Kupffer cells, the probable seeding of the regenerating liver with foreign macrophages, and the alteration of the liver microenvironment so that it becomes favorable to erythropoiesis.[68,69]

Planimetric measurements made from photographs of specimens taken at 2500× are much more accurate than 970× measurements because the cell borders are more distinct. Although the relative areas of the various cell types are similar regardless of the magnification used, a periodicity in the parenchymal cell response that was not seen at 970× is noted at 2500× (Figure 27-10). It is probable that the increases in area observed for the parenchymal cells are due to additional nuclear material per cell as regeneration progresses.[70,71] Since nuclear changes of this nature have not been reported for Kupffer cells,[32] we feel that the buildup of intracellular materials associated with both lysosomal and secretory activity account for the difference in area of the Kupffer cells. The rise and fall of relative area values for the Kupffer cells (Figure 27-10) roughly approximate those of Ep production following hepatectomy, bilateral nephrectomy, and hypoxia (Figure 27-1). It is of interest to note that when hepatectomy is followed by bilateral nephrectomy and the livers examined 24 hr later, Kupffer cell areas have increased considerably when

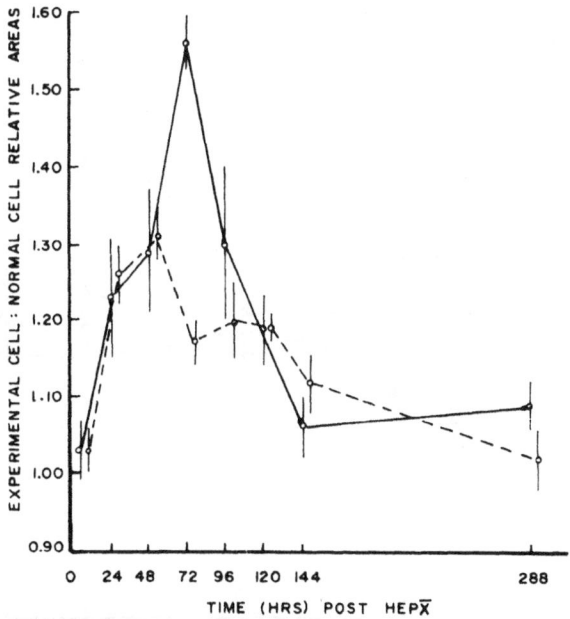

Figure 27-11 Mean relative areas of hepatic cells ±1 SEM at various times after partial hepatectomy and nephrectomy. Tracings are made from photographs taken at 970× after a 2× enlargement. Experimental to normal cell ratios significantly higher than 1 are indicative of hypertrophy. Solid line, Kupffer cells; broken line, Parenchymal cells.

204 Extrarenal Erythropoietin

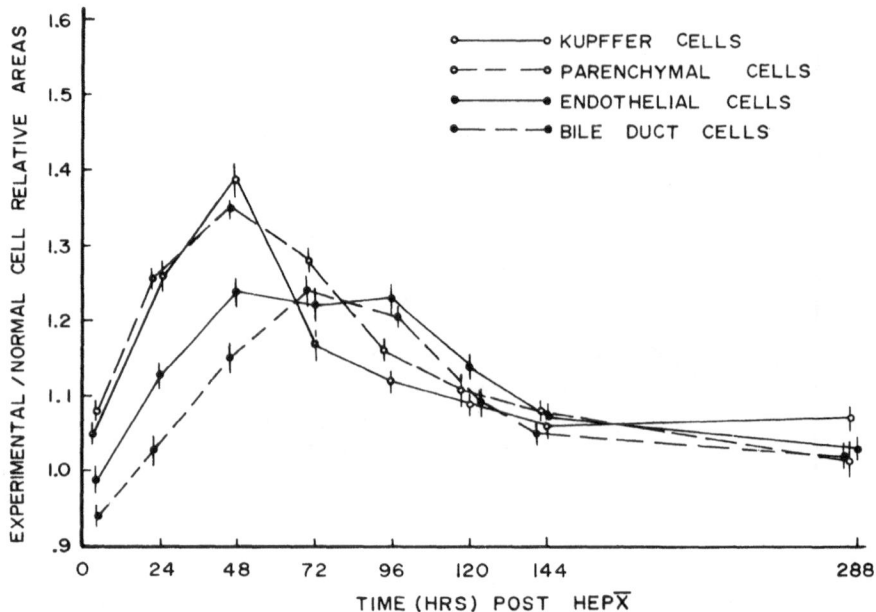

Figure 27-12 Mean relative areas of hepatic cells ±1 SEM at various times after partial hepatectomy, nephrectomy, and hypoxia. Tracings are made from photographs taken at 2,500×. Experimental to normal cell ratios significantly higher than 1 are indicative of hypertrophy.

compared to those following hepatectomy alone whereas parenchymal cell elements are not affected (Figure 27-11). If hepatectomy and bilateral nephrectomy are followed by hypoxia, Kupffer cell relative areas decrease considerably when compared to hepatectomized and bilaterally nephrectomized groups (Figure 27-12) whereas parenchymal cells show no significant variation. Some increase is also noted in Kupffer cell to parenchymal cell ratios as well as in Kupffer cell numbers per constant area after hepatectomy, bilateral nephrectomy, and hypoxia when compared to hepatectomy alone (Figure 27-13). The

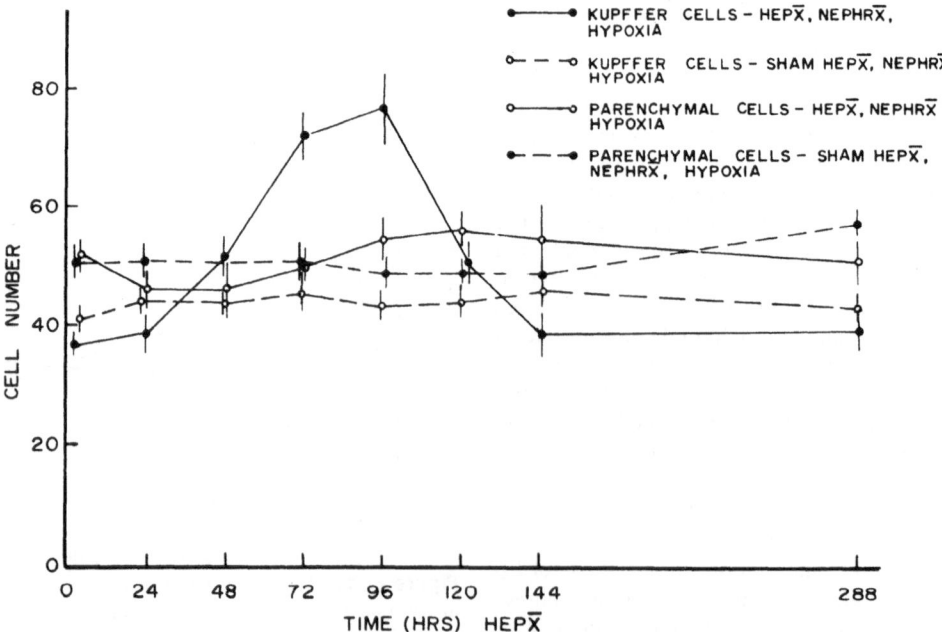

Figure 27-13 Mean number of hepatic cells ±1 SEM counted using a Kellner ocular with a constant area of interest at 400× and at various times after hepatectomy, nephrectomy, and hypoxia or after sham hepatectomy, nephrectomy, and hypoxia.

27-2, 27-3, 27-4). If there is an 18-hr delay between bilateral nephrectomy and a 6-hr exposure to hypoxia, the effect is diminished. Similar patterns hold for the relative areas (Table 27-3) as well as the numbers of hepatic cells per constant area (Table 27-4). It is suggested that bilateral nephrectomy and hypoxia exert a synergistic effect in the activation of hepatic regeneratory mechanisms and the production of Ep.

Radionuclide Studies of Hepatic Regeneration

Methods of imaging the liver by using a pinhole collimator and a gamma camera have previously been described.[73] The radiopharmaceutical most widely used for this purpose is ^{99m}technetium sulfur colloid (TSC).[74] TSC is rapidly taken up by the RES, especially the Kupffer cells in the liver.[75,76] Among the advantages of TSC are a low radiation exposure and an easily collimated 140 keV gamma emission.[74] Scanning techniques can be used *in vivo* to quantitate liver mass using either colloidal gold[77] or TSC.[78] Scintillation scanning of the liver using TSC has been used to evaluate liver regeneration in humans[79] and in animals.[80] This method has proved useful in the detection of metastatic malignancies,[81,82] hepatic abscesses

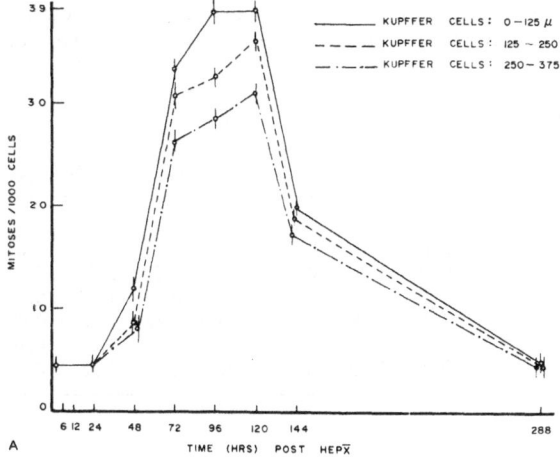
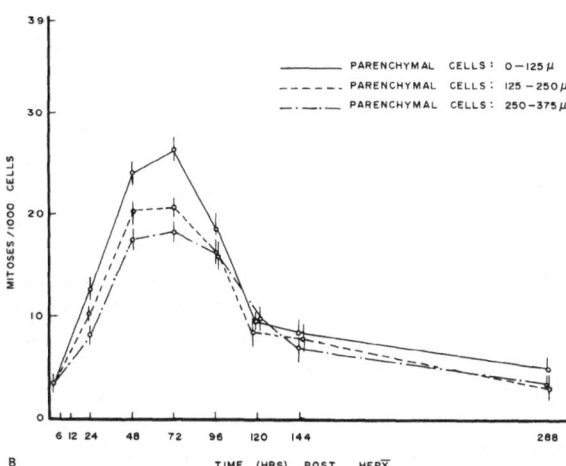

Figures 27-14A and B Mean number of dividing Kupffer (A) and parenchymal (B) 1,000 total cells ±1 SEM at different distances from the hepatic central vein and at various times after hepatectomy, nephrectomy, and hypoxia as evaluated by tritiated thymidine autoradiography.

percentage of cells in division is not affected by the imposition of nephrectomy and hypoxia on hepatectomy although the peak Kupffer cell labeling period is expanded (Figure 27-14). Also, nephrectomy and hypoxia do not alter the pattern of relative areas (Figure 27-15) or cell division (Figure 27-14) at various distances from the central vein when compared to the livers of rats subjected to hepatectomy alone (Figures 27-6 and 27-7). Hypoxia and nephrectomy alone exert some effects on liver cells. The incidence of dividing Kupffer cells is increased somewhat by 6 hr of hypoxia but decreased after a longer (24 hr) hypoxic stimulus (Table 27-2).[72] Parenchymal cells on the other hand show higher divisions rates in the 24-hr hypoxia group when compared to the 6-hr hypoxia group. The highest degree of activity is noted when nephrectomy is followed by 6 hr of hypoxia (Tables

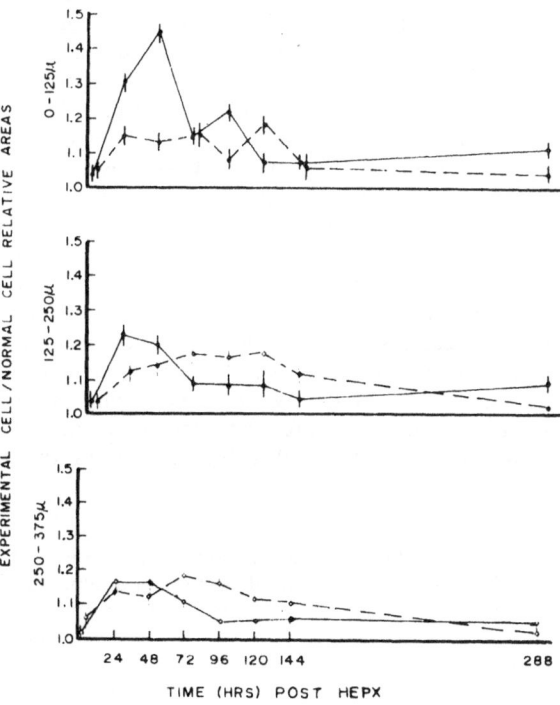

Figure 27-15 Mean relative areas ±1 SEM for hepatic cells at different distances from the hepatic central vein and at various time intervals after hepatectomy, nephrectomy, and hypoxia. Tracings are made from photographs taken at 970× after a 2× enlargement. Experimental to normal cell ratios significantly higher than 1 are indicative of hypertrophy. Solid line, Kupffer cells; broken line, Parenchymal cells.

Table 27-2 Mean number of dividing cells per 1000 total cells ±1 standard error of the mean after nephrectomy, hypoxia, or combined nephrectomy and hypoxia as evaluated by tritiated thymide autoradiography at various distances from the hepatic central vein.

	Kupffer Cells			Parenchymal Cells		
	0–125μ	125–250μ	250–375μ	0–125μ	25–250μ	250–375μ
Normal	2.13 ± 0.18	2.09 ± 0.06	2.11 ± 0.19	1.15 ± 0.08	1.13 ± 0.13	1.16 ± 0.11
Nephrex (6 hr)	2.05 ± 0.12	1.99 ± 0.11	2.05 ± 0.11	1.19 ± 0.13	1.13 ± 0.12	1.16 ± 0.05
Nephrex (24 hr)	2.33 ± 0.10	2.16 ± 0.05	2.20 ± 0.08	3.40 ± 0.41	3.19 ± 0.20	2.92 ± 0.14
Hypoxia (6 hr)	2.30 ± 0.08	2.31 ± 0.35	2.20 ± 0.12	1.59 ± 0.81	1.75 ± 0.41	1.40 ± 0.23
Hypoxia (24 hr)	2.06 ± 0.12	2.15 ± 0.09	2.12 ± 0.09	2.08 ± 0.19	1.98 ± 0.18	1.70 ± 0.08
Nephrex (18 hr) and Hypoxia (6 hr)	2.40 ± 0.18	2.22 ± 0.08	2.20 ± 0.10	2.71 ± 0.25	2.54 ± 0.07	2.43 ± 0.10
Nephrex and Hypoxia (6 hr)	4.24 ± 0.16	4.00 ± 0.29	4.01 ± 0.30	2.10 ± 0.17	2.50 ± 0.17	2.58 ± 0.43

and occlusions,[83,84] and other hepatopathies.[85,86] In the adolescent rat, TSC counts/cm² of hepatic tissue show steady increases 24 to 72 hr after hepatectomy when scans are taken 4 min after injection of TSC (Figure 27-16). When compared to normal livers, TSC counts/cm² then decline by 120 to 144 hr and reach normal values by 288 hr post-hepatectomy. Immediately after hepatectomy, the remaining liver assumes the form of a convoluted mass of tissue and although increasing in weight, it retains this shape up to 48 hr post-hepatectomy (Figure 27-17). The median lobe then begins a rapid period of growth and both the right and left lobes grow ventrolaterally until normal liver appearance is attained. This occurs 120 to 144 hr after hepatectomy in the young rat (Figure 27-2). When sham hepatectomy is followed by bilateral nephrectomy and hypoxia for 6 hr, there is a substantial rise in TSC counts/cm² when compared to those groups subjected to sham hepatectomy alone (Figure 27-16). Some elevation in TSC uptake is noted after either long term hypoxia (180 hr at 0.4 atm) or bilateral nephrectomy alone (Table 27-5). Short-term hypoxia (6 hr at 0.4 atm) does not have a significant effect on TSC uptake. When hepatectomy is followed by bilateral nephrectomy and hypoxia, TSC counts/cm² are considerably higher than in those groups experiencing only hepatectomy (Figure 27-16). These results are to be expected if hepatic TSC uptake is dependent on both numbers and activity of the Kupffer cells. When Kupffer cell relative areas and numbers (Figures 27-5, 27-10, 27-13) are compared to hepatic TSC uptake, it is seen that total TSC counts/cm² are proportional to both the number and activity of Kupffer cells. Partial hepatectomy results in a marked reduction in liver mass and total liver blood flow is reduced.[87,88] The increased perfusion rate per gm of hepatic tissue[89] is accompanied by a rise of only about 25 to 30% in the portal venous pressure.[90] Since hepatic extraction efficiency is dependent on hemodynamic factors,[91] it was expected that TSC uptake/cm² of hepatic tissue would be highest initially after hepatectomy and would drop thereafter, until liver mass is restored (Figure 27-2), but this was not the case. TSC counts/cm² increased gradually to a peak at 72 hr after hepatectomy, declining to slightly elevated levels 96 to 288 hr post hepatectomy. In our experiments, partial hepatic ischemia, brought about by portal vein ligation after hepatectomy, did little to alter TSC counts. Likewise, 30% hepatectomy accounted only for a slight elevation in the TSC counts. These findings agree with studies performed in isolated perfused livers

Table 27-3 Relative areas evaluated by planimetry of various cell types in the liver after bilateral nephrex, hypoxia, or combined bilateral nephrex and hypoxia (970×).

	Kupffer Cells			Parenchymal Cells		
	0–125μ	125–250μ	250–375μ	0–125μ	125–250μ	250–375μ
Nephrex (6 hr)	1.08 ± 0.04	1.05 ± 0.02	1.04 ± 0.06	1.10 ± 0.06	1.04 ± 0.03	1.03 ± 0.04
Nephrex (24 hr)	1.10 ± 0.03	1.07 ± 0.03	1.07 ± 0.02	1.18 ± 0.05	1.16 ± 0.01	1.15 ± 0.03
Hypoxia (6 hr)	1.15 ± 0.01	1.13 ± 0.03	1.12 ± 0.01	1.03 ± 0.01	1.05 ± 0.03	1.04 ± 0.01
Hypoxia (24 hr)	1.20 ± 0.03	1.22 ± 0.06	1.14 ± 0.03	1.01 ± 0.05	1.06 ± 0.04	1.04 ± 0.06
Nephrex (18 hr) and Hypoxia (6 hr)	1.19 ± 0.01	1.12 ± 0.04	1.10 ± 0.01	1.09 ± 0.03	1.03 ± 0.02	1.05 ± 0.02
Nephrex and Hypoxia (6 hr)	1.28 ± 0.03	1.22 ± 0.04	1.09 ± 0.07	1.05 ± 0.06	1.08 ± 0.04	1.07 ± 0.03

Table 27-4 Relative numbers of various cells types in the hepatic central vein area after nephrex, hypoxia, or combined nephrex and hypoxia.

	Kupffer Cells	Parenchymal Cells
Nephrex (6 hr)	38.45 ± 3.90	53.98 ± 4.71
Nephrex (24 hr)	44.73 ± 2.61	54.86 ± 4.93
Hypoxia (6 hr)	43.12 ± 4.39	55.71 ± 5.31
Hypoxia (24 hr)	35.63 ± 3.89	56.76 ± 4.98
Nephrex (18 hr) and Hypoxia (6 hr)	44.99 ± 5.38	54.57 ± 3.69
Nephrex and Hypoxia (6 hr)	51.62 ± 2.34	55.09 ± 4.56

Table 27-5 Technetium sulfur colloid uptake by liver after unilateral nephrectomy, bilateral nephrectomy, hypoxia, or combined bilateral nephrectomy and hypoxia.

	Counts/cm²/4 min Imaging	SEM
Bilateral nephrex (24 hr)	5,252	689
Unilateral nephrex (24 hr)	4,337	135
Bilateral nephrex (6 hr)	4,688	126
Unilateral nephrex (6 hr)	3,623	140
Hypoxia (0.4 atm) (7 hr)	2,906	246
Hypoxia (0.4 atm) (180 hr)	5,429	390
Combined bilateral nephrectomy and hypoxia (0.4 atm) (7 hr)	11,668	611

after hepatectomy.[47] In those experiments, $CrPO_4$ uptake by the liver was maximal at 72 hr after hepatectomy, remaining elevated for up to 90 days.

Studies in Rats after Multiple Hepatectomies

Repeated hepatectomies were carried out in adolescent rats using the procedure described above. The accumulation of fat has been demonstrated in the early stages of hepatic regeneration after single or multiple hepatectomies.[32,92] This produces a hardening of the hepatic tissue and renders further surgery less manageable. The removal of only the right and median hepatic lobes during the first hepatectomy makes possible the left hepatic lobectomy performed 96 hr later. Peripheral blood hematocrit values remained relatively unchanged in all rats undergoing multiple hepatectomy. The point-slopes of a percent regeneration vs. time post-hepatectomy graph (Figure 27-2) reveal that the most active period of hepatic growth in the singly hepatectomized adolescent rat is 24 to 48 hr whereas in the adult singly hepatectomized rat it is 72 to 96 hr after hepatectomy. Analysis of the response to repeated hepatectomies revealed that, initially, these livers regenerated more rapidly than in rats subjected to a single hepatectomy. It is interesting to note that tissue mass never completely returns to normal within the period

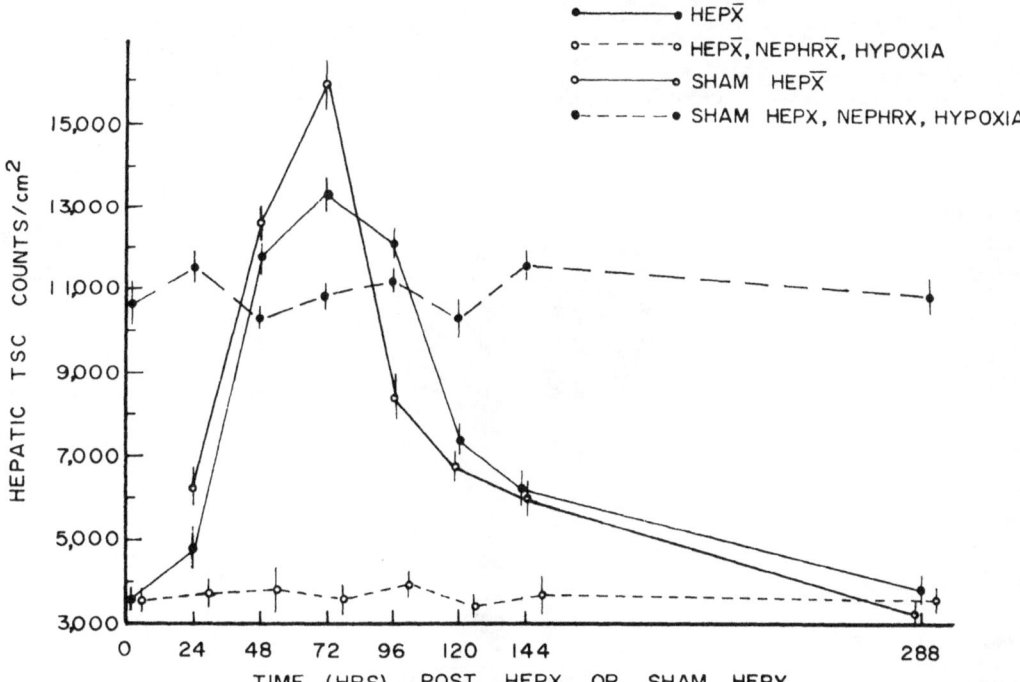

Figure 27-16 Mean hepatic TSC counts/cm² ±1 SEM at various times after hepatectomy, sham hepatectomy, hepatectomy, nephrectomy, and hypoxia; or sham hepatectomy, nephrectomy, and hypoxia.

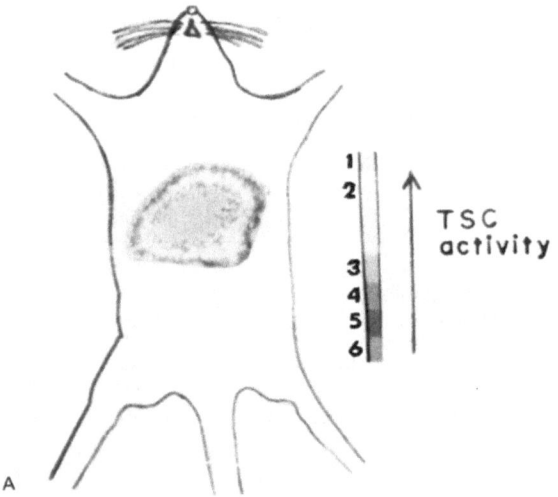

Figure 27-17 Polaroid scintigraphs of normal and regenerating liver taken at various time intervals after subtotal hepatectomy. **A.** Diagram depicting the gradation of TSC activity (1–6) in hepatic phagocytic cells. Subsequent scintigraphs (B–G) are keyed in this manner.

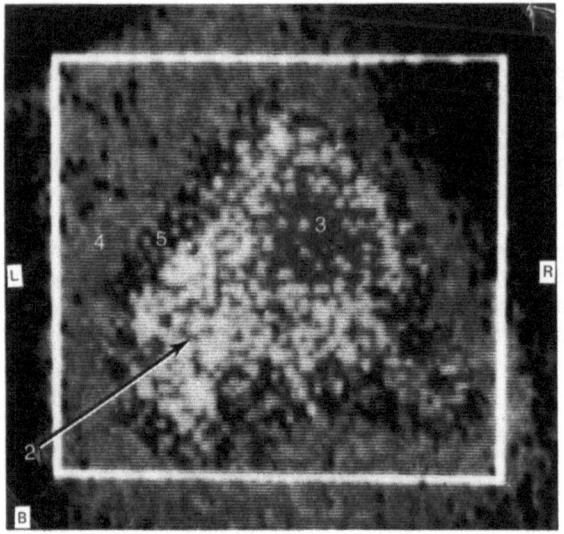

Figure 27-17 B. Normal.

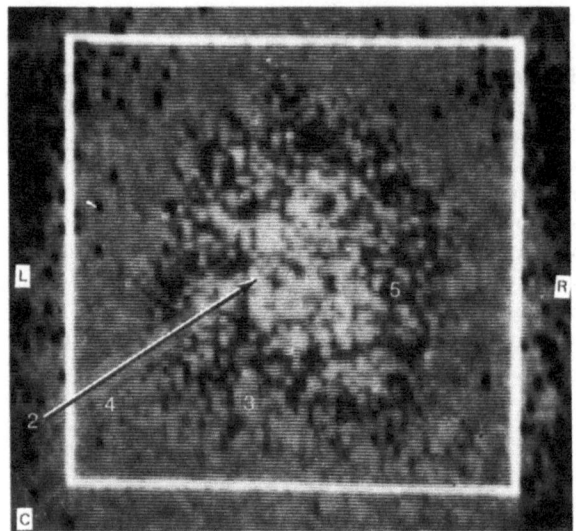

Figure 27-17 C. 24 hr post hepx.

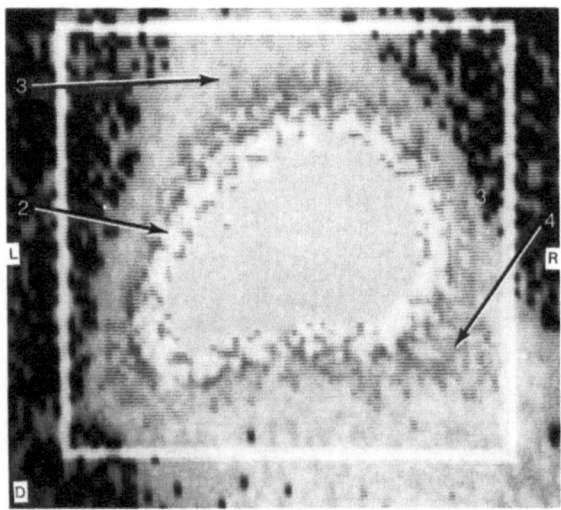

Figure 27-17 D. 48 hr post hepx.

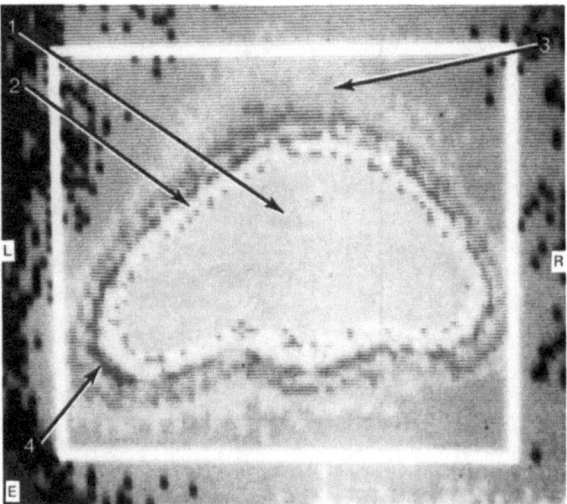

Figure 27-17 E. 72 hr post hepx.

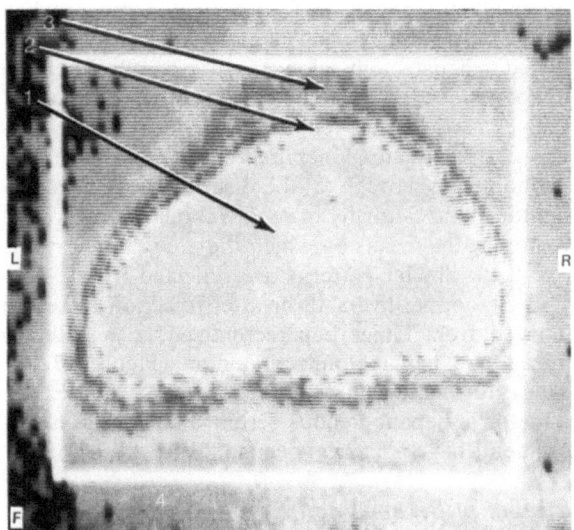

Figure 27-17 F. 96 hr post hepx.

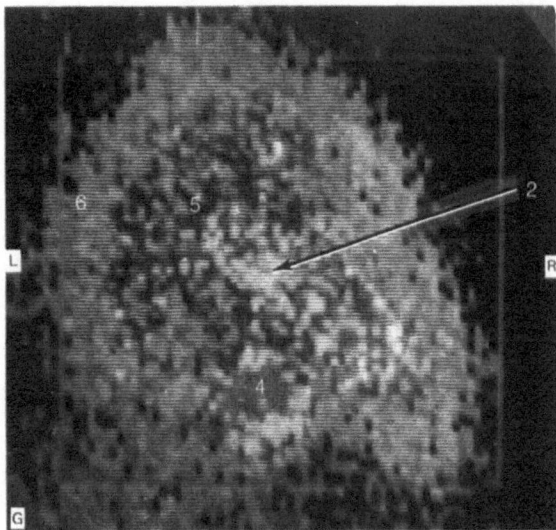

Figure 27-17 G. 120 hr post hepx.

of observation in rats undergoing multiple hepatectomy. With each successive hepatectomy, the initial growth rate is greater than in the preceding hepatectomy but the peaks are not so high (Figure 27-2). Presumably, the tissue subjected to multiple hepatectomy loses much of its capacity to regenerate following each successive hepatectomy. The localized necrosis, multiple abscesses, and fibrous scarring that accompany multiple hepatectomy[92] may account for this phenomenon. At various intervals after multiple hepatectomies, nephrectomy, and hypoxia, serum Ep levels are elevated compared to normal, anephric, hypoxic rats but Ep levels are considerably lower than those obtained for the same intervals after single hepatectomy, bilateral nephrectomy, and hypoxia (Figure 27-1 and Table 27-6).

Studies on a Serum Factor(s) Triggering Hepatic Production of Ep

Many experiments have been performed to demonstrate a possible humoral mechanism controlling liver regeneration. Classic experiments have been carried out in the parabiotic animal. One normal animal is surgically joined to another animal whose liver has been removed. Mitotic activity in the liver cells of the normal parabionts were shown to be elevated when compared to normal partners of sham hepatectomized parabionts.[56,57,93] Serum from hepatectomized rats has the capacity to stimulate fetal liver mitoses[94] and DNA synthesis.[95] Several workers have been unable to confirm these reports.[96,97] Normal liver cell cultures treated with serum from partially hepatectomized rats[98] or humans[99] showed an increased number of mitoses per total cells when compared to cultures treated with sham-hepatectomized serum. Serum from hepatectomized rats can stimulate livers already regenerating[100,101] and perfusion of isolated rat livers with blood from partially hepatectomized rats has also resulted in augmented numbers of liver mitoses.[47,102]

Recent evidence also supports a humoral system for kidney hypertrophy following unilateral nephrectomy.[103,104] The presence of a factor found in the portal vein has been reported to stimulate hepatic regeneration[105,106] particularly the hypertrophy of parenchymal cells.[107] Others have shown that in fowl regeneration is not dependent on portal blood flow rate.[108] The spleen has been reported to have a positive influence on regeneration,[109] but others have found that splenectomy, which decreases the liver blood flow by 38%, actually stimulates regeneration.[110] Although serum from hepatectomized animals enhances the growth of hepatic cells, it may also

Table 27-6 Ep response at various time intervals after multiple hepatectomies.

Number of Hepatectomies	Time (hr) after Last Hepatectomy	Ep (IU) ± SEM
2	24	0.15 ± 0.03
2	48	0.23 ± 0.02
2	72	0.13 ± 0.02
2	120	0.18
2	144	0.12
3	48	0.30 ± 0.04
3	72	0.25 ± 0.02
4	72	0.19 ± 0.04

The data for Ep represent equivalent units of Ep derived from the standard dose-response curves for the international reference preparation (IRP) of human urinary Ep, assayed in the exhypoxic polycythemic mouse.

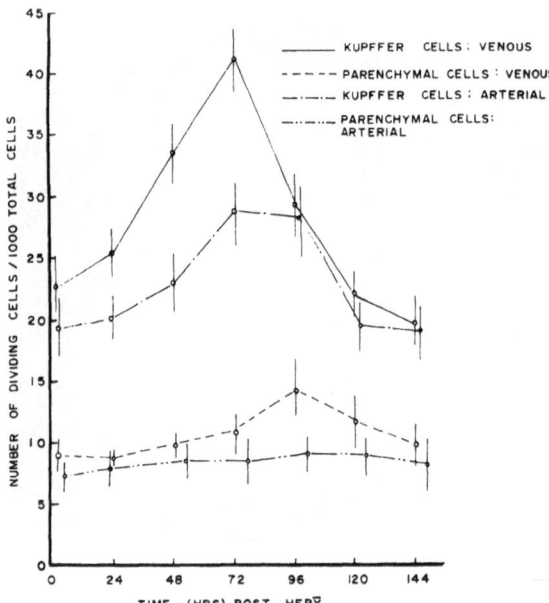

Figure 27-18 Mean number of dividing hepatic cells ±1 SEM in normal rats 18 hr after injection of either venous or arterial serum obtained from hepatectomized donor rats as determined by tritiated thymidine autoradiography.

contain a generalized growth factor that stimulates mitoses in other tissues as well.[111]

Effects of Serum Factor on Hepatic Histology

The regenerating liver responds to bilateral nephrectomy and hypoxia with increased levels of Ep (Figure 27-1) and if liver regeneration is mediated by a humoral mechanism, then serum from hepatectomized animals, if it does indeed stimulate hepatic cells in the normal animal, should also augment Ep production. This was found to be the case. In the typical experiment, serum from donor rats is obtained from either the dorsal aorta or the hepatic vein at different times after hepatectomy and 2 ml are injected intravenously. Stimulatory activity in venous as well as in arterial serum was noted on Kupffer cell elements whereas parenchymal cells revealed only slight stimulation after the injection of venous serum (Figure 27-18) when compared to control groups injected with saline or normal rat serum. These effects have been previously reported.[106] With venous serum, the numbers of mitoses were significantly higher in the recipients' Kupffer cells after injection of 48 to 72 hr post-hepatectomy serum and in parenchymal cells after injection of 96 hr post-hepatectomy serum (Figure 27-18) when compared to saline or normal rat serum injected controls. When serum donors were hepatectomized and later bilaterally nephrectomized 24 hr before collection of serum, twin peaks were observed for the recipients' Kupffer cells whereas parenchymal cells showed only slight increases when compared to groups in which the donors were subjected to hepatectomy alone (Figure 27-19). When the donors' kidneys were removed, the differences noted between arterial and venous serum disappeared (Figure 27-19). Bilateral nephrectomy may cause this phenomenon through (a) the disappearance of an inhibitor found normally in renal venous serum, or (b) removing the means by which the growth factor(s) is excreted. Similar patterns are reflected in TSC uptake by normal livers 18 hr after injection of serum derived from either hepatectomized or hepatectomized and nephrectomized donors (Figure 27-20). This is further indication of the priming of the RES by a factor in hepatic venous serum. Planimetric measurements have not as yet been made on these groups.

Effects of Serum Factor on Serum Ep of Recipient Rats

A significant increase in Ep production is noted 18 hr after injection of serum from hepatectomized animals into normal animals subjected to nephrectomy and exposure to hypoxia when compared to the effects seen with serum from sham hepatectomized or normal rats, or with saline (Figure 27-21). Ep levels in normal rats after the injection of serum derived from hepatectomized donors and followed by nephrec-

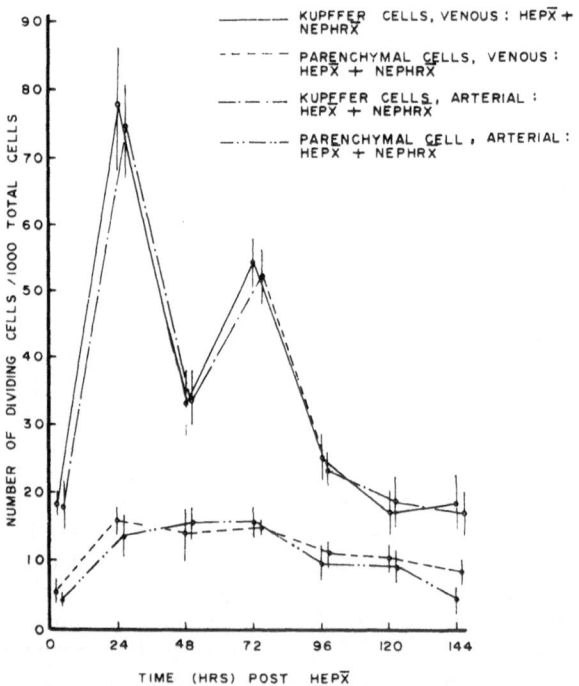

Figure 27-19 Mean number of dividing hepatic cells ±1 SEM in normal rats 18 hr after injection of either venous or arterial serum taken from donor rats that were hepatectomized and subjected to bilateral nephrectomy 24 hr before collection of serum.

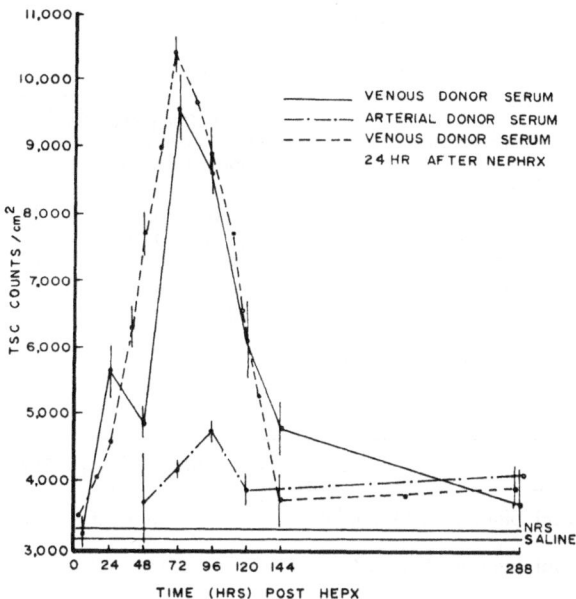

Figure 27-20 Hepatic TSC counts/cm² 18 hr after injection of donor serum derived from either the hepatic vein or the dorsal aorta of rats at various times after partial hepatectomy.

tomy and exposure to hypoxia are similar to those observed in hepatectomized animals after nephrectomy and hypoxia (Figure 27-1), although the peak Ep production is considerably lower.

Hepatic venous serum is considerably more potent in producing this Ep response than arterial serum. Hypoxia, however, is necessary for this Ep response; if absent, Ep levels are barely detectable. When both the liver and kidneys are removed from the donors, there is no significant difference between arterial and venous serum in evoking the Ep response (Figure 27-22), although a double peak is observed similar to that noted in the autoradiographic studies (Figure 27-19). When donor serum taken after hepatectomy and bilateral nephrectomy is injected into a recipient that is bilaterally nephrectomized immediately before injection and 18 hr later rendered hypoxic for 6 hr, the greatest extent of Ep production is noted (Figure 27-21). Although there is greater variance in Ep values within this group, the peak is similar to that observed in the hepatectomized, nephrectomized, hypoxic animal (Figure 27-1). Since bilateral ureter ligation in either the hepatectomized or sham-hepatectomized rat has little effect on either Ep production or cellular changes (Table 27-7), an effect of uremia is probably to be discounted. Removal of the kidneys augments Ep production as well as Kupffer cell changes in all groups studied. Normal serum in-

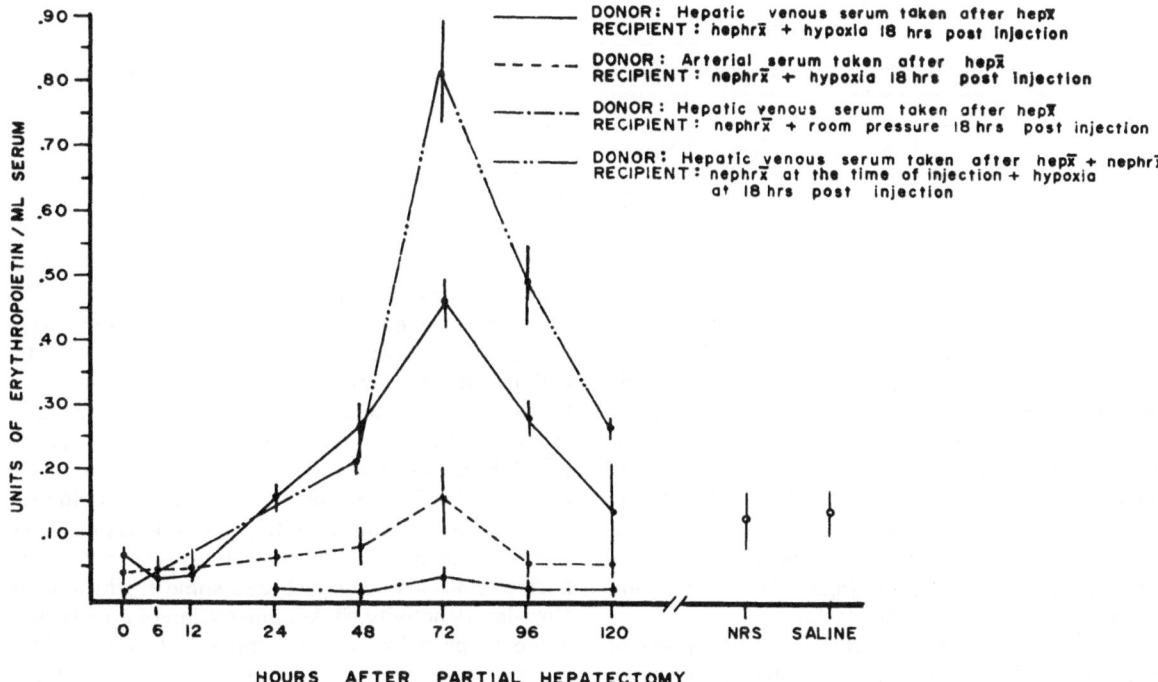

Figure 27-21 Serum Ep levels ±1 SEM for normal recipient or anephric recipient rats 18 hr after injection of 2 ml of serum derived from either the hepatic vein or the dorsal aorta of donor rats at various times after either hepatectomy or combined hepatectomy and nephrectomy. The Ep assay is described in the text (ref. 43).

Table 27-7 Ep assay and cellular measurements after various control procedures.

	Ep Response in Controls		Ep Response in Controls
Donor Material	Ep (IU) ± SEM in Recipients 18 hr after Injection, Nephrex, and 6 hr Hypoxia	Donor Material; Time after Hepx	Ep (IU) ± SEM in Recipients 18 hr after Injection of Serum, Nephrex, and 7 hr Ambient Pressure
Normal Rat Serum	0.07 ± 0.02	24 hr	0.02 ± 0.01
Saline	0.07 ± 0.01	48 hr	0.01 ± 0.01
Ep (1.1 IU)	0.02 ± 0.01	72 hr	0.03 ± 0.01
		96 hr	0.02 ± 0.02
		120 hr	0.02 ± 0.01

^{59}Fe (%) Incorporation after Various Control Procedures

Protocol		Ep (% RBC-^{59}Fe inc) ± SEM
Donor of Serum (2 ml)	Recipient	
48 hr post hepx, bilat. ureter ligation	Normal—18 hr after inj, bilat. nephrex & 6 hr hypoxia	9.66 ± 1.01
72 hr post hepx, bilat. ureter ligation	Normal—18 hr after inj, bilat. nephrex & 6 hr hypoxia	13.15 ± 0.64
48 hr post hepx, bilat. ureter ligagion	Normal—18 hr after inj, bilat. ureter ligation & 6 hr hypoxia	13.77 ± 0.85

Standards

0.9%	NaCl	1.37 ± 0.30
0.05	IRP U Ep	3.55 ± 0.54
0.20	IRP U Ep	9.44 ± 0.91
0.80	IRP U Ep	19.86 ± 2.29

Cell Measurements after Bilateral Ureter Ligation			Experimental/Normal Cell Relative Areas		
Time Post Ligation	Incidence of Mitoses/1000 Cells 0–125μ from the Hepatic Central Vein		Time Post Ligation	Incidence of Mitoses/1000 Cells 0–125μ from the Hepatic Central Vein	
	Kupffer Cells	Parenchymal Cells		Kupffer Cells	Parenchymal Cells
0 hr	3.60 ± 1.08	2.01 ± 1.29	0 hr	1.01 ± 0.03	1.03 ± 0.02
6 hr	3.06 ± 1.28	2.20 ± 0.88	6 hr	0.97 ± 0.06	1.01 ± 0.04
24 hr	2.99 ± 0.76	2.09 ± 1.01	24 hr	1.07 ± 0.02	1.04 ± 0.03

hibitors of hepatic regeneration have previously been reported.[112–115] The normal course of regeneration is impeded by the daily administration of normal plasma[112] or normal serum[114] whereas if a parabiotic union is made between a hepatectomized rat and two normal partners, mitoses in the hepatectomized animal are considerably lower than in non-parabiosed hepatectomized animals.[115] Fetal liver extracts can stimulate *in vitro* erythropoiesis.[116] Since the regenerating liver bears some structural and functional similarities to the fetal liver (Figure 27-9), the Ep-inducing mechanisms may be similar.

The liver at one time was thought to be the site of inactivation of Ep[17] but it has been shown that when livers are perfused with mixtures containing Ep, a significant decrease in Ep activity was not observed.[117] Ep levels in the plasma of rats treated with phenylhydrazine[118] or carbon tetrachloride[119] were elevated when liver tissue was damaged. Similar Ep activity was not seen when treated animals had sustained no hepatic injury. Hepatic injury due to carcinoma results in many cases in increased levels of Ep.[120] Renal disease may also bring about increased levels of Ep.[121,122] Reports of specific endocrine receptor changes, especially to growth hormone, have been demonstrated in regenerating liver.[123] Growth hormone has also demonstrated some erythropoietic stimulatory activity.[124] Administration of growth hormone to partially hepatectomized rats results in a prolonged wave of DNA synthesis[125] and can stimulate the repair of livers injured by carbon tetrachloride administration.[126] Although hypophysectomy

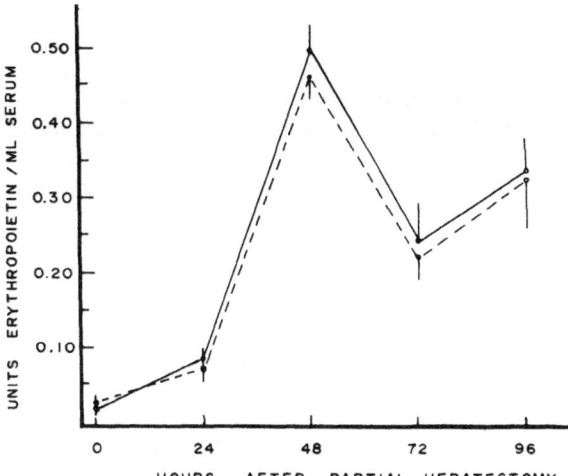

Figure 27-22 Serum Ep levels ± 1 SEM for normal recipient rats 18 hr after injection of 2 ml of serum derived from either the hepatic vein or the dorsal aorta of donor rats at various times after partial hepatectomy and nephrectomy.

decreases parenchymal cell polyploidy after hepatectomy, it has little effect on regeneration as a whole.[71]

Summary

Subtotal hepatectomy evokes an elevated erythropoietin (Ep) response in anephric rats subjected to hypoxia when compared to sham-hepatectomized, anephric hypoxic groups. Hepatectomy also induces numerous hepatocellular changes, among these being an increased Kupffer cell to parenchymal cell ratio, pronounced elevation in relative areas as well as in the % mitoses of Kupffer cells when compared to parenchymal cells. The values vary depending on the distance of the cells from the hepatic central vein. The ability of the Kupffer cell to accumulate radioactive technetium sulphur colloid (TSC) also increases during the course of regeneration. This is, most likely, related to the increased numbers and activity of the Kupffer cells.

Hepatic Ep production decreases with increasing age as does the hepatic regenerative potential. Erythroblastic nests have been noted in livers 24 hr after hepatectomy. In this connection, partial liver removal may lead to an alteration in the hepatic microenvironment, rendering it again favorable for erythropoiesis. Ep production varies in animals undergoing multiple hepatectomies but in all cases the peak values attained parallel the period of maximal tissue proliferation.

Humoral factor(s) involved with liver regeneration have previously been cited, although the precise nature of these growth principles is as yet unclear. Our present experiments indicate that the serum of hepatectomized donor rats is capable of stimulating mitotic activity, as well as inducing changes in the morphology of the hepatic cells of normal rats. Serum from hepatic venous blood is considerably more potent in eliciting these responses than is serum of arterial origin. The structural changes evoked are most conspicuous in the Kupffer cell population. Normal rats also respond to the injection of serum from hepatectomized donors with increased levels of serum Ep when the recipients are nephrectomized and rendered hypoxic 18 hr after serum injection. Removal of the kidneys from the hepatectomized donors augments further this Ep response to the serum in recipient rats and the differences noted between venous and arterial serum are abolished. Indeed, when the kidneys are removed from hepatectomized donors and serum is administered to nephrectomized recipients, the highest serum Ep levels are seen ater an 18 hr delay and exposure to hypoxia.

Conspicuous alterations occur in the Kupffer cell elements of recipient animals after injection of serum from donors that had been subjected to either hepatectomy or combined hepatectomy and nephrectomy. These changes include an increased number of mitoses, elevated TSC uptake, and an increase in the cellular areas when compared to controls. The serum factor is clearly liver-derived, since ligation of the portal vein did not diminish its effects and very little activity was noted when serum from the systemic circulation was used. In addition, the presence of the kidney has a suppressive effect on this factor, since bilateral nephrectomy in either donor or recipient rats results in increased serum Ep levels in the recipient when compared to non-nephrectomized animals. Our present concept depicting renal–hepatic interrelations in Ep production is shown in Figure 27-23. Kupffer cells have been shown to elaborate hematopoietically active substances and they may represent

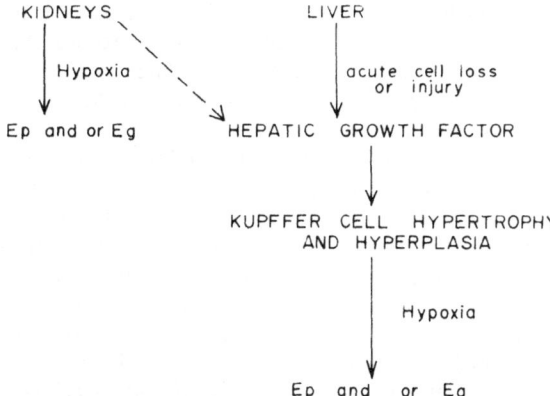

Figure 27-23 Figure indicates interrelations between kidneys and liver in the production of an hepatic growth factor and Ep. Solid line, stimulatory; broken line, inhibitory.

the cellular site of origin of Ep. In addition, we propose that another hepatic factor, distinct from Ep, is necessary for the hepatic production of this latter hormone. It has been demonstrated that a principle in fetal liver augments *in vitro* erythropoiesis.[116] The *in vivo* mechanism which we have described may be similar in nature.

Acknowledgment

This study was supported by grant 5 R01HLB03357-20 from the National Heart, Lung and Blood Institute, National Institutes of Health, USPHS. The authors wish to thank D. J. Birnbach, J. A. Piliero, A. Ronquillo, and M. Roy for their help on this project.

References

1. Krantz SB, Jacobson LO: Erythropoietin and the Regulation of Erythropoiesis. Chicago, University of Chicago Press, 1970.
2. Gordon AS: Erythropoietin, in *Vitamins and Hormones*. New York, Academic Press, *31:*105-174, 1973.
3. Jacobson LO, Goldwasser E, Fried W, et al.: Role of the kidney in erythropoiesis. *Nature 179:*633-634, 1957.
4. Mirand EA, Prentice TC, Slaunwhite WR: Current studies on the role of erythropoietin on erythropoiesis. *Ann NY Acad Sci 77:*677-702, 1959.
5. Naets JP: in Williams PC (ed): *Hormones and the Kidney*. New York, Academic Press, pp. 75-86, 1963.
6. Fried W, Kilbridge T, Krantz S, et al.: Studies on extrarenal erythropoietin. *J Lab Clin Med 73:*244-248, 1969.
7. Schooley JC, Mahlmann LJ: Evidence for the *de novo* synthesis of erythropoietin in hypoxic rats. *Blood 40:*662-670, 1972.
8. Schooley JC, Mahlmann LJ: Erythropoietin production in the anephric rat. I. Relationship between nephrectomy, time of hypoxic exposure, and erythropoietin production. *Blood 39:*31-38, 1972b.
9. Gordon AS, Kaplan SM: Erythrogenin (REF), in Fisher JW (ed): *Kidney Hormones,* Vol II, London, Academic Press, pp. 187-229, 1977.
10. Gordon AS, Zanjani ED, Peterson EN, et al.: Studies on fetal erythropoiesis, in Gordon AS, Condorelli M, Peschle C (eds): *First Int Conf on Hematop Regul of Erythropoesis*. Milan, Ill Ponte, pp. 188-204, 1972.
11. Carmena AO, Howard D, Stohlman F Jr: Regulation of erythropoietin. XXII. Production in the newborn animal. *Blood 32:*376-382, 1968.
12. Wang F, Fried W: Renal and extrarenal erythropoietin production in male and female rats of various ages. *J Lab Clin Med 79:*181-186, 1972.
13. Schooley JC, Mahlmann LJ: Extrarenal erythropoietin production by the liver in the weanling rat. *Proc Soc Exptl Biol Med 145:* 1081-1083, 1974.
14. Peschle C, Marone G, Genovese A, et al.: Erythropoietin production by the liver in fetal neonatal life. *Life Sci 17:*1325-1330, 1975a.
15. Kaplan SM, Weinberg SR, Gordon AS: Erythropoietin production in the neonatal rat. Sites of erythrogenin production. *Biol Neonate 24:*145-150, 1974.
16. Zanjani ED, Poster J, Burlington H, et al.: Liver as the primary site of erythropoietin formation in the fetus. *Lab Clin Med 89:*640-644, 1977.
17. Burke WT, Morse BG: Studies on the production and metabolism of erythropoietin in rat liver and kidney, in Jacobson LO, Doyle M (eds): *Erythropoiesis*. New York, Grune & Stratton, pp. 111-119, 1962.
18. Reissmann KR, Nomura T: Erythropoietin formation in isolated kidneys and liver, in Jacobson LO, Doyle M (eds): *Erythropoiesis*. New York, Grune & Stratton, 1962.
19. Katz R, Cooper GW, Gordon AS, et al.: Studies on the site of production of erythropoietin. *Ann NY Acad Sci 149:*120-127, 1968.
20. Fried W: The liver as a source of extrarenal erythropoietin production. *Blood 40:*671-677, 1972.
21. Mirand EA, Murphy GP: Renal erythropoietic (RDF) assays in anephric and intact baboons following varied erythropoietic stimulus. *J Med 2:*192-200, 1971.
22. Kaplan SM, Rothmann SA, Gordon AS, et al.: Extrarenal sites of erythrogenin production. *Proc Soc Exptl Biol Med 143:*310-313, 1973.
23. Peschle C, Marone G, Genovese A, et al.: Relation between the reticuloendothelial system (RES) and extrarenal erythropoietin production, in Nakao K, Fisher JW, Takaku F (eds): *Erythropoiesis: Proceedings of the 4th International Conference on Erythropoiesis*. Baltimore, University Park Press, 1974.
24. Peschle C, Marone G, Genovese A, et al: Increased erythropoietin production in anephric rats with hyperplasia of the reticuloendothelial system induced by colloidal carbon or zymosan. *Blood 47:*325-337, 1976a.
25. Peschle C, Marone G, Genovese A, et al.: Hepatic erythropoietin: Enhanced production in anephric rats with hyperplasia of Kupffer cells. *Br J Haematol 32:*105-111, 1976b.
26. Haurani FI, O'Brien R: The erythropoietic ef-

fect of a reticuloendothelial blocking agent. *J Res 13:*126–133, 1973.
27. Carson EJ, Roy M, Naughton BA, et al.: The effect of reticuloendothelial overload on extrarenal erythrogenin and erythropoietin production. Proceedings of the 19th annual meeting. *Am Soc Hematol:*107, 1976.
28. Joyce RA, Chervenick PA: Stimulation of granulopoiesis by liver macrophages. *J Lab Clin Med 56:*112–117, 1975.
29. Chervenick PA, LoBuglio AF: Human blood monocytes: Stimulators of granulocyte and mononuclear colony formation *in vitro*. *Science 178:*164–167, 1972.
30. Gruber DF, Zucali JR, Wlekinski J, et al.: Temporal transition in the site of rat erythropoietin production. *Exp Hematol* (in press).
31. Higgins GM, Anderson RM: Experimental pathology of the liver: Restoration of the liver of the white rat following partial surgical removal. *Arch Pathol 12:*186–202, 1931.
32. Bucher NR: Regeneration of mammalian liver. *Intern Rev Cytol 15:*245–300, 1963.
33. Morley CG: Humoral regulation of liver regeneration and tissue growth. *Persp Biol Med 17:*411–428, 1974.
34. Stenger RJ, Confer DB: Hepatocellular ultrastructure during liver regeneration after subtotal hepatectomy. *Exp Mol Pathol 5:*455–474, 1965.
35. Anderson WM, Grundholm A, Sells BH: Modification of ribosomal proteins during liver regeneration. *Biochem Biophys Res Comm 62:*669–676, 1975.
36. Naughton BA, Kaplan SM, Roy M, et al.: Hepatic regeneration and erythropoietin production in the rat, in *Prog 19th Ann Meet Am Soc Hematol*, p. 106, 1976.
37. Naughton BA, Kaplan SM, Roy M, et al.: Hepatic regeneration and erythropoietin production in the rat. *Science 196:*301–302, 1977a.
38. Naughton BA, Piliero JA, Kruger RE, et al.: Age related variations in hepatic regeneration and erythropoietin production in the rat. *Am J Anat 149:*431–438, 1977b.
39. Anagnostou A, Schade S, Barone J, et al.: Effects of partial hepatectomy on extrarenal erythropoietin production in rats. *Blood 50:*457–462, 1977.
40. Eränkö O: in Quantitative Methods in Histology and Microscopic Histochemistry. Boston, Little, Brown, pp. 57–73, 1955.
41. Edgren J, Laasonen L, Kock B, et al.: Kidney function and compensatory growth of the kidney in living kidney donors. *Scand J Urol Nephrol 10:*134–136, 1976.
42. Millonig G: The advantages of a phosphate buffer for OsO_4 solutions in fixation. *J Appl Physics 32:*1637, 1961.
43. Camiscoli, JF, Weintraub AH, Gordon AS: Comparative assay of erythropoietin standards. *Ann NY Acad Sci 149:*40–45, 1968.
44. Becker FF, Lane BP: Regeneration of the mammalian liver I. Autophagocytosis during differentiation of the liver cell in preparation for cell division. *Am J Pathol 47:*783–801, 1965.
45. Bucher NR, Glinos AD: The effect of age on regeneration of rat liver. *Cancer Res 10:*324–332, 1950.
46. Straube RL, Patt HM: Regeneration following partial hepatectomy as a function of age in the mouse. *Fed Proc 20:*286, 1961.
47. Leong GF, Pessotti RL, Braeur RW: Liver function in regenerating rat liver. $CrPO_4$ colloid uptake and bile flow, *Am J Physiol 197:*880–886, 1959.
48. Grisham JW: A morphologic study of DNA synthesis and cell proliferation in regenerating rat liver: Autoradiography with thymidine-H^3. *Cancer Res 22:*842–849, 1962.
49. Edwards JL, Koch A: Parenchymal and littoral cell proliferation during liver regeneration. *Lab Invest 13:*32–43, 1964.
50. Edwards JL, Smith SW, West-mark ER, et al.: Interrelations of DNA synthesis and cell division in normal and regenerating liver. *Fed Proc 18:*475, 1959.
51. Schultze B, Oehlert W: Autoradiographic investigation of incorporation of ^{3}H-Tdr into cells of the rat and mouse. *Science 131:*737–738, 1960.
52. Wilson ME, Stowell RE, Yokoyama HO, et al.: Cytological changes in regenerating mouse liver. *Cancer Res 19:*561–565, 1959.
53. Sigel B, Baldia LB, Dunn MR, et al.: Humoral control of liver regeneration. *Surg Gynecol Obstet 124:*1023–1031, 1967.
54. Johnson HA: Liver regeneration and the "critical mass" hypothesis. *Am J Pathol 57:*1–15, 1969.
55. Harkness RD: The spatial distribution of dividing cells in the liver of the rat after partial hepatectomy. *J Physiol 116:*373–379, 1952.
56. Bucher NR, Scott JF, Aub JC: Regeneration of the liver in parabiotic rats. *Cancer Res 11:*457–465, 1951.
57. Wenneker AS, Sussman N: Regeneration of liver tissue following partial hepatectomy in parabiotic rats. *Proc Soc Exptl Biol Med 76:*683–686, 1951.
58. Bernhard W, Rouiller C: Close topographical relationship between motochondria and ergostoplasm of liver cells in a definite phase of cellular activity. *J Biophys Biochem Cytol 2:*73–78, 1956.
59. Oberling C: The structure of cytoplasm. *Internat Rev Cytol 8:*1–31, 1959.
60. Bartok I, Viragh S: Zur Entwicklung und Dif-

ferenzierung des Endoplasmatischen Retikulums in den Epithelizellan der Regenerierden Leber. *Z Zellforsch Mikroscop Anat 68:*741–754, 1965.
61. Naughton BA, Arce J, Liu P, et al.: Ultrastructural analysis of hematopoiesis in regenerating liver. Submitted for publication, 1977.
62. Silini G, Pozzi LV, Pona S: Studies on the haematopoietic stem cells of mouse fetal liver. *J Embryol Exp Morph 17:*303–312, 1967.
63. Sjögren U: Erythroblastic islands and extramedullary erythropoiesis in chronic myeloid leukemia. *Acta Haemat 55:*272–276, 1976.
64. Metcalf D, Moore MAS: Haematopoietic Cells. Amsterdam, North Holland Publishing, 1972.
65. Ben-Ishay Z: Selective eradication or rat bone marrow-erythroid cells by administration of cytosine arabinoside: Observations on resumptive phase of erythropoiesis. *Scand J Haematol 14:*361–368, 1975.
66. Ben-Ishay Z: Reticular cells in erythroid islands: Their erythroblastophagocytic function. *J Res 16:*340–346, 1974.
67. Stephenson JR, Axelrad AA, McLeod DL, et al.: Induction of colonies of hemoglobin-synthesizing cells by erythropoietin *in vitro*. *Proc NAS 68:*1942–1946, 1971.
68. Trentin JJ: Influence of hematopoietic organ stroma (hematopoietic inductive microenvironment on stem cell differentiation, in Gordon AS (ed): *Regulation of Hematopoiesis*. New York, Appleton-Century-Crofts, pp. 159–184, 1970.
69. Weis, L: The hematopoietic microenvironment of the bone marrow: An ultrastructural study of the stroma in rats. *Anat Rec 186:*161–184, 1976.
70. Hoffman J, Himes MB, Klein A, et al.: Responses of the liver to injury. *AMA Arch Pathol 60:*10–18, 1955.
71. Geschwind I, Alfert M, Schooley C: Liver regeneration and hepatic polyploidy in the hypophysectomized rat. *Exptl Cell Res 15:*232–235, 1958.
72. Talarico KS, Feller DD, Neville ED: Effects of hyperoxia and hypoxia on mitosis in the normal and regenerating rat liver. *Proc Soc Exptl Biol Med 131:*430–434, 1969.
73. Whang KS, Fish MB: Evaluation of hepatic photoscanning with radioactive colloidal gold. *J Nucl Med 6:*494–499, 1965.
74. Harper PV, Lanthrop KA, Jiminez F, et al.: Technetium 99m as a scanning agent. *Radiology 85:*101–109, 1965.
75. Croft DN: Diagnostic Uses of Radioisotopes in Medicine. London, Hospital Medical Publications, pp. 61–71, 1969.
76. Freeman LM, Johnson PM: Clinical Scintillation Imaging. New York Harper & Row, 1975.
77. Baker RR, Tohru M, Wagner HN: A quantitative method of measuring liver regeneration in dogs. *J Surg Res 7:*578–590, 1967.
78. Leach KG, Karran SJ, Wisbey ML, et al.: *In vivo* assessment of liver size in the rat. *J Nucl Med 16:*380–385, 1975.
79. Samuels LD, Grosfeld JL: Serial scans of liver regeneration after hemi hepatectomy in children. *Surg Gynecol Obstet 131:*453–457, 1970.
80. Karran SJ, Leach KG, Blumgart LH: Assessment of liver regeneration in the rat using the gamma camera. *J Nucl Med 15:*10–16, 1974.
81. McAfee JG, Ause RG, Wagner HN: Diagnostic value of scintillation scanning of the liver. *AMA Arch Int Med 116:*95–100, 1965.
82. Uszler JM, Swanson LA: Focal nodular hyperplasia of the liver: Case report. *J Nucl Med 16:*831–832, 1975.
83. Staab EV, Hartman RC, Parrott JA: Liver imaging in the diagnosis of hepatic venous thrombosis in paroxysmal nocturnal hemoglobinuria. *Radiology 117:*341–348, 1975.
84. Spencer RP: Radionuclide liver scans in tumor detection. *Cancer Suppl 37:*475–479, 1976.
85. Geslien GE, Pinsky SM, Roth RK, et al.: Sensitivity and specificity of /sup 99m/ Tc-sulfur colloid liver imaging in diffuse hepatocellular disease. *Radiology 118:*115–119, 1976.
86. Lopez-Majano V: Indications for liver scintigraphy. *Int Surg 59:*362–366, 1974.
87. Lowrance P, Chanutin A: The effect of partial hepatectomy on the blood volume of the white rat. *Am J Physiol 135:*606–608, 1942.
88. Benacerraf B. Bilbey D, Biozzi G, et al.: The measurement of liver blood flow in partially hepatectomized rats. *J Physiol 136:*287–293, 1957.
89. Saba TM: Liver blood flow and intravascular colloid clearance alterations following partial hepatectomy. *J Res 7:*406–417, 1970.
90. MacKenzie RJ, Furnival CM, O'Keane MA, et al.: The effect of hepatic ischaemia on liver function and the restoration of liver mass after 70 per cent partial hepatectomy in the dog. *Br J Surg 62:*431–437, 1975.
91. Brauer RW, Leong GF, McElroy RF, et al.: Circulatory pathways in the rat liver as revealed by P^{32} chromic phosphate colloid uptake in the isolated perfused liver preparation. *Am J Physiol 184:*593–598, 1956.
92. Ingle DJ, Baker BL: Histology and regenerative capacity of liver following multiple partial hepatectomies. *Proc Soc Exptl Biol Med 95:*813–816, 1957.
93. Moolten FL, Bucher NL: Regeneration of rat liver: Transfer of humoral agents by cross circulation. *Science 158:*272–274, 1968.
94. Paul D, Leffert H, Sato G, et al.: Stimulation of DNA and protein synthesis in fetal rat liver cells

by serum from partially hepatectomized rats. *PNAS 69:*374–377, 1972.
95. Morley CG, Kingdon HS: The regulation of cell growth. I. Identification and partial characterization of a DNA synthesis stimulating factor from the serum of partially hepatectomized rats. *Biochem Biophys Acta 308:*260–275, 1973.
96. Islami AH, Pack GT, Hubbard JC: The humoral factor in regeneration of the liver in parabiotic rats. *Surg Gynecol Obstet 108:*549–554, 1959.
97. Rogers AE, Shaka JA, Pechet G, et al.: Absence of a bloodborne factor affecting liver regeneration in paired and triplet parabiotic rats. *Fed Proc 20:*287, 1961.
98. Hayes DM, Tedo I, Matsushima Y: Stimulation of *in vitro* growth of rat liver cells with calf serum drawn following partial hepatectomy. *J Surg Res 9:*133–137, 1969.
99. Demetriou A, Seifter E, Levanson S: Effect of sera obtained from normal and partially hepatectomized rats and patients on the growth of cells in tissue culture. *Surgery 76:*779–785, 1974.
100. Adibi S, Paschkis KE, Cantarow A: Stimulation of liver mitosis by blood serum from hepatectomized rats. *Exptl Cell Res 18:*396–398, 1959.
101. Stich HF: Regulation of mitotic rate in mammalian organisms. *Ann NY Acad Sci 90:*603–609, 1960.
102. Levi JU, Zeppa R: Source of the humoral factor that initiates hepatic regeneration. *Ann Surg 174:*364–370, 1971.
103. Preuss HG, Terryi EF, Keller AI: Renotropic factor(s) in plasma from uninephrectomized rats. *Nephron (Basel) 7:*459–470, 1970.
104. Dicker SE, Morris C, Shipolini R: Regulation of compensatory kidney hypertrophy by its own products. *J Physiol 269:*687–706, 1977.
105. Alston WC, Thomson RY: Humoral and local factors in liver regeneration. *Cancer Res 23:* 901–909, 1963.
106. Fisher B, Szuch P, Fisher ER: Evaluation of humoral factor in liver regeneration utilizing liver transplants. *Cancer Res 31:*322–330, 1971.
107. Max MH, Price JB, Takeshige K, et al.: The role of factors of portal origin in modifying hepatic regeneration. *J Surg Res 11:*590–593, 1972.
108. Thomson RY, Clarke AM: Role of portal blood supply in liver regeneration. *Nature 208:*392–393, 1965.
109. Starzyl TE, Francavilla A, Halgrimson CG, et al.: The original hormone nature and action of hepatotrophic substances in portal venous blood. *Surg Gynecol Obstet 137:*179–199, 1973.
110. Bollman L: Liver. *Ann Rev Physiol 23:*183–206, 1961.
111. Paschkis KE, Goddard J, Cantarow A, et al.: Stimulation of growth by partial hepatectomy. *Proc Soc Exptl Biol Med 101:*184–186, 1959.
112. Smythe RL, Moore RO: A study of possible humoral factors in liver regeneration in the rat. *Surgery 44:*561–571, 1958.
113. Jackson B, Bohnel E: Effect of serum injections on body weight and liver regeneration of partially hepatectomized rats. *Fed Proc 21:*301, 1962.
114. Leong GF, Grishom JW, Hole B: Effect of rapid "total" exchange transfusion on hepatic DNA synthesis in partially hepatectomized rats. *Fed Proc 22:*192, 1963.
115. Pechet GS, Rogers AE, MacDonald RA: Inhibitory humoral factors and liver regeneration. *Fed Proc 22:*192, 1963.
116. Salvatorelli G, Smith J: Effet de diverses fractions d'extraits de foie embryonnaire de poulet sur l'erythropoiese *in vitro. C R Hebd Seances Acad Sci Ser D Sci Natur (Paris) 272:*869–872, 1971.
117. Fisher S, Roheim PS: Role of the liver in the inactivation of erythropoietin. *Nature 200:*899–900, 1963.
118. Jacobsen EM, Davis AK, Alpen EL: Relative effectiveness of phenylhydrazine treatment and hemorrhage in the production of an erythropoietic factor. *Blood 11:*937–945, 1956.
119. Prentice TC, Mirand EA: Effect of acute liver damage plus hypoxia on plasma erythropoietin content. *Proc Soc Exptl Biol Med 95:*231–234, 1957.
120. Gordon AS, Zanjani ED, Zalusky R: A possible mechanism for the erythrocytosis associated with hepatocellular carcinoma in man. *Blood 35:*151–157, 1970.
121. Nixon RK, O'Rourke W, Rupe CE, et al.: Nephrogenic polycythemia. *Arch Intern Med 106:*797–802, 1960.
122. Mann DL, Gallagher NI, Donati RM: Erythrocytosis and primary aldosterism. *Ann Intern Med 66:*335–340, 1967.
123. Leffert H, Alexander NM, Faloona G, et al.: Specific endocrine and hormonal receptor changes associated with liver regeneration in adult rats. *PNAS 72:*4033–4036, 1975.
124. Peschle C, Rappaport IA, Sasso GF, et al.: Mechanism of growth hormone (GH) action on erythropoiesis. *Endocrinol 91:*511–517, 1972.
125. Cater DB, Holmes BE, Nee LK, et al.: The effect of growth hormones upon cell division and nucleic acid synthesis in the regenerating liver of the rat. *J Biochem 66:*482–486, 1957.
126. Post J, Himes MB, Klein A, et al.: Responses of the liver to injury. *AMA Arch Pathol 64:*284–289, 1957.

28 *In Vitro* Aspects of Erythropoietin Production

James R. Zucali and Edwin A. Mirand

Introduction

Erythropoietin (Ep) is a hormone active in the regulation of erythropoiesis; it is believed to be produced mainly in the kidney of adult animals and man.[1-3] Jacobson et al.[1] postulated the kidney to be the sole source of erythropoietin and Naets[2] substantiated Jacobson's findings in the dog. While the kidney plays a major role as a site for erythropoietin production or activation in the adult, the question arises: where in the kidney is erythropoietin produced and are there extrarenal sites for Ep? Fisher et al.[4] and Frenkel et al.[5] using a fluorescent antibody technique, localized erythropoietin in the glomerular tuft of anemic sheep kidneys. Busuttil et al.[6,7] reported localization of erythropoictin in the epithelial cells of the glomerulus in anemic human kidneys and hypoxic dog kidneys. In addition, Gruber et al.[8] using indirect immunofluorescent staining for erythropoietin, found fluorescent staining of glomerular cells in the kidneys of hypoxic adult rats (Figure 28-1) whereas no staining was seen in the kidneys of rats 1 or 2 weeks of age. This fluorescent staining was correlated with results of erythropoietin bioassay indicating that the kidney was not involved in erythropoietin elaboration at 1 or 2 weeks of age (Figure 28-2). While this study offers some credence to the use of immunofluorescence techniques, it is still debatable whether the glomerulus is a site of erythropoietin production or storage.

Studies in rodents by Mirand et al.[9-11] in 1957 and later studies by others[12-14] using a variety of animal species have confirmed that there is an extrarenal source for erythropoietin. Nathan et al.[15] reported the persistence of erythropoiesis in anephric man without the detection of erythropoietin in the plasma. However, Mirand et al.[16-18] demonstrated the presence of high titers of plasma erythropoietin in a renoprival state prior to homotransplantation. This discrepancy in results may be due to the circadian or rhythmic nature of extrarenal Ep levels noted in man since these variations have been found for up to 160 days after nephrectomy in over 16 patients.[19]

The liver has been implicated as a source of extrarenal erythropoietin[20-22] and Fried[23] has demonstrated an ablation of extrarenal erythropoietin if nephrectomy is combined with hepatectomy prior to hypoxia. These studies on the importance of the liver have been substantiated by Schooley[24,25] in both weanling rats and lead administered rats. In more recent studies, Peschle et al.[26] have implicated the reticuloendothelial system (RES) with extrarenal Ep production. They found a correlation between potentiation of extrarenal Ep production and hyperplasia of the RES with colloidal carbon and zymosan. Naughton et al.[27] reported that the regenerating liver produces erythropoietin in response to hypoxia and that hepatic Ep production is dependent on the stage of regeneration with the highest levels being produced 48 to 72 hr post hepatectomy, the period of greatest proliferation and increase in liver mass.

Additional support for the liver's involvement in erythropoietin production in man comes from studies by Mirand and Murphy[28] on Ep production in human liver disease. They describe elevated Ep levels in the plasma, urine, and tumor extracts of hepatocellular carcinoma patients. It is interesting to note that outside of renal tumors the most common tumors associated with elevated Ep levels and erythrocytosis are hepatic tumors.

In neonate and weanling rats, less dependence on the kidney for erythropoietin production has been shown; thus, extrarenal erythropoietin is important in the regulation of erythropoiesis. Carmena et al.[29] have compared Ep production in neonatal and adult rats. These authors found that during the first 15 days of life, bilateral nephrectomy did not abolish Ep production after exposure to hypoxia whereas it did affect Ep production in the adult animal. Bilateral nephrectomy has also been shown to have little affect

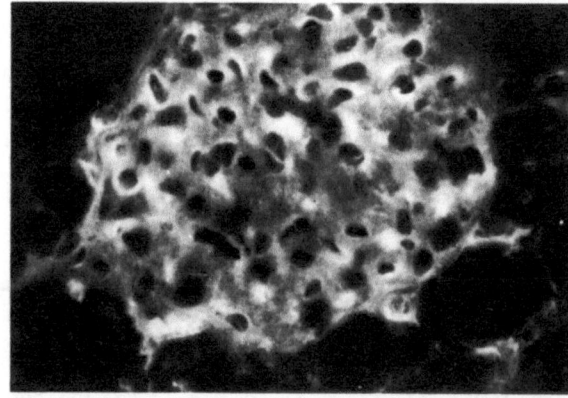

Figure 28-1 Kidney section of an adult rat subjected to hypoxia. Heavy immunofluorescent staining is visible throughout the glomerular area. ×700. From *Exp Hematol* 5:399–407, 1977.

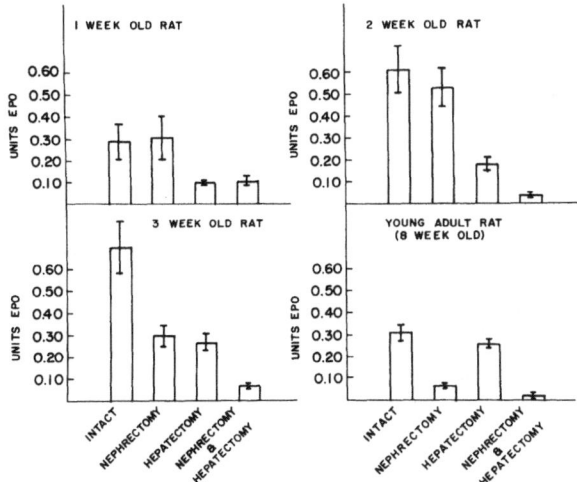

Figure 28-2 Effect of nephrectomy, hepatectomy, and nephrectomy plus hepatectomy on erythropoietin production of various aged rats stimulated by 6 hr of hypoxia (0.35 atm of air). Units of erythropoietin were extrapolated from ^{59}Fe uptake values to erythropoietin standard assay values. From *Exp Hematol* 5:399–407, 1977.

on the erythroid compartment of the bone marrow and Schooley and Mahlmann[24] have reported that removal of a substantial portion of the liver in weanling rats significantly reduced the production of Ep induced by hypoxia. In addition, Gruber et al.[8] have recently shown that there is a temporal transition from liver to kidney as the major site of Ep production as the rat matures. Thus, the liver appears to be the primary site of Ep production during the first 2 weeks of life; but, by the third week, the kidneys began producing Ep and by 8 weeks, they were the major organs of Ep production (Figure 28-2). Once again, these studies implicate the liver as a major source of extrarenal erythropoietin.

Fetal erythropoiesis has been shown to operate independently of maternal erythropoiesis. Thus, polycythemia of pregnant rats suppressed maternal but not fetal erythropoiesis.[30] Zanjani and Gordon[31,32] in studies on fetal goats demonstrated that induction of anemia in pregnant goats resulted in elevated fetal erythropoietin levels at a time when no increase could be detected in maternal erythropoietin levels. They were also able to show that Ep production in the fetus was not affected by nephrectomy of the fetus and/or mother, suggesting an extrarenal origin of erythropoietin in the fetus.

A promising approach to the problem of erythropoietin production appears to lie in *in vitro* culture studies. Early studies have dealt with attempts to obtain Ep from kidney cell cultures. Ozawa[33] cultured rabbit bone marrow cells with and without kidney cells and measured ^{59}Fe incorporation into heme. He reported that cultures to which kidney cells had been added and which were then exposed to an anoxic environment had increased heme synthesis. McDonald et al.[34] established a bovine kidney cell monolayer in specially modified Spinner flasks and exposed the cells to a low O_2 environment. These authors found that the culture media contained elevated Ep levels when measured in the polycythemic mouse assay. Burlington et al.[35] also reported that long-term cultures of goat renal glomeruli produced erythropoietin and Sherwood and Goldwasser[36] recently reported on the production of erythropoietin by human renal carcinoma cells in culture. While these studies look promising, they have not been easily reproducible [37,38]; thus kidney cells in culture require further study.

Discussion

In a series of reports, we[39–41] have demonstrated the biosynthesis of an erythropoietically active substance by a fetal mouse liver culture system (Table 28-1). This material resembles erythropoietin by giving a dose-response in the hypertransfused, plethoric mouse bioassay and its *in vivo* activity was suppressed by an antibody to erythropoietin and destroyed by prior incubation with the enzyme neuraminidase (Table 28-2). In addition, fetal liver conditioned media was shown to restore erythropoiesis in the hypertransfused, lethally irradiated mouse reconstituted with bone marrow (Table 28-3). We have had repeated success with collecting erythropoietically stimulating material from over 100 different fetal liver cultures, thus demonstrating the reproducibility of this culture system. Congote[42] in studies with human fetal liver cells has also reported on the erythropoietic effect of liver conditioned media. He found that the liver conditioned media stimulates ^{59}Fe incorporation into heme associated with hemoglobin from freshly plated fetal liver

Table 28-1 Erythropoietin production over a 21-day period by fetal mouse liver cultures.[a]

Groups	Materials Injected	^{59}Fe Incorporation (%)
1	Saline	0.19 ± 0.01[b]
2	0.05 unit Erythropoietin	0.54 ± 0.17
3	0.20 unit Erythropoietin	3.93 ± 0.68
4	0.80 unit Erythropoietin	16.41 ± 2.85
5	Control Media	0.23 ± 0.08
6	Culture Media day 1	0.27 ± 0.01
7	Culture Media day 7	3.72 ± 0.01
8	Culture Media day 14	8.59 ± 1.50
9	Culture Media day 21	12.78 ± 1.19

[a] From Gruber et al. *Exp Hematol* 5:392, 1977.
[b] Mean ± S.E.

220 *In Vitro* Aspects of Erythropoietin Production

Table 28-2 Inactivation of culture media erythropoietin by anti-erythropoietin and neuraminidase.

Groups	Materials Injected	^{59}Fe Incorporation (%)
1	Saline	0.17 ± 0.01^a
2	0.20 unit Erythropoietin	4.26 ± 1.17
3	Control Media	0.19 ± 0.04
4	Culture Media (1×)	0.89 ± 0.08
5	Culture Media (2.5×)	2.79 ± 0.26
6	Culture Media (5×)	5.72 ± 1.06
7	(Culture Media (5×) + anti-Ep^b) + GARGG^c	0.21 ± 0.03
8	(anti-Ep + GARGG) + Culture Media (5×)	5.03 ± 0.89
9	(Culture Media (5×) + 200 units Neuraminidase) + 80°C	0.22 ± 0.08
10	(200 units Neuraminidase + 80°C) + Culture Media (5×)	4.83 ± 0.58
11	Fetal Liver Extract	0.15 ± 0.02

^{a} Mean ± SEM.
^{b} Anti-erythropoietin
^{c} Goat anti-rabbit gamma globulin.

erythroid cells. Thus, the liver appears to be an important site of Ep production as well as a site of erythropoiesis in the fetus.

In additional studies, Zucali et al.[43] demonstrated that fetal mouse liver, obtained from day 13 to day 19 of gestation, was able to produce both erythropoietin (Figure 28-3) and colony-stimulating factor (CSF) (Figure 28-4) but not thrombopoietin (TSF) (Table 28-4). Maximum Ep activity was detected in media collected from day 14 and day 15 fetal liver cultures, whereas maximum CSF was found in media collected from day 16 fetal liver cultures. Colony-stimulating factor was also reported to be produced by adult liver macrophages in culture.[44] Thus, liver macrophages are an important source of CSF. McDonald et al.[45] have demonstrated that embryonic kidney cells produce thrombopoietin but not erythropoietin in culture, implying that the fetal kidney is important in the production of thrombopoietin whereas the fetal liver is the important site of erythropoietin production.

In an attempt to identify the cells responsible for

Table 28-4 Thrombopoietin assay results.

	Dose per Mouse	^{35}S Incorporation × 10³ (%)	Control (%)
Day 13^a	20 mg Protein	3.21 ± 0.24^b	101
Day 14	20 mg Protein	3.57 ± 0.10	113
Day 15	20 mg Protein	3.22 ± 0.26	101
Day 16	20 mg Protein	3.15 ± 0.36	99
Day 17	20 mg Protein	2.47 ± 0.28	78
Day 18	20 mg Protein	3.88 ± 0.29	122
Day 19	20 mg Protein	3.88 ± 0.29	122
Control Media	20 mg Protein	3.17 ± 0.21	—
Saline		3.59 ± 0.30	—
TSF Std.^{c}	1.5 ml	6.52 ± 0.34	182
Ep Std.^{d}	1.0 unit	4.81 ± 0.51	134

From Zucali et al. *Exp Hematol* 5:385, 1977.
^{a} Day of gestation.
^{b} Mean ± SEM.
^{c} Thrombopoietin standard (Partially purified preparation from kidney cell culture media; 21.2 mg protein/ml).
^{d} Erythropoietin standard (Step 1 sheep plasma erythropoietin).

Table 28-3 Spleen colonies in polycythemic mice after irradiation and a bone marrow transplant.

Materials Injected	Mean Colony Count per Spleen		Spleen Colony Differential		
	Gross	Microscopic	Erythroid	Granuloid	Mixed
1 0.50 unit Ep^a	16.8 ± 1.4^d	17.6 ± 0.6	12.0 ± 0.4	3.8 ± 0.4	1.9 ± 0.3
2 (0.50 unit Ep + 0.06 ml anti-Ep^b) + 0.25 ml GARGG^c	8.2 ± 1.4	7.6 ± 0.4	0	5.6 ± 0.4	1.9 ± 0.1
3 1.0 ml Fetal liver culture media	10.8 ± 1.8	12.2 ± 0.6	4.4 ± 0.4	5.1 ± 0.3	2.5 ± 0.1
4 (1.0 ml Fetal liver culture media + 0.06 ml anti-Ep) + 0.25 ml GARGG	5.8 ± 0.6	5.8 ± 0.4	0	4.9 ± 0.3	0.8 ± 0.2

From Zucali et al. *Exp Hematol* 5:103, 1977.
Colonies were counted 9 days after 800 rad whole body irradiation and injection of 10⁵ bone marrow cells.
^{a} Erythropoietin.
^{b} Anti-erythropoietin.
^{c} Goat anti-rabbit gamma globulin.
^{d} Mean ± SEM.

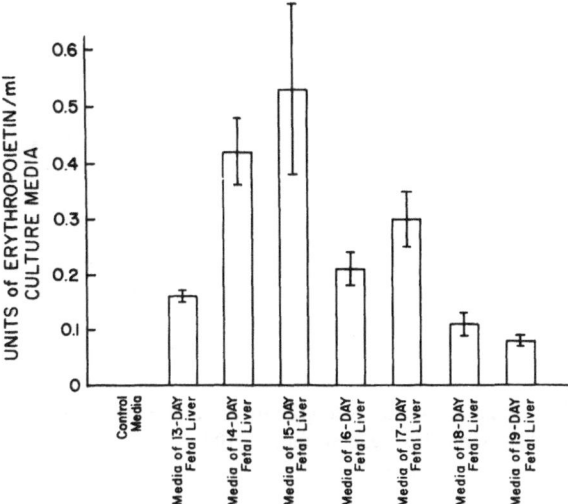

Figure 28-3 *In vitro* production of erythropoietin over 21 culture days by fetal mouse liver (gestation 13 to 19 days). All culture media were concentrated 5 times before assay. Vertical bars = ±1 SEM. From *Exp Hematol* 5:385–391, 1977.

the production of Ep in fetal mouse liver, Gruber et al.[46] applied an immunofluorescent technique to the cultured cells. These authors report that the only cells demonstrating the blue-green fluorescent staining were cells ingesting carbon particles (Figure 28-5). Thus, this study implicates the liver macrophages or Kupffer cells as being involved in Ep production or storage. However, caution is advised since the antibody used was not purified and immunoflu-

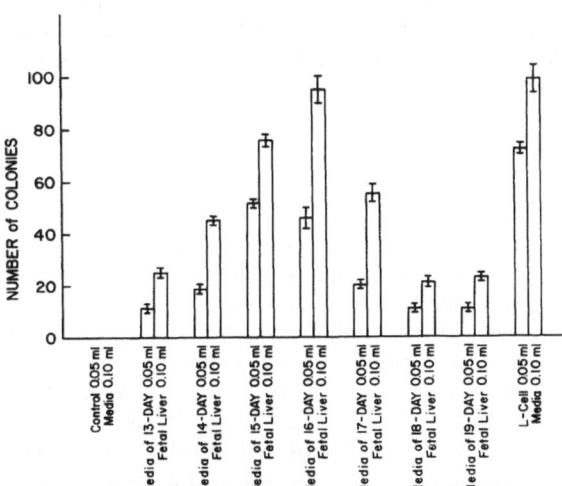

Figure 28-4 Number of colonies as measure of colony stimulation by media from cultures of fetal mouse liver (13 to 19 days gestation). All culture media were concentrated 5 times before assay. Vertical bars = ±1 SEM. From *Exp Hematol* 5:385–391, 1977.

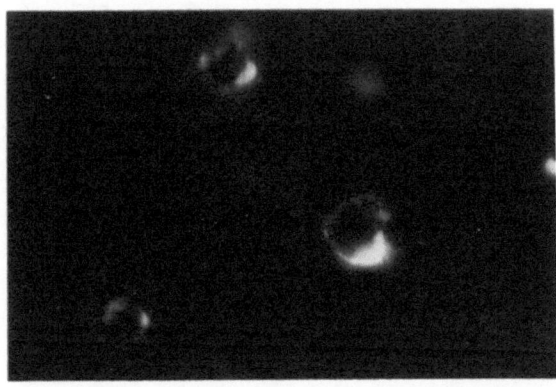

Figure 28-5 Immunofluorescent staining for erythropoietin in a 21-day culture of fetal mouse liver. Ingested carbon granules identifying liver macrophages or Kupffer cells are seen within specifically stained cells. × 540. From *Exp Hematol* 5:392–398, 1977.

orescence implicates only the presence but not the production of Ep. In contrast, Congote[42,47] suggests that the fetal liver parenchymal cells are responsible for the erythropoietic activity seen in fetal liver conditioned media. But no mention is made of his method of isolating parenchymal cells for culture. Therefore, the exact cell or cells in fetal liver respon-

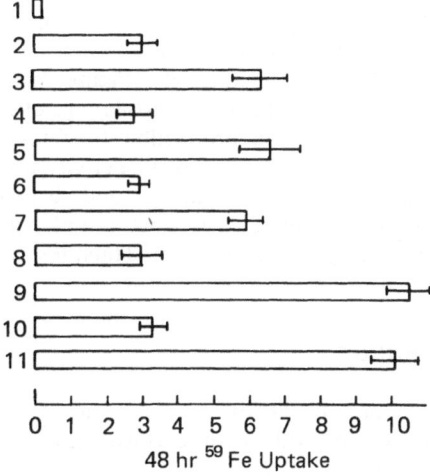

Figure 28-6 Effect of testosterone hemisuccinate on erythropoietin production by fetal mouse liver cultures. Testosterone was added on day 10 of culture. Horizontal bars = ±1 SEM. *1.* Control Media; *2.* Normal Culture Media, day 10; *3.* Normal Culture Media, day 20; *4.* 10^{-6}M Testosterone Culture Media, day 10; *5.* 10^{-6}M Testosterone Culture Media, day 20; *6.* 10^{-8}M Testosterone Culture Media, day 10; *7.* 10^{-8}M Testosterone Culture Media, day 20; *8.* 10^{-10}M Testosterone Culture Media, day 10; *9.* 10^{-10}M Testosterone Culture Media, day 20; *10.* 10^{-12}M Testosterone Culture Media, day 10; *11.* 10^{-12}M Testosterone Culture Media, day 20. From *Br J Haematol* 35:639, 1977.

222 In Vitro Aspects of Erythropoietin Production

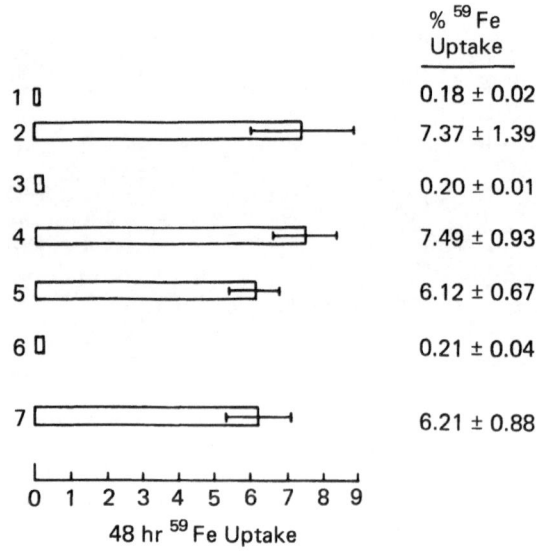

Figure 28-7 Neutralization of the erythropoietic effect induced by testosterone by anti-erythropoietin. *1.* Saline; *2.* 0.20 u Ep; *3.* Control Testosterone (10^{-10}M) Media; *4.* Control Testosterone (10^{-10}M) Media + 0.20 u Ep; *5.* Culture Media with 10^{-10}M Testosterone; *6.* Culture Media with 10^{-10}M Testosterone + (anti = Ep + Goat Anti-Rabbit Gamma Globulin (GARGG; *7.* Culture Media with 10^{-10}M Testosterone + (Anti-EP + GARGG). From *Br J Haematol 35:*639, 1977.

sible for Ep production remain to be elucidated and further studies requiring isolation or cloning of a particular cell type are required.

Finally, Zucali and Mirand[48] (Figure 28-6) have shown that testosterone, over a dose range of 10^{-12} to $10^{-10}M$, stimulated Ep production by fetal mouse liver in culture. This erythropoietic activity was completely neutralized with anti-Ep, and testosterone alone did not stimulate or enhance the activity seen with erythropoietin (Figure 28-7). Thus, testosterone

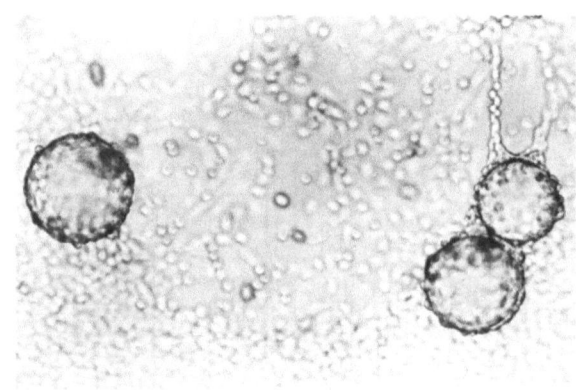

Figure 28-8 Monolayer growth of fetal mouse cells on DEAE–Sephadex bead surface. × 100.

acted directly on the cells producing Ep, but whether the stimulatory effect was due to an increase in the number of cells capable of producing Ep or on the amount of Ep produced per cell was not determined. Estradiol, on the other hand, had no observable effect on Ep production by fetal mouse liver cells.

Studies are now underway to purify the Ep obtained in culture and to use it in an anemic, uremic rat model resembling the condition of chronic renal disease in man.[49] Such a study should demonstrate the feasibility of using culture-produced Ep in clinical situations. Also, attempts are being made to scale-up our fetal liver culture system employing a relatively new microcarrier culture technique.[50-52] This system (Figure 28-8) brings together the characteristics of both monolayer culture and suspension culture while retaining the advantages of both. With the larger surface area afforded by the use of a microcarrier, a larger cell yield is feasible, allowing for the production of greater quantities of cell products such as erythropoietin.

References

1. Jacobson LO, Goldwasser E, Fried W, et al.: Role of the kidney in erythropoiesis. *Nature* 179:633–634, 1957.
2. Naets JP: The role of the kidney in erythropoiesis. *J Clin Invest* 39:102–110, 1960.
3. Stohlman F Jr: Erythropoiesis. *New Engl J Med* 267:342–348, 1962.
4. Fisher JW, Taylor G, Porteous DD: Localization of erythropoietin in glomeruli of sheep kidney by the fluorescent antibody technique. *Nature* 205:611–612, 1965.
5. Frenkel EP, Suki W, Baum J: Some observations on the localization of erythropoietin. *Ann NY Acad Sci* 149:292–293, 1968.
6. Busuttil RW, Roh BL, Fisher JW: The cytological localization of erythropoietin in the human kidney using the fluorescent antibody technique. *Proc Soc Exp Biol Med* 137:327–330, 1971.
7. Busuttil RW, Roh BL, Fisher JW: Localization of erythropoietin in the glomerulus of the hypoxic dog kidney using a fluorescent antibody technique. *Acta Haematol* 47:238–242, 1972.
8. Gruber DF, Zucali JR, Wleklinski J, et al: Temporal transition in the site of rat erythropoietin production. *Exp Hematol* 5:399–407, 1977.
9. Mirand EA, Prentice TC: Presence of plasma erythropoietin in hypoxic rats with or without kidney(s) and/or spleen. *Proc Soc Exp Biol Med* 96:49–51, 1957.
10. Mirand EA, Prentice TC, Slaunwhite WR: Cur-

rent studies on the role of erythropoietin on erythropoiesis. *Ann NY Acad Sci* 77:677–702, 1959.
11. Mirand EA: Studies on erythropoietin in tamarins. *Invest Urol* 2:579–588, 1965.
12. Piliero SJ, Medici PT, Gordon AS: Production of the circulating erythropoietic factor in hypoxic endocrine-deficient rats. *Proc 7th Cong Intern Soc Hemat*, 1958.
13. Gallagher MI, McCarthy JJ, Lange RD: Erythropoietin production in uremic rabbits. *J Lab Clin Med* 57:281–289, 1961.
14. Naets JP: Erythropoiesis in nephrectomized dogs. *Nature* 181:1134–1135, 1958.
15. Nathan DG, Schupak E, Stohlman F Jr, et al: Erythropoiesis in anephric man. *J Clin Invest* 43:2158–2165, 1964.
16. Mirand EA, Murphy GP, Steeves RA, et al: Extra-renal production of erythropoietin in man. *Acta Haematol* 39:359–365, 1968.
17. Mirand EA: Renal and extrarenal relations to erythropoietin production in animals and man. *S Afr Med J* 42:462–466, 1968.
18. Mirand EA, Murphy GP, Steeves RA, et al: Erythropoietin activity in anephric, allotransplanted, unilaterally nephrectomized and intact man. *J Lab Clin Med* 73:121–128, 1969.
19. Mirand EA, Murphy GP: Erythropoietin activity in anephric humans, in Gordon AS, Condorelli M, Peschle C (eds): *Regulation of Erythropoiesis*. Milano, Italy, Il Ponte, 1971.
20. Burke WT, Morse BS: Studies on the production and metabolism of erythropoietin in rat liver and kidney, in Jacobson LO, Doyle M (eds): *Erythropoiesis*. New York, Grune & Stratton, 1962.
21. Reissman KP, Normura T: Erythropoietin formation in isolated kidneys and liver, in Jacobson LO, Doyle M (eds): *Erythropoiesis*. New York, Grune & Stratton, 1962.
22. Fried W, Kilbridge T, Krantz S, et al: Studies on extrarenal erythropoietin. *J Lab Clin Med* 73:244–248, 1969.
23. Fried W: The liver as a source of extrarenal erythropoietin. *Blood* 40:671–676, 1972.
24. Schooley JC, Mahlmann LJ: Extrarenal erythropoietin production by the liver in the weanling rat. *Proc Soc Exp Biol Med* 145:1081–1083, 1974.
25. Schooley JC, Mahlmann LJ: Hepatic erythropoietin production in the lead-poisoned rat. *Blood* 43:425–428, 1974.
26. Peschle C, Marone G, Genovese A, et al: Hepatic erythropoietin: Enhanced production in anephric rats with hyperplasia of Kupffer cells. *Br J Haematol* 32:105–111, 1976.
27. Naughton BA, Kaplan SM, Roy M, et al: Liver regeneration and erythropoietin production in the rat. *Science* 196:301–302, 1977.
28. Mirand EA, Murphy GP: Erythropoietin alterations in human liver disease. *NY State J Med* 71:860–864, 1971.
29. Carmena AO, Howard D, Stohlman F Jr: Regulation of erythropoiesis. XXII. Erythropoietin production in the newborn animal. *Blood* 32:376–382, 1968.
30. Jacobson LO, Marks EK, Gaston EO: Studies on erythropoiesis. XII. The effect of transfusion-induced polycythemia in the mother on the fetus. *Blood* 14:644–653, 1959.
31. Zanjani ED, Gidari AS, Peterson EN, et al.: Humoral regulation of erythropoiesis in the fetus, in Cross KW, Dawes GS, Nathaniels PW, Comline KS (eds): *Foetal and Neonatal Physiology*. Cambridge, England, Cambridge University Press, 1973.
32. Zanjani ED, Peterson EN, Gordon AS, et al: Erythropoietin production in the fetus: role of the kidney and maternal anemia. *J Lab Clin Med* 83:281–287, 1974.
33. Ozawa S: Erythropoietin from the kidney cells cultured *in vitro*. *Keio J Med* 16:193–203, 1967.
34. McDonald TP, Martin DH, Simmons ML, et al: Preliminary results of erythropoietin production by bovine kidney cells in culture. *Life Sci* 8:949–954, 1969.
35. Burlington H, Cronkite EP, Reincke U, et al: Erythropoietin production in cultures of goat renal glomeruli. *Proc Nat Acad Sci* 69:3547–3550, 1972.
36. Sherwood JB, Goldwasser E: Erythropoietin production by human renal carcinoma cells in culture. *Endocrinol* 99:504–510, 1976.
37. Zucali JE: personal observations, 1977.
38. Lange RD: personnel communications, 1977.
39. Zucali JR, Stevens V, Mirand EA: In vitro production of erythropoietin by mouse fetal liver. *Blood* 46:85–90, 1975.
40. Zucali JR, Mirand EA: Biosynthesis of erythropoietin by mouse fetal liver in culture. *Blood Cells* 1:485–496, 1975.
41. Zucali JR, McGarry MP, Mirand EA: Stimulation of erythropoiesis in a grafted animal by mouse fetal liver culture media. *Exp Hematol* 5:103–108, 1977.
42. Congote LF: Regulation of fetal liver erythropoiesis. *J Steroid Biochem* 8:423–428, 1977.
43. Zucali JR, McDonald TP, Gruber DF, et al: Erythropoietin, thrombopoietin and colony stimulating factor in fetal mouse liver culture media. *Exp Haematol*, 5:385–391, 1977.
44. Joyce RA, Chervenick PA: Stimulation of granulopoiesis by liver macrophages. *J Lab Clin Med* 86:112–117, 1975.
45. McDonald TP, Clift R, Lange RD, et al: Thrombopoietin production by human embryonic kidney cells in culture. *J Lab Clin Med* 85:59–66, 1975.

46. Gruber DF, Zucali JR, Mirand EA: Identification of erythropoietin producing cells in fetal mouse liver cultures. *Exp Hematol,* 5:392–398, 1977.
47. Congote LF, Solomon S: On the site of origin of erythropoietic factors in human fetal tissues. *Endocrin Res Commun* 1:495–504, 1974.
48. Zucali JR, Mirand EA: Effect of testosterone and oestradiol on erythropoietin production *in vitro.* *Br J Haematol* 35:639–645, 1977.
49. Anagnostou A, Barone J, Kedo A, et al: Effect of erythropoietin on the red cell mass of uremic and non-uremic rats. *Blood* 48:978, 1976.
50. Van Wezel Al: Growth of cell-strains and primary cells on micro-carriers in homogeneous culture. *Nature* 216:64–65, 1967.
51. Van Hemert P, Kilburn DG, van Wezel AL: Homogeneous cultivation of animal cells for the production of virus and virus products. *Biotechnol Bioeng* 11:875–885, 1969.
52. Horng C-B, McLimans W: Primary suspension culture of calf anterior pituitary cells on a micro-carrier surface. *Biotechnol Bioeng* 17:713–732, 1975.

29 Renal and Extrarenal Erythropoietin Production in Anemia

A. J. Erslev, J. Caro, and E. Kansu

It has been shown in anephric animals and man that small amounts of extrarenal erythropoietin is generated in response to severe anemia or hypoxia. Although this response is inadequate to do more than maintain a hematocrit of 10 to 20%, it seems possible that a stimulated production of extrarenal erythropoietin could replace renal production in anephric patients and provide a new therapeutic approach to patients with impaired renal endocrine function.

In order to investigate this possibility, renal and extrarenal erythropoietin production were studied in adult male Wistar rats rendered anemic by bleeding or phenylhydrazine. The results are summarized in Figures 29-1 and 29-2.

The results indicate that phenylhydrazine-treated rats were probably slightly more hypoxic than bled rats. However, their renal and extrarenal erythropoietin production was far in excess of what could be caused by this small difference, and must be related to a specific effect of phenylhydrazine.

It has been suggested that extrarenal erythropoietin is produced by the reticuloendothelial (RE) tissue, and phenylhydrazine administration is indeed associated with a pronounced increase in splenic size due to RE hyperplasia. However, simultaneous splenectomy did not significantly change the rate of extrarenal erythropoietin production. Even after 6 or more weeks of biweekly injections of phenylhydrazine and after the spleen had increased almost ten times in size, extrarenal erythropoietin production both in the presence and in the absence of the spleen was about the same as after a single injection of phenylhydrazine. Consequently, it seems more likely that an organ other than the RES is responsible for extrarenal erythropoietin production.

A number of studies have implicated the liver as this organ and recent observations on the regenerating liver have suggested that rapid proliferation of hepatocytes promotes erythropoietin production. Phenylhydrazine affects liver metabolism by causing hemoglobinemia with secondary uptake and catabolism of hemoglobin by the parenchymal cells. Such an hepatic effect could underly the pronounced extrarenal erythropoietin production observed after phenylhydrazine injections. The concomitant in-

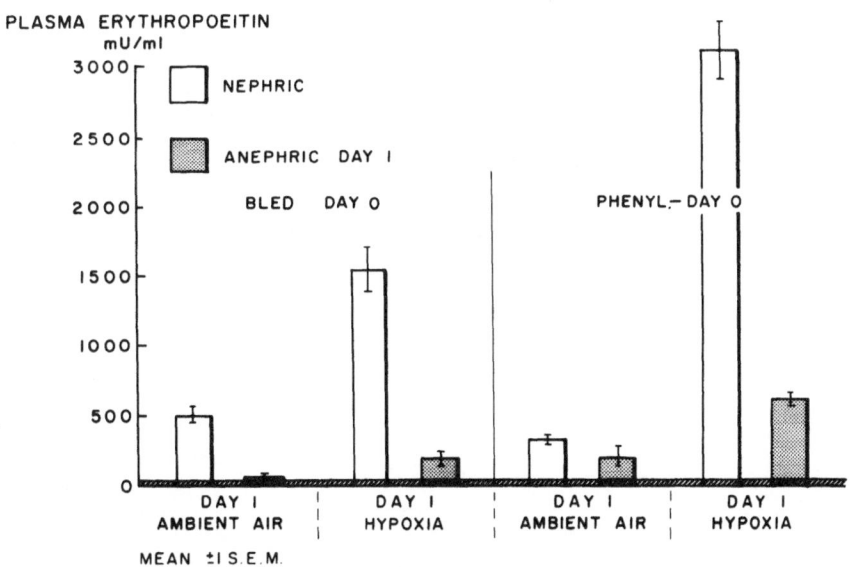

Figure 29-1 The left panel lists the mean ± 1 SEM response of hematocrit, hemoglobin, absolute reticulocyte count, oxygen tension of a peritoneal air pocket and plasma erythropoietin to an acute bleeding of 25 ml per kg body weight. The right panel lists the corresponding response to an injection of 60 mg phenylhydrazine per kg body weight.

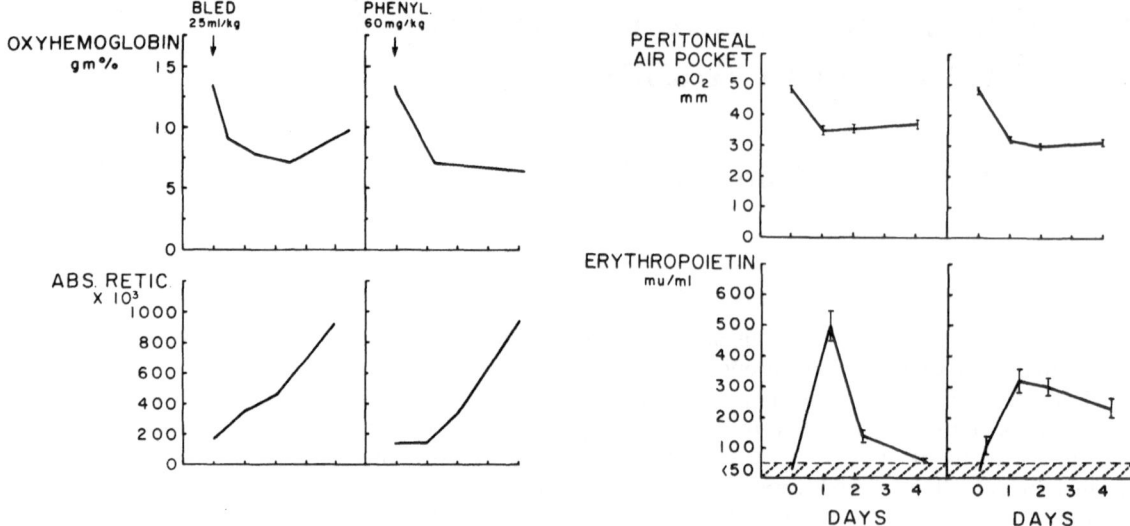

Figure 29-2 The bars in the left panel depict the mean ± 1 SEM of plasma erythropoietin in rats bled on day 0 and nephrectomized or left intact on day 1 followed by exposure to 6-hr ambient air or 6-hr 0.4 atm pressure. The bars in the right panel depict the corresponding values for rats given phenylhydrazine on day 0.

crease in renal erythropoietin production may be related to simultaneous catabolism of free hemoglobin by the renal tubular cells. However, it is also possible that the hepatic effect of phenylhydrazine underlies both extrarenal and renal erythropoietin production, as suggested previously by Gordon and his co-workers.

These possibilities are being studied in an attempt to unravel the biogenesis of extrarenal erythropoietin and to develop an alternate pathway for erythropoietin production in patients with renal disease.

Acknowledgment

This work was supported in part by USPHS Grant 4612.

30 Recent Advances in Erythropoietin Physiology

C. Peschle, M. C. Magli, C. Cillo, F. Lettieri, F. Pizzella, G. Migliaccio, A. Soricelli, G. Scala, G. Mastroberardino, and G. F. Sasso

Introduction

It is well established that the rate of erythropoiesis, both under normal and tissue hypoxia conditions, is largely or totally regulated by erythropoietin (Ep).[1-3] It is therefore of crucial interest to elucidate the mechanisms underlying both renal and extrarenal Ep production. Recent studies from our laboratory that focused on these aspects are presented here.

Section I Hypophyseal Modulation of Hypoxic Ep Production

Extensive evidence indicates that under physiological conditions the endocrine system plays a major role in the regulation of erythropoiesis.[1-3] Early after removal of the pituitary, the metabolic rate declines rapidly and then levels off.[4] This leads to a parallel decrease of the erythropoietic activity,[5] which causes a slower, gradual reduction of red cell mass (RCM) values.[4,6] The latter parameter tapers off at approximately 3 to 4 months after hypophysectomy. Thus the initial, rapid decline of O_2 consumption is thereafter compensated by a slower, progressive drop of O_2 transport. Since it is conceded that Ep is the major regulator of physiological red cell production,[1-3] hypophysectomy apparently causes a sequential decline of O_2 demand, Ep activity, erythropoietic rate, and RCM values.

The hypophyseal influence on normal erythropoiesis is clearly mediated by pituitary and target hormones. In hypophysectomized rats, regression of the anemia is observed after combined treatment with GH, ACTH, corticosteroids, thyroid hormones, and testosterone.[7] These agents apparently operate, at least partially, via elevation of Ep activity.[8-13] However, androgens[14,15] and corticosteroid derivatives[16] may also exert a direct stimulatory effect on erythropoiesis. Of further interest is that, in contrast with other target hormones, estrogens induce an inhibitory influence on red cell production[17]; thus, administration of physiological amounts of estradiol in mice or rats causes respectively a decline of either the number of erythroid colony-forming units in marrow[18] or the level of hypoxic Ep activity in serum.[19]

It is generally accepted that under hypoxic conditions the hypophysis is not essential for Ep production.[1] The present studies indicate that this concept may be partially reconsidered. In this regard, hypophysectomized rats showed a marked reduction of the Ep response to hypoxia, which was applied at different time intervals (i.e., from 2 weeks through 7 months) after ablation of the pituitary. Thus, it is postulated that during hypoxia pituitary and target hormones play a significant role in the modulation of Ep production.

Materials and Methods

Male Sprague-Dawley rats weighting between 90 and 150 gm at the time of hypophysectomy were employed. The animals, made available within 1 week after the operation, were maintained on a normal diet of laboratory pellets and tap water *ad libitum* supplemented with 3% sucrose.

1. *Erythropoietic activity in serum of hypophysectomized rats exposed to a bout of hypoxia (0.42, 0.40 or 0.35 atm air for 6 to 24 hr).* In the first series of experiments rats weighing 140 to 150 gm at the time of hypophysectomy were exposed to hypoxia (0.42 or 0.35 to 0.37 atm air/6 hr) at 2, 4, 6, 8, or 12 weeks after the operation. Sham-operated controls of the same age were included only in experiments with milder hypoxia (i.e., 0.42 atm air). A minimum of three to four rats per group were employed. Blood was collected by cardiac puncture under light ether anesthesia immediately after the end of the hypoxic period. Heparinized blood was pooled and centrifuged at 2,000 rpm for 20 min. The plasma was then collected and stored at $-20°C$ until assayed.

In the second series of experiments, rats weighing 90 to 100 gm at the time of hypophysectomy were exposed to hypoxia (0.40 or 0.35 atm air/6 hr) at 3.5, 4, 4.5, 5, or 7 months after the operation. Sham-operated controls of the same age were included in all experiments. A minimum of three to four rats per group were employed. Bleeding, collection, and storage of serum were performed as indicated above.

In the third series of experiments, rats weighing 90 to 100 gm at the time of hypophysectomy were exposed to severe hypoxia (0.37 atm air) for 6, 12, or 24 hr at 4 or 12 weeks after the operation. Sham-operated controls were included in both experiments.

228 Recent Advances in Erythropoietin Physiology

Each group comprised a minimum of three to four rats. Bleeding, collection, and storage of serum were performed as indicated above.

2. Assay of erythropoietic activity in hypoxic rat serum in exhypoxic polycythemic mice. CF 1, female mice weighing 20 to 25 gm, maintained on a diet of standard laboratory pellets and tap water *ad libitum*, were employed to assay the erythropoietic activity in all series of experiments.

In the first experiment, the mice were exposed to intermittent hypoxia (0.42 atm air/18 hr per day) up to a total of approximately 220 hr. Test sera or standard Ep* were injected intraperitoneally on day 6 and 7 post-hypoxia.[20] On day 8 the animals received intraperitoneally 0.5 μCi ^{59}Fe citrate. In the second and third series of experiments, the mice, exposed to an equivalent hypoxic stimulus up to a total of 220 hr, were given test materials and standard Ep** subcutaneously on days 3 and 4, radioiron intravenously on day 5.[21,22] In both assays, 48-hr % RBC-^{59}Fe incorporation values were determined. A minimum of five to six mice per group were employed. Results are expressed in terms of either 48-hr % RBC-^{59}Fe incorporation values or equivalent units of the IRP I or II of Ep. Mice with a final hematocrit of less than 56% were discarded.

3. Evaluation of RCM values at different time intervals after hypophysectomy or sham operation. In rats weighing 130 to 150 gm at the time of either hypophysectomy or sham operation, RCM values were evaluated starting from 2 weeks through 12 months after the ablation, according to the ^{59}Fe-labelled red blood cells dilution technique.[6] A minimum of three rats per group were employed here.

Results

The present results indicate that the Ep activity evoked by a 6 to 24 hr bout of hypoxia is significantly lower in hypophysectomized rats than in sham-operated controls (Figures 30-1, 30-2, 30-3, 30-4 and unpublished observations).

Progressive increase of the Ep response to a 6-hr bout of hypoxia is associated with extension from 2 weeks up to 2 to 4 months of the time interval between hypophysectomy and hypoxia (i.e., from either nondetectable activity or 0.08 to 0.16 IU of Ep per ml of serum up to 0.1 or 0.3 IU, following respectively 0.42 or 0.37 to 0.35 hypoxia) (Figures 30-1 and

---

* Sheep plasma Ep Step I (Connaught Medical Research Lab., Toronto, Canada), standardized against the International Reference Preparation (IRP) I or II (Division of Biological Standards, National Institute for Medical Research, London, England).

** Sheep plasma Ep Step III (Connaught Medical Research Lab., Toronto), standardized against the IRP II.

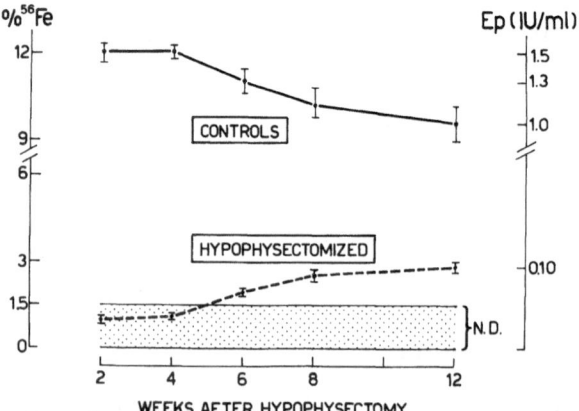

Figure 30-1 Mean ± SEM of % RBC-^{59}Fe incorporation values in assay mice and corresponding IU of Ep/ml of serum from hypophysectomized (— — —) or sham (———) rats subjected to hypoxia (0.42 atm air/6 hr) at different time intervals after the operation.

30-2). As previously mentioned, ablation of the hypophysis induces initially a "relative plethora" (i.e., normal RCM values with low O_2 demand). Thereafter, RCM values gradually decline (Figure 30-5). In 1- to 3-month hypophysectomized rats, the gradual increase of hypoxic Ep activity is thus apparently mediated by the simultaneous decline of RCM values (i.e., of the "relative plethora"). It cannot be excluded, however, that significant modifications of Po_2 50 values in hypophysectomized versus sham-operated rats may also underlie this phenomenon. Further studies are currently in progress in an attempt to elucidate this particular aspect.

Feigin and Gordon[23] showed that, in 2-month hypophysectomized rats, exposure to mild hypoxia (0.54 atm air) did not enhance the erythroid activity, whereas a considerable elevation of this parameter

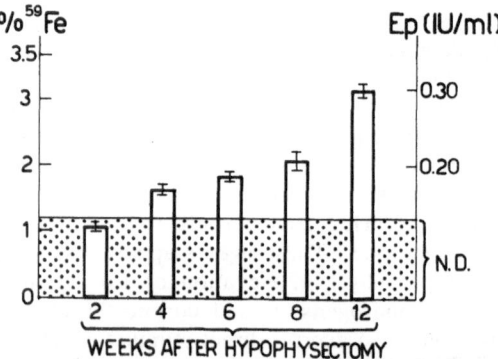

Figure 30-2 Mean ± SEM of % RBC-^{59}Fe incorporation values in assay mice and corresponding IU of Ep/ml of serum from hypophysectomized rats subjected to hypoxia (0.35 to 0.37 atm air/6 hr) at different time intervals after hypophysectomy.

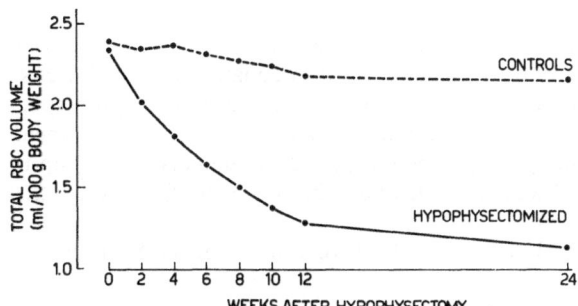

Figure 30-3 Ep levels (IU/ml of serum) in rats subjected to hypoxia (0.40 atm air/6 hr) from 3.5 through 7.0 months after hypophysectomy (———) or sham operation (— — —).

was observed after a more severe stimulus (0.42 atm air). In line with these findings, the present studies indicate that in 2-month hypophysectomized rats significant Ep production is evoked by severe hypoxia.

It is of crucial interest that in 3- to 7-month hypophysectomized rats hypoxic Ep activity is lower than in sham-operated controls (Figures 30-3 and 30-4). In these animals, the "relative plethora" is negligible or absent: accordingly Ep levels, although showing some fluctuations, apparently reach plateau levels (i.e., 0.2 to 0.4 or 0.2 to 0.7 after respectively 0.40 or 0.35 to 0.37 hypoxia). Furthermore, the hypothesis of Ep production by the hypophysis and/or its target glands is obsolete. Thus, the reduced Ep response in these hypophysectomized rats suggests that during hypoxia the activity of pituitary and/or target hormones (except for estrogens) plays a significant role

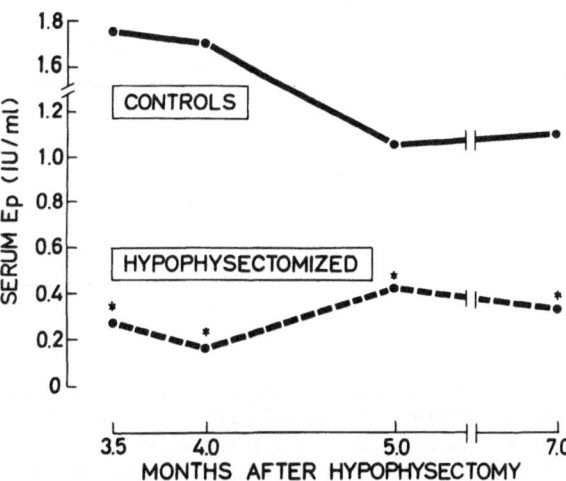

Figure 30-4 Ep levels (IU/ml of serum) in rats subjected to hypoxia (0.35 atm air/6 hr) from 3.5 through 7.0 months after hypophysectomy (— — —) or sham operation (———).

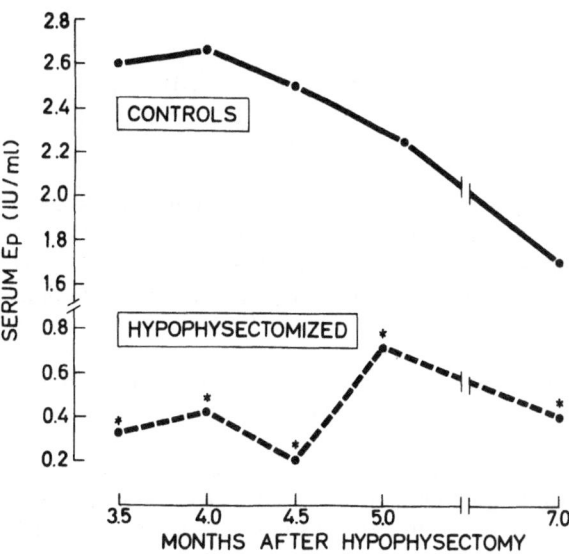

Figure 30-5 Mean values of total RBC volume (ml/100 gm body weight) at different time intervals after hypophysectomy or sham operation.

in Ep production. In line with this concept, evidence has been presented indicating that several of these hormones potentiate serum Ep activity under normal or hypoxic conditions.[8-13]

The mechanisms underlying the hypophyseal influence on hypoxic Ep levels are not yet elucidated. Evidence has been presented indicating that the activity of erythrogenin (Eg, also termed pro-Ep: cfr. Section III), in the kidney of T_4-treated normal mice or GH-injected hypophysectomized rats, was significantly more elevated than in vehicle-treated controls.[12,13] Preliminary studies from our laboratory further suggest that renal pro-Ep activity is barely detectable in the kidneys of hypoxia-exposed, hypophysectomized rats. Thus, it may be envisioned that hypophyseal and target hormones enhance hypoxic Ep production via renal pro-Ep. In this regard the possibility should be considered that these agents, with particular reference to testosterone, exert a stimulatory effect on the production of this factor. This action may be mediated via renal tissue hypoxia, thus triggering the O_2-sensor mechanism regulating Ep production.

Conclusion

In conclusion these studies indicate that in male rats the hypophysis (i.e., hypophyseal and target hormones, with the exception of estrogens) modulate Ep production under hypoxic conditions, apparently via a permissive enhancement of pro-Ep production and/or activity in the kidney.

Section II Additive Role of Kidney and Liver in Ep Production: Transition from Hepatic to Renal Ep Activity

In anephric rats, subtotal hepatectomy abolishes almost completely the Ep response to hypoxia,[24-26] thus indicating that the liver is the main site of extrarenal Ep production. In line with these findings, elevated levels of pro-Ep or Eg have been observed in hepatic and splenic homogenates from nephrectomized animals subjected to hypoxia,[25,27] thereby suggesting that the biogenesis of extrarenal Ep is mediated by hepatic and possibly splenic pro-Ep.

In the intact rat, however, the role of the liver in Ep production is not yet elucidated. Subtotal hepatectomy in adult animals either abolished partially[28] or did not modify the Ep response to hypoxia.[24] The interaction(s) between liver and kidney in the biogenesis of Ep is also obscure. In adult rodents the liver may represent a source for the serum factor interacting with Eg[28]; this concept, however, has not been substantiated conclusively. In the neonatal period, the Ep response to hypoxia, although virtually unmodified in 1-hr anephric animals,[29] showed a sharp reduction in 24-hr nephrectomized rats.[30] This phenomenon might indicate that, although neonate Ep is not derived from the kidneys, the renal function exerts a significant influence on extrarenal sources for Ep.

In an attempt to elucidate these aspects, Ep production in sham-operated, nephrectomized and/or hepatectomized rats at various ages was evaluated. It is concluded that liver and kidney play an additive role in this function. Hepatic Ep, although prevalent in the neonate rat, is progressively obscured by massive renal Ep production, which is initiated in the late neonatal period and peaks in the young, adult animal.

Materials and Methods

Ep levels in serum of sham-operated, nephrectomized, hepatectomized, hepatectomized-nephractomized rats exposed to a standard bout of hypoxia (0.45 atm air/6 hr). Male Wistar rats were employed. All the operations were performed by the ventral route under light ether anesthesia. Eighty to ninety % hepatectomy was carried out according to a slight modification of the procedure by Higgins and Anderson.[31] The animals were exposed to hypoxia starting 1 hr after the operation. Control rats were maintained at room pressure. Heparinized blood was collected by cardiac puncture immediately after the end of the hypoxic period. The plasma obtained from each group was pooled and stored at −20°C until utilized.

Each group consisted of a minimum of 16 to 18 neonate, 8 weanling, or 5 adult rats. The following groups were selected: (a) neonate rats, 1 or 2 weeks of age (mean body weight, 15 or 30 to 35 gm respectively), (b) weanling animals, 3.5 weeks of age (mean body weight, 60 gm), (c) adult animals, from 5 up to 10 weeks of age (mean body weight, from 100 to 270 gm).

For assay of Ep levels in serum see Section I.

Results

The present studies provide an evaluation of hepatic and renal Ep production in neonatal, weanling and adult rats.

Figures 30-6 and 30-7 indicate that, in the early neonatal (i.e., 1 week of age) period, the liver is a major source of Ep. Thus, Ep levels were sharply reduced in subtotally hepatectomized animals, as compared to either sham-operated or nephrectomized controls. In the weanling-adult rat (starting from 3.5 weeks of age), hepatectomy plus nephrectomy abolished almost completely the Ep response to hypoxia (Figure 30-6), thereby indicating that the liver still represents the main site of extrarenal production of Ep. In this regard, the maximal hepatic Ep production occurs at 3.5 weeks (72% of total activity in sham-operated animals). It is also noteworthy that extrarenal (i.e., hepatic) Ep activity is gradually reduced, i.e., from 0.20 to 0.08 IU Ep/ml of serum in 3.5 week or 10-week-old rats, respectively.

On the other hand, renal Ep is virtually absent in

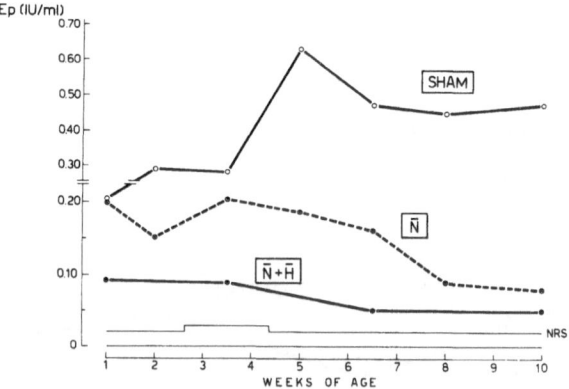

Figure 30-6 Ep levels (IU/ml) in serum of either sham-operated (SHAM), nephrectomized (N) or hepatectomized-nephrectomized rats (N + H) exposed to a standard bout of hypoxia (0.45 atm air/6 hr starting 1 hr after the operation). Upper borderline of shaded area indicates Ep levels in serum of normal, control rats maintained at room pressure (normal rat serum: NRS). From Peschle et al., *Am J Physiol,* 230:845, 1976.

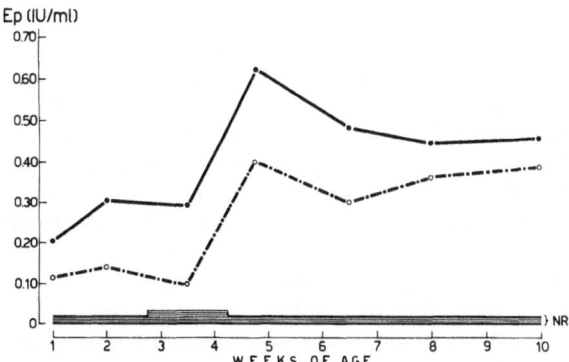

Figure 30-7 Ep levels (IU/ml) in serum of either sham-operated (———) or subtotally hepatectomized rats (– – –) of different ages subjected to a standard bout of of hypoxia (0.45 atm air/6 hr starting 1 hr after the operation). Upper borderline of shaded area indicates Ep levels in serum of normal, control rats maintained at room pressure (normal rat serum: NRS), as approximately evaluated by means of the standard dose-response curve between 0.02 and 0.04 IU of Ep (IRP II). From Peschle et al., *Am J Physiol*, 230:845, 1976.

1-week-old neonate rats (Figure 30-6). The Ep response curve in hepatectomized animals (Figure 30-7) indicates that the renal Ep production is "switched on" between 2 and 4 weeks of age, peaks at 5, and is thereafter maintained at sustained, elevated levels up to 10 weeks. The mechanism underlying the 5-week

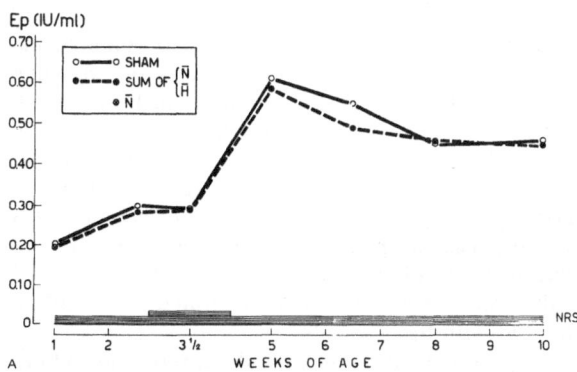

Figure 30-8A Ep levels (IU/ml) in serum of sham-operated rats (SHAM, ———) or sum of Ep activity in serum of nephrectomized (N̄) and subtotally hepatectomized rats (H̄) (– – –). Rats were subjected to a standard bout of hypoxia (0.45 atm air/6 hr starting 1 hr after operation) at different ages. Ep levels in anephric 1-week-old rats (N̄: ⊗) were not added to corresponding values in hepatectomized animals, since at this particular age the Ep response in nephrectomized and sham-operated rats is strictly equivalent. Upper borderline of shaded area indicates Ep levels in serum of normal, control rats maintained at room pressure (normal rat serum: NRS). From Peschle et al., *Am J Physiol*, 230:845, 1976.

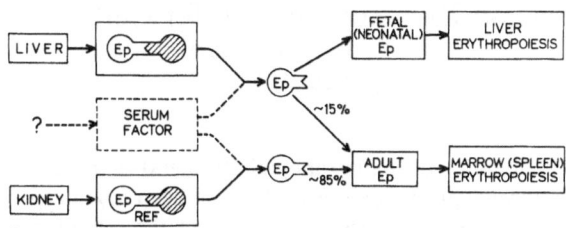

Figure 30-8B A model for Ep production in both fetal-neonatal (above) and adult age (below). Pro-Ep (i.e., Ep linked with a carrier molecule through its biologically active site), synthetized first in the liver and then also in the kidney, is activated to Ep (possibly via interaction with a serum factor) and finally exerts its stimulatory action on the erythroid tissue. The % contribution of renal and hepatic Ep in the overall Ep production in adult life is also indicated.

peak has not been clarified. In this regard, initiation of substantial androgen secretion, thus enhancing Ep activity,[8] is an attractive hypothesis.

The Ep production curve in sham-operated rats is approximately equivalent to that calculated by addition of corresponding values in hepatectomized or nephrectomized animals (Figure 30-8A). This apparently indicates that liver and kidney play an additive role in Ep production, from the late neonatal period throughout the adult life. Hepatic Ep, prevalent in neonate animals, is gradually obscured in the weanling-adult rat by both progressive initiation of massive renal Ep production and gradual decrease of liver Ep activity (Figure 30-8B).

Several corollaries of the present studies are worthy of consideration. Neonatal erythropoiesis in the rat corresponds to the fetal stage of erythrocyte production in the majority of mammalian species. This may allow extrapolation of these results in the neonate rat to Ep production mechanisms in mammalian fetuses (Figure 30-8B).

These studies possibly reconcile previous reports on Ep production in hepatectomized, adult rats. In line with the observations reported by Katz et al.,[28] it is apparent that Ep activity is reduced following subtotal hepatic ablation. However, this difference is both barely significant and progressively reduced in older rats, thus possibly explaining the opposite results reported by Fried.[24]

Conclusion

In conclusion, these studies originally indicated that Ep derives from two functionally distinct and additive sources, the kidney and the liver, possibly via synthesis of renal and hepatic pro-Ep respectively (Figure 30-8B).

232 Recent Advances in Erythropoietin Physiology

Section III Role of Kupffer Cells in Hepatic Ep Production

The identity of the intrahepatic tissue(s) involved in extrarenal production of Ep has not been elucidated. In the present studies hyperplasia of the reticuloendothelial system (RES) in the liver is correlated with enhanced extrarenal Ep production, thus apparently indicating that the hepatic RES is a source for extrarenal Ep.

Materials and Methods

Studies with colloidal carbon. In the first series of experiments, Wistar rats received intravenously 1.0 ml/250 gm of body weight of either colloidal carbon (C 11/1431 A, Pelikan Werke, Hannover) or its vehicle (4.3% fish glue and 1% carbolic acid in H_2O). These agents were injected daily from days 1 through 4. On day 5, the animals were subjected to the operations (sham operation, bilateral nephrectomy, ligation of ureters or 80 to 90% hepatectomy) under light ether anesthesia and then to a bout of hypoxia (0.45 atm air/6 hr), starting 1 hr after the operation. Three further groups of rats were primed with colloidal carbon or its vehicle daily from days 0 through 3, sham-operated, nephrectomized or ureter-ligated on day 4 and subjected to hypoxia on day 5, simultaneously with the above groups. Subtotal hepatectomy was performed according to a slight modification of the procedure by Higgins and Anderson.[31]

In a second series of studies involving five separate experiments, Wistar rats, primed with colloidal carbon, its vehicle, or saline daily from days 1 through 4, were subjected to a 6-hr bout of hypoxia on day 5, starting 1 hr after bilateral nephrectomy.

Studies with zymosan. One group of Wistar rats received zymosan intravenously (Sigma Co., St. Louis, Missouri) at the dosage of 5 mg/200 gm of body weight, once on days 1 and 2, twice on day 3; the control group was given an equivalent volume of physiological saline. On day 4, three subgroups from each group were nephrectomized, ureter-ligated, or sham-operated and then subjected to hypoxia (0.45 atm air/6 hr), starting 1 hr after the operation.

In four similar experiments, zymosan- or vehicle-treated rats were subjected to nephrectomy and a 6-hr bout of hypoxia, starting 1 hr after the operation.

Studies with gadolinium. Wistar rats were injected intravenously with either 0.4 or 2 mg/200 gm of body weight of gadolinium chloride (Sigma Co.) on day 1, nephrectomized or ureter-ligated on either day 2 or 6 and finally exposed to hypoxia (0.45 atm air/6 hr), starting 1 hr after the operation.

For assay of erythropoietic activity see Section I.

Morphological and physiological studies of hepatic and splenic RES in anephric or sham-operated rats primed with colloidal carbon or zymosan. Wistar rats received either (a) colloidal carbon from days 1 through 4, (b) zymosan from days 2 through 4, (c) gadolinium on day 4, as indicated above, or (d) their vehicles, injected on corresponding days. On day 5, the animals were sham-operated or nephrectomized. The phagocytic index was evaluated on the basis of colloidal carbon clearance values. Thus, 16 mg of carbon/100 g of body weight were injected 1 hr after the operation. Blood samples were obtained under pentobarbital anesthesia prior to and 3, 6, 9, 12, or 15 min after the injection. The logarithm of carbon concentration was plotted against time, and the value of the phagocytic index was calculated according to Biozz et al.[32]

In similar experiments, liver and spleen were ablated from colloidal carbon-, zymosan-, gadolinium- or vehicle-treated rats subjected to nephrectomy or sham operation and then to hypoxia (0.45 atm air/6 hr) starting 1 hr after the operation. The ablated organs were weighed and standard histological examination of the RES was performed.

Although results are not presented here, liver weight values were significantly more elevated in anephric rats primed with colloidal carbon or zymosan than in vehicle-treated controls.[22] On the other hand, no increase of this parameter was observed after gadolinium administration.[22] Histological examination of colloidal carbon- and zymosan-livers showed that the weight increase was associated with marked hyperplasia of the RES.* No colloidal materials were observed in hepatocytes or anywhere outside of Kupffer cells, even at the highest magnification. It was also observed that zymosan or colloidal carbon administration induced, respectively, a large or slight augmentation of spleen weight and a corresponding level of hyperplasia of splenic RES. Hyperplasia of splenic or hepatic RES was not induced by gadolinium administration. Equivalent results were observed in sham-operated rats similarly treated with these agents.

It is of relevance that colloidal carbon and gadolinium induced an inhibitory effect on the phagocytic capacity of anephric rats, whereas zymosan exerted an opposite influence.[22] Equivalent results were observed in sham-operated rats receiving these agents.

As indicated in Figure 30-9A, priming of either sham-operated or hepatectomized rats with colloidal carbon did not modify significantly the Ep response to a bout of hypoxia. However, an equivalent treatment of anephric, hypoxic rats induced a three- to

---

* Hyperplasia of hepatic and splenic RES causing augmentation of liver and spleen weight is a well-established phenomenon (see ref. 22).

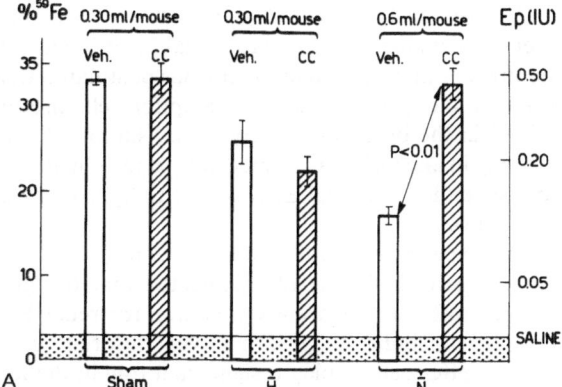

Figure 30-9A Mean ± SEM of % RBC-^{59}Fe incorporation values in assay mice and corresponding IU of Ep in serum from sham, subtotally hepatectomized (H̄) and nephrectomized (N̄) rats primed with colloidal carbon (CC) or its vehicle (Veh.) and subjected to a standard bout of hypoxia (0.45 atm air/6 hr starting 1 hr after the operation). From Peschle et al., *Blood*, 47:325, 1976.

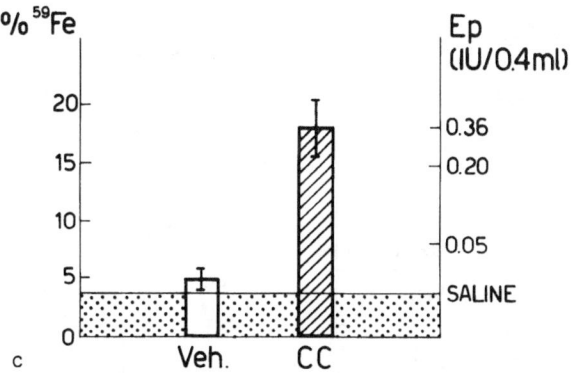

Figure 30-9C Mean ± SEM of % RBC-^{59}Fe incorporation values in assay mice and corresponding IU of Ep/0.4 ml of serum from nephrectomized rats primed with colloidal carbon (CC) or its vehicle (Veh.) and subjected to hypoxia (0.45 atm air/6 hr starting 1 hr after the operation).

fivefold increase of serum erythropoietic activity over control levels (Figures 30-9A–C).

The erythropoietic activity in colloidal carbon-treated, anephric rats was totally neutralized following incubation with anti-Ep and GARGG,[22] thus indicating the presence of elevated Ep levels in the serum. No significant difference was observed between hypoxia-induced Ep levels in ureter-ligated rats primed with either colloidal carbon or its vehicle.[22] This precluded the possibility that toxic factors accumulated after kidney ablation and constituted a factor influencing Ep production.

On the other hand, the kinetics of exogenous Ep was not significantly modified by colloidal carbon,[22] thereby showing that this agent enhances Ep activity in serum via augmentation of its production.

It should further be emphasized that colloidal carbon potentiated Ep levels in the serum of anephric rats subjected to hypoxia (0.45 atm air/6 hr) starting 24 hr after nephrectomy (Figure 30-10). In this regard, a 24-hr extension of the time interval between nephrectomy and initiation of the hypoxic stimulus gradually abolished the Ep response to a 6-hr hypoxic exposure.[20] Thus, the potentiating effect of colloidal carbon on the Ep response to hypoxia in 24-hr anephric rats conclusively indicated that this agent acted on extrarenal sources for Ep.

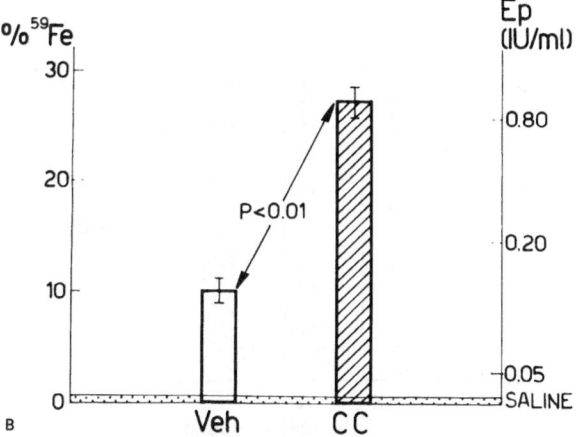

Figure 30-9B Mean ± SEM of % RBC-^{59}Fe incorporation values in assay mice and corresponding IU of Ep/ml of serum from nephrectomized rats primed with colloidal carbon (CC) or its vehicle (Veh.) and subjected to hypoxia (0.40 atm air/6 hr starting 1 hr after the operation). From Peschle et al.: in Nakao K, Fisher JW, Takaku F (eds): *Erythropoiesis*. Tokyo, University of Tokyo Press, 1975.

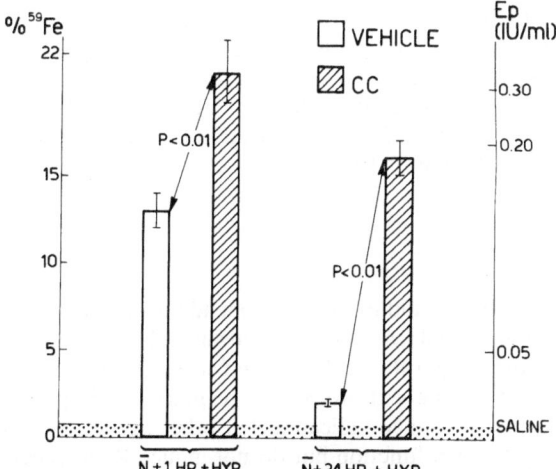

Figure 30-10 Mean ± SEM of % RBC-^{59}Fe incorporation values in assay mice and corresponding IU of Ep/ml of serum from nephrectomized rats primed with colloidal carbon (CC) or its vehicle (Veh.) and subjected to hypoxia (0.40 atm air/6 hr starting 1 or 24 hr after the operation). From Peschle et al., *Blood*, 47:325, 1976.

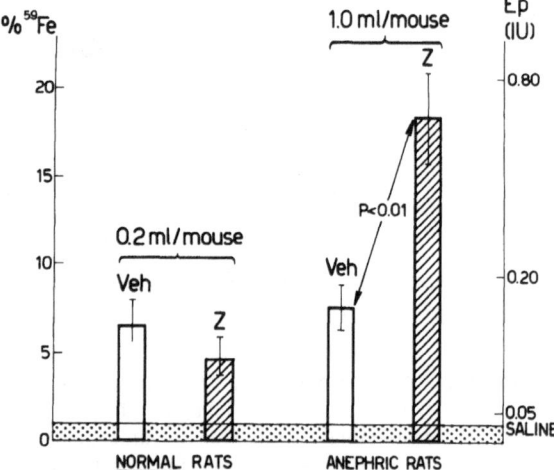

Figure 30-11 Mean ± SEM of % RBC-^{59}Fe incorporation values in assay mice and corresponding IU of Ep in serum from normal or nephrectomized rats primed with zymosan (Z) or its vehicle (Veh) and subjected to hypoxia (0.45 atm air/6 hr, starting 1 hr after the operation). From Peschle et al.: in Nakao K, Fisher JW, Takaku F(eds): *Erythropoiesis*. Tokyo, University of Tokyo Press 1, 1975.

As shown in Figure 30-11, priming of anephric rats with zymosan potentiated the Ep response to hypoxia. It is noteworthy that the magnitude of this enhancing effect was similar to that induced by colloidal carbon. No potentiation of the Ep response to hypoxia was induced by zymosan treatment in ureter-ligated rats.[22] Furthermore, this agent did not modify the Ep response to hypoxia in sham-operated rodents (Figure 30-11). Thus, the influence of zymosan on both renal and extrarenal Ep production was similar to that of colloidal carbon.

Figure 30-12 shows that administration of gadolinium did not modify the Ep response to hypoxia in nephrectomized rats, as well as in ureter-ligated controls. Similar results, although not presented here, were obtained after administration of a lower dosage of gadolinium (0.4 mg).

Results

The present studies indicate that administration of either colloidal carbon or zymosan induced simultaneously hyperplasia of liver and spleen RES and elevation of the extrarenal Ep response to hypoxia. Further evidence indicated that other agents, which modify RES function but do not induce RES hyperplasia (i.e., gadolinium, etc.), did not potentiate the extrarenal Ep activity following hypoxia. Thus, a direct correlation was established in anephric rats between hyperplasia of hepatosplenic RES and enhanced Ep production, indicating that the RES plays a significant role in this function.

As previously indicated, it is conceded that the liver is the major source of extrarenal Ep.[24] In view of the present findings it may be further postulated that in nephrectomized animals the Kupffer cells are involved in the biogenesis of Ep; thus, subtotal hepatectomy in anephric rats primed with zymosan almost abolished the extrarenal Ep response to hypoxia (Figure 30-13).

The experiments with fluorescein-conjugated anti-Ep serum by Fisher et al.[33] indicated that the glomerular tufts are apparently a source for renal Ep. The present studies suggest that liver "littorial" cells (i.e., Kupffer cells)[34] play a significant role in the hepatic production of Ep. In anephric animals primed with these agents the phagocytic activity of the RES was not correlated with the extrarenal Ep levels. Under the present experimental conditions, we have confirmed that colloidal carbon or zymosan depressed or stimulated respectively the phagocytic activity of the RES. However, both agents enhanced the extrarenal production of Ep. This lack of correlation is not at variance with the concept postulated here, that is, generation of Ep by the liver RES. Distinct metabolic pathways may underlie phagocytic activity and Ep production in Kupffer cells, resulting in the lack of correlation between these functions.

It was considered possible that hyperplasia of the RES caused increased production of Ep through hypoxia of hepatocytes via compression of sinusoids by the Kupffer cells. However, histological observations showed that hyperplasia of the RES did not lead to sinusoidal compression.[22]

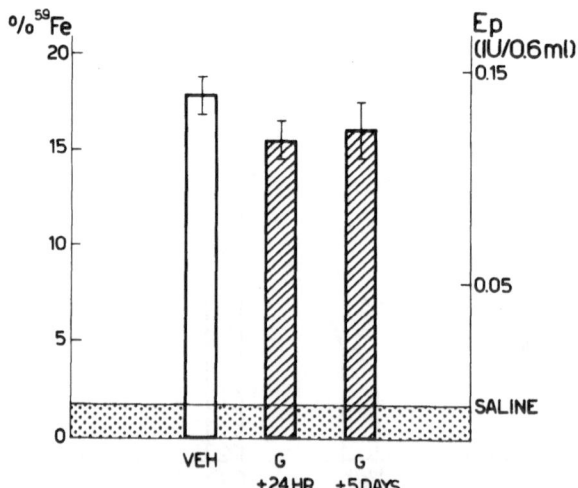

Figure 30-12 Mean ± SEM of % RBC-^{59}Fe incorporation values in assay mice and corresponding IU of Ep/0.6 ml of serum from rats primed with gadolinium (G) or its vehicle (Veh), nephrectomized 24 hr or 5 days after G treatment and subjected to hypoxia (0.45 atmospheres of air/6 hr, starting 1 hr after the operation). From Peschle et al., *Blood*, 47:325, 1976.

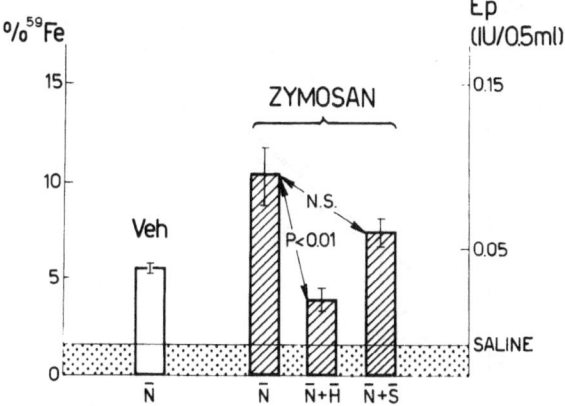

Figure 30-13 Mean ± SEM of % RBC-^{59}Fe incorporation values in assay mice and corresponding IU of Ep/0.5 ml of serum of rats primed with zymosan or its vehicle (Veh), subjected to nephrectomy ($\bar{N}$), associated or not with either subtotal hepatectomy ($\bar{H}$) or splenectomy ($\bar{S}$), and finally exposed to hypoxia (0.45 atm of air/6 hr, starting 1 hr after the operation). Modified from Peschle et al., *Br J Haematol*, 32:105, 1976.

Potentiation of the Ep response to hypoxia by colloidal carbon or zymosan, although demonstrated in anephric rats, was not observed in sham-operated animals. This phenomenon is possibly related to lack of a significant influence of these agents on glomerular cells. It should also be noted that pro-Ep or Eg activity in liver and spleen, although elevated in anephric rats exposed to hypoxia, is not detected in sham-operated controls,[25,27] thus indicating that the extrarenal source of Ep is "switched on" when renal production is shut off. The possibility also exists that in sham-operated animals the large amount of Ep derived from the kidney obscured the enhancing action of these agents on extrarenal Ep activity.

In anephric rodents similar levels of Ep activity were observed following administration of either lead acetate,[26] colloidal carbon or zymosan. In this regard, since the former agent modifies RES function,[35] its stimulatory effect on the extrarenal sources for Ep may be also mediated by the RES. Recently, enhanced Ep activity has been observed in hypoxic, anephric animals with liver regeneration[36,37] since Kupffer cell hyperplasia occurred under these conditions,[36] these studies apparently confirm the role of hepatic RES in extrarenal Ep production.

Conclusion

Finally, the present experimental studies might prove of clinical significance, as reduced Ep production is a major factor in renal anemia.[38] Thus, administration of nontoxic agents inducing RES hyperplasia may prove beneficial in the management of this condition.

Conclusive demonstration of the role played by Kupffer cells in extrarenal Ep production requires further investigations, such as the culture of Kupffer cells and immunofluorescent observations with anti-Ep serum, currently in progress in our laboratory. However, these studies provide the first indication that in nephrectomized rats colloidal agents enhancing RES proliferation sharply potentiate hepatic Ep production.

Section IV The Biogenesis of Renal Ep: The Pro-Ep–Ep System

The biogenesis of renal Ep, although subjected to extensive investigation, still represents a negative research field. In this regard, Gordon and co-workers[39] postulated that it involves interaction of a renal enzyme, termed the renal erythropoietic factor (REF) or erythrogenin (Eg), with a serum substrate (i.e., a precursor of Ep) possibly synthetized by the liver.[28,40] However, several aspects of this interesting concept are still unclarified. In this regard, a major controversy is the inability to generate large amounts of Ep in the REF + serum incubation mixture.[40] This phenomenon is hardly in line with the enzyme–serum substrate concept.

Most of the experiments reported by the New York group may suggest alternative concepts for the biogenesis of Ep. In this regard, it may be postulated that the REF represents a pro-Ep factor, which is activated on incubation with normal serum.[21,41] Strong evidence favoring this concept is presented here. Alternatively, the possibility cannot be excluded that REF preparations contain both the active Ep molecule and an inhibitor of Ep; the latter might be inactivated following incubation with normal serum, thus leading to restoration of the biological activity of Ep in the REF solution. However, preliminary incubation of REF with rabbit anti-Ep and goat anti-rabbit gammaglobulin (GARGG) does not modify the Ep activity generated on subsequent incubation of the REF with normal rat serum (NRS).[42]

It has been recently reported[43] that occasional REF preparations (REF II) exerted a direct erythropoietic activity when injected intraperitoneally in ex-hypoxic polycythemic mice. This intriguing finding raises further questions on the nature of the standard type of REF (REF I), which is biologically inactive when administered by itself.

In the studies reported here, a first series of experiments were designed to clarify the relationship between the REF II preparation and Ep. Furthermore, conversion of standard REF (REF I) to REF II was obtained by a variety of physiochemical procedures. Finally, evidence was obtained indicating that in REF I plus serum mixtures the biologically active Ep molecule derives from the REF and not from the serum.

Materials and Methods

Extraction and incubation of REF. All REF preparations were made from the kidneys of hypoxic 150 to 200 gm male Wistar rats according to a slight modification of the method by Zanjani et al.[43] for REF I preparation.[21,25,44] NRS was prepared and the incubations (45 min) were carried out according to previous procedures.[21,25,44]

For assay of Ep activity see Section I.

Figure 30-14 indicates that, although standard REF (REF I) has no direct erythropoietic effect, occasional REF preparations (REF II) exert a direct stimulatory action on red cell production, which is not significantly enhanced upon incubation with NRS. Furthermore, some REF preparations (REF I + REF II) show a direct erythropoietic activity, which is further enhanced following incubation with NRS.

It has been indicated that incubation of REF II with anti-Ep and GARGG resulted in complete neutralization of its direct erythropoietic activity, which was not restored by subsequent incubation with NRS(44). The activity in the REF II + NRS mixture was also fully neutralized after incubation with anti-Ep and goat anti-rabbit gammaglobulin (GARGG). In

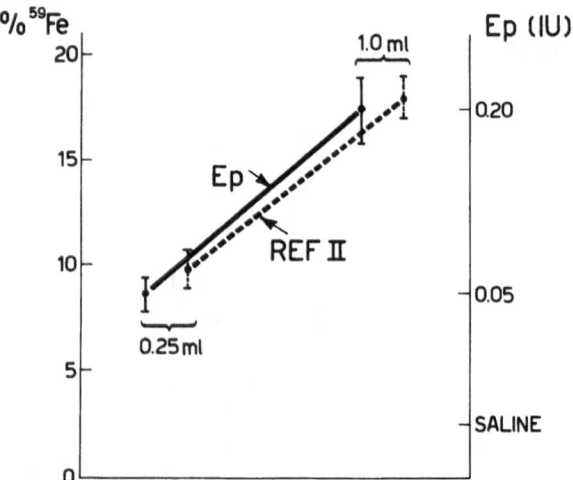

Figure 30-15 Parallelism of dose/response regression lines following injection of graded amounts of Ep or REF II in assay mice. The erythropoietic activity was evaluated on the basis of % RBC-^{59}Fe incorporation values (mean ± SEM) or corresponding IU of Ep. From Peschle et al.: in Nakao K, Fisher JW, Takaku F. *Erythropoiesis*. Tokyo, University of Tokyo Press, 1975.

line with these findings the dose-response regression line of REF II is strictly parallel to that of Ep, as indicated in Figure 30-15. It is therefore apparent that REF II cannot be distinguished from Ep, either immunologically or biologically. It is concluded that REF II is equivalent to the circulating, biologically active Ep.

As shown in Figure 30-16A–B, the REF I preparations were often converted to REF II following either

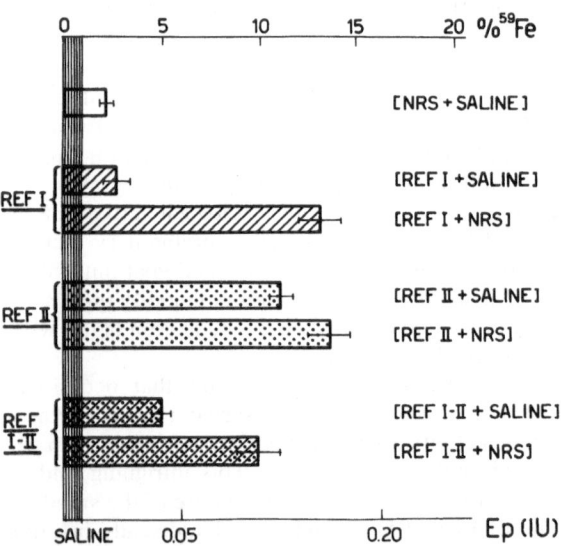

Figure 30-14 Percent RBC-^{59}Fe incorporation values (mean ± SEM) and corresponding IU of Ep in assay mice receiving "REF I", "REF II" or "REF I-II" preparations after incubation with either saline or NRS. From Peschle et al.: in Nakao K, Fisher JW, Takaku F. *Erythropoiesis*. Tokyo, University of Tokyo Press, 1975.

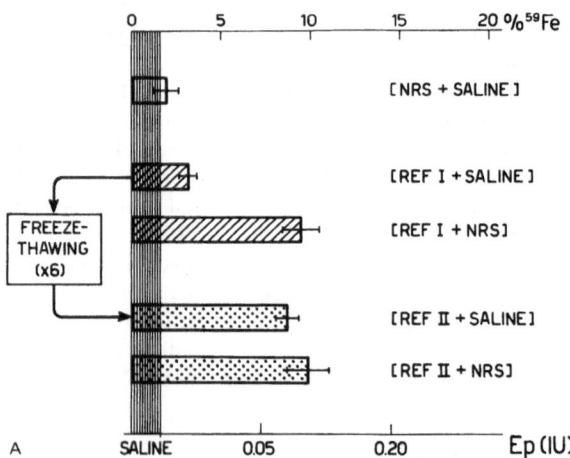

Figure 30-16A Conversion of REF I to REF II activity without incubation with NRS after repeated freeze-thawing (× 6). The erythropoietic activity in assay mice was evaluated on the basis of % RBC-^{59}Fe incorporation values (mean ± SEM) or corresponding IU of Ep. From Peschle et al.: in Nakao K, Fisher JW, Takaku F. *Erythropoiesis*. Tokyo, University of Tokyo Press, 1975.

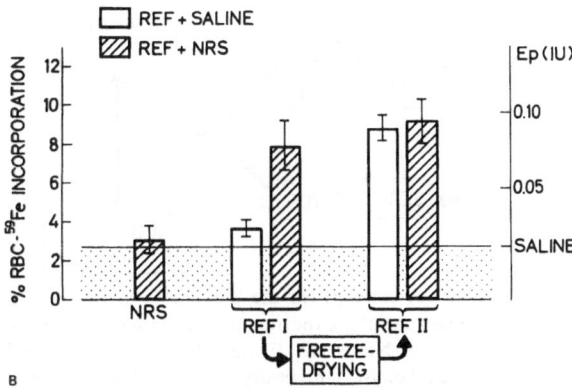

Figure 30-16B Conversion of REF I to REF II activity without incubation with NRS after repeated freeze-drying (× 2). The erythropoietic activity in assay mice was evaluated on the basis of % RBC-^{59}Fe incorporation values (mean ± SEM) or corresponding IU of Ep.

repeated freeze-thawing or freeze-drying. Similarly, conversion of REF I to REF II was observed following a prolonged period of storing at −20°C.[44] It should be emphasized that both types of conversion did *not* involve incubation with NRS. Furthermore, a similar phenomenon (i.e. REF I to REF II conversion *without* incubation with NRS) has been occasionally observed following a variety of physiochemical procedures, as column chromatography, and so forth (unpublished observations).

Figure 30-17 shows that 0.01 or 0.1 ml of rabbit anti-human urinary Ep respectively neutralized up to 0.12 IU of rat or rabbit Ep. In this regard, it is of crucial interest that 0.08 IU of Ep/2 ml of incubation mixture generated following addition of normal rat serum with rabbit REF I were not neutralized on further addition of 0.01 of anti-Ep and GARGG (Figure 30-18). On the other hand an equivalent activity generated following incubation of normal rabbit serum and rat REF I was totally neutralized by further addition of the same amount of anti-Ep and GARGG (Figure 30-18). Although the results are not presented here, a similar phenomenon was observed on incubation of these materials (rabbit or rat serum + rabbit or rat REF I, respectively) with anti-Ep, without further addition of GARGG.

Results

These studies suggest that the REF I is a pro-Ep molecule that is synthetized by the hypoxic kidney. Strong evidence in favor of this concept derives from two lines of observations: (a) the REF I was converted to Ep (i.e., REF II) either on incubation with normal serum or by means of a variety of experimental procedures (prolonged storing at −20°C, repeated freeze-thawing, freeze-drying, column chromatography, and so forth). (b) Incubation of rat or rabbit REF I with rat or rabbit serum resulted in the generation of Ep deriving from the REF I solution and not from the normal serum. In an attempt to partially rec-

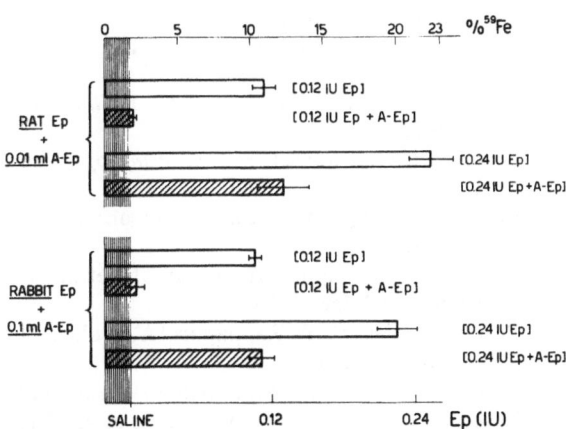

Figure 30-17 Incubation of 0.01 or 0.1 ml of rabbit anti-human urinary Ep with either 0.12 or 0.24 IU of respectively rat or rabbit serum Ep. The incubation was carried out for 30 min at 37°C in a water bath with constant shaking. The precipitate after centrifugation (2,000 rpm for 15 min) was discarded and the supernatant fluid injected in assay mice. The erythropoietic activity was evaluated on the basis of % RBC-^{59}Fe incorporation values (mean ± SEM) or the corresponding IU of Ep. From Peschle and Condorelli, *Science*, 190:910, 1975.

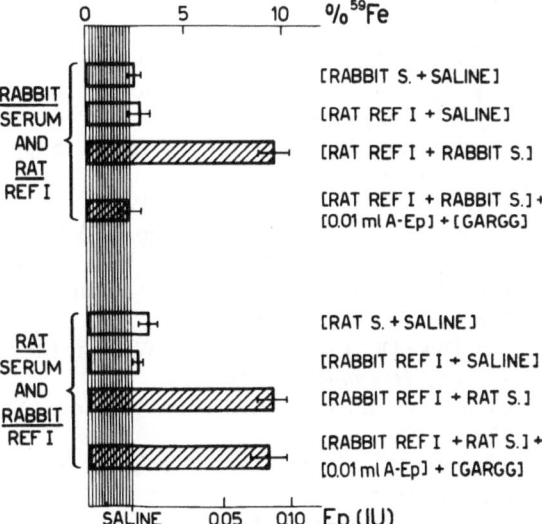

Figure 30-18 Incubation of either rabbit normal serum with rat REF I, or rat serum with rabbit Ref I. The (REF + serum) mixtures were further incubated with 0.01 ml of anti-Ep serum (A-Ep) and GARGG. The activity in the final mixtures was evaluated in assay mice, on the basis of % RBC-^{59}Fe incorporation values (mean ± SEM) or corresponding IU of Ep. From Peschle and Condorelli, *Science*, 190:910, 1975.

oncile the Gordon concept with the present results, it may be envisioned that the REF solution contains both the enzyme and its substrate.

It is tentatively concluded that the REF I–Ep system[21,44] is apparently similar to the proinsulin-insulin one.[45] In both cases a precursor of the hormone is synthesized in a specific organ, represented respectively by kidney and pancreas. Furthermore, in the extracts of these organs, both a precursor molecule and the biologically active hormone (REF I and Ep, proinsulin and insulin, respectively) can be demonstrated.

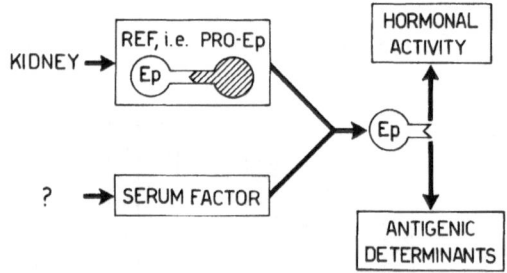

Figure 30-19 A concept for the biogenesis of Ep, postulated on the basis of the experiments presented here. From Peschle et al.: in Nakao K, Fisher JW, Takaku F (eds): *Erythropoiesis*. Tokyo, University of Tokyo Press, 1975.

Conclusion

In conclusion, these studies provide strong evidence indicating that the biogenesis of renal Ep involves inactivation of a kidney-produced pro-Ep molecule by a serum factor (Figure 30-19). The REF II activity is due to the biologically active Ep. The standard REF (REF I) described by Gordon et al.[39] is apparently constituted by a pro-Ep molecule. Furthermore, it must be postulated that in the pro-Ep complex the antigenic determinants of Ep are masked by the carrier molecule, thus preventing recognition of the pro-Ep by anti-Ep serum and of Ep by anti-REF serum.[46] It is also apparent that the pro-Ep is not stored in large amounts in the kidney and has a high turnover rate during hypoxia. However, further investigation is required to assess the physiological role of the serum component interacting *in vitro* with the pro-Ep molecule to generate Ep.

Acknowledgments

This work was supported by grants from EURATOM, Bruzelles (No. 159–76–7–BIOI); Volkswagen Foundation, Hannover; and CNR, Rome (No 75.01009.65 and 76.01467.04).

We express our appreciation to Mr. P. Ciaglia and P. Barba for their excellent technical help.

References

1. Krantz SB, Jacobson LO: Erythropoietin and the Regulation of Erythropoiesis. Chicago, University of Chicago Press, 1970.
2. Gordon AS: Erythropoietin. *Vit Horm* (Suppl) 31:105, 1973.
3. Peschle C: *Br J Haematol* 31:69, 1975.
4. Evans ES, Contopoulos AN, Simpson ME: *Endocrinology* 60:403, 1957.
5. Fried W, Plzak LF, Jacobson LO, et al.: *Proc Soc Exp Biol Med* 94:237, 1957.
6. Van Dyke DC, Contopoulos AN, Williams BS, et al.: *Acta Haematol* 11:203, 1954.
7. Crafts RC, Meineke HA: *Ann NY Acad Sci* 77:501, 1959.
8. Mirand EA, Gordon AS, Wenig J: *Nature* 206:270, 1965.
9. Alexanian R: *Blood* 33:564, 1969.
10. Fisher JW, Roh BL, Halvorsen S: *Proc Soc Exp Biol Med* 126:97, 1967.
11. Jepson IH, McGarry EE: *Blood* 39:239, 1970.
12. Peschle C, Zanjani ED, Gidari AS, et al.: *Endocrinology* 89:609, 1971.
13. Peschle C, Rappaport IA, Sasso GF, et al.: *Endocrinology* 91:511, 1972.
14. Byron JW: *Blood* 40:198, 1972.
15. Peschle C, Magli MC, Cillo C, et al.: *Life Sci,* 21:773, 1977.
16. Golde DW, Eersch N, Cline MJ: *J Clin Invest* 57:57, 1976.
17. Dukes PP, Goldwasser E: *Endocrinology* 69:21, 1961.
18. Peschle C, Magli MC, Cillo C, et al.: *Life Sci* 21:1303, 1977.
19. Peschle C, Rappaport IA, Sasso GF, et al.: *Endocrinology* 92:358, 1973.
20. Peschle C, Sasso GF, Rappaport IA, et al.: *J Lab Clin Med* 79:950, 1972.
21. Peschle C, Condorelli M: *Science 190*:910, 1975.
22. Peschle C, Marone G, Genovese A, et al.: Relation between the reticulo-endothelial system (RES) and extrarenal erythropoietin (Ep) production, in Nakao K, Fisher JW, Takaku F (eds): *Erythropoiesis*. Tokyo, University of Tokyo Press, 1975; Peschle C, Marone G, Genovese A, et al.: *Blood* 47:325, 1976; Peschle C, Marone G, Genovese A, et al.: *Br J Haematol* 32:105, 1976.
23. Feigin WM, Gordon AS: *Endocrinology* 47:364, 1950.

24. Fried W: *Blood 40:*671, 1972.
25. Peschle C, D'Avanzo A, Rappaport IA, et al.: *Nature 243:*539, 1973.
26. Schooley JC, Mahlmann LJ: *Blood 43:*425, 1974.
27. Kaplan SM, Rothman SA, Gordon AS, et al.: *Proc Soc Exp Biol Med 143:*310, 1973.
28. Katz R, Cooper GW, Gordon AS: *Ann NY Acad Sci 149:*120, 1968.
29. Carmena AO, Howard D, Stohlman F: *Blood 32:*376, 1968.
30. Schooley JC, Mahlmann LJ: *Blood 39:*31, 1972.
31. Higgins CM, Anderson RM: *Arch Pathol 12:*186, 1931.
32. Biozzi G, Benacerraf B, Stiffel C, et al.: *CR Soc Biol 148:*431, 1954.
33. Fisher JW, Taylor G, Porteous DD: *Nature 205:*611, 1965.
34. Kelly LS, Brown BA, Dobson EL: *Proc Soc Exp Biol Med 110:*555, 1962.
35. Trejo RA, Di Luzio NR, Loose LD, et al.: *Exp Mol Pathol 17:*145, 1972.
36. Naughton BA, Kaplan SM, Roy M, et al.: *Science 196:*301, 1977.
37. Anagnostou A, Schade S, Barone J, et al.: *Blood 50:*457, 1977.
38. Erslev AJ: Anemia of chronic renal failure, Hematology. Edited by WJ Williams, E Bentler, AJ Erslev, RW Rundles-McGraw-Hill, New York, 1972.
39. Gordon AS, Cooper GW, Zanjani ED: *Sem Hematol 4:*337, 1967.
40. Zanjani ED, Contrera JF, Gordon AS et al.: *Proc Soc Exp Biol Med 125:*505, 1971.
41. Kuratowska Z, Lewartovski B, Lipinski B: *J Lab Clin Med 64:*226, 1964.
42. Schooley JC, Zanjani ED, Gordon AS; *Blood 35:*276, 1970.
43. Zanjani ED, Gidari AS, Gordon AS, et al.: Further studies on the mechanism of erythropoietin formation by the kidney, in Gordon AS, Condorelli M, Peschle C (eds): *First International Conference on Hematopoiesis. Regulation of erythropoiesis.* Il Ponte, Milan, 1972.
44. Peschle C, Marone G, Genovese A, et al.: Studies on erythrogenin (Eg)-erythropoietin (Ep) system, in Nakao K, Fisher JW, Takaku F (eds): *Erythropoiesis.* Tokyo, University of Tokyo Press, 1975.
45. Steiner DF: Proinsulin and insulin biosynthesis, in Loubetieres A, Renold AE (eds): *Pharmacokinetics and Mode of Action of Oral Hypoglycemic Agents.* Milan, Il Ponte, 1969.
46. McDonald TP, Zanjani ED, Lange RD, et al.: *Br J Haematol 20:*113, 1971.

Discussion (Chapters 27–30)

Dr. Peschle: We have been intrigued by the role of the spleen in Ep production. We had a negative experience, as you had, in the splenectomy experiments: no difference was observed in regard to Ep production between the splenectomized and the non-splenectomized animal. Recently, we have tried to induce massive proliferation of spleen RES with prolonged injection of zymosan. Only under these particular conditions did splenectomy apparently decrease extrarenal Ep production.

Dr. Erslev: I wonder if your splenic proliferation is greater after zymosan than after 6 weeks of phenylhydrazine? This drug caused an increase in size to ten times normal.

Dr. Peschle: I don't remember exactly the amount of increase of the spleen weight after zymosan. However, this is not only a quantitative matter: massive amounts of hemolyzed red cells may well interfere in Ep production. As it was mentioned earlier, the *in vitro* experiments can certainly lead to artifacts; it is fair to say now that the *in vivo* ones can be awfully difficult to interpret. I also emphasize that not all macrophages besides Kupffer cells, must necessarily be involved in Ep production.

Dr. Zanjani: I guess when we are talking about extrarenal erythropoietin production, the fetus may be the best model, because there, nearly all the erythropoietin is produced extrarenally. In the fetus, if you take the spleen out, it doesn't really affect the production of erythropoietin at all, and this is true if you remove the kidney as well. So I think, in the fetus at least, the data support your view that the spleen may not be an important agent in extrarenal production of erythropoietin.

Dr. Fisher: I think everyone here agrees that the spleen is not very important in extrarenal erythropoietin production. Can splenic proliferation be equated with Kupffer cell proliferation? Are we really permitted to make an association between splenic proliferation and Kupffer cell function?

Dr. Zanjani: I think the differences that exist between liver and spleen are that the production of proteins in the liver are just enormous compared to the spleen, and the proximity of the cell and where it's located may be an important point in all of these considerations. For example, we know that cells cannot become red cells or any of the blood cell types unless they sit in an appropriate microenvironment, so that may be the reason why Kupffer cells may be very important in the liver regarding erythropoiesis.

Dr. Fisher: The spleen still appears to be Galen's organ of mystery. Dr. Erslev, your work on the air pockets reminds me of our frequent reference to the oxygen sensor in the kidney. I remember when Dr. Linus Pauling was visiting with us a few years ago at Tulane; I explained the renal oxygen sensor concept to him, and he immediately asked what is the site of the oxygen sensor in the kidney? Is it an enzyme? It stimulated us to look for some site within the kidney that may be responsive to an oxygen deficit and our search led us to the prostaglandins system. I wonder if phospholipase, which is a very early step in the synthesis of prostaglandins, is triggered by hypoxia to increase the arachidonate pool and leads to an increase in the prostaglandins which are tissue hormones. This could be the first step in the production of erythropoietin. We must search further for this very sensitive oxygen receptor in the kidney.

Dr. Gordon: Dr. Erslev, you went into this problem a little with your medullary-cortical theory. Would you want to say a little about that?

Dr. Erslev: No, because we really have no new data on it. Certainly, the most frequent association within appropriate erythropoietin production is the renal cyst; and the renal cyst is made up by tubular epithelium. Consequently, I'm more inclined to believe that this erythropoietin-producing and oxygen-sensing apparatus is in the medulla rather than in the cortex. However, this is very indirect reasoning.

Dr. Zucali: If I may make a note of caution. When people start talking about looking at increased numbers of cells versus increased amount of cell product, it could be that the reason that one sees increased cell product is that one has more cells that are making the product, but it also could be that each cell is making more of the product.

Dr. Erslev: I agree completely.

Dr. Sassa: When red cells are hemolyzed, the first thing that happens in the liver is the induction of microsomal heme oxygenase. This enzyme induction also occurs in the kidney. Other potent inducers of microsomal heme oxygenase are certain metals including cobalt. Cobalt is known to produce erythropoietin in the kidney. I wonder whether someone has studied cobalt effect in the liver.

Dr. Erslev: We're looking at phenobarbital and at a number of other substances that we think may in some way induce microsomal activity. Cobalt can do it, but we are not really looking at cobalt because it produces both renal and extrarenal erythropoietin. We would like to find other compounds that would stimulate microsomal activity.

Dr. Sassa: If I may add, phenobarbital is an inducer of cytochrome P-450 but microsomal heme oxygenase is not related to cytochrome P-450. If you want to look for another compound that induces erythropoietin production in the liver but not in the kidney, other metallic inducers on heme oxygenase would be interesting to try. For example, tin has been reported by Maines and Kappas to be a potent inducer of heme oxygenase. Since tissue anoxemic effect of cobalt goes hand in hand with the production of erythropoietin and the induction of heme oxygenase, I wonder whether those metals that cause heme oxygenase induction can induce hepatic erythropoietin production.

Dr. Erslev: It sounds like a very useful approach and I hope you get some ideas about compounds which we could try.

Dr. Rifkind: Speaking as a noncombatant in this problem, and sort of responding to the trend that Dr. Sassa raised, I just wonder whether biogenetically, the machinery for this response was designed to respond to phenylhydrazine or to bleeding?

Dr. Erslev: Well, no, I disagree. I think phylogenetically the stem cells are responding to hemolysis. When a cell dies it is engulfed and lysed and the feedback is sensing what is being destroyed, not what is being bled.

Dr. Trobaugh: After Dr. Rifkind's last question, I'm not sure I want to ask mine. In thinking of oxygen sensors, which as we all consider are within the kidney, in the arterial system, how much does the Po_2 of the portal vein and the hepatic vein blood change in the face of anemia?

Dr. Erslev: My recollection is that the venous blood coming out of the liver has a Po_2 of about 40 milliliters, very much the same as for most other organs. The Po_2 of venous blood coming from the kidney is much higher, 80 milliliters, and the Po_2 of venous blood coming out of the heart much lower, almost zero.

Dr. Trobaugh: I just wonder if it changes very much, nearly as much as the arterial oxygen capacity changes.

Dr. Fisher: Yes, this is an important question and one that Kurt Reissman and I used to debate. I have searched for references to the A-V oxygen difference in the kidney during anemia and have not found a report on this subject. I rather doubt that it has been done in the liver.

Dr. Erslev: You know, it's always very dangerous to say it has not been done, because one can always find an obscure author who did it before. In this case it is a Swedish author named Aperia (*Acta Physiol Scand* 75:353, 1969), who has done some studies on the A-V oxygen differences across the kidney in anemia and in normal conditions. There was a very small difference between the two conditions, too small to be significant.

Dr. Peschle: I am delighted by this presentation, which is fully in line with our findings on the liver RES as a source for Ep. Several aspects of your studies on anephric rats with liver regeneration show a striking similarity with our observations in anephric animals treated with colloidal carbon or zymosan, published 1 or 2 years ago. First, the hyperplasia of Kupffer cells during liver regeneration is very similar, on a quantitative basis, to that induced by colloidal carbon and zymosan. Second, the increase of hepatic Ep production in the regenerating phase reproduces exactly that observed after treatment with these agents. Finally, I agree that the potentiation of hepatic Ep production via hyperplasia of Kupffer cells is inhibited by the kidney. In this regard, we reported that, in the intact animal, priming with colloidal carbon or zymosan did not enhance Ep production; we envisioned therefore that the kidney partially suppresses liver Ep production, as you have suggested here.

Dr. Murphy: Dr. Gordon, your results in adult rats that indicate that hepatic erythropoietin is generated only in the absence of kidneys are very exciting. How can we reconcile these data with those of the adult sheep in which we have demonstrated copious amounts of erythropoietin in hepatic lymph of adult sheep with intact and functional kidneys? You will recall that erythropoietin was found in both renal and hepatic lymph following hemolytic anemia induced by phenylhydrazine.*

Dr. Gordon: There is the possibility here that phenylhydrazine, in addition to inducing anemic hypoxia, has injured the kidney, resulting in an interference with the ability of the kidney to exert its inhibitory influence on extrarenal Ep production. This is pure conjecture but it is the only explanation I can come up with at this time.

Dr. Zanjani: Dr. Peschle, I feel that estrogen should be included in your model on the endocrine regulation of erythropoiesis, but on the inhibitory side.

Dr. Peschle: Sure. This is exactly what I meant: that model was focused on hormones stimulating erythropoiesis.

Dr. Zanjani: I have several questions on the pro-Ep concept, as well as the REF-Ep conversion. The antiserum against REF, prepared by Drs. Lange and McDonald, did not neutralize Ep. Is this finding compatible with your pro-Ep concept? Second question. The rabbit anti-Ep serum you employed neutralized Ep produced by rat REF but not that generated by rabbit REF which was neutralizing titer of the anti-Ep against rat and rabbit Ep re-

---

* Murphy MJ, et al.: *Exp Hematol* 5:41, 1977.

spectively? Third one: you showed that it is possible to convert REF I to REF II (that is, REF to Ep) without incubation with serum. How did you exactly perform these experiments?

Dr. Peschle: In regard to the first question our concept postulates that, in the pro-Ep molecule, both the biologically active site and the antigenic determinants of Ep are masked by the Ep carrier. Thus, the anti-REF serum cannot recognize Ep; and also, the anti-Ep serum cannot neutralize REF. Second question. The neutralizing titers of rabbit anti-Ep against respectively rat or rabbit Ep had a ratio of at least 10:1. Similar results were reported by Schooley and Garcia in regard to their rabbit anti-Ep. In the REF experiments, the rat REF + rabbit serum mixture was incubated with the amount of anti-Ep necessary to neutralize the rat Ep generated in the mixture. However, the same volume of anti-Ep did not significantly neutralize a similar amount of rabbit Ep, generated via incubation of rabbit REF + rat serum, i.e., the 10% neutralization of rabbit Ep by anti-Ep was not monitored by the bioassay. Third question. The conversion of REF (pro-Ep) to Ep, in the absence of serum, can be observed after a variety of physicochemical manipulations—in some experiments, after repeated freeze-thawing. In this case, the original REF was subdivided into 2 aliquots: let us call them bottle A and B. Bottle B was subjected to repeated freeze-thawing ($\times 6$); bottle A was not touched. Thereafter, the contents of both A and B were incubated with either serum or saline, and finally assayed: bottle A contained REF (pro-Ep), bottle B Ep. A conversion from pro-Ep to Ep was thus induced by repeated freeze-thawing; these experiments have an inescapable logic.

Dr. Meyers: What happens if you incubate REF I from rat with mouse serum?

Dr. Peschle: I have not done this experiment. Dr. Zanjani observed no species specificity when incubating REF and normal serum from different mammalian species.

Dr. Zanjani: That is true.

Dr. Meyers: In light of that, why, when you inject your REF 1 into your assay animal, does it not have erythropoietic activity?

Dr. Peschle: This is an intriguing aspect. Of course, REF may react differently with normal serum *in vitro* than with polycythemic serum in assay mice. However, you should ask this point to Dr. Zanjani, who performed these experiments.

Dr. Zanjani: I don't know if we did. Anyway, the concept was that REF got inactivated by parenteral injection.

Appendix 1 *In Vitro* Modulation of Human Erythropoiesis by Lymphocyte-Rich Fractions of Blood

Esmail D. Zanjani, Benet Nomdedeu, John Rinehart, Bobby J. Gormus and Manuel E. Kaplan

The plasma clot culture system originally described by Stephenson et al.,[1] improved[2] and modified for use with human bone marrow,[3] was used to assess erythroid colony formation by human bone marrow (CFU-e), peripheral blood mononuclear cells (BFU-e), and the role of peripheral blood mononuclear cells in erythropoiesis. The latter involved the co-culture of bone marrow cells and blood mononuclear elements (PBL). The culture method for CFU-e was the same as described by Tepperman et al.[3] In co-culture studies, bone marrow cells (6×10^5 cells) and different concentrations of PBL (0.5 to 6×10^5 cells) were admixed at the start of the culture and the mixture distributed in microtiter wells within 30 to 60 min of mixing. The culture procedure for BFU-e was similar to CFU-e except that semipurified bovine thrombin (Grade 1, Sigma Chemical Co., St. Louis, Missouri) at a final concentration of 1 unit / 1.1 ml was substituted for beef embryo extract,[4] and the incubation time extended to 13 days.

Preparation of PBL for Co-culture Studies

Venous blood was collected in heparinized glass tubes, and diluted 1:1 with Ca^{2+}- and Mg^{2+}-free Hank's balanced salt solution (HBSS, GIBCO, Grand Island, New York). Five ml of the diluted blood was carefully layered by means of a sterile disposable syringe, fitted with a 19-gauge needle, on top of 3 ml of Ficoll-Hypaque (Pharmacia, Ltd., Piscataway, New Jersey) in a 10 ml glass tube and spun at 400 g for 30 min at 20°C. Cells at interface were aspirated using sterile glass Pasteur pipets, pooled in sterile plastic tubes (Falcon Plastics, Oxnard, California) and washed three times with HBSS by centrifugation at 4,000 rpm in a Serofuge for 2 to 3 min. The cells were resuspended in minimum essential medium (GIBCO, Grand Island, New York) made 15% with heat-inactivated fetal calf serum (FCS, GIBCO, Grand Island, New York), and used.

Preparation of T- and B-Cell Enriched Fractions from PBL for Co-Culture Studies

Sheep red blood cells (SRBC, Colorado Serum Co., Denver, Colorado) were washed three times with HBSS to remove preservative. Appropriate numbers of PBL and sheep red cells were mixed in a glass tube (50 to 100 SRBC/lymphocytes) in the presence of heat-inactivated AB serum and incubated at 37°C for 5 min. The mixture was centrifuged at 500 rpm for 5 min at room temperature, and the tube incubated at 4°C for at least 1 hr. The cells were subjected to gentle agitation and 6 ml of the mixture layered on top of 3 ml of Ficoll-Hypaque in a 10 ml glass tube. After centrifugation (400 g, 30 min, 20°C), the interface (nonrosetted) cells were removed, washed two times with HBSS, resuspended in 15% FCS-MEM solution and designated as B-cell enriched fraction. The "rosetted" cells (at bottom of tube) were resuspended in a 1:5 solution of phosphate buffered saline and distilled water and allowed to stand at room temperature for 45 sec. The mixture was spun (Serofuge) at 4,000 rpm for 2 min, and the cells washed two times with HBSS, resuspended in 15% FCS-MEM solution and designated as T-cell enriched fraction.

Preparation of FI-FIV Fractions of Blood

For studies of BFU-e proliferation, human peripheral blood was separated into four cell fractions. Blood was drawn from the antecubital fossa as follows: 0.2 ml of heparin (200 units, 1,000 units/ml) was drawn into a 50 ml syringe. After a tourniquet was placed on the upper arm, a 19-gauge butterfly needle was placed in the vein and 50 ml of blood was drawn into the heparinized syringe. Blood was transported to the laboratory in the heparinized syringes at room temperature and was used within 30 min. All laboratory procedures were performed in a laminar flow hood utilizing materials sterilized by autoclaving and media sterilized by filtration (.45 micron filters).

Ficoll-Hypaque (F-H) gradients were prepared as follows: 15 ml of F-H was placed in 50 ml conical polypropylene tubes. Blood was diluted 1:1 with cold Seligman's balanced salt solution (SBSS) then 20 ml of the diluted blood was poured into 30 ml syringes (plunger removed) with a 19-gauge needle in place. The needle and attached syringe were placed over the F-H tubes so that gravity induced the flow of diluted blood down the side of the tubes over the F-H. Gradients were then centrifuged in a swinging bucket centrifuge at 400 g for 25 min, 15°C. Cells at the interface between the F-H and diluted plasma or fraction I were then removed with a 10 ml syringe and a 19-gauge needle, placed in 50 ml polycarbonate tubes, and centrifuged at 400 g for 10 min at 4°C. Supernates were then decanted and the cell buttons were resuspended by vigorous agitation. Cells were then diluted in 40 ml of ice cold SBSS and centrifuged at 200 g for 10 min at 4°C. This step was repeated before the cells were resuspended in 20 ml of SBSS and placed on ice. An aliquot of this cell suspension was then diluted 1:10 in a white blood cell pipet using 1% acetic acid and the diluted cells were counted in a Neubquer hemocytometer. A second aliquot was used to determine the percentage of monocytes and lymphocytes present: 0.1 ml of cell suspension and 0.1 ml of Hank's balanced salt solution (HBSS) with 20% fetal calf serum were placed in cytocentrifuge tubes, centrifuged for 5 min at 400 rpm. Slides were then air dried, fixed in methanol for 30 sec, air dried, and stained for 10 min with Giemsa. Slides were again air dried and differential cell counts were performed. The percentage of monocytes and lymphocytes determined by routine morphology correlated well with Sudan black positivity and latex phagocytosis (±3%). Cell counts and cell differentials were performed as described on all subsequent cell fractions obtained.

The interface cells of fraction I were then centrifuged again at 200 g for 10 min. Then using the cell count and differential counts just described, these cells were resuspended to adjust the monocyte concentration to 1.5×10^6/ml of RPMI-1640 with 20% fresh autologous sera; 3.0 ml of this suspension was placed onto 60-mm culture dishes and incubated for 1.0 hr at 37°C, 5% CO_2 in air with 100% humidity. Nonadherent cells were then mobilized by swirling and decanted, and the dishes were washed by placing 2 ml of HBSS onto each culture dish, swirling, and decanting. This procedure was repeated six times. Nonadherent cells or fraction II cells were pooled, stored on ice for 30 to 40 min, and then further processed as described in the next paragraph. Adherent cells or fraction IV were immediately removed from washed culture dishes as follows: the final HBSS wash was decanted from plates and 3 ml of SBSS containing 30 mM lidocaine and 10% fresh autologous sera were added to each plate and incubated at room temperature for 15 min. Adherent cells were removed by this treatment or by gently streaming SBSS onto residual cells utilizing a Pasteur pipet and suction bulb. These suspended cells were then pooled in 35 ml polycarbonate tubes and immediately washed three times by centrifugation (200 g for 10 min) with iced SBSS.

Pooled fraction II or nonadherent cells were centrifuged at 200 g for 10 min and resuspended to 10^6 cells/ml of autologous plasma in 50-ml polypropylene tubes. Then 25 mg/ml of carbonyl iron was mixed with the cell suspension and the mixture was rotated (20 rpm) for 30 min; 15 ml of this mixture was then layered over 15 ml of F-H as previously described and centrifuged at 400 g for 30 min at room temperature. Resultant interface cells or fraction III cells were then washed twice in RPMI-1640 by centrifuging at 200 g for 10 min at 4°C.

Materials

50 ml conical polypropylene tubes and 60 mm culture dishes (Falcon Plastics, Oxnard, California), polycarbonate tubes, 30- and 50 ml (Fischer Scientific Co., Pittsburgh, Pennsylvania), Nalge Filters (Nalge-Sybran Corp., Rochester, New York) were used in these procedures.

F-H was prepared as follows: Ficoll-400 (Pharmacia Chemicals, Piscataway, New Jersey) was dissolved in distilled water to 9%. Hypaque-50% (Winthrop Labs., New York, New York) was diluted to 34% with distilled water. Then 24 parts of 9% Ficoll and 10 parts of 34% Hypaque were mixed and sterilized by filtering (0.45 μ pore size). Specific gravity of the resulting mixture was 1.075 to 1.080 at 20°C. SBSS was prepared in our laboratory by dissolving NaCl (13.3 gm), $NaCO_2CH_3$ (3 gm), Dextrose (2 gm), $NaHCO_3$ (1.4 gm), KCl (0.4 gm), KH_2PO_4 (0.2 gm) and NaH_2PO_4 (0.1 gm) in 2 liters of distilled water and sterilizing by filteration. Xylocaine, 30 mM was prepared by mixing 8.75 ml lidocaine (40 mg/ml Astra Pharmaceutical Products, Worchester, Massachusetts) and 41.25 ml SBSS in the presence of 0.7 ml of 7.5% $NaHCO_3$. The pH was 7.4. Carbonyl iron powder (GAF Corp., New York, New York) was weighed into 0.5 mg aliquots, placed in glass tubes, and sterilized by heating to 180°C for 1 hr. The following were used as supplied commercially. Sodium heparin, 1,000 units/ml (Abbot Labs., Chicago, Illinois), fetal calf serum (Reheis Chemical Co., Phoenix, Arizona), RPMI-1640 (Associated Biomedic Systems, Inc., Buffalo, New York), HBSS (GIBCO, Grand Island, New York).

References

1. Stephenson JR, Axelrad AA, McLeod DL, et al: Induction of colonies of hemoglobin-synthesizing cells by erythropoietin *in vitro*. *Proc Natl Acad Sci 68:*1542, 1971.
2. McLeod DL, Shreeve MM, Axelrad AA: Improved plasma culture system for production of erythrocytic colonies *in vitro:* quantitative assay method for CFU-e. *Blood 44:*517, 1974.
3. Tepperman AD, Curtis JE, McCulloch EA: Erythropoietic colonies in cultures of human marrow. *Blood 44:*659, 1974.
4. Clarke BJ, Housman D: Characterization of an erythroid precursor cell of high proliferative capacity in normal peripheral blood. *Proc Natl Acad Sci 74:*1105, 1977.

Appendix 2

James W. Fisher, Yasahico Ohno, Bruno Modder, Franciszek Przala, Gregory D. Fink, and Dennis M. Gross

Plasma Clot Erythroid Colony-Forming Cell Bone Marrow Culture Technique*

Normal (sham) or uremic rabbits are exsanguinated via cardiac puncture and one femur removed rapidly using an aseptic technique. Both ends of the femur are cut off and the bone marrow flushed into 10 ml of cold collection medium. Each 100 ml of collection medium contains 10 ml of MEM-Hank's balanced salt solution (10 × concentrated GIBCO, Grand Island, New York), 1 ml of MEM nonessential amino acid solution (100 × GIBCO), 1 ml of MEM sodium pyruvate solution (100 × GIBCO), 1 ml of L-glutamine (100 × GIBCO), 1.0 ml of 7.5% $NaHCO_3$ solution, 2 ml fetal calf serum (heat inactivated, GIBCO), 10,000 units penicillin and 5 μg streptomycin in sterilized distilled water. The cell clumps are broken up by gently pipetting the suspension eight to ten times. The cells are then washed three times with cold collection medium and finally the number of nucleated cells is determined using a cell counting chamber. Disposable microtiter plates are sterilized by UV-irradiation for 2 hr and 0.1 ml of culture medium containing 5×10^4 nucleated cells plus 0.9 ml media containing 20% fetal calf serum, 10% of a beef embryo extract solution (SIGMA, St. Louis, Missouri, GIBCO) in NCTC-109, 10% of a L-asparagine solution in NCTC-109, Erythropoietin (0.2 unit), 40% NCTC-109 (Microbiological Assoc., Bethesda, Maryland) and 10% bovine citrated plasma (I.S.I. Biol., Cary, Illinois) is pipetted into each microtiter well (Cook Lab. Prod., Alexandria, Virginia). Four wells for each determination are performed and allowed to clot at room temperature. The microtiter plates are placed into sterile 100 mm petri dishes containing a small dish filled with water, and incubated in humidified atmosphere of 95% air and 5% CO_2 for 4 days at 37°C. The plasma clots are then removed from the microtiter wells, placed on a microscopic glass slide, dried with filter paper, fixed with 5% glutaraldehyde in phosphate buffer solution, stained with benzidine and counterstained with Giemsa solution. A drop of immersion oil is placed on each fixed plasma clot and benzidine-positive erythroid colonies that consist of at least eight cells are scored over the entire clot under a microscope with a 100× magnification. For each value the mean ± SEM of three to four clots is determined. Figure 16-3 shows the type of CFU-e dose-response curve seen in the normal rabbit bone marrow culture to erythropoietin. Note that the linear part of our dose-response curve with increasing concentrations of human urinary erythropoietin is between 0.1 and 1.0 unit/ml of culture medium. A plateau effect is seen in our system between 1 and 2 units of erythropoietin. Note in Figure 16-4 that when the rabbit bone marrow is cultured over a 96-hr period that no erythroid colonies are formed until 48 hr when a linear increase in CFU-e is seen up to 84 hr. A plateau in CFU-e is seen between 84 and 96 hr in culture. As seen in Figure 16-5, when the number of bone marrow cells plated is plotted against the number of erythroid colonies formed no colonies are seen until 2×10^4 cells/plate is reached. The regression line is linear between 2×10^4 and 10×10^4 cells/plate.

* Modification of the method of McLeod DL et al: *Blood* 44: 517–534, 1974.

Sources of Media and Other Materials Used in the Plasma Clot Culture

MEM (Hanks, 10×)	Grand Island Biological Co., Grand Island, New York
MEM nonessential amino acid sol. (100×)	GIBCO
MEM sodium pyruvate sol. (100×)	GIBCO
L-glutamine (100×)	GIBCO
$NaHCO_3$ sol.	GIBCO
Fetal calf serum (heat inactivated)	GIBCO
Beef embryo extract	GIBCO
NCTC-109	Microbiological Associates
L-asparagine	SIGMA, St. Louis, Missouri

Petri dishes (100 mm)	FALCON, Oxnard, California
Glutaraldehyde	SIGMA
Bovine citiated plasma	I.S.I. Biological, Cary, Illinois
Disposable microtiter plates	Cook Lab. Prod., Alexandria, Virginia
Erythropoietin (human urinary Ep) (Batch M-7—8.8 μ/mg)	Supplied by: The Blood Resources Branch Division of Blood Diseases and Resources National Heart, Lung and Blood Institute Bethesda, Maryland

The erythropoietin preparations are sterilized by passage through a Millipore filter (type HA-0.45 μm) before being added to the culture. Approximately 25% of the erythropoietin is lost by passage through the Millipore filter.

Plasma Clot Erythropoietic Burst Forming Cell Bone Marrow Culture Technique*

Cells were cultured at a concentration of 5×10^5 bone marrow cells/ml in the culture medium of the same composition as used for the plasma clot erythroid colony forming cells (CFU-e) described earlier with the exception that the concentration of Erythropoietin (Ep) was 1.0 unit per ml of culture medium. 0.1 ml of this complete culture medium (5×10^4 cells) (containing 0.1 unit of Ep) was incubated for 8 or 10 days at 37°C in a humidified atmosphere in 5% CO_2 in air. 0.1 unit of Ep in NCTC-109 was added to the surface of each clot on first, second, third, and fourth days. (Each well contains a total of 0.5 unit of Ep.) Benzidine-positive erythropoietic bursts which consist of 1,000 cells or more are scored over the entire clot using a microscope with a $100 \times$ magnification.

* Modification of the method of Heath DS et al: *Blood* 47:777–792, 1976.

Appendix 3   Methylcellulose Clonal Cell Culture of Murine and Human Erythropoietic Precursors

Makio Ogawa

Methylcellulose clonal cell culture of murine and human erythropoietic cells is based on the method described by Iscove et al.[1] Culture mixtures were prepared in 16 ml Falcon tissue culture tubes. The following is an example for preparing 5 ml of marrow culture mixture:

For culture of murine cells, pipet 0.5 ml of marrow cells (1 to 2×10^6 nucleated cells/ml), 1.5 ml of fetal bovine serum, 0.5 ml of bovine serum albumin, 0.5 ml of mercaptoethanol (1×10^{-3} M), and 0.5 ml of erythropoietin (20 units/ml). Add 1.5 ml of methylcellulose stock solution (2.7%) using a 15-gauge needle and a 6 ml syringe. Shake vigorously and plate 1 ml of the mixture into a 35 mm Lux standard nontissue culture dish (#5221R) (Flow Laboratories, Inc., Rockville, Maryland) using a disposable 6 ml syringe fitted with a 15-gauge needle. Incubate in a humidified atmosphere flushed with 5% CO_2 in air. For culture of human erythropoietic cells, mercaptoethanol may be replaced by α-medium (Flow Laboratories, Inc.). Furthermore, concentrations of erythropoietin and duration of incubation should be adjusted according to the maturational stages of the precursors to be examined.[2-4]

The following is a more detailed account of each ingredient:

Marrow Cells: Cells are suspended in α-medium.
Methylcellulose: MeC (4,000 centipoise) is purchased from Fisher Scientific Co., Fair Lawn, New Jersey. To prepare stock solution, add 500 ml of boiling deionized water to 27 gm of MeC powder in a 2 liter flask and cool at room temperature to approximately 40°C with continuous stirring on a large magnetic stirrer. Add 500 ml of chilled, double-strength α-medium that would gel the solution immediately. Stir continuously in a cold room for 48 hr and aliquot in 50 to 100 ml volume and store in a freezer.

Fetal Calf Serum: Careful screening for a proper lot is needed. We purchase fetal calf serum from Flow Laboratories, Inc.

Bovine Serum Albumin: Dissolve Cohn's fraction V (Calbiochem, San Diego, California) in deionized water in a final concentration of 10% (w/v), deionized with analytical-grade Ion Exchange Resin (AG 501-X8[D]) (Bio-RAD Labs, Richmond, California), sterilize with filtration, and store in a freezer in aliquots of 2.5 ml per tube. On thawing, add 0.1 ml of 7% $NaHCO_3$ solution to a tube before use in culture.

Mercaptoethanol: Prepare mercaptoethanol fresh in α-medium on the day of experiment.

Erythropoietin: Dissolve the step III preparation of sheep plasma erythropoietin purchased from Connaught Labs, Ltd., Willowdale, Ontario, Canada, in α-medium and dialyze against α-medium by changing the medium three times every 8 hr. Dialysis for more than 48 hr partially inactivates the EPO activity.

References

1. Iscove NN, Sieber F, Winterhalter KH: *J Cell Physiol 83:*309, 1974.
2. Gregory CJ: *J Cell Physiol 89:*389, 1976.
3. Gregory CJ, Eaves AC: *Blood 49:*855, 1977.
4. Ogawa M, MacEachern MD, Avila L: *Am J Hematol 3:*29, 1977.

Appendix 4 Technique for Culture of Murine BFU-e

G. Wagemaker

Culture Conditions

Large erythroid colonies, termed bursts, become apparent after 6 days of incubation in semisolid cultures containing erythropoietin (EP), and α-medium, (lacking nucleosides), supplemented with 10% (v/v) horse serum, 10% (v/v) fetal calf serum, 1% (w/v) bovine serum albumin (BSA), 3×10^{-5} M/l^{-1} egg lecithin, 7×10^{-6} M/l^{-1} transferrin saturated with FeCl$_3$, 10^{-7} M/l^{-1} Na$_2$SeO$_3 \cdot$ 5 H$_2$O, and 10^{-4} M/l^{-1} 2-mercaptoethanol, using 0.8% (w/v) methylcellulose as a semisolidifying agent. Bursts were scored at various intervals of incubation, and could be easily detected in unstained cultures by their specific growth pattern, small cell size, and red color, using an inverted microscope (Zeiss) at a magnification of 79×. The cultures were incubated at 37°C in an atmosphere of 10% CO$_2$ in air and 100% humidity. This culture method is a slight modification of the methods employed by Iscove and Sieber,[1] Iscove,[2] and Guilbert and Iscove.[3]

Reagents

All reagents were of the purest grade available. Methylcellulose (Methocel A4M Premium, The Dow Chemical Co.), purified human transferrin (Behringwerke AG), Na$_2$SeO$_3 \cdot$ 5H$_2$O (Merck) and egg lecithin (Merck) were prepared in solutions as described by Iscove[4] and by Guilbert and Iscove.[3] BSA was exhaustively dialyzed, lyophilized, and dissolved in α-medium. Solutions of 10^{-1} M 2-mercaptoethanol (Merck) in α-medium were stored at 4°C for not more than a week.

Sera

Combinations of fetal calf serum (Flow) and horse serum (Flow) were selected on the basis of optimal support of erythroid burst formation (see Chapter 7).

Erythropoietin

Erythropoietin used was step III, lot no. 3013-16, spec. act. 7 units/mg^{-1}, prepared from anemic sheep plasma (Connaught Medical Research Lab., Willowdale, Ontario), or human urinary erythropoietin (pool H31-TaLSL), collected and concentrated by Centro de Estudios Farmacologicos y de Principios Naturales, Buenos Aires, Argentina, further processed and assayed by Hematology Res. Lab., Children's Hospital of Los Angeles, California, and authorized for distribution by the Divison of Blood Diseases and Resources of the National Heart Lung and Blood Institute, NIH, Bethesda, Maryland, spec.act. 18, $6 \pm 1,2$ units/mg^{-1}.

Preparation of Cell Suspensions

Mouse bone marrow cells were flushed from femurs with α-medium. A single cell suspension was obtained by pipetting and sieving through a nylon gauze. Nucleated cell counts were made by hemocytometer after dilution with Türk's solution.

Cultures

Cultures were prepared by adding the following volumes of ingredients:

  1.6 ml or 0.6 ml of a 2.0% stock solution of methylcellulose
  0.4 ml of 0.15 ml horse serum
  0.4 ml or 0.15 ml fetal calf serum
  0.4 ml or 0.15 ml of a mixture containing BSA, transferrin, lecithin, selenite and 2-mercaptoethanol at 10× final concentrations in 4 medium
  0.4 ml of 0.15 ml of EP at various concentrations
  0.4 ml or 0.15 ml bone marrow cells at various concentrations
  0.4 ml of 0.15 ml of various additives or α-medium
  4.0 ml   1.50 ml

for respectively 1 ml cultures in 32 mm petri dishes (Corning) of 0.25 ml cultures in 15 mm petri dishes (Costar). The tubes (Falcon) containing the initial culture mixture were thoroughly shaken by hand prior to plating by syringe and needle.

References

1. Iscove NN, Sieber F: *Exp Hematol 3:*32, 1975.
2. Iscove NN: *Cell Tissue Kinet 10:*323, 1977.
3. Guilbert LJ, Iscove NN: *Nature 263:* 594, 1976.
4. Iscove NN: *In vitro* culture of hemopoietic cells. Van Bekkum DW, Dicke KA (eds): Rijswijk, Radiobiological Institute TNO, 1972, p. 459.

Appendix 5 Technique for BFU and CFU-e Cultures

N. G. Testa

Preparation of Mouse Bone Marrow Cell Suspension

Femoral bone marrow cells are obtained by repeated flushing of the bone cavity with α-medium (GIBCO) using a syringe attached to a 23-gauge needle. The cells are suspended in ice cold α-medium containing 2% fetal calf serum, counted in a hemocytometer and plated as soon as possible, usually within one hr.

Plating and Incubation of Bone Marrow Cells

Each 1.0 ml culture contains

   0.40 ml 2.2% methylcellulose
   0.30 ml fetal calf serum
   0.10 ml bovine serum albumin solution
   0.01 ml transferrin solution
   0.01 ml β-mercaptoethanol solution
   0.01 ml sodium selenite solution
   0.05 ml erythropoietin
   0.12 ml cell suspension

The components are stored separately, and are combined at the time of plating. Usually 2.5 ml of the mixture is prepared to plate duplicate 1 ml cultures, as the viscosity of the methylcellulose makes some waste unavoidable. Two ml aliquots are drawn with a syringe, and plated in 35 mm diameter plastic petri dishes (Corning). The dishes are enclosed in a 100 mm diameter petri dish containing a third 35 mm dish, filled with water. The cultures are incubated at 37°C, 9 days for BFU, and 2 days for CFU-e, in a sealed box in an atmosphere of 5% CO_2 in air. Incubation in an atmosphere of 5% CO_2, 5% O_2 in nitrogen increases plating efficiency up to 3 times.

Stock Solutions

α-medium is purchased in powder, and prepared according to the manufacturer's instructions, but osmolarity is reduced to 280 mOs. Single strength medium is kept at 4°C and used within a month.

Methylcellulose: 4,000 CPS (Dow Chemicals) is prepared as a 2.2% solution as follows: a flask with 1L double distilled water heated to just below boiling point is placed on a magnetic stirrer and 44 gm of methylcellulose powder is dispersed on the surface. The mixture is allowed to boil for one min. When it has cooled to room temperature, an equal volume of $2\times$ concentrated α-medium is added. The mixture is placed on a magnetic stirrer at 4°C overnight, and then kept at 4°C for 1 to 2 days. It can be stored frozen for several weeks until needed. Before use, 1 ml of 200 mM solution of glutamine per 40 ml is added (C. Gregory, personal communication).

Bovine serum albumin (Fraction V, Sigma) is prepared according to the technique of Worton et al.[1]

Fetal calf serum: (Tissue Culture Supplies). FCS is stored frozen and used within 14 days after thawing. Great variation in the capacity to support BFU and CFU-e growth is found between batches. Some batches were found to support CFU-e, but not BFU growth.

Supplements: 10^{-2} M solution of 2-Mercaptoethanol (BDH Chemicals). 3.4×10^4 M solution of human transferrin (Grade II, Sigma) half saturated with $FeCl_3$, and 10^{-5} M solution of sodium selenite are prepared in alpha medium, sterilized by filtration and kept frozen until needed.

Erythropoietin: Step III sheep plasma Ep (Connaught) is dissolved in alpha medium at the desired concentration, and stored frozen. Usually, 2 units per ml of culture are used for the BFU assay, and 0.5 unit for the CFU-e assay.

Scoring: Colonies are recognized by their characteristic morphology, and are scored directly under $75\times$ magnification.

Reference

1. Worton RG, McCulloch EA, Till JE: *J Cell Physiol 74:*171, 1969.

Bibliography

Gilbert LJ, Iscove NN: *Nature (London) 263:*594, 1976.
Iscove NN, Sieber F: *Exp Hematol 3:*32, 1975.
Testa NG, Dexter TM: *Differentiation* (in press) 1977.

Appendix 6   Growth of Murine and Human Erythroid Colonies *In Vitro*

Noelle Bersch and David W. Golde

Materials and Preparation of Solutions

Centrifuges: Sorvall GLC-1 for routine cell centrifugation, 1,000 rpm for 10 min. IEC-PRJ for centrifugation of human bone marrow specimens in Wintrobe tubes, 2,700 rpm for 10 min.

Microscope: Leitz Diavert for enumeration of erythroid colonies. $10\times$ objective for counting, verification with 20 and $32\times$ objectives.

Cytocentrifuge: Shandon Eliott Cytospin

Falcon culture tubes: #2054 (12×75 mm) for plating methylcellulose; #2037 (16×125 mm) for diluting drugs and hormones.

Petri dishes: Lux #5217 (35×10 mm with 2 mm grid) for enumeration of erythroid colonies (Microbiological Associates).

Centrifuge tubes: Corning #25310, 15 ml, conical with screw cap, for preparation of cell suspensions following Wintrobe tube separations.

Nalgene filter units: Nalge #120-0020 (0.2μ) for media filtering.

PBS (phosphate-buffered saline): Dulbecco's $10\times$, calcium- and magnesium-free (GIBCO #420), for erythropoietin suspension and some drug and hormone dilutions.

Pen-strep: GIBCO #514, 10,000 units/ml streptomycin, 10,000 mg/ml penicillin.

Erythropoietin: Sheep plasma, step III, Connaught Laboratories, Willowdale, Ontario, Canada. Human urinary erythropoietin obtained from the Blood Diseases Branch, National Heart and Lung Institute, National Institutes of Health, Bethesda, Maryland. Both preparations are diluted with PBS, $1\times$.

Ammonium chloride lysing solution: 1 ml Tris buffer pH 7.65 plus 9 ml ammonium chloride added to cell pellet for 10 min at 37°C. The pellet is then centrifuged and resuspended in fresh medium.

Fetal calf serum: GIBCO, non-heat activated. Several lots must be tested to select the one that produces the highest cloning efficiency.

α-thioglycerol: Calbiochem #59525, 90% B grade. A stock solution of 10^{-2} M is stable for 1 month but fresh solutions are usually made every 3 weeks as follows: In a fume hood, 0.113 ml of α-thioglycerol is added to 99.887 ml of $1\times$ α-medium without fetal calf serum and filtered through a 0.2μ Nalge filter.

α-Medium: α-Modified Eagle's with Earle's salts and glutamine, without ribosides, deoxyribosides, or sodium bicarbonate, obtained from Flow Laboratories, Catalog #10-311-22. Ten-liter packets are dissolved in 5 liters double-distilled water, filtered through 0.45μ Nalge filters, then aliquoted into 500 ml bottles and refrigerated. This preparation is termed α-$2\times$. Prior to use, 29 ml of GIBCO 7.5% sodium bicarbonate is added to the 500 ml of α-$2\times$ and the entire volume is filtered through a 0.2μ Nalge filter.

To prepare 100 ml α-$1\times$ (20% fetal calf serum), combine 40 ml double-distilled water, 40 ml α-$2\times$, 20 ml fetal calf serum, 1 ml pen-strep.

Methylcellulose: Dow Chemical Corporation Methocel-A4M premium, lot #MN022358A. A 2.3% stock solution is made as follows: Weigh out 6.9 g methylcellulose and add it to a sterile 500 ml bottle. Add 150 ml hot double-distilled water, cap and shake vigorously for 5 min. Then add 150 ml of cold double-distilled water, cap and shake vigorously for another 5 min and refrigerate overnight. The next day the preparation is warmed in a 37°C water bath before it is autoclaved (with the caps loose) for 30 min. Let cool at room temperature for several hours after autoclaving before refrigerating overnight. The next day the solution should flow evenly and smoothly and be ready for use. At least four lots of methylcellulose should be tested and the clearest one selected for evaluation in the CFU-e assay. The clearest methylcellulose usually produces the highest number of erythroid colonies.

Methylcellulose culture medium (100 ml):

35 ml 2.3% methylcellulose
33 ml $2\times$ α-medium
1 ml pen-strep
1 ml 10^{-2} M α-thioglycerol
30 ml fetal calf serum

Procedure

Murine hematopoietic cells (fetal liver, spleen, and bone marrow) are obtained aseptically and single-cell suspensions prepared in complete α-$1\times$ medium with 20% fetal calf serum. Human bone marrow is aspirated into preservative-free heparin and passed through progressively smaller bore needles until it easily traverses a #25 needle. The suspension is centrifuged in a Wintrobe tube (2,700 rpm for 10 min) and the buffy coat removed and put into complete α-1 $\times$ medium and centrifuged 1,000 rpm for 10 min. The cell pellet is lysed with Tris–NH_4Cl and then washed once in complete α-medium and resuspended in fresh medium.

Each culture is set up in an individual #2054 Falcon tube prior to plating in Lux petri dishes. The sequence of addition of ingredients to the tubes is usually as follows:

Erythropoietin (added in less than 100-λ volumes)
Other additives (hormones, drugs, etc.)
Methylcellulose culture medium
Cells

The total volume in the tube should be approximately 1.5 ml so that on transferring the mixture to the dish a consistent 1 ml volume may be delivered after accounting for adherence of the methylcellulose in pipets and tubes. For this procedure glass pipets are used. After adding the cells to the tubes, the solution is gently mixed by mouth pipetting; the tube is then vortexed at high speed, allowed to stand for about 1 min, and the contents transferred to a culture dish.

The plates are incubated in a completely humidified atmosphere of 5% CO_2 in air at 37°C. Erythroid colonies are enumerated with an inverted phase microscope, counting all clusters of eight or more hemoglobinized cells. The erythroid nature of the cells can be confirmed by picking individual colonies or cytocentrifuging the entire contents of the plate and staining with benzidine. Murine CFU-e are enumerated after 48 hr and human CFU-e at 8 days. The same system may be used for bursts and intermediate colonies by increasing the dose of erythropoietin and the time of incubation.

Appendix 7 Method for Culturing Dog Bone Marrow Cells for Erythroid Colony Formation

John W. Adamson, N. Lin, and Faith Shiota

Method for Culturing Dog Bone Marrow Cells for Erythroid Colony Formation

Preparation of Cells

Random bred dogs are anesthetized by the intravenous injection of Surital (thiamylal sodium; Parke-Davis). Up to 500 mg per dog are injected, depending on the size of the animal. About two-thirds of the dose is given as a bolus injection and the remainder administered depending on the state of sedation obtained.

An area over the posterior iliac crest or the proximal end of the femur or humerus is shaved and the skin cleansed first with an iodine solution followed by 70% alcohol. Marrow cells are aspirated using a sterile 6-inch needle inserted into the shaft of the humerus or femur, or a short bone marrow aspiration needle inserted into the iliac crest. Two to five ml of marrow are obtained by aspirating in 0.5 ml volumes as the needle is gradually withdrawn. The usual yield is approximately 150×10^6 nucleated cells. A 10 ml syringe containing 0.1 ml of preservative-free heparin (1,000 units/ml) is used for aspiration. It is important to wet the walls of the syringe with the heparin prior to aspirating cells to prevent premature clotting.

After the cells have been aspirated they are mixed in the syringe to completely distribute the heparin and then transferred, in a sterile laminar air flow hood, to a 15 ml plastic tube (Falcon; 17×100 mm, #2057) containing 5 ml of Hank's balanced salt solution (BSS; Microbiological Associates) without fetal calf serum (FCS). The tube is capped, inverted several times and then spun in a refrigerated (4°C) centrifuge for 10 min at 1,000 rpm. The fatty layer is removed by aspiration from the top along with the supernatant. A buffy coat layer (approximately 1 ml) is removed by Pasteur pipet and transferred to a 5 ml black screw-top tube (Falcon; #2027). Four ml of BSS containing 2% FCS (Grand Island Biological Co., Grand Island, New York) is added, the tube inverted several times and the cells spun again as described above. The buffy coat is now harvested as cleanly as possible and put into a sterile 3 ml conical pyrex tube to which 2 ml of BSS with 2% FCS are added. The cells are aspirated up and down repeatedly with a 1.0 ml plastic tissue culture pipet (Falcon; #7522) and then spun again as described above. A small (0.2 to 0.3 ml) buffy coat is removed and resuspended in 3 to 5 ml of α-medium (Microbiological Associates) containing 2% FCS in a 12×75 mm plastic tube (Falcon; #2058). An aliquot is removed, diluted in an isotonic solution containing 0.1% trypan blue, and the cells counted in a hemocytometer. The dye-excluding cell concentration is then adjusted such that 0.4 ml of the suspension contains the total number of cells desired in a final culture volume of 1 ml.

Preparation of Media

Bovine serum albumin (BSA) is obtained in powdered form as Fraction V from Sigma Chemical Co. An appropriate solution of 40% by weight (50 gm in 91 ml sterile distilled H_2O) is made and deionized two times against 5 gm of Bio-Rad mixed bed resin (AG501-X8(D); 20 to 50 mesh) as described by Worton et al.[1] The deionized albumin is then made isotonic by adding $10 \times$ Dulbecco's phosphate buffered saline (Microbiological Associates). The solution is centrifuged in 40 ml screw-capped conical shaped tubes at 3,000 rpm for 30 min at 4°C. The albumin solution is made up to 10% concentration and then filtered through a 0.22 μ filter using a 120 ml Nalgene Filter unit (Nalge Sybron Corp.), distributed into 17×100 mm plastic tubes (Falcon; #2057), and stored at −20°C. A volume of 0.1 ml of this is used in the final volume of 1 ml culture medium. Beef embryo extract (Grand Island Biological Co.) is diluted as recommended by the manufacturer to 10 ml and then 50 ml of α-medium added. One-tenth ml is then used in a final volume of 1 ml of culture. FCS is heat-inactivated at 56°C for 30 min in a water bath, aliquoted into small volumes (10 to 12 ml) and stored at −20°C in plastic (Falcon; #2057) tubes. Different batches of FCS are tested for their growth promoting properties and then large volumes of a suitable batch stocked. A total of 0.3 ml FCS is used in the final culture of 1 ml α-Medium (Flow Laboratories) is obtained in powder form with glutamine but without $NaHCO_3$. Two gm of $NaHCO_3$ is added per liter and the medium stored at 4°C. Medium is rotated every 3 months to insure no loss of activity. Ribosides and deoxyribosides (10 mg/liter each of adenosine, deoxyadenosine, deoxycytidine, deoxyguanosine, guanosine, thymidine, uridine, and cytidine) are dissolved in stock medium. Two μl/ml of a penicillin-streptomycin solution (Microbiological Associates; 5,000 units penicillin and 5,000 μg streptomycin/ml stock) is added/ml when the final cell culture medium is made along with 5λ/ml of a stock concentration of 10^{-3} β-mercaptoethanol (βME). The βME is obtained as 14 M from Matheson, Coleman and Bell, diluted with BSS and stored at $-20°C$ in a 1 M concentration.

All of the above materials are added together with cells and allowed to sit on ice until final distribution into the culture dishes.

Preparation of Culture

Erythroid colonies are assayed using a modification of the method of Stephenson et al.[2] The appropriate number of Lux 10 × 35 mm petri dishes (Catalog #5221) are set out, the tops labeled, and to each is added 0.1 ml of bovine citrated plasma (Grand Island Biological). Sheep plasma erythropoietin (Step III material, Connaught Laboratories, Willowdale, Ontario, Canada) is added in the appropriate concentration. The erythropoietin is dissolved in BSS with 0.1% polyethylene glycol in a concentration of 100 units/ml. The solution is sterilized by passing it through a 0.22μ Selas Flotronics Filter. Appropriate amounts up to 100 μl are added using sterile, calibrated glass capillary pipets. If enhancing factors are to be tested, then microliter quantities (never greater than 10 μl) are added to the appropriate dishes. At this point, 0.9 ml of cell suspension in culture medium is added to the dishes, each dish is swirled in order to mix the contents, and the total material allowed to "firm" for 5 to 10 min on a perfectly level area. A minimum of three dishes is set up for each experimental point. No more than six dishes are loaded at a time prior to swirling in order to make sure that uniform mixing occurs before clotting of the contents of the dish. The dishes are then placed in a 37°C, high humidity tissue culture incubator (National Appliance Corp.) gassed with a mixture of 5% CO_2 and 95% air.

The cultures are removed after 48 to 60 hr of incubation. Erythroid colonies appear as small, grape-like clusters of orange cells through the inverted microscope. Such a colony from normal dog marrow cultures is shown in Figure 1. Only colonies containing more than eight cells are counted. To facilitate count-

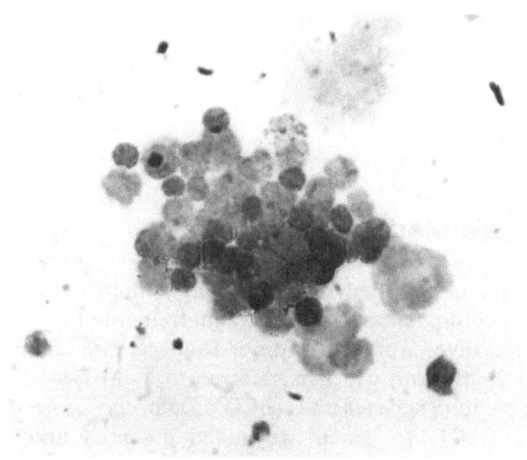

Figure 1 A benzidine-stained erythroid colony grown from normal dog marrow in plasma clot culture. The colony contains approximately 32 cells and was obtained after 48 hr incubation.

ing and allow storage of dishes, 0.5 ml of 5% glutaraldehyde in phosphate buffered saline is added to each dish at the time of removal from the incubator. Counting accuracy can be confirmed by removing the clots from the dish, affixing them to a glass slide, blotting them dry, and then staining with benzidine.

Acknowledgment

This work was supported by designated research funds of the Veterans Administration and research grant AM-19410 of the National Institutes of Health.

References

1. Worton RG, McCulloch EA, Till JE: Physical separation of hemopoietic stem cells forming colonies in culture. *J Cell Physiol* 74:171–182, 1969.
2. Stephenson JR, Axelrad AA, McLeod DL, Shreeve MM: Induction of colonies of hemoglobin-synthesizing cells by erythropoietin in vitro. *Proc Natl Acad Sci USA* 68:1542–1546, 1971.

Appendix 8 Serum-Free Culture of Hemopoietic Cells

L. J. Guilbert and N. N. Iscove

Culture Medium

The use of a suitably enriched and osmotically correct medium is essential for serum-free growth.[1] We have evolved a modification of Dulbecco's modified Eagle's medium which is referred to as M-DMEM. The medium is based on GIBCO's Dulbecco's formulation H-21, to which are added essential amino acids, vitamins, sodium selenite, thiogycerol, and HEPES buffer. The final osmolarity of the revised formula is 280 mOsM. The medium is at the correct pH (7.4) in a 5 to 6% CO_2 atmosphere. Details are given in reference 2.

Methylcellulose

A detailed description of the preparation and handling of methylcellulose is given in reference 3.

Albumin, Transferrin, Lecithin and Cholesterol

Details of the preparation of these materials are given in reference 2. For erythroid and granulocytic colonies from mouse bone marrow, optimum concentrations are bovine serum albumin 1%, transferrin 300 μg/ml, and lipid concentrations as indicated in reference 2. Both lecithin and cholesterol are required for colony formation by CFU-e in the absence of serum.

References

1. Guilbert LJ, Iscove NN: Partial replacement of serum by selenite, transferrin, albumin and lecithin in haemopoietic cell cultures. *Nature* 263: 594–595, 1976.
2. Iscove NN, Melchers F: Complete replacement of serum by albumin, transferrin and soybean lipid in cultures of lipopolysaccharide-reactive B lymphocytes. *Submitted for publication.*
3. Iscove NN, Schreier MH: Clonal growth of cells in semi-solid or viscid media, in Lefkovits I, Pernis B (eds): *Research Methods in Immunology.* New York, Academic Press, Chapter 28, 1978.

Appendix 9 *In Vitro* Culture of Erythroid Precursors: Materials and Methods

F. E. Trobaugh, Jr., Umi Sawada, and S. S. Adler

In our laboratories, colony-forming units–erythroid (CFU-e) and burst-forming units–erythroid (BFU-e) are cultured in an erythropoietin containing medium made semisolid with methylcellulose. The cultures are performed in 1 ml amounts in 35 mm plastic tissue culture dishes maintained in a fully-humidified atmosphere containing 5% CO_2. The techniques for both CFU-e and BFU-e are similar and are detailed below. These techniques are based on those described by Iscove, et al.[1,2]

I. Materials
   A. Semisolid Tissue Culture Medium
      2.5 ml of semisolid culture medium contains the following:
      1. Fetal calf serum (FCS)   0.75 ml
         (See section E)
      2. α-Medium (single strength)   0.55 ml
         (See section C)
      3. Stock methylcellulose solution   1.0 ml
         (See section B)
      4. Erythropoietin solution   0.1 ml
         (See section D)
      5. Suspension of cells in   0.1 ml
         α-medium (See section C)

   Instructions for preparing the culture medium and setting it up for culture are detailed below in the Methods section.

   B. Stock Methylcellulose Solution
      The stock methylcellulose solution contains the following:
      1. Methylcellulose (Dow Methocel,   10 g
         A4M premium, type MC, premium
         grade, 4,000 CPS, Dow Chemical Co.)
      2. α-Thioglycerol, mol wt, 108.2,   0.1 mg
         (Cal Biochem., Cat. No. 59525)
      3. Sterile distilled water (tissue   233 ml
         culture grade, triple distilled
         in glass)
      4. Double strength α-medium   245 ml
         (See section C)

   Instructions for preparing this stock solution are given in the Methods section below.

   C. α-Medium(s)
      1. The α-medium used to prepare the stock methylcellulose solution is double strength, prepared from powder cell culture medium (alpha modification of Eagle's minimal essential medium with Earle's salts and L-Glutamine and without sodium bicarbonate, Flow Laboratories, Catalog No. 1-F-094). This powder is dissolved in an appropriate volume of sterile distilled water to make a double strength solution. Sodium bicarbonate is then added in the amount of 3.75 gm/liter.
      2. α-medium, single strength, MEM α-medium, without ribonucleosides and deoxyribonucleosides, GIBCO MO7-2561, is used:
         (a) to suspend test cells
         (b) to add to stock methylcellulose solution in preparing the 2.5 ml of culture medium.

   D. Erythropoietin
      1. Dr. Eugene Goldwasser of the University of Chicago has kindly supplied our erythropoietin (human, highly purified, mean potency circa 1,140 μ/mg protein). The fraction used is a side fraction of the sulfopropyl-sephadex chromatograph step in purification.[3]
      2. Bovine serum albumin (BSA), Cohn fraction V, powder, 96–99% albumin, Sigma Chemical Co., No. A-4503, Lot No. 77C-0162.
      3. Sterile isotonic saline, sterile, nonpyrogenic 0.9% (w/v) sodium chloride solution without preservative (Irrigation).

   A description of the preparation and storage of working amounts of erythropoietin is provided below in Methods.

   E. Tissue Culture Dishes
      These are 35 × 10 mm plastic tissue culture dishes with a 2-mm grid on their bottom surfaces. (Lux Scientific Corp., Newbury Park, California 91320, Cat. No. 5217).

   F. Fetal Calf Serum (FCS)
      This is "Fetal bovine serum, mycoplasma-free, Flow Laboratories, Catalog No. 4-055M, Lot No. 4055985." Five liters of this serum were obtained in January 1976, and

have been stored in their original 500 ml containers at $-20°C$ in a mechanical refrigerator.

G. Incubator

We use a National Appliance Company two-compartment incubator, Model No. 3312. This is a water-jacketed incubator with two separate compartments. Each compartment has two doors: (a) an inner glass door, and (b) an outer insulating door. Full humidification is achieved by leaving water standing in the bottom of the incubator. The inside of the glass doors is continuously dripping wet and the run-off is collected and returned to the bottom of the incubator by a little gutter system designed by National.

The 5% CO_2 in the incubator's atmosphere is maintained by a National CO_2 Control Master, Model No. 3636; this has proven quite satisfactory. The unit maintains any selected concentration of carbon dioxide in the incubator. The CO_2 concentration is monitored continuously and is reported by a direct meter readout; CO_2 is automatically injected into the incubator as needed. To check on the unit's accuracy and stability, approximately once a week the concentration of CO_2 in a 30 ml sample of air from the incubator is measured on a blood gas machine in the respiratory laboratory. This control unit has reduced the use of CO_2 from approximately 200 liters per week to less than 10 liters per week. As the unit has no fan in the incubator, there is no evident drying of the cultures. We have tested the growth of CFU-e in tissue culture dishes placed directly in the incubator with that in tissue culture dishes maintained in covered petri dishes containing additional water and have detected no difference in the growth.

II. Methods

A. Preparation of Methylcellulose Stock Solution

This solution contains the ingredients listed in 1-B. Methylcellulose is present in the concentration of 2.09% (w/v).

1. Measure out 233 ml fresh distilled water and place in a sterile 1 liter Erlenmeyer flask. Cover the flask loosely with aluminum foil and heat over an open flame, bringing the solution to a boil. Put a sterile stirring magnet in the flask. Pour the 10 gm of Methocel into the flask, replace the foil cover, and agitate the flask vigorously in order to suspend *completely* the Methocel powder. Replace the flask over the flame and bring to the boiling point again, being careful that the mixture does not boil over; remove the flask from the flame just at the point when boiling begins. Place the flask on a magnetic sitrring unit at room temperature and allow to cool, stirring vigorously all the time.

2. Add the 0.01 ml of α-thioglycerol and mix thoroughly.

3. Measure out the 245 ml of freshly thawed or freshly prepared double strength α-medium.

4. When the Methocel solution has reached 40 to 45°C, add the double strength alpha medium and mix thoroughly. Replace the foil cover and continue slow stirring in a 4°C cold room overnight. Aliquot the solution into sterile screw-capped 100 ml bottles, about 80 ml per bottle. Store at $-20°C$ or colder. After each bottle has been thawed for use, it can then be stored at 4 to 6°C.

B. Preparation of Cell Suspension

1. Mouse

a. Marrow. A femur is dissected free of its attachments under sterile conditions and both femoral epiphyses are removed. A 22-gauge needle on a sterile tuberculin syringe containing α-medium with 2% FCS is inserted tightly into the marrow cavity so that the femur is stuck on the needle. Using the syringe as a handle, the open end of the femur is inserted into 2 ml of single strength α-medium with 2% FCS contained in a Falcon 2001 plastic test tube. The marrow cells are ejected from the femur and a cell suspension is prepared by repeatedly ejecting and aspirating the cell suspension through the femoral cavity.

b. Spleen. Under sterile conditions, the spleen of each animal is removed and weighed. A weighed portion of the spleen is placed in a Thomas syringe and cut into many small pieces with a pair of fine scissors. The small pieces are then sieved through three layers of stainless steel bolting cloth, of 88, 200, and 30 mesh (openings: 0.201 mm, 0.086 mm, and 0.68 mm, respectively), to form a suspension of single cells in 10 ml of single strength α-medium with 2% FCS contained in a 100 ml glass beaker.

2. Human Marrow

Human marrow cells are aspirated by standard techniques, usually from a posterior iliac spine. Approximately 1 ml of sterile isotonic saline containing 100 units heparin per ml is aspirated into a 20 ml syringe; the entire interior surface of the syringe is coated with the heparin solution and all excess heparin solution is expelled. Approximately 5 ml of marrow is aspirated into the heparin-coated syringe, mixed well, and transferred to a sterile culture tube (Falcon 2001). A monocell suspension is prepared by repeatedly aspirating and ejecting this suspension into and out of a 5 ml syringe through a 22-gauge needle. The red blood cells are then separated from the white cells by either of two methods:

a. The culture tube is allowed to stand at a modest slant, about 20°C from the vertical, at room temperature, until the red cells have settled. This usually requires 1 to 2 hr.
b. The aspirate is added to an equal volume of α-medium containing 2% FCS and centrifuged at 1,000 rpm (150 RCF) for 3 min.

Following either of these procedures, the leukocyte-rich plasma is harvested and the concentration of nucleated cells determined. The desired concentration of cells is then obtained by the addition of α-medium containing 2% FCS.

C. Incorporation of Cell Suspension in Semisolid Tissue Culture Medium
1. In a 15 ml sterile plastic test tube (Falcon 2001), mix the following:
   a. Fetal calf serum   0.75 ml
   b. α-medium, single strength 0.55 ml
   c. Erythropoietin solution   0.1 ml
2. To the above, add 1 ml of the stock methylcellulose solution, and mix for circa 1 sec on a vortex mixer. Let the mixture stand while the suspension of test cells is prepared.
3. Prepare the test cell suspension, adjusting the cell concentration so that the number of cells to be contained in 2.5 ml of final concentration is contained in 0.1 ml of α-medium with 2% FCS.
4. Add 0.1 ml of the test cell suspension to the test tube containing the above ingredients, and mix on a vortex mixer for approximately 0.5 sec. During the process of mixing the cell suspension into the methylcellulose solution on the vortex mixer, bubbles develop. By allowing the tube to stand 3 to 5 min, these bubbles will slowly rise and disappear.

D. Plating the Cell Suspension in Tissue Culture Dishes
Take up as much as possible of the 2.5 ml solution through a 16-gauge needle into a plastic 3 ml syringe, get rid of any bubbles, and place 1 ml of the solution in the center of a 35 mm plastic tissue culture dish, lifting the cover only far enough to insert the needle. Tilt the dish carefully until all of the bottom and the adjacent sides are wet with the viscous culture solution.

Incubate the dishes at 37°C in a fully-humidified atmosphere containing approximately 5% CO_2.

E. Preparation and Storage of Working Amounts of Erythropoietin (Ep)
We started with 8,000 units of erythropoietin (Ep) in approximately 7 mg of lyophilized protein. This protein was dissolved in 8 ml of saline containing 0.1% bovine serum albumin (BSA). The 0.1% solution of BSA is prepared by dissolving 0.1 gm powdered serum albumin in 100 ml 0.9% (w/v) sterile, nonpyrogenic sodium chloride solution which contains no preservative. The BSA is not sterile; after preparing the solution, sterilize it by filtering it through a Millipore filter with 0.45μ pore size (HAWPO 1300, Millipore Corp.). One ml of the diluted material was placed in each of seven plastic tubes (Falcon 2001) and stored at −20°C in a mechanical freezer.

The remaining 1 ml of erythropoietin solution was diluted with 0.1% BSA and saline and divided in such a manner that some tubes contained 43.75 units Ep in 0.5 ml solution, providing 0.1 ml Ep solution containing 8.75 units Ep for each of five culture tubes. As each culture tube contains 2.5 ml of a cell suspension in semisolid tissue culture medium, each ml of culture medium contains 3.5 units Ep, or 3.5 units per plate. Other tubes were prepared that contained 4.375 units Ep in 0.5 ml solution, which is the amount of Ep needed to prepare five tubes of culture medium from which ten plates, each containing 0.35 unit Ep, may be prepared. All solutions are stored at −20°C in plastic tubes (Falcon 2001).

F. Assay of Specific Erythroid Precursors
The basic culture techniques for CFU-e and BFU-e (both human and mouse) are the same. Each requires different amounts of erythropoietin and different culture times (detailed below), but all are cultured in the same type of plastic tissue culture dishes at 37°C in a fully-humidified atmosphere containing approxi-

260   Appendix 9

Experiment no. _____   Counted by _____   Date plated _____

Date CFU-e examined _____

Date BFU-e examined _____

Culture Procedure: 2.5 ml of culture solution is prepared by mixing fetal calf serum—0.75 ml, erythropoietin solution—0.1 ml, alpha-medium (single strength)—0.55 ml, stock methylcellulose solution—1.0 ml, and the test cell suspension—0.1 ml. These constituents are mixed at room temperature, drawn into a 3 ml disposable syringe. Bubbles are eliminated, and 1 ml is placed in each of two plastic petri dishes (Lux, cat. no. 5217). CFU-e are examined after 2 days of culture in a fully-humidified atmosphere of approximately 5% CO_2 in air, BFU-e after 7-10 days.

Mem-Alpha-Medium: Without ribonucleosides, without deoxyribonucleosides (cat. no. M07-2561, Grand Island Biological Company).

Stock Methylcellulose: Methylcellulose powder (Methocel, 4,000 cps)—60 gm., sterile distilled water—1400 ml, double strength alpha-medium—1470 ml, alpha-thioglycerol (Calbiochem, m.w. 108.2)—0.06 ml.

Plate Number	Group or Patient	Description or Diagnosis	Cells Plated		E-CSA		CFU-e Plate #				BFU-e Plate #			
			Source	#	Source	Amt.	1	2	Mean	Comments	1	2	Mean	Comments

mately 5% CO_2. In the incubator, the medium should be at a pH of 7.2 to 7.4, effecting a red-orange to orange color of the phenol red indicator. We use an Olympus inverted microscope to visualize and count the colonies.

1. CFU-e

    All suspensions assayed for CFU-e are plated in 1 ml of medium which contains 0.35 unit erythropoietin. These are examined using $100\times$ magnification, and all cell clusters containing eight or more cells are counted.

    a. Mouse. Cell clusters formed from mouse CFU-e are counted 44 to 48 hr after plating. Mouse erythroblasts are tiny and are clearly distinguishable from leukocytes. If the identity of cells composing a colony is in doubt, the colony may be aspirated, placed on a slide, and stained according to the techniques described in Section H below. Mouse marrow cells or spleen cells are plated at a concentration of 100,000 cells per ml.

    b. Human. Human marrow cells are plated at a concentration of 100,000 cells per ml and their colonies are counted on day 7.

2. BFU-e

    Assay of cell suspensions for BFU-e are cultured in a medium containing 3.5 units of erythropoietin per ml. The colonies are counted using $40\times$ magnification.

    a. Mouse. The colonies formed from mouse BFU-e are assayed on days 8 through 9. Mouse marrow cells or spleen cells are plated at a concentration of 100,000 cells per ml.

    b. Human. The colonies formed from the BFU-e are counted on day 21. Human marrow cells are plated in a concentration of 100,000 cells per ml.

G. Cleaning of Glassware

Glassware is made of hard glass, most of it pyrex, and is cleaned using a special tissue culture detergent, "$7\times$, Linbro Scientific, Inc., Division of Flow Laboratories, Hamden, Connecticut." Following the cleaning with this detergent solution, the glassware is rinsed several times in tapwater, then with distilled water. The glassware is sterilized by dry heat at 170°C for 60 min.

H. Staining Cells in a Colony

The identity of cells in a colony can be determined by removing the colony from methylcellulose, and staining it first with a benzidine stain, followed by a Giemsa counterstain. Removing the colony is a tricky business that requires a steady hand and considerable practice. The procedure is as follows:

1. A colony is aspirated from the methylcellulose into a very fine tipped Pasteur pipet equipped with a rubber bulb. Such pipets are not available commercially and must be prepared by drawing the tip from a standard Pasteur pipet until it has an internal diameter of approximately 0.1 mm. It helps to have this tip drawn at a right angle to the rest of the pipet.
2. Eject the colony from the pipet onto a slide. Mark the location of the colony; this can be done by drawing a circle on the bottom of the slide with a diamond-tipped marker.
3. Allow the colony to air dry.
4. Fix the colony by immersing the slide in absolute methanol for 5 min. The methanol should be free of water; absorbed water can be removed from the methanol by a molecular sieve (Sephadex, G-25-80, beads for gel filtration, particle size: 20 to 80μ, Sigma Chemical Co., St. Louis, Missouri).
5. Using a fine Pasteur pipet, apply a small drop of water to the colony. Allow the water to stand for 30 sec and remove the water by aspirating it into the pipet. The methylcellulose dissolves very rapidly and can be removed in this manner, permitting the cells to be stained.
6. Air dry the slide.
7. Stain with a benzidine stain (4) as follows:
    a. Stock staining solutions:
        (1) Benzidine solution:
            (a) Benzidine base   200 mg
            (b) Methanol   10 ml
        This 2% w/v solution can be stored indefinitely if protected from light. Benzidine is a carcinogen and you should avoid contact with it.
        (2) Peroxide solution:
            (a) Methanol   5.6 ml
            (b) Distilled water   4.2 ml
            (c) Hydrogen peroxide, 30% 0.25 ml
        This is a 0.75% H_2O_2 solution. It will remain usable for at least a week if stored at 4 to 6°C.
    b. Staining methods:
        (1) Cover the slide with the benzidine solution for five minutes and remove the solution from the slide by tipping the slide. DO NOT RINSE.
        (2) Immediately cover the slide with the peroxide solution for 1 min; rinse thoroughly but gently with distilled water.
        (3) Counterstain the slide with 5 to 10% Giemsa solution (1 drop stock Giemsa stain per ml distilled water) for 3 to 5 min.
        (4) Wash carefully with distilled water and dry.

            In this stain, all the red cells will be a brown-orange in color, the nuclei will be dark blue, and the cytoplasm of white cells will be as usually seen in a Giemsa-stained slide.

References

1. Iscove ND, Sieber F, Winterhalter KH: Erythroid colony formation in cultures of mouse and human bone marrow: analysis of the requirement for erythropoietin by gel filtration and affinity chromatography on agarose-concanavalin A. *J Cell Physiol 83:*309–320, 1974.
2. Iscove NN, Sieber F: Erythroid progenitors in mouse bone marrow detected by macroscopic colony formation in culture. *Exp Hematol 3:*32–43, 1975.
3. Miyake T, Kung C K-H, Goldwasser E: Purification of Human Erythropoietin *J Biol Chem 252:*5558–5564, 1977.
4. Borsook H, Ratner K, Tattrie B: Studies on erythropoiesis. II. A method of segregating immature from mature adult rabbit erythroblasts. *Blood 34:*32–41, 1969.

Appendix 10 Culture of Erythroid Stem Cells from Murine and Human Marrow and Blood

Martin J. Murphy, Jr. and Maureen E. Sullivan

Preparation of 1× or 2× Alpha (α) Medium
Materials
1 package of α-medium without nucleosides (Flow Laboratories, Inc.).
600 mg Penn G
1 gm Streptomycin
22 gm $NaHCO_3$
Nalgene filter (0.45μ) #245-0045

Method
The first three items are dissolved in 10 liters of sterile distilled water at room temperature and immediately before filtering, 22 gm $NaHCO_3$ is added. Aliquots are dispensed in 100 ml bottles and refrigerated at 4°C. The 2× α-medium is prepared exactly as above except that 5 liters of sterile distilled water are added instead of 10 liters.

Preparation of Bovine Serum Albumin (BSA)
Materials
BSA-Fraction V, B-grade, (Calbiochem, San Diego, California).
Mixed Bed Resin #AG 501-X 8 (D), analytical grade (Biorad Lab, Richmond, California).
Sterile 10 ml serological pipet
2× α-medium (prepared)
(3) 125 ml Erlenmeyer flasks (sterile)
100 ml sterile graduated cylinder
Sterile gauze (4 × 4)

Method
To 44.2 ml of sterile distilled water, pour without agitation, 10 gm of BSA and refrigerate overnight at 4°C. The following morning at room temperature, add 1 gm of mixed bed resin which is bluish in color. Intermittently swirl for mixing. When 90% of the resin changes to a golden-yellow color, place a piece of sterile gauze over the mouth of the flask and pour the BSA into a fresh sterile 125 ml Erlenmeyer flask to which another 1 gm of the mixed bed resin had been added. This is similarly treated with intermittent swirling for 2 to 3 hr until the resin is golden. The resin is again filtered with another sterile gauze cloth into a 100 ml sterile graduated cylinder. The volume is noted and an equal volume of 2× α-medium added and mixed. The BSA is then sterilized by Nalgene filtration (0.45μ), after which aliquots of 2.5 ml BSA/sterile 5 ml tubes are stored frozen. For usage, a BSA tube is thawed in a 37°C water bath and 0.1 ml of a sterile 7% $NaHCO_3$ solution is added just before use. Once thawed, BSA is not returnable to the freezer. If not entirely used up, the remainder is discarded.

Preparation of 9% Ficoll-Metrizoate
Materials
Ficoll 400 (Pharmacia Fine Chemicals)
Sodium metrizoate (Accurate Chemical & Scientific Corp.).
Sterile 250 ml graduated cylinder
Sterile 10 ml serological pipet
Sterile 200 ml beaker
Test tubes #2001. 17 × 100 mm with caps (Kramer Scientific, NY)
Nalgene filter 0.45μ #245-0045

Method
1. 13.5 gm of Ficoll is dissolved in 150 ml sterile distilled water in a sterile 200 ml beaker (solution A).
2. A 20 ml vial of sodium metrizoate is added to 25 ml sterile distilled water in a 250 ml sterile graduated cylinder (solution B).
3. 108 ml of the dissolved Ficoll (solution A) is added to the sodium metrizoate (solution B).

Sterilization with Nalgene filtration (0.45μ) and aliquots of 3 ml in 17 × 100 mm sterile tubes are stored at 4°C for up to 1 month.

Preparation of Methylcellulose (2% and 2.67%)
Materials
Methocel A4M (Dow Chemical Co.)
Large sterile magnetic flea
Large magnetic stirrer
Sterile 3 liter Erlenmeyer flask with a sterile stopper
Sterile distilled water
2× α-medium (at room temperature), prepared
Sterile 100 ml bottles

Method

20 gm Methocel = 2% solution
26.7 gm Methocel = 2.67% solution

The gram weight desired of methylcellulose is put into a sterile, 3 liter Erlenmeyer flask that has a sterile stopper and already containing a sterile magnetic flea on a large magnetic stirrer. To prevent frothing, stirring is initiated while 500 ml of sterile boiling distilled water is gently poured down the side of the flask. Stirring continues at room temperature until the flask gradually cools to room temperature (this may take an hour). When the flask is no longer warm to the touch, 500 ml of 2× α-medium, which had been allowed to come to *room temperature* is added to the flask without frothing. The flask is stoppered and transferred to a cold room (4°C) where stirring continues for 48 hr. The solution is then sterilely aliquoted into sterile 100 ml bottles. The bottles are stored frozen for up to 6 months. Prior to using, 2 ml of stock glutamine solution is added and vigorously shaken.

Preparation of L-(+) Glutamine Solution
Materials

L-(+) glutamine (Eastman Kodak Co.)
Sterile distilled water
Nalgene filter (0.45μ) #245-0045
Sterile test tubes #2003, 12 × 75 mm (Kramer Scientific)
Sterile 10 ml serological pipets

Method

2.72 gm of L-glutamine is dissolved in 80 ml sterile distilled water at room temperature. This is filtered for sterilization through a Nalgene filter (0.45μ) in the laminar flow hood and 2 ml aliquots are collected in 12 × 75 mm sterile tubes and stored at 4°C for up to 6 months. Once the stopper is removed the tube is not to be returned to storage. It is either used entirely or the remains discarded.

Preparation of 10^{-3} M Mercaptoethanol
Materials

Mercaptoethanol, mol wt 78 (Pierce Chemical Co., Rockford, Illinois).
1× α-medium (prepared)
Sterile test tubes #2001, 17 × 100 mm with caps
Sterile 10 ml serological pipets

Method

A 1 M solution of mercaptoethanol is prepared with 1× α-medium, e.g., 0.78 ml mercaptoethanol and 9.22 ml 1× α-medium (sterile). This 1 M solution is diluted three times 1:10. Namely, 1 ml of the 1 M solution is added to 9 ml of 1× α-medium. This solution is now 10^{-1} M. Therefore, after three 1 to 10 dilutions with 1× α-medium the molarity of the working solution is 1 × 10^{-3} M. This is aliquoted in 10 ml volumes in 17 × 100 mm sterile tubes, and is stored at 4°C for up to 3 months.

Preparation for Processing Heparinized Peripheral Blood
Materials

10 to 15 ml heparinized blood
3 tubes Ficoll-metrizoate (9%)
1× α-medium (prepared)
Pasteur pipets (sterile) with rubber bulbs
Test tubes #2001, 17 × 100 mm

Method

With the aid of a sterile Pasteur pipet, 3 ml of heparinized blood is carefully layered down the sides of each (3) 17 × 100 mm test tubes containing 3 ml of a 9% Ficoll-Metrizoate solution which has been brought to room temperature prior to use. The tubes are stoppered and centrifuged for 30 min at 2,000 rpm. After centrifugation, the interface of cells is harvested and twice washed in 1× α-medium and the nucleated cells counted. The final working concentration is brought to 3 × 10^6 nucleated cells/ml.

Procedure for the Processing of Mouse Marrow Cultures
Materials

BDF$_1$ mice (10 to 15 weeks old)
Tuberculin syringes and 21-gauge needles (sterile disposable)
26-gauge needle
1× α-medium
Mercaptoethanol 10^{-3} M
Erythropoietin (20 units/ml)
2.67% methylcellulose
Sterile gauze sponges
Sterile scissors and forceps

Method

Mice are killed by cervical dislocation. With the aid of a tuberculin syringe and under sterile conditions, the diaphyseal contents of both femora are harvested in a 35 mm petri dish containing 4 ml of 1× α-medium. A single cell suspension is obtained by repeated aspiration with a 1 ml syringe and 26-guage needle. Following this, the suspension is centrifuged for 5 min at 2,000 rpm; the supernatant decanted and the cell pellet resuspended in 1 ml of 1× α-medium for cell enumeration. The final working cell concentration is 1 × 10^6/ml. Follow instructions for culturing of human marrow, however, use 1.5 ml of 2.67% methylcellulose; 0.5 ml 10^{-3} M 2-mercaptoethanol and 0.5 ml of 20 units/ml of erythropoietin. For murine hematopoietic cells the scoring of CFU-e is done on the second day following incubation and BFUe on the ninth day.

Procedure for Processing of Human Marrow Cultures
Materials

> Heparinized marrow (1 to 3 ml depending on the diagnosis)
> Test tubes: #2001 (17 × 100 mm with caps)
> #2003 (12 × 75 mm with caps, Kramer Scientific, N.Y.)
> Sterile Pasteur pipets with rubber bulbs
> 1 ml and 5 ml serological pipets (sterile)
> Ficoll metrizoate
> 1× α-medium

Method
Heparinized bone marrow is centrifuged for 5 min at 2,000 rpm at room temperature. The buffy coat is removed and added to a 17 × 100 mm test tube containing 3 ml of 1× α-medium. The suspension is thoroughly mixed and layered on 3 ml of Ficoll-Metrizoate. The tube is centrifuged for 30 min at 2,000 rpm then the interface is harvested with a sterile Pasteur pipet and transferred to another tube with 1× α-medium in which the cells are washed twice. After the final wash, the supernatant is aspirated and the pellet resuspended in 1 ml of 1× α-medium for counting.

Preparation of 7% NaHCO₃ Solution
Materials

> 3.5 gm NaHCO₃
> 50 ml sterile distilled water
> Sterile 10 ml serological pipet
> Nalgene filter (0.45μ) #245-0045
> Sterile test tubes #2001 (17 × 100 mm with caps)

Method
The gram weight of NaHCO₃ is dissolved in 50 ml sterile distilled water. This is filtered for sterilization through a Nalgene filter (0.45μ) in a laminar flow hood and aliquots of 10 ml are collected in 17 × 100 mm sterile tubes. These are stored at 4°C for up to 1 month. Any one tube may be used repeatedly provided the cultures are not contaminated and as long as the solution is stored in the refrigerator.

Preparation for Cell Counting
Materials

> Sterile Pasteur pipets with rubber bulbs
> Parafilm, American Can Co.
> WBC diluting pipet
> Hemacytometer (Neubauer)
> 3% acetic acid

Method
One drop of the thoroughly mixed cell suspension is transferred to a piece of parafilm by use of a sterile pipet and with the aid of a WBC diluting pipet, 3% acetic acid and a hemacytometer, the count is made and adjusted so that the final working concentration is 1 × 10⁶ nucleated cells/ml.

Final Preparation for Cell Culturing
Materials

> Human Urinary Erythropoietin (Ep 10 units/ml)
> Ep of known specific activity is made up in 1 × α-medium to 20 units/ml stock solution. This is Millipore filtered (0.45 μ) and stored frozen until used.
> 2% Methylcellulose with L (+) glutamine added
> Bovine Serum Albumin (BSA), 10% (prepared)
> Fetal Calf Serum (FCS) (Flow Laboratory)
> 5 ml and 10 ml syringes, sterile (B-D Glaspak, disposable)
> 15- and 18-gauge needles
> 35 mm Lux petri dishes #5221-R (Lux Scientific Corp.)
> 1 ml, 5 ml, 10 ml sterile serological pipets
> 1× α-medium (prepared)
> 7% NaHCO₃ (prepared)
> Optilux petri dishes #1001 (Falcon)
> CO₂ incubator (37°C)

Method
Two 17 × 100 mm test tubes are routinely used for each experiment. One is labelled "Experimental" and the other "Control." With the use of a 15-gauge sterile needle and 5 ml syringe, 2 ml of the 2% methylcellulose is added to each tube; 1.5 ml fetal calf serum (i.e., each batch of FCS tested for optimum growth efficiency); 0.5 ml bovine serum albumin (BSA), to which 0.1 ml 7% NaHCO₃ is added just before use; and 0.5 ml of the cell concentration at 1 × 10⁶/ml. Finally, 0.5 ml of 10 units of Ep is added to the "experimental" tube while the same amount of 1× α-medium is added to the "control." The tubes are stoppered, vortex-mixed and the cells allowed to settle for 2 to 3 min before aliquots of 1 ml are added

to 35 mm petri dishes with 18-gauge needles and 5 ml syringes. Two of these 35 mm dishes are enclosed in a 100 × 15 mm Optilux petri dish, with an uncovered 35 mm dish containing sterile distilled water and incubated at 37°C with 5% CO_2. For human hematopoietic cells the scoring of CFU-e is performed on the seventh day following incubation and BFU-e on the 14th day. For proper identification the modified benzidine staining technique is performed.* Cytocentrifuge preparations are also made and counter-stained, Wright-Giemsa for morphology.

* Gallicchio V: *Lab Med* 6:15, 1975.

Appendix 11 Assay for the Commitment of Murine Erythroleukemia Cells to Differentiate

Richard A. Rifkind, Eitan Fibach, and Paul A. Marks

The rate at which murine erythroleukemia cells (MELC) initiate terminal differentiation *in vitro* may be profoundly modulated, either to increase or decrease the rate, by exposure in culture to a variety of chemical inducers or inhibitors of induced differentiation.[1,2,3] The commitment of MELC to differentiate is defined as the ability of the cell to express differentiation, by the accumulation of erythrocyte-specific products (such as hemoglobin) after removal of the inducing reagent. Commitment can be assayed, at the single cell level, by transfer of cells from suspension culture with inducer to a semisolid cloning medium free of inducer.[4]

Commitment Assay
Materials

Eagle's Basal Medium
Fetal Calf Serum
Inducing Agent, e.g., Hexamethylene bisacetamide (5 mM final conc.)
Dimethylsulfoxide (280 mM final conc.)
Sodium butyrate (1 to 1.5% final conc.)
37 mm Falcon plastic petri dishes
Methylcellulose, 4,000 centipoise
Dulbecco's Modified Eagle's Medium, 1× and 2× concentration

Procedure

MELC at 10^5 cells/ml are cultured in suspension in medium consisting of Eagle's Basal Medium supplemented with fetal calf serum (final concentration 15%; pretested for ability to support growth and induced differentiation of MELC), penicillin and streptomycin, and the selected chemical inducer at the concentration to be tested. Cultures are maintained at 37°C in a humidified incubator under 5% CO_2; at selected periods of time aliquots are removed, counted, washed in inducer-free medium and inoculated into inducer-free semisolid medium.

The semisolid medium is composed of Dulbecco's Modified Eagle's Medium supplemented with methylcellulose (final concentration 1.44%, v/v) and fetal calf serum (final concentration 15%, v/v). To prepare this methylcellulose medium, 1.8 gm of methylcellulose is autoclaved in a flask with a magnetic stirring bar; while still hot 50 ml of double distilled, autoclaved hot water are added, the mixture stirred for 30 min and 50 ml of 2× concentrated Dulbecco's Modified Eagle's Medium are added. This mixture is stirred at 4°C for 2 days. The serum supplement is added just prior to use. MELC washed free of inducer are suspended at 5×10^3 cells/ml in the semisolid medium, then dispersed in one ml aliquots into petri dishes, and examined under the inverted tissue culture microscope to insure that the cells are distributed singly and uniformly.

After a total of 4 to 5 days of culture (suspension plus semisolid) the petri dishes are scored for their content of colonies and the content of hemoglobinized cells by the benzidine reaction.

Benzidine Reaction for Committed Colonies
Materials

0.2% benzidine dihydrochloride in 0.5 M acetic acid
30% hydrogen peroxide (superoxide)

Procedure

The staining reagent consists of 100 ml of the benzidine solution to which 0.4 ml of hydrogen peroxide is added just prior to use. One ml of this reagent is added to each petri dish and 5 min later the dish is scored for the numbers (proportion) of colonies which are uniformly benzidine-unreactive (colorless), uniformly benzidine-reactive (blue) and colonies containing both reactive and unreactive cells (mixed colonies containing both differentiated, hemoglobin-containing and undifferentiated cells).

References

1. Tanaka M, Levy J, Terada M, Breslow R, Rifkind RA, Marks PA: *Proc Natl Acad Sci (USA)* 73:1003, 1975.
2. Reuben RC, Wife RL, Breslow R, Rifkind RA, Marks PA: *Proc Natl Acad Sci (USA)* 73:862, 1976.
3. Yamasaki H, Fibach E, Nudel U, Weinstein IB, Rifkind RA, Marks PA: *Proc Natl Acad Sci (USA)* 74:3451, 1977.
4. Fibach E, Reuben RC, Rifkind RA, Marks PA: *Cancer Res 37*:440, 1977.

Appendix 12 Materials and Methods for Mouse Friend Virus-Infected Cells

Shigeru Sassa, Joel L. Granick, Harvey Eisen, and Wolfram Ostertag

Growth of Friend Virus-Infected Cells

A clone of Friend erythroleukemic cells (745A) was obtained from Drs. C. Friend and W. Scher of Mt. Sinai School of Medicine, New York, New York; T3-Cl-2 cells from Dr. P. Leder, National Institutes of Health, Bethesda, Maryland. Clones 745A, T3-Cl-2 and F4N are hemoglobin-inducible when they are treated with dimethylsulfoxide (DMSO). F4N+2, F4+, and F4D-5 are subclones of F4N that have been selected for their ability to grow in the presence of DMSO. K2, donated from Dr. Y. Ikawa, Cancer Institute, Tokyo, Japan, is a subclone of T3-Cl-2, which is also resistant to DMSO. Ma, U91, and U99 were derived from clone 745A and resistant to DMSO treatment. These clones were used either directly without preliminary cloning, or after subcloning in the presence of DMSO.

Cells were routinely diluted twice weekly to a final concentration of 5×10^4 cells/ml in 25 cm^2 plastic flasks (Falcon No. 3013), containing 5 ml of a modified Ham's F12 medium[1] supplemented with 10% heat-inactivated fetal bovine serum. Cultures were incubated at 37°C in a 5% CO_2 atmosphere.

Fluorometric Assay of Heme

Cell suspensions containing 10^5 cells were transferred to a 1 ml disposable glass tube (6 × 50 mm, Kimble Products). The tubes were centrifuged at 680 × g for 5 min. Cell pellets were washed once with 500 µl of Earle's buffer devoid of Ca^{2+} and Mg^{2+} (pH 7.4) by centrifugation. 500 µl of 2 M oxalic acid was then added to the pelleted cells. The suspension was mixed, and the mixture was immediately heated for 30 min at 100°C by inserting the lower half of the tube into an aluminum block in a dry-bath. In this way, evaporation from each tube was minimized and kept relatively uniform. A tissue blank was run to check for the presence of endogenous porphyrins in the cell; tissue blanks were prepared which contained the same amount of cells in 2 M oxalic acid, but they were not heated. Standards were made by adding hemin solution prepared in 0.01 N KOH-50% methanol to 2 M oxalic acid solution, then heating as previously.

After cooling, tubes were inserted into a semimicrocell holder having four appropriate holes. Fluorescence was determined with a Hitachi-Perkin Elmer MPF4 fluorescence spectrophotometer equipped with a red sensitive 777-01 photomultiplier using a 400 nm light as the exciting source. Fluorescence emission was determined at 662 nm. Calibration of the spectrofluorometer was carried out using coproporphyrin solution (10^{-7} M) in 0.5 N perchloric acid-50% methanol.

The fluorescence generated from heme was linear between 10^{-9} M and 10^{-6} M. The lower limit of detection of heme by this method was 0.5 pmol. Fluorescence yield was linear to cell concentration up to approximately 2×10^5 cells per assay for both untreated and DMSO-treated cells, but it became nonlinear at cell concentrations greater than 2×10^5. It is therefore essential not to exceed 2×10^5 cells per assay.

Assay for δ-aminolevulinic Acid (ALA)-synthase Activity

The activity of ALA-synthase was determined by a minor modification of the radiochemical assay of Strand et al.[2] using 10^7 cells per assay. The incubation mixture contained 100 mM glycine, 50 mM Tris-HCl, pH 7.4, 20 mM $MgCl_2$, 10 mM EDTA, 1 mM pyridoxal 5'-phosphate, 1 mM dithiothreitol, 10 mM Na malonate, 5 mM Na DL-malate, 5 mM Na arsenate, 2.5 µg/ml antimycine A, 25 mM ATP, 1 mM coenzyme A, 12.5 mM Na succinate, cells, 2 units of succinyl CoA synthetase (1 unit = 1 µmol succinohydroxamate/mg protein, 37°C, hr), and 5 µCi of (1,4-^{14}C)succinic acid (sp.act. 24 mCi/mmol) per assay. The final volume of the assay mixture was adjusted to 400 µl and quickly frozen and thawed in dry-ice acetone three times; the mixture was then incubated for 20 min at 37°C. The reaction was stopped with one volume of 8% TCA. ALA was isolated by using Dowex column chromatography (Dow Chemical Co., Midland, Michigan) after the addition of carrier unlabeled ALA. Enzyme activity is expressed as pmol ALA formed/10^6 cells, hr at 37°C.

Assay for ALA-dehydratase Activity:

The activity of ALA-dehydratase was assayed spectrophotometrically by the micromethod of Sassa et al.[3] Cell suspension equivalent to 5×10^5 cells was pipetted into each of three disposable glass tubes (6×50 mm, Kimble Products) and centrifuged at $680 \times g$ for 5 min. After removal of the supernatant, 50 µl of 50 mM sodium phosphate buffer (pH 5.8) was added into the first tube which serves as the control. In the other two duplicate tubes, 50 µl of the same buffer containing 5 mM ALA and 20 mM dithiothreitol was added. The pH of the final mixture was 6.3. The mixture was freeze-thawed once in liquid nitrogen or in dry ice-acetone and the tubes were incubated for 1 hr at 37°C in a water-bath. The reaction was terminated by the addition of 150 µl of 6% trichloracetic acid containing 0.1 M $HgCl_2$, then centrifuged at $1,000 \times g$ for 5 min. 150 µl of the supernatant was removed and mixed with an equal volume of the modified Ehrlich reagent and the absorbance of the porphobilinogen (PBG)-Ehrlich color salt was determined spectrophotometrically using an EmM of 61 at 553 nm.[4] Enzyme activity is expressed as nmol PBG formed/10^6 cells, hr, at 37°C.

Assay for Uroporphyrinogen-I Synthase Activity:

This enzyme assay was carried out by the microfluorometric method of Sassa et al.[5] using 5×10^5 cells per assay. Cell pellet equivalent to 5×10^5 cells was mixed with 25 µl of 100 µM porphobilinogen (PBG) in a 6×50 mm glass tube. The mixture was freeze-thawed once prior to the incubation. Incubation was made at 37°C, for 1 hr in a subdued light. The reaction was terminated by cooling down tubes in ice; immediately followed by the addition of 300 µl of ethylacetate-acetic acid. Then 300 µl of 0.5 N HCl was added and the mixture was shaken vigorously. The mixture was allowed to stand in room light for 30 min, and then fluorescence in the lower aqueous phase was determined using a HITACHI-PERKIN ELMER MP4 fluorescence spectrophotometer. An excitation wavelength of 400 nm and an emission wavelength of 654 nm was used. The rate of enzyme reaction was monitored by incubating a control human blood sample in the same manner as described above. The control human blood was a normal origin and stored in small aliquots in liquid nitrogen.

Assay for Ferrochelatase Activity

This enzyme activity was indirectly assessed in cultured cells by incubating them with ^{59}Fe-citrate. Cells ($\sim 10^7$ cells) were incubated with ^{59}Fe-citrate (0.5 µCi/ml) for 20 hr, at 37°C, in 5% CO_2. Then hemin was isolated from washed cell suspensions using acid-cyclohexanone method.[6] It was found unnecessary to preincubate ^{59}Fe-citrate with mouse transferrin; thus the tracer was added directly to culture medium containing 10% fetal bovine serum. Cyclohexanone extracts were counted in a gamma counter. The results obtained in this manner were proportional to the heme assay data using the fluorometric method.

Microflowphotometric Assay for Benzidine-positive Cells

One hundred µl of cell suspension was mixed with an equal volume of benzidine-H_2O_2 reagent, and incubated at room temperature for 2 min. Benzidine-H_2O_2 reagent was made freshly as follows: 5 ml of solution containing 10 mg benzidine-HCl, 25 µl of glacial acetic acid was filtered through a 0.45 µ mesh millipore filter prior to use. Then 200 µl of 3% H_2O_2 was added to the benzidine solution. 1.6 ml of Earle's buffer devoid of Ca^{2+} and Mg^{2+} was added at the end of 2 min incubation. The mixture was assayed in a Cytograf Model 6300A, using absorption on X axis, and scatter on Y axis. A typical example of benzidine positive cells are shown in Fig. 22-4. The results obtained by this method is in good agreement with the benzidine stain carried out manually; however, it allowed fast and reproducible results.

Hemoglobin Fractionation

Approximately 50×10^6 cells were collected by centrifugation. Cell lysates were prepared by repeated freeze-thawing of washed cells in 1 ml Earle's buffer salt solution devoid of Ca^{2+} and Mg^{2+}. Lysates were then centrifuged at $10,000 \times g$ for 20 min. The supernatant was removed and dialyzed overnight against 0.01 M phosphate buffer (pH 7.0). The dialysate was loaded to a CM-Sephadex column pre-equilibrated with 0.01 M phosphate buffer (pH 7.0). The column was washed 2 times with 3 bed volumes of 0.01 M phosphate buffer (pH 7.0) and hemoglobin was eluted with 0.01 M phosphate buffer (pH 8.0). Concentration of hemoglobin treated with CO was determined by absorbtion at 418 nm using an extinction coefficient of 138 for mouse DBA/2J hemoglobin. In those experiments in which tracers were used, aliquots of the hemoglobin eluates were either directly used for counting (with ^{59}Fe), or were mixed with Aquasol-2 for scintillation counting (with ^{3}H). In some experiments heme and globin from the hemoglobin eluates were split by acid-cyclohexanone prior to counting.

Preparation of Radioactive Hemin

^{59}Fe-hemin was prepared by injecting ^{59}Fe-citrate intraperitoneally to mice (100 µCi/mouse) that had been treated with phenylhydrazine (40 mg/kg body weight) for 5 days.

Red cells were collected 24 hr after injection of the tracer, and hemin was isolated by crystallization from hot glacial acetic acid saturated with NaCl.[7] Crystalline hemin was washed with 0.1 N HCl a few times and has proven to be free of porphyrins by thin layer chromatography and free of the tracer ^{59}Fe by counting. The specific activity of the ^{59}Fe-hemin preparation was 4.4×10^5 CPM/mg. ^{14}C-Hemin was prepared by incubating ^{14}C-δ-aminolevulinic acid (ALA) with mouse reticulocytes obtained from mice carrying genetic hemolytic anemia (nb/nb) at 37°C, for 6 hr. ^{14}C-Hemin was isolated as described as above. The specific activity of the ^{14}C-hemin preparation was 15.3×10^5 CPM/mg.

References

1. Sassa S, Kappas A: Induction of δ-aminolevulinate synthase and porphyrins in cultured liver cells maintained in chemically defined medium. Permissive effects of hormones on the induction process. *J Biol Chem* 252:2428–2436, 1977.
2. Strand LJ, Swanson AL, Manning J, et al.: Radiochemical microassay of δ-aminolevulinic acid synthetase in hepatic and erythroid tissues. *Anal Biochem* 47:457–470, 1972.
3. Sassa S, Granick S, Bickers DR, et al.: Studies on the inheritance of human erythrocyte δ aminolevulinate dehydratase and uroporphyrinogen synthetase. *Enzyme* 16:326–333, 1973.
4. Mauzerall D, Granick S: The occurrence and determination of δ-aminolevulinic acid and porphobilinogen in urine. *J Biol Chem* 219:435–446, 1956.
5. Sassa S, Granick S, Bickers DR, et al.: A microassay for uroporphyrinogen-I synthetase, one of three abnormal enzyme activities in acute intermittent porphyria, and its application to the study of the genetics of this disease. *Proc Natl Acad Sci USA* 71:732–736, 1974.
6. Krantz SB, Gallien-Lartigue O, Goldwasser E: The effects of erythropoietin upon heme synthesis by marrow cells *in vitro*. *J Biol Chem* 238:4085–4090, 1963.
7. Fischer H: Hemin. $C_{34}H_{32}O_4N_4FeCl$. *Org Syn* 3:442–446, 1955.

Appendix 13 Materials and Methods for FV-P and Rauscher Virus Infection

H. J. Seidel and Uta Opitz

Virus preparation
RLV was serially passaged in Balb/c mice. Mice were infected i.p. with 0.2 ml of a 10% cell free homogenate of leukemic spleens (2 to 3 gm in weight) in saline after 15 min centrifugation at 2,000 rpm at 4°C. RLV from the serum of infected mice was prepared as follows: 2 to 3 weeks after i.p. infection mice were anesthesized with ether and blood was taken with a 1 ml plastic syringe from the axillary vessels and centrifuged for 20 min at 3,000 rpm. The supernatant serum was diluted 1:1 with TNE (Tris 0.01 M, NaCl 0.15 M, EDTA 0.001 M, pH 8.3) and loaded on a 15 ml cushion of 20% glycerol in TNE. Centrifugation for 60 min at 98,000 g in a SW 27 rotor in a Beckman Ultracentrifuge. The sediment was resuspended in 2 ml of TNE, diluted to 30 ml with TNE and recentrifuged for 60 min at 98,000 g. The sediment was diluted in 2 ml TNE and frozen at -70°C in 0.5 ml volumes. For infection of mice these samples were thawed, diluted 1:5 in saline, and 0.2 ml were injected i.p.

FV-P
An NB tropic strain of the polycythemia inducing Friend virus was kindly provided by Prof. W. Schäfer (Tübingen). It was serially passaged in NMRI and DBA/2 mice. It was prepared in the same way as RLV. All steps were done at a temperature between 0° to 4°C.

Mice
Female NMRI (Südd. Versuchstierfarm/Tuttlingen) and DBA/2 (Zentralinstitut für Versuchstiere/Hannover) of 18 to 22 gm weight were used. 10 animals per cage (Macrolon No. 2) were kept in artificial light 12 hr daily. They were fed commerical pellets and water *ad libitum*.

^{59}Fe incorporation
0.25 μCi of ^{59}Fe citrate (specific activity 18 μCi/μg Fe) was injected i.v. after at least 30 min incubation in normal mouse serum diluted 1:1 with saline. After 6 hr the spleens and femora of one group were removed and kept in Bouin's solution, peripheral blood was taken from another group after 48 hr, and 0.2 ml were diluted in 2 ml of Drabkin's solution. All samples and two to four standards (0.25 μCi) were kept until the end of the experiment and counted in a γ-counter. The total radioactivity of spleen, femur and peripheral blood was expressed in % of the activity injected. The blood volume was assumed to be 6% of the body weight.

Preparation of cell suspensions
The mice, four to six per experimental point, were killed by cervical dislocation. Spleens were weighed and spleens and femora were kept in α-medium (Flow) containing 2% of FCS (Seromed) at room temperature before cell preparation. Spleens were teased with forceps in a plastic petri dish, transferred to a plastic tube and suspended with needles, 1-, 12-, eventually 18-gauge. After 5 min of sedimentation of coarse particles the suspension was transferred to a new plastic tube and counted in a Coulter counter. Appropriate cell concentrations were obtained by dilution with α-medium and 2% FCS. The femora were cut at both ends and the cells were flushed out using a syringe with a 12-gauge needle. A single cell suspension was obtained by the procedure as described. All steps were done at room temperature and only plastic material (tubes, syringes, and pipets) was used throughout.

CFU-e culture procedure
The method described by Iscove and Sieber[1] was used. Powder for 1 liter α-medium was diluted in 750 ml aqua dest., supplemented with 100 mg/liter penicillin (Grünenthal) and 100 mg/liter streptothenat (Grünenthal) and mixed for 1 hr on a magnetic stirrer. The pH was adjusted to 7 with sodium bicarbonate (7.5%, Flow). The medium was then filled up to 1 liter and sterilized by filtration through a Sartorius filter (0.22μ) and stored at 4°C. Before use 0.2 ml glutamin 200 mM (Flow) per 20 ml α-medium was added. Double concentrated medium was prepared in analogy.

Methylcellulose
Methocel MC 4,000 CP purum (Fluka) was used since methylcellulose 4,000 cps from Serva did not allow CFU-e growth since one year. 470 ml aqua dest. were boiled for 2 min, 20 gm of MC were added and mixed avoiding clumping. The suspension was further boiled for 4 min and then placed on a magnetic stirrer at 30° to 40°C. 470 ml of cold double concentrated α-medium was added and mixed. After 24 hr stirring at 4°C 100 ml portions were frozen at -20°C. Before use

it is thawed and distributed with 2 ml plastic syringes without needle into culture tubes.

Fetal calf serum

Different serum batches (Seromed) have to be tested and normally 1 out of 10 batches gave good CFU-e growth. The serum was stored at $-20°C$ in 20 ml aliquots to avoid repeated thawing and freezing.

Erythropoietin Step III (Connaught)

A stock solution of 50 units/ml in α-medium with 2% FCS is stored at $-20°C$ in plastic tubes. Depending on the batch of Ep 0.2 to 0.4 unit/ml gave optimal CFU-e growth.

α-Thioglycerol (Fluka)

A stock solution of 10^{-2} M in α-medium is stored up to 3 weeks at 4°C. Before use a 1:4 dilution in medium with 2% FCS is prepared.

Assay mixture	Final concentration
1 ml 2% MC	0.8%
0.75 ml FCS	30.0%
0.1 ml Ep	0.2 to 0.4 unit/ml
0.1 ml α-Thioglycerol	10^{-4} M
0.3 ml α-Medium with 2% FCS	
0.25 ml Cells 1 to 3 $\times$ 10^6/ml	1 to 3 $\times$ 10^5/ml

Only plastic material is used. The final volume of 2.5 ml is mixed and 1 ml is plated into a 3.5 cm Greiner plastic petri dish with a 2 ml syringe and 1-gauge needle. For each experimental point four plates were set up. Incubation for 48 hr at 37°C in a humidified atmosphere containing 5% CO_2 (Forma Scientific Incubator).

The CFU-e colonies growth was dependent on the batch of Ep, MC, and FCS. Normally there were 350 to 400 CFU-e/10^5 bone marrow cells and 100 to 200/10^5 spleen cells. The CFU-e concentration varied also with the strain of mice.

Erythroid colonies with more than eight small cells (erythroblasts) were scored without staining at a magnification of 100 with a Zeiss Invertoscop or Leitz Diavert. To check scoring accuracy colonies were transferred to slides with a Pasteur pipet, fixed in methanol for 10 min. and stained for hemoglobin according to Borsook et al.[2]

BFU-e culture procedure

The assay system as described for CFU-e growth was used with the following alterations: Ep step III was dialyzed against saline and α-medium for 48 hr at 4°C (2 ml Ep against 1 liter of saline for 24 hr and against 1 liter of α-medium). The Ep dose was 3 units/ml. Some batches of FCS that gave poor growth for CFU-e and good growth for BFU-e were used.

Feeding of BFU-e cultures with 50 μl double strength α-medium at days 2 and 5 after onset of the cultures improved BFU-e colony growth of FV-P infected cell cultures.

BFU-e cell cultures were incubated for 10 days at 37°C under identical conditions as CFU-e cultures. Normally there were 13 to 25 BFU-e/10^5 bone marrow cells, but depending on the batches of Ep and FCS up to 50 BFU-e per 10^5 bone marrow cells were found. BFU-e colonies containing more than 200 cells in a loose or tight aggregate with CFU-e-like colonies in the periphery were counted. Colonies transferred to slides, fixed with methanol, and stained with the Pappenheim method showed erythropoietic cells at different stages of maturation.

References

1. Iscove NN, Sieber F: Erythroid progenitors in mouse bone marrow detected by macroscopic colony formation in culture. *Exp Hematol* 3:32–43, 1975.

2. Borsook H, Ratner K, Tattric B: Studies on erythropoiesis. II. A method of segregating immature from mature adult rabbit erythroblasts. *Blood* 34:32–41, 1969.

Index

A

A-Ep, *see* Anti-erythropoietin serum
AA, *see* Anemia, aplastic
Actinomycin D, 178
Adenosine monophosphate, cyclic, *see* cAMP
Adenyl cyclase, 95–98
   isoproterenol and, 125
β-Adrenergic activation of erythropoiesis, 105–109
β_2-Adrenergic
   activation of CFU-e, 108–109
   agonist drugs, sites of action for, 113–114
Adrenergic agonists, 82
Agar culture system, 34–35
AIA, *see* Allylisopropylacetamide
AIP, *see* Porphyria, acute intermittent
ALA (δ-Aminolevulinic acid), 156
   -dehydratase (ALA-D), 151–154
      activity, assay for, 269
      in acute intermittent porphyria, 157–158
      heme biosynthesis and, 135–141
   -synthase (ALAS), 125, 127
      activity, assay for, 268
      in acute intermittent porphyria, 156–158
      heme biosynthesis and, 135–141
Albumin fraction V, bovine serum (BSA), 32, 248
   preparation of, 254–255, 262
Albuterol, 96
   erythroid colony formation and, 108–109
Allylisopropylacetamide (AIA), 125
α-Amanitin, 164–165
δ-Aminolevulinate
   dehydratase, *see* ALA-dehydratase
   synthase, *see* ALA-synthase
δ-Aminolevulinic acid, *see* ALA
AMP, cyclic, *see* cAMP
Androgens, 82
   aplastic anemia and, 127–128
   erythroid colony formation and, 110–111
   erythropoietin and, 123
Androstanes, 110
Anemia
   aplastic (AA), 11, 12–17, 39
      androgen therapy and, 127–128
      granulocyte colonies and, 102
      lymphocyte-mediated suppression in, 100–102
   congenital-hypoplastic, *see* Diamond-Blackfan syndrome
   erythropoietin production in, 225–226
   induction, 46
   of renal insufficiency, 181–187

Anti-erythropoietin, 180, 220
   serum (A-Ep), 88–91
Antilymphocyte globulin, 17
Antithymocyte globulin, 14
Aplasia
   mechanisms of, 17
   pure red cell (PRCA), 12
L-Asparagine, 32

B

B-cell enriched fraction, 243
B-lymphocyte precursor cell, *see* CFU-B
Benzidine
   -positive cells, microflowphotometric assay for, 269
   reaction, 129–130
      for committed colonies, 266
   solution, 261
BFA, *see* Burst feeder activity
BFU, *see* BFU-e
BFU-e (erythroid progenitors) (erythropoietic burst-forming units), 1, 21
   in anemia of renal insufficiency, 182–187
   assay, 260
      for human and murine, 22
      method for murine, 34
   cultures, technique for, 243, 251, 272
      murine, 249
   determinations with lymphocyte-rich fractions of blood, 9–10
   DMSO and, 23–26
   endotoxin and, 75
   ERC and, 44
   and erythropoietin, 125
      dependence, 179–180
      independence, 3
      responsiveness, 47–51
   in Friend or Rauscher virus-infected mice, 144–147
   HLCM and, 53–54
   after hypertransfusion, 60
   in vitro culture of, 257–261
   kinetics, 86
   in long-term
      bone marrow cultures, 73
      cultures, 68–70
   lymphocyte-mediated suppression and, 100–102
   migration from bone marrow to spleen, 75
   miniaturized methylcellulose culture system and, 28–30
   mouse adherent peritoneal exudate cells and, 190

274 Index

BFU-e (continued)
  pH and, 42
  plasma clot techniques for, 104–106
  pool, testosterone propionate and, 88
  production *in vitro*, 72–74
  following Rauscher virus infection, 172–175
  after RBC transfusion, 87–88
  red blood cell extracts and, 64–66
  suicide rates of, 76
  in suspension, 80
  testosterone and, 112
Blood
  mononuclear cells, 9–10
  peripheral
    CFU-e colonies in, 126
    preparation for processing heparinized, 263
  preparation of FI-FIV fractions of, 243–244
Bone marrow (BM)
  BFU-e migration to spleen from, 75
  cell freezing, 22–23
  cells, 1
    BFA and, 49–51
    concentration and burst formation of, 50
    for erythroid colony formation, method of culturing dog, 254–255
  cultures
    long-term, 73
    plasma clot erythroid colony-forming cell, 246–247
    plasma clot erythropoietic burst-forming cell, 247
    procedure for processing human, 264
    procedure for processing mouse, 263–264
  functions, cryopreservation of, 21–26
  reconstitution, 17
  suspensions, 31–34
  transplantation, 17, 25–26
BPA, *see* Burst-promoting activity
BrdU, *see* 5-Bromo-2'-deoxyuridine
5-Bromo-2'-deoxyuridine (BrdU), 136
BSA, *see* Albumin fraction V, bovine serum
BUN in anemic-uremic patients, 181–185
Burst
  colony versus, 40
  -enhancing activity
    in HLCM, 53–54
    in mouse serum, 52–53
  erythroid, *see* Erythroid bursts
  feeder activity (BFA), 47–51
    bone marrow cells and, 49–51
  formation, cell concentration and, 50
  -forming
    cell bone marrow culture technique, plasma clot erythropoietic, 247
    units, erythropoietic, *see* BFU-e
  -promoting activity (BPA), 4–6
  size, 73–74

Busulfan, 112–114
Butoxamine, 96–97
  erythroid colony formation and, 106, 108
Butyric acid, 152, 159

C

cAMP (cyclic adenosine monophosphate), 82, 95–98
  db-, 95
  in murine erythroleukemia, 130, 132
Carbon, colloidal, 232–233, 241
Carbon tetrachloride, 212
Carrageenans, 193
Cascade versus polycistronic model, 140
Catecholamines, erythroid colony growth and, 95–98
CBP, *see* Plasma, citrated bovine
cDNAs, *see* DNAs, complementary
Cell
  counting, preparation for, 264
  culturing, final preparation for, 264–265
  -mediated regulation of erythropoiesis, 18
  proliferation, hormones and, 123
  staining, in colony, 261
  suspension, 79
    preparation of, 249, 258–259
Centrifugation experiment, density, 50
CFU-B (B-lymphocyte precursor cell), in long-term culture, 68–70
CFU-c (colony-forming units in culture) (granulocyte/macrophage progenitors), 3, 21
  assay for human and murine, 22
  busulfan and, 114
  DMSO and, 23–26
  endotoxin and, 75
  enumeration after hypertransfusion, 59, 60
  ERC and, 44
  fetal calf serum and, 79
  in Friend or Rauscher virus-infected mice, 144–147
  HLCM and, 53–54
  following Rauscher virus infection, 172–175
  after RBC transfusion, 87–88
  in suspension, 80
  testosterone propionate and, 90
CFU-e (erythroid colony-forming units) (erythroid precursors), 1, 21
  β_2-adrenergic activation of, 108–109
  in anemia of renal insufficiency, 181–187
  assay, 260
    for human and murine, 22
    method for murine, 31–34
  busulfan and, 113–114
  culture method for, 243, 251, 271
  determinations with lymphocyte-rich fractions of blood, 8–9
  DMSO and, 24–25
  dose-response curve, 246
  EEC versus, 120–121

enumeration after hypertransfusion, 58–59, 60
ERC and, 44
erythropoietin dependence and, 179–180
estradiol benzoate and, 92–93
fixation of, 33
in Friend or Rauscher virus-infected mice, 144–147
growth hormone and, 123, 124
in vitro culture of, 257–261
kinetics, 86
in long-term culture, 68–70
lymphocyte-mediated suppression and, 100–102
miniaturized methylcellulose culture system and, 28–30
mouse adherent peritoneal exudate cells and, 190
nephrectomy and, 181–187
in peripheral blood, 126
pH and, 42
plasma clot techniques for, 104–106
in polycythemia vera, 119–121
proliferation, peripheral blood lymphocytes and, 11–18
following Rauscher virus infection, 172–175
after RBC transfusion, 87–88
red blood cell extracts and, 64–66
scoring of, 33–34
serum requirement of, 4
staining of, 33
steroid pretreatment and, 110–111
testosterone
and 5α-DHT effects on, 111–112
propionate and, 88
tritiated thymidine and, 3
CFU-m (megakaryocytic progenitors), 31
assay method for, 34–35
enumeration after hypertransfusion, 59, 61
fixation, 35
in long-term culture, 68–70
staining, 35
CFU-s (multipotential stem cells) (pluripotential stem cells), 1, 3
in anemia of renal insufficiency, 186
busulfan and, 112–114
cryopreservation of, 39–40
enumeration after hypertransfusion, 59, 61
ERC and, 44
estradiol benzoate and, 92
in Friend or Rauscher virus-infected mice, 144–147
in long-term culture, 68–70
repression, 8
in suspension, 80
testosterone propionate and, 90–91
CHA (congenital-hypoplastic anemia), *see* Diamond-Blackfan syndrome
Chalone, erythrocyte, *see* Erythrocyte chalone
Chlorambucil, 113
Cholera enterotoxin, 96, 98
Cholesterol, 256

Chromatin proteins, non-histone (NHCP), 162–163
Cobalt, 240
Cocaine, 130
Colony
benzidine reaction for a committed, 266
burst versus, 40
-forming cell bone marrow culture technique, plasma clot erythroid, 246–247
-forming cells
continuum of, 77
cryogenic preservation of, 21–26
-forming units
in culture, *see* CFU-c
erythroid, *see* CFU-e
staining cells in, 261
-stimulating factor (CSF), 46, 194, 220
Creatinine in anemic-uremic patients, 181–185
Cryopreservation
of bone marrow functions, 21–26
of CFU-s, 39–40
of colony-forming cells, 21–26
CSF, *see* Colony-stimulating factor
Culture
long-term, 68–70
method, erythrocytic colony, 33
osmolarity of, 42
Cytochrome P-450, 128, 241

D

Density centrifugation experiment, 50
Deoxycortisol, 82
Dexamethasone, 82
erythropoietin and, 124
5α-DHT, *see* 5α-Dihydrotestosterone
Diamond-Blackfan syndrome, 10–11, 12–17
Dibutyryl cyclic adenosine monophosphate, *see* cAMP, db-
5α-Dihydrotestosterone (5α-DHT), 110–113
CFU-e and, 111–112
Diiodo-L-thyronine, 97
Dimethylacetamide (DMA), erythroid differentiation by, 150–153
Dimethylformamide (DMF), erythroid differentiation by, 150–153
Dimethylsulfoxide (DMSO) (Me$_2$SO), 21, 130–131
BFU-e and CFU-e and, 23–26
cryoprotective effects of, 22–26
erythroid differentiation and, 149–154
Friend
cells and, 159
virus-transformed erythroleukemia cells and, 135–141
DMA, *see* Dimethylacetamide
DMF, *see* Dimethylformamide
DMSO, *see* Dimethylsulfoxide
DNAs, complementary, 159

Dobutamine, erythroid colony formation and, 105–108

E

Eagle's minimum essential medium, 31
EB, *see* Estradiol benzoate
EEC, *see* Erythroid colonies, endogenous
Eg, *see* Erythrogenin
Endotoxin, 127
    BFU-e and CFU-e and, 75
Enterotoxin, cholera, 96, 98
Enucleation of cells, 177
Ep, *see* Erythropoietin
Epinephrine, 96
EPO, *see* Erythropoietin, sheep plasma
ERC, *see* Erythropoietin-responsive cells
Erythroblastic islands, 189, 201
Erythroblasts, macrophages and, 201
Erythrocyte chalone, 66
Erythrocytes, enucleated, 177
Erythrogenin (Eg) (Pro-Ep), 229, 235, 241–242
Erythroid
    burst formation, horse serum and fetal calf serum and, 51–52
    bursts, characteristics of, 46–47
    colonies, endogenous (EEC), 118–121
        CFU-e versus, 120–121
    colonies *in vitro*, 252–253
    colony formation
        androgens and, 110–111
        method for culturing dog bone marrow cells for, 254–255
    colony-forming
        cell bone marrow culture technique, plasma clot, 246–247
        units, *see* CFU-e
    colony growth, hormonal influences on, 95–98
    differentiation
        in Friend virus-transformed cells, 149–154
        by hemolysates, 150–151
    functions, early, 141
    precursors, *see also* CFU-e
        assay of specific, 259–260
        in vitro culture of, 257–261
    progenitors, *see* BFU-e
    stem cells, culture of, 262–265
Erythroleukemia, murine, 129–132
    cell (MELC), 129–132
        differentiation, 226
        interferon and, 159–170
Erythropoiesis, *see also* Hemopoiesis
    β-adrenergic activation of, 105–109
    cell-mediated regulation of, 18
    erythroid stem cell kinetics and, 86–88
    erythropoietin-independence of early, 3–6
    estradiol benzoate and, 91–93
    fetal, 219
    glucortocosteroids and, 81–82
    hormonal modulation of, 81–84
    in vitro, 1
    inhibitors of, 64–67
    Kupffer cells and, 240
    levels and testosterone propionate, 89
    lymphocyte-rich fractions of blood and, 8–18, 243–244
    macrophages and, 189–190
    neonatal, 231
    following Rauscher virus infection, 172–175
    testosterone
        propionate and, 88–91
        and steroid metabolites and, 109–115
    uremic toxins and, 181–182
Erythropoietic
    burst-forming
        cell bone marrow culture technique, plasma clot, 247
        units, *see* BFU-e
    colony-forming cells, cryogenic preservation of, 21–26
    factor, renal (REF), 235–238, 241–242
        extraction and incubation of, 236–237
    precursors, methylcellulose clonal cell culture of, 248
Erythropoietin (ESF) (Ep), 1, 8, 44, 95, 103, 218
    androgens and, 123
    BFU-e and, 125
    bioassay, 103–104
        plasma, 104
    dependence, CFU-e and BFU-e and, 179–180
    dexamethasone and, 124
    endogenous erythroid colonies and, 118–121
    erythroid differentiation by, 150–153
    and estradiol, 124–125
        benzoate, 92
    extrarenal, 194–214, 218, 225–226
    Friend cells and, 38
    during hepatic regeneration, 198
    hypoxia and, 227–229
    -independence
        of early erythropoiesis, 3–6
        Rauscher virus and, 147
    methylcellulose and, 78
    physiology, recent advances in, 227–238
    preparation and storage of working amounts of, 259
    production
        additive role of kidney and liver in, 230–231
        in anemia, 225–226
        hepatic, 232–235
        hypophyseal modulation of, 227–229
        in vitro aspects of, 218–222
        in liver, 194
        spleen in, 240

Rauscher virus and, 177
receptors, 6
-responsive cells (ERC), 44, 59
-responsiveness in BFU-e, 47–51
serum factors triggering hepatic production of, 209–211
sheep plasma (EPO), 22
sites of action for, 113–114
sterilization of, 42
after subtotal hepatectomy, 197–198
testosterone and, 221–222
thyroid hormone and, 124, 126
ESF, see Erythropoietin
Estradiol
benzoate (EB)
erythropoiesis and, 91–93
erythropoietin activity and, 92
erythropoietin and, 124–125
Estrogen, 82, 241
Exudate cells, mouse adherent peritoneal, 189–190

F

F-H, see Ficoll-Hypaque
FCS, see Serum, fetal calf
Ferrochelatase, 136
activity, assay for, 269
Ficoll-Hypaque (F-H) gradients, 244
Ficoll-metrizoate, preparation of, 262
Fixation
CFU-e, 33
CFU-m, 35
glutaraldehyde, 33
methanol, 33
Fluorometric assay of heme, 268
FLV, see Friend leukemia virus
Fractionation, hemoglobin, 269
Friend
cells
DMSO and, 159
enucleation of, 177
erythropoietin and, 38
growth and cell cycle of unstimulated, 163
treated with interferon, tumorigenicity of, 168
tumorigenicity of, 160
leukemia virus (FLV), 160
-virus induced polycythemia, polycythemia vera versus, 128
virus-infected cells
growth of, 268
materials and methods for mouse, 268–270
virus infection (FV-P), 143–147
materials and methods for, 271–272
virus-transformed
cells, erythroid differentiation in, 149–154
erythroleukemia cells, 135–141
FV-P, see Friend virus infection

G

Gadolinium, studies with, 232, 234
Gammaglobulin, goat anti-rabbit (GARGG), 233, 235–237
Globin
glycine incorporation into, 137–138
mRNAs, 159, 161–162, 165, 169
Glucocorticosteroids, erythropoiesis and, 81–82
Glutamine solution, preparation of, 263
Glutaraldehyde fixation, 33
Glycine incorporation into globin, 137–138
GM-CSF, see Granulocyte-macrophage colony-stimulating factors
Gonadotropin, human chorionic, 82
Granulocyte
colonies with aplastic anemia, 102
-macrophage
colony-stimulating factors (GM-CSF), 6
precursor cell assay in long-term culture, 68–70
progenitors, see CFU-c
Granulocytes after hypertransfusion, 60
Granulopoietic colony-forming cells, cryogenic preservation of, 21–26
Growth hormone, 82–83
CFU-e and, 123, 124

H

Hank's balanced salt solution, 31
Hb, see Hemoglobin
Hematocrit
in anemic-uremic patients, 183–185
during hepatic regeneration, 198
Hematopoiesis, hypertransfusion and, 58–63
Hematopoietic inductive microenvironment (HIM), 8
Heme
biosynthesis, 135–141
fluorometric assay of, 268
iron uptake and, 75
oxygenase, microsomal, 240–241
pathway enzymes and hemin, 179
synthesis, inhibitors of, 181
Hemin, 136–141
erythroid differentiation by, 150–153
heme pathway enzymes and, 179
preparation of radioactive, 270
solution, preparation of, 149–150
Hemoglobin (Hb)
formation, 137–138
fractionation, 269
synthesis and interferon, 163–164
Hemolysates
erythroid differentiation by, 150–151
preparation of, 150
Hemopoiesis, see also Erythropoiesis
humoral regulation of, two-stage model of, 5–6

Hemopoietic
    cells, serum-free culture of, 256
    stem cell pools, 143–147
Hepatectomy, 195
    sham, 206
    studies in rats
        after multiple, 207–209
        after subtotal, 197–207
Hepatic
    production of erythropoietin, 209–211, 232–235
    regeneration
        erythropoietin during, 198
        microscopic analysis in, 198–205
        radionuclide studies of, 205–207
Hexamethylene bisacetamide (HMBA), 129–131, 162
HIM, see Hematopoietic inductive microenvironment
HLCM, see Leukocyte conditioned medium, human
HMBA, see Hexamethylene bisacetamide
Hormonal
    influences on erythroid colony growth, 95–98
    modulation of erythropoiesis, 81–84
Hormones
    cell proliferation and, 123
    hypophysectomy and, 123
HS, see Serum, horse
³HTdR, see Thymidine, tritiated
Humoral regulation of hemopoiesis, two-stage model of, 5–6
Hypertransfusion, hematopoiesis and, 58–63
Hypogammaglobulinemia, 16
Hypophyseal modulation of erythropoietin production, 227–229
Hypophysectomy, 227–229
    hormones and, 123
Hypoxanthine, 162
    erythroid differentiation by, 150–153
Hypoxia, 44
    erythropoietin and, 227–229

I

IF, see Interferon
Ig production, 16
In vivo versus *in vitro* observations, 125–126, 127
Incubator, 258
Interferon (IF), 160
    doses, 168–169
    hemoglobin synthesis and, 163–164
    immune, 160
    murine erythroleukemic cells and, 159–170
    tumorigenicity of Friend cells treated with, 168
Iodo-L-tyrosine, 97
Iron uptake, heme and, 75
Isoproterenol, 96–97
    adenylate cyclase and, 125
    erythroid colony formation and, 105–107

K

Kidney
    interrelations with liver, 213
    role in erythropoietin production, 230–231
Kupffer cells, 198–206
    erythropoiesis and, 240
    in hepatic erythropoietin production, 232–235

L

Lecithin, 256
Leukocyte conditioned medium, human (HLCM), 46
    burst-enhancing activity in, 53–54
Liver
    erythropoietin production in, 27
    interrelations with kidney, 213
    regeneration, see Hepatic regeneration
    role in erythropoietin production, 230–231
Lymphocyte
    -mediated suppression, 100–102
    -rich fractions of blood, 243–244
        erythropoiesis and, 8–18
Lymphocytes, 100–102
    peripheral blood (PBL), 11–18, 243
        antithymocyte globulin and, 14

M

MA, see β-Mercaptoethylamine
Macrophages
    erythroblasts and, 201
    erythropoiesis and, 189–190
Marrow, bone, see Bone marrow
Media, collection, 31–32
α-Medium, 252, 257
    preparation of, 262
Megakaryocytic progenitors, see CFU-m
MELC, see Erythroleukemia cell, murine
Mercaptoethanol preparation, 263
β-Mercaptoethylamine (MA), 178
Me₂SO, see Dimethylsulfoxide
Metaproterenol, 96
Methanol fixation, 33
Methylcellulose, 251, 252, 271–272
    clonal cell culture, 248
    erythropoietin and, 78
    method, miniaturized, 28–30
    preparation of, 262
    stock solution, 257
        preparation of, 258
    technique, 1
Microflowphotometric assay for benzidine-positive cells, 269
Monocytes, 15

Mononuclear cells
  blood, 9–10
  peripheral, 100–102; see also Lymphocytes
mRNAs, globin, 159, 161–162, 165, 169
Multipotential stem cells, see CFU-s

N

NCTC-109, 32
Nephrectomy, 181–187, 194
Neuraminidase, 220
NHCP, see Chromatin proteins, non-histone
Norepinephrine, 96

O

Osmolarity of culture, 42
Ouabain, 130
Oxygenase, microsomal heme, 240–241

P

Parenchymal cells, 198–206
PBG, see Porphobilinogen
PBL, see Lymphocytes, peripheral blood
Perchloric acid (PCA), 178
Peroxide solution, 261
pH, BFU-e and CFU-e and, 42
Phentolamine, 96–97
Phenylephrine, 96
Phenylhydrazine, 52–53, 212, 225–226, 240, 241
Plasma
  citrated bovine (CBP), endogenous erythroid colonies and, 118
  clot
    culture system, see Plasma culture system
    erythroid colony-forming cell bone marrow culture technique, 246–247
    erythropoietic burst forming cell bone marrow culture technique, 247
    techniques for CFU-e and BFU-e, 104–106
  culture system, 1, 28, 79
    assay method for murine BFU-e, 34
    megakaryocytic progenitors and, 31–35
Pluripotential stem cells, see CFU-s
Pokeweed Conditioned Medium (PWCM), 5
Polycistronic versus cascade model, 140
Polycythemia
  hypertransfusion, 58
  induction, 46
  inhibitors, 192
  vera (PV), 118–121
    CFU-e in, 119–121
    Friend-virus induced polycythemia versus, 128
Polyvinylpyrrolidone (PVP), 21
Porphobilinogen (PBG), 156
Porphyria, acute intermittent (AIP), 156–158

Practolol, 96–97
  erythroid colony formation and, 106–107
PRCA, see Aplasia, pure red cell
Pregnanes, 110
Pregnanolone, 127
Pro-Ep, see Erythrogenin
Proerythroblast-like cells, 175
Progesterone, 82
Prolactin, 82–83
Propranolol, 96–97
  erythroid colony formation and, 106–107
Prostaglandins E, 82, 190
Protoporphyrin-IX (PROTO), 156–157
PV, see Polycythemia vera
PVP, see Polyvinylpyrrolidone
PWCM, see Pokeweed Conditioned Medium

R

Radioiron, 58
Rauscher virus (RLV)
  erythropoietin
    administration and, 177
    independence and, 147
  infection, 143–147, 172–175
  materials and methods for, 271–272
RCE, see Red blood cell extracts
Receptors, erythropoietin, 6
Red blood cell extracts (RCE), 64–66, 75
REF, see Erythropoietic factor, renal
Renal
  erythropoietic factor (REF), see Erythropoietic factor, renal
  insufficiency, anemia of, see Anemia of renal insufficiency
RES, see Reticuloendothelial system
Reticulocytes, release of, 114
Reticuloendothelial system (RES), 194
Retroviruses, 160
RLV, see Rauscher virus
RO-20-1724, 96

S

Salbutamol, erythroid colony formation and, 105–109
Scoring of CFU-e, 33–34
Serum
  factors triggering hepatic production of erythropoietin, 209–211
  fetal calf (FCS), 32, 248, 251, 257–258, 272
    CFU-c and, 79
    endogenous erythroid colonies and, 118–120
    erythroid burst formation and, 51–52
    preparation of, 254–255
  -free culture of hemopoietic cells, 256
  horse (HS), erythroid bust formation and, 51–52

Serum (continued)
  inhibitors in anemia of renal insufficiency, 181–187
  mouse, burst-enhancing activity in, 52–53
Silicon particles, 193
Somatomammotropin, human chorionic, 83
Spectrin
  accumulation, 162
  studies, 178
Spleen
  BFU-e migration from bone marrow to, 75
  cell suspensions, 31
  in erythropoietin production, 240
Staining
  of CFU-e, 33
  of CFU-m, 35
  methods, 261
Stem
  cell
    kinetics, erythroid, erythropoietin and, 86–88
    pools, hemopoietic, 143–147
  cells
    erythropoietin responsive, 59
    multipotential, see CFU-s
    pluripotential, see CFU-s
Sterilization of erythropoietin, 42
Steroid
  metabolites, erythropoiesis and, 109–115
  pretreatment, erythroid colony formation and, 110–111
Steroids
  in cultured chick embryo liver cells, 127
  sites of action for, 113–114
Suicide rates of BFU-e, 76
Suspension, cell, see Cell suspension

T

T-cell enriched fraction, 243
T cells, 76
Technetium sulfur colloid (TSC), 196, 205–213
Terbutaline, erythroid colony formation and, 105–109
Testosterone
  BFU-e and, 112
  CFU-e and, 111–112
  erythropoiesis and, 109–115
  erythropoietin and, 221–222
  propionate (TP)
    CFU-c and, 90
    CFU-s and, 90–91
    erythropoiesis and, 88–91
    erythropoietin levels and, 89
Tetracaine, 130
12-0-Tetradecanoyl-phorbol-13-acetate (TPA), 131
Tetramethylurea (TMU), erythroid differentiation by, 150–153
α-Thioglycerol, 252, 272
Thrombocytes after hypertransfusion, 60
Thrombopoietin (TSF), 220
Thymidine, tritiated (^{3}HTdR), 3
Thyroid hormones, 82
  erythroid colony growth and, 95–98
  erythropoietin and, 124, 126
Thyronine, diiodo-L-, 97
L-Thyroxine, 97
TMU, see Tetramethylurea
TP, see Testosterone propionate
TPA, see 12-0-Tetradecanoyl-phorbol-13-acetate
TSC, see Technetium sulfur colloid
TSF, see Thrombopoietin
Tumorigenicity of Friend cells, 160
  treated with interferon, 168
Tyrosine, iodo-L-, 97

U

U 91 cells, 151–154
U 99 cells, 152–154, 178, 179
Uremia, 181
  inhibitors, 192
Uremic toxins and erythropoiesis, 181–182
Uroporphyrinogen-I (URO) synthase (URO-S), 141, 151–154
  activity, assay for, 269
  in acute intermittent porphyria, 156–158
  heme biosynthesis and, 135–141

Z

Zymosan, 240, 241
  studies with, 232–235